# OPTOELECTRONICS
## An Introduction to Materials and Devices

Electronics and VLSI Circuits

Senior Consulting Editor
**Stephen W. Director**, Carnegie Mellon University

Consulting Editor
**Richard C. Jaeger**, Auburn University

# OPTOELECTRONICS
## An Introduction to Materials and Devices

**Jasprit Singh**
*University of Michigan*

**The McGraw-Hill Companies, Inc.**
New York  St. Louis  San Francisco  Auckland  Bogotá  Caracas
Lisbon  London  Madrid  Mexico City  Milan  Montreal  New Delhi
San Juan  Singapore  Sydney  Tokyo  Toronto

# OPTOELECTRONICS
An Introduction to Materials and Devices
International Editions 1996

1 2 3 4 5 6 7 8 9 0 CWP FC 9 8 7 6

The editors was Lynn Cox;
the production supervisor was Louise Karam.
The book design and all illustrations were done by Teresa Singh;
the jacket was designed by Teresa Singh.

.
ISBN 0-07-057650-5

Library of Congress Catalog Card Number: 95-78648


**When ordering this title, use ISBN 0-07-114727-6**


Printed in Singapore

## About the Author

Jasprit Singh received his Ph.D. in solid state physics from the University of Chicago. He has carried out research in solid state electronics at the University of Southern California, Wright Patterson Air Force Laboratories, and the University of Michigan, Ann Arbor, where he is currently a Professor in the Department of Electrical Engineering and Computer Science. His research interests cover the area of semiconductor materials and their devices for information processing. He is also the author of *Physics of Semiconductors and Their Heterostructures*, McGraw-Hill (1993), *Semiconductor Devices: An Introduction*, McGraw-Hill (1994) and *Semiconductor Optoelectronics: Physics and Technology,* (1995).

*To*
*Nilu and Nihal*

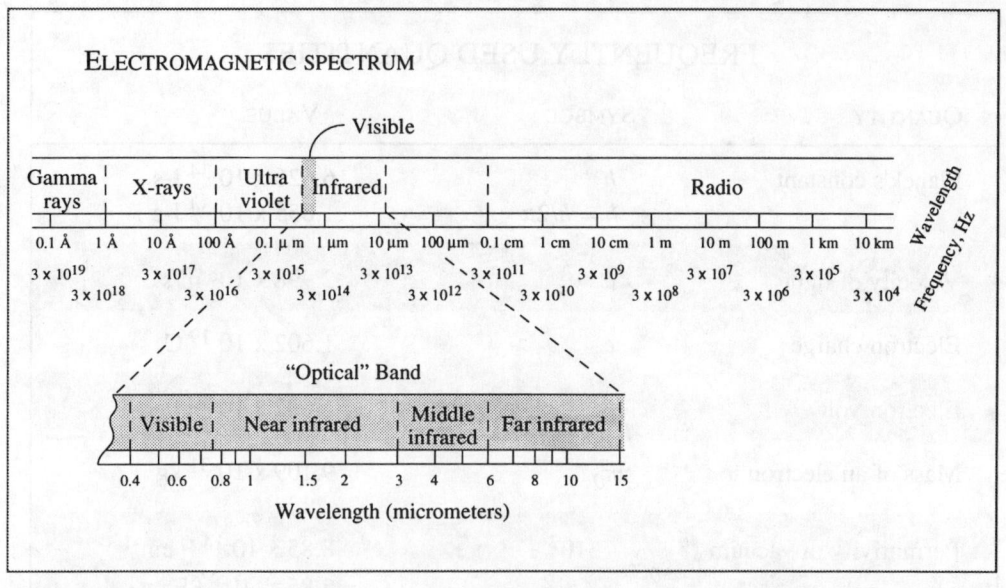

## ELECTROMAGNETIC SPECTRUM

Visible

| Gamma rays | X-rays | Ultra violet | Infrared | Radio |

| 0.1 Å | 1 Å | 10 Å | 100 Å | 0.1 μm | 1 μm | 10 μm | 100 μm | 0.1 cm | 1 cm | 10 cm | 1 m | 10 m | 100 m | 1 km | 10 km |

Wavelength

$3 \times 10^{19}$  $3 \times 10^{17}$  $3 \times 10^{15}$  $3 \times 10^{13}$  $3 \times 10^{11}$  $3 \times 10^{9}$  $3 \times 10^{7}$  $3 \times 10^{5}$

$3 \times 10^{18}$  $3 \times 10^{16}$  $3 \times 10^{14}$  $3 \times 10^{12}$  $3 \times 10^{10}$  $3 \times 10^{8}$  $3 \times 10^{6}$  $3 \times 10^{4}$

Frequency, Hz

### "Optical" Band

| Visible | Near infrared | Middle infrared | Far infrared |

0.4   0.6   0.8   1        1.5   2        3   4      6   8   10   15

Wavelength (micrometers)

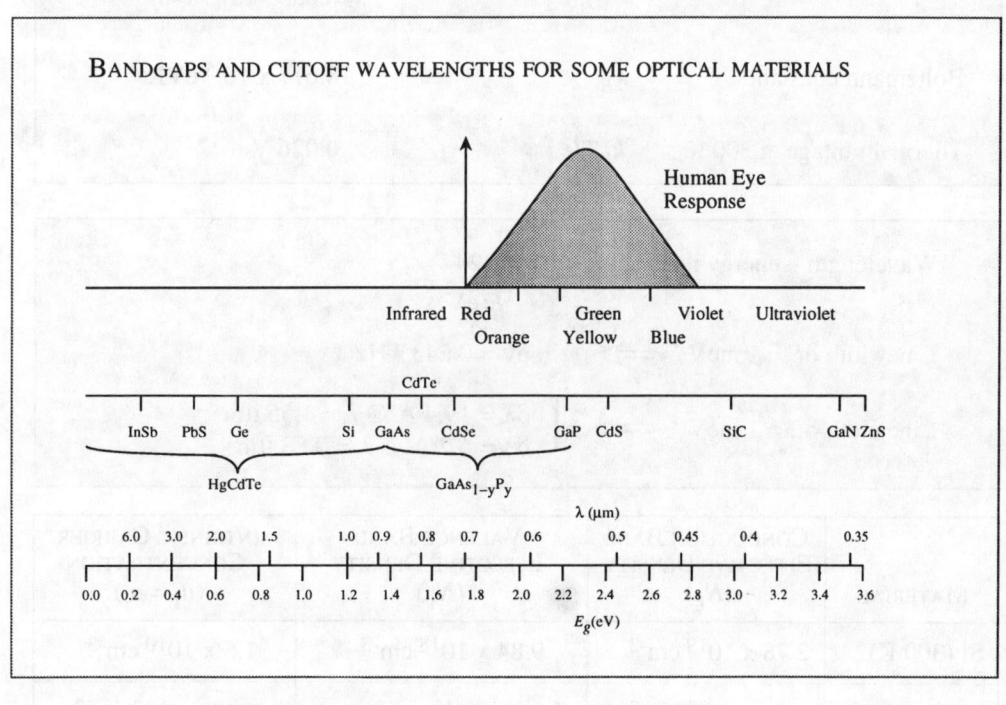

## BANDGAPS AND CUTOFF WAVELENGTHS FOR SOME OPTICAL MATERIALS

Human Eye Response

Infrared   Red   Green   Violet   Ultraviolet
           Orange   Yellow   Blue

CdTe

| InSb | PbS | Ge | Si | GaAs | CdSe | GaP | CdS | SiC | GaN ZnS |

HgCdTe                              $GaAs_{1-y}P_y$

$\lambda$ (μm)

6.0   3.0   2.0   1.5        1.0   0.9   0.8   0.7      0.6        0.5      0.45      0.4        0.35

0.0   0.2   0.4   0.6   0.8   1.0   1.2   1.4   1.6   1.8   2.0   2.2   2.4   2.6   2.8   3.0   3.2   3.4   3.6

$E_g$ (eV)

# FREQUENTLY USED QUANTITIES

| QUANTITY | SYMBOL | VALUE |
|---|---|---|
| Planck's constant | $h$ | $6.626 \times 10^{-34}$ J-s |
| | $\hbar = h/2\pi$ | $1.055 \times 10^{-34}$ J-s |
| Velocity of light | $c$ | $2.998 \times 10^8$ m/s |
| Electron charge | $e$ | $1.602 \times 10^{-19}$ C |
| Electron volt | eV | $1.602 \times 10^{-19}$ J |
| Mass of an electron | $m_0$ | $9.109 \times 10^{-31}$ kg |
| Permittivity of vacuum | $\varepsilon_0 = \dfrac{10^7}{4\pi c^2}$ | $8.85 \times 10^{-14}$ F cm$^{-1}$ $= 8.85 \times 10^{-12}$ F m$^{-1}$ |
| Boltzmann constant | $k_B$ | $8.617 \times 10^{-5}$ eVK$^{-1}$ |
| Thermal voltage at 300 K | $k_B T/e$ | 0.026 V |

Wavelength – energy relation: $\lambda(\mu m) = \dfrac{1.24}{E(eV)}$

Linewidth $\delta E = 1$ meV $\Longrightarrow$ $\delta \nu = 0.243$ THz

Linewidth $\delta E = 1$ meV $\Longrightarrow$ $\begin{cases} \delta\lambda = 19.4 \text{ Å @ } \lambda = 1.55 \text{ }\mu m \\ \delta\lambda = 6.2 \text{ Å @ } \lambda = 0.88 \text{ }\mu m \end{cases}$

| MATERIAL | CONDUCTION BAND EFFECTIVE DENSITY ($N_c$) | VALENCE BAND EFFECTIVE DENSITY ($N_v$) | INTRINSIC CARRIER CONCENTRATION ($n_i = p_i$) |
|---|---|---|---|
| Si (300 K) | $2.78 \times 10^{19}$ cm$^{-3}$ | $9.84 \times 10^{18}$ cm$^{-3}$ | $1.5 \times 10^{10}$ cm$^{-3}$ |
| Ge (300 K) | $1.04 \times 10^{19}$ cm$^{-3}$ | $6.0 \times 10^{18}$ cm$^{-3}$ | $2.33 \times 10^{13}$ cm$^{-3}$ |
| GaAs (300 K) | $4.45 \times 10^{17}$ cm$^{-3}$ | $7.72 \times 10^{18}$ cm$^{-3}$ | $1.84 \times 10^6$ cm$^{-3}$ |

# Contents

# 2 LIGHT PROPAGATION IN MEDIA                                  41

# 3 LIGHT PROPAGATION IN WAVEGUIDES 78

# 5 TRANSPORT AND OPTICAL PROPERTIES OF SEMICONDUCTORS

**177**

# 6 LIGHT DETECTION AND IMAGING                                           **236**

# 7 THE LIGHT EMITTING DIODE                                               **294**

# 8 THE LASER DIODE                                                      338

# 9 MODULATION AND DISPLAY DEVICES     388

# 10 OPTICAL COMMUNICATION SYSTEMS DEVICE NEEDS — **438**

# 11 FABRICATION AND PROCESSING OF DEVICES 468

# PREFACE

The last decade of the twentieth century is seeing unprecedented changes in human experience. These changes are occuring in the area of technology, politics, international trade, environmental conciousness and even a reevaluation of the meaning of the quality of life. As a result powerful new market forces are being unleashed spawning new industries. As the world becomes one big market place, the importance of information—fast, accurate and in a comprehensible form—has become paramount. Serving this global market are the information processing systems built upon computer networks and communication. It is indeed difficult to imagine any modern industry without access to computers and communication to be a key player in the changing market. The computer and communication systems are themselves evolving rapidly, thanks to a number of key "enabling technologies." These enabling technologies are loosely classified under the categories of semiconductor electronic devices, optoelectronics, and new developments in software. This book deals with the general area of optoelectronics.

Optoelectronics—the technology dealing with information processing with light—has been around for a long time. However, over the last fifteen years or so it has acquired a new potency, providing us products that could not have been possible without optoelectronics. Examples of these products are the compact disc players, optical communication systems, laptop portable computers, personal video cameras, laser printers, and so on. In all of these products, the optoelectronic component— whether it be a semiconductor laser, optical fiber, charge coupled device, or a liquid crystal display—has played a key role in bringing the product to the market place. It is widely expected that due to the increasing importance of information processing, optoelectronics-based products will have a global market share in 2010 which is comparable to the integrated- circuit-based market share. This would mean a twenty-fold increase in the market from the value of about ten billion dollars in 1995. It is quite clear that given the importance of optoelectronics, the electrical engineering curricula should cater to this area. Due to the historic status of semiconductor devices, in most electrical engineering schools, while transistors, diodes etc., are discussed at the undergraduate level, the discussion of optoelectronic devices is mostly avoided. This has been changing recently, although most schools still offer optoelectronic courses only to advanced graduate students.

My intention in writing this introductory textbook on optoelectronics is two-fold. Firstly, I wanted to develop a text that could be used for seniors and beginning graduate students who have had a course in semiconductor devices and one in electromagnetics or optics. Such a course, say, with a title like "An Introduction to Optoelectronics," could be offered in a setting where beginning graduate students and qualified seniors could take it together. One may argue that some texts in the market could already satisfy the needs for such a course. This brings me to the second reason for this text. I wanted to develop a text that addresses the technical areas that are responsible for present and future products. Most texts on introductory optoelectronics address the topics of semiconductor devices such as light emitting diodes, lasers, and detectors and focus on the technology that has led to the optical communication systems. Topics such as amorphous silicon thin film transistors, charge coupled devices, liquid crystal cells etc., are either skipped over completely or barely mentioned. Given that the optical communication systems, while of great importance, only account for about ten percent of the optoelectronic market, it is only fair that an introductory course on optoelectronics should discuss the other devices mentioned above in detail comparable to the traditional devices. I have tried to present such a treatment in this book.

Having discussed the motivations for this book, let me also express some concerns that may be faced by students (especially seniors who wish to learn about optoelectronics) and instructors. I have encountered these concerns in my classroom teaching and in discussions with colleagues. These concerns arise mainly from the notion that the level of quantum mechanics needed for appreciating optoelectronics is much higher than that needed for semiconductor electronic devices. This is not entirely true, since the concepts of semiconductor bandstructure, effective mass, mobility, etc., are all based on very advanced quantum mechanics. However, historically these concepts have been given to the undergraduate students without full justification—an approach that is extremely valuable. The exciting area of electronic devices is brought within the grasp of an undergraduate student by making him or her take a leap of faith by accepting the concepts of bandgaps, conduction band, valence band, mobility, etc., in semiconductors. I have used a similar approach in this text. Concepts of optical emission and absorption are introduced using intuitive arguments, rather than detailed quantum mechanical derivations. Once the student accepts these concepts, he or she can then learn about detectors, LEDs, lasers, display devices, etc. Given the fact that most electrical engineers will not take the two or three advanced quantum mechanics courses needed to fully understand electron-phonon or electron-photon interactions, this approach is entirely justified. This approach then allows the student to appreciate the excitement of optoelectronic devices.

This manuscript was typed by Ms. Izena Goulding, to whom I am extremely grateful. The figures, cover design, and the formatting of this book were done by Teresa Singh, my wife. She also provided the support without which this book would not be possible. I am also indebted to my colleagues and the administration at the University of Michigan for providing the atmosphere and the physical resources to make book writing possible in these times of fast paced research.

Finally, I want to thank Ms. Lynn Cox and Mr. George Hoffman, my editors at McGraw-Hill, for providing me with encouragement and support for this project. I am particularly grateful to the excellent team of referees who provided critical comments on this book. Their feedback was of great benefit to me. I gratefully acknowledge Professors Thomas A. DeMassa (Arizona State University), Joe C. Campbell (University of Texas), R. P. Kenan (Georgia Institute of Technology), David Brady ((University of Illinois), A. Safaai-Jazi (Virginia Polytechnic Institute and State University), James Coleman (University of Illinois), Marek Osinski (University of New Mexico), Karl Hess (University of Illinois), and Dr. Daniel Renner (Ortel Corp.).

A solution manual is available to the instructors wishing to use this textbook. Please write on your department stationary to McGraw-Hill to get a copy.

Jasprit Singh

# INTRODUCTION

## I.1 THE INFORMATION AGE

As we move into an intensely competitive and global economy, timely and accurate information is becoming of paramount importance. Of course, information has always been of great importance. Imagine the value of the knowledge of an oncoming hurricane or an advancing enemy. Such knowledge has been of importance for thousands of years. However, because of the enormous improvements in the computer and communication technologies, information affects every aspect of our lives. It is hard to imagine a world without computers, satellites, undersea fiber networks, televisions, fax machines, laser printers, and a myriad of other information processing tools. Whether we like it or not, most of the livelihoods of workers in industrially developed countries are intimately tied to the ability to access and process information.

The survival of today's industries depends not only on how good the products their workers make are, but how well the companies can respond to today's global market. For example, what good is a product without a customer? And very often these days the customer may be thousands of miles away. Information about market demands from thousands of miles away are as critical as the product itself. Thus, information access, information manipulation, and information communication are becoming increasingly important. As a result, the scientists and engineers who design and build information processing devices carry an enormous burden on their shoulders. Regardless of what the information is, we want to process it faster, with fewer inaccuracies, at a lower cost, with a system which consumes less space, etc. This text will address the information processing devices that fall under the category of *optoelectronics*. In principle, a vast range of devices can be classified under optoelectronics. These may include the neon light displays, the electric bulb and the television screen. However, we will focus only on those devices that have emerged over the past twenty years or so as enabling technologies in the information age. We will discuss this aspect further in a later section.

The modern age of information processing has been ushered in by the electronic devices, particularly the mass-produced high-density semiconductor devices. These devices process information at blinding speeds, crunching numbers at speeds of millions of instructions per second. Electronic devices are deeply entrenched in any information processing system, and for good reason. Can optoelectronic devices, which exploit light and electrons, make inroads into the domain of electronic devices? Over the last decade, optoelectronic devices have started to make an impact on the information processing scene. This impact has been felt most in the area of information communication, information storage (and retrieval) and information display. To understand the challenges facing optoelectronics and the potential payoffs, we examine the demands placed upon devices that are to be successful in information processing.

## I.2   DEMANDS OF THE INFORMATION AGE

As noted in the previous section, we live in an age where acquiring, manipulating, and transmitting information are of utmost importance. In Fig. I.1 we show some of the important functions that need to be carried out in order to survive in the information age. Let us briefly examine these functions and see what sort of requirements they put on devices.

**Information Reception/Detection**

This is, of course, one of the most important functions in an information system. For example, in our own case we receive information about the world we live in by our eyes, nose, skin, ears, and tongue. Our five senses allow us to obtain important information about our surroundings, and this information is conveyed to our main processing unit—our brain. Devices which hope to serve as sensors/detectors must have a well-defined response to an external input. They must convert the input information into a form which can be used for further processing.

**Information Enhancement/Amplification**

Very often the information that is received is of either very poor quality, or is too "weak" to be directly useful. In such cases, the information must be amplified or enhanced. We often ask people to repeat themselves more loudly, since we cannot hear them well. Many hearing impaired people need hearing aids which can amplify the sound coming in. Thus, amplifiers are an essential part of an information processing system. To be able to amplify information, the device must have the very important characteristic of *gain*, i.e., a small change in input should result in a large change in output.

**Information Manipulation**

This is, of course, the most important and "intelligent" aspect of information processing. When some information comes in, much of it may be redundant or in a form

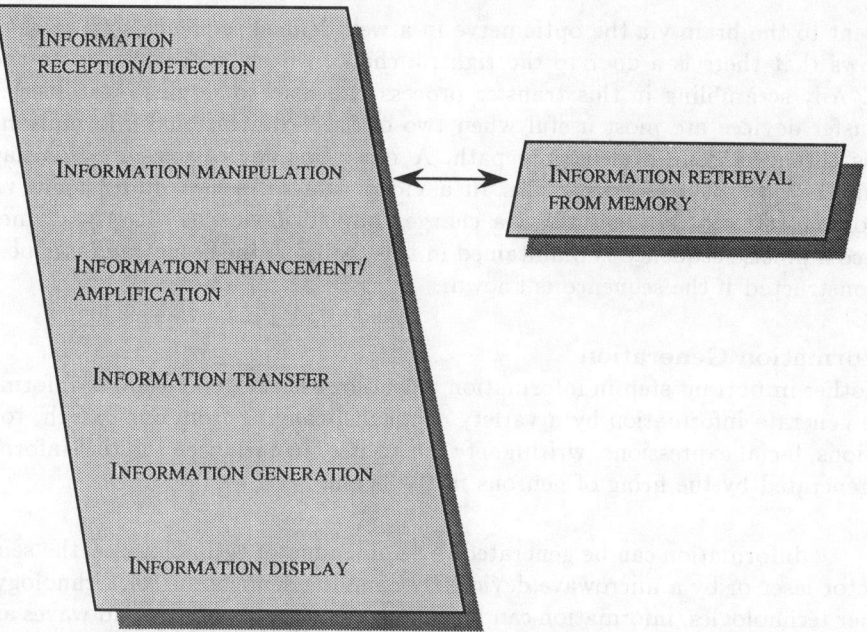

Figure I.1: Necessary functions to survive in the "information age."

that is not appropriate. Invariably, it has to be manipulated, which may mean carrying out processes like addition, subtraction, division, multiplication, comparison with previous information, extracting a "signature" of the information, etc.

### Memory

Memory is obviously essential in an information processing system. The process of learning, comparing, selecting, and reusing information all require memory. The memory device should be able to store information by, perhaps, changing the state of the device, and then one should be able to retrieve the information (i.e., be able to WRITE/READ). This page that you are reading is a form of a memory—perhaps the most influential kind of memory in the history of humankind.

Memory is an area where a number of important technologies are used. Optical memories based on a plastic disc (the compact disc), magnetic tape memories, magnetic bubble memories, semiconductor memories, etc., all form important spokes of the memory technology.

### Information Transfer

An important function in an information processing system is to be able to transfer the information or the results obtained after manipulating the information into a storage or memory. For example, when an image is seen by our eye, the information is sent to the brain via the optic nerve in a well-defined sequence so that the brain knows that there is a door to the right, a chair in front, a blackboard to the left, etc. Any scrambling in this transfer process can lead to serious disabilities. Such transfer devices are most useful when two or three dimensional information must pass through a one dimensional path. A most popular case where this happens is in the case of a video camera. In a video camera, a two dimensional view is recorded and sent sequentially via charge coupled devices (CCDs) to a memory. Since a precise sequence is maintained in this transfer, the scene can later be easily reconstructed if the sequence is known.

## Information Generation

Another important step in information technology is the generation of information. We generate information by a variety of means, ranging from our speech, to hand actions, facial expressions, writing, etc. Of course, in each case the real information is generated by the firing of neurons in the brain.

Information can be generated in semiconductor technology by the semiconductor laser or by a microwave device. By coupling semiconductor technology with other technologies, information can be generated in the form of sound waves as well.

## Information Display

The saying, "A picture is worth a thousand words," seems to be more and more valid as the amount of available information becomes greater and greater. Often in our daily life, a single facial expression conveys more information than any speech or writing could. Displaying information is extremely important and has great impact on human experience. Consider the enormous sum of money spent by companies on advertisements. Displays need not just be pictures—they can be words conveying information as well. Display technology is one of the fastest growing technologies in recent years. Nations and companies vie fiercely to obtain an edge in display technology. New display technologies, such as high density television (HDTV), flat panel displays, programmable transparencies, etc., hold keys to the economic success of many companies. Graphic workstations have already transformed the lives of designers of houses, automobiles, and microelectronic chips. Semiconductor technology has coupled extremely well with liquid crystal technology to produce displays. Also, for active displays and light sources, semiconductor devices such as LEDs and laser diodes serve an important need.

# I.3 OPTOELECTRONICS: AN ENABLING TECHNOLOGY

The reason optoelectronics has and will continue to make an impact in the information processing age is embodied in the term "enabling technology." Consider a gallium arsenide laser diode. Mass produced for the CD player it costs a couple of dollars. However, without the laser diode, there would be no CD player and CD music market—a market worth billions of dollars. Thus when one looks at the complete picture, it becomes clear why highly market driven companies would make laser diodes.

The laser diode also drives another massive market. This is the telecommunication market. The laser diode and the optical fiber have combined to enable the telecom market to explode over the last ten years or so. The enormous data handling of the fiber with the remarkable capabilities of the laser have created entirely new markets in the communication markets. Thus while by itself, the laser diode—an ultimate "high tech" in most people's minds—has an embarrassingly small market, its true impact is enormous.

Another important example of enabling technology is the flat panel display based on liquid crystal cells. Without this technology the laptop computer would not be possible. This technology is introducing important new products and has the potential to replace conventional TV and computer display screens.

The light emitting diode is also creating new products in the area of displays (with brightness far greater than what is possible by liquid crystal displays)—products that could replace the 100-year-old technology of light bulbs.

Finally, let us not forget the charge coupled devices (CCDs) which have brought us the home video camera. Once again the CCD impact has been far greater than what one might think of if the devices were to be looked upon in isolation. Not only has the home video camera been made possible by CCDs, but the professional video cameras used in TV studios are also increasingly based on CCD technology.

In the mid 1990's the optoelectronic market is only about ten billion dollars, according to the Optoelectronic Industry Development Assiciation (OIDA). Most of this market is due to displays (about 90%). However, OIDA estimates that over fifty billion dollars worth of products depends upon optoelectronics. It is projected that the optoelectronic component of the market will reach one hundred billion dollars by 2010 and its impact on the market will be at least three times greater. (See, for example, *Business Week*, special issue on "The Information Revolution," McGraw-Hill, New York, 1994). This important market is projected to be dominated by: i) light emitting devices based on semiconductors—the light emitting diode and the laser diode; ii) light detection and imaging devices, such as photodiodes and charge coupled devices; iii) light transmission devices, such as optical fibers and dielectric waveguides; and vi) light display devices, such as liquid crystals. Of course, new optoelectronic devices having information manipulation capabilities comparable or even better than the electronic transistor may also make their mark.

In this book we will examine the physics and technology of the four categories of devices mentioned above. We will also discuss briefly some state-of-the-art devices that are being examined in the research labs around the world.

## I.4  ELECTRONIC DEVICES: SOME CRACKS IN THE ARMOR

Electronic devices have dominated the modern information processing systems. Essentially all the demands of information processing can be met by electronic devices. Field effect transistors (FETs) and bipolar junction transistors (BJTs) provide high gain and are extremely fast. They are widely used for microwave devices as amplifiers and oscillators, digital switches, and memories. Two terminal Gunn diodes, tunnel diodes, IMPATTS, etc., are used for signal generations at hundreds of gigahertz.

Electronics does, however, have some vulnerable spots. The electronic circuits are formed by connecting devices to each other using metallic inter-connects. This limits the inter-connectivity of the devices, and system architecture calling for massive inter-connections are very difficult to implement. An optics-based system would not have such problems.

Another difficulty faced by electronics is in the transmission of information over very long distances. For such transmissions (e.g., telecommunications), cables made from metals (e.g., copper) are needed. The system is quite expensive, cannot carry a large number of information channels, and requires repeaters after a kilometer or so because of severe signal decay. This is an area where optoelectronics has made a most significant impact. This has been made possible by the trio of laser diodes, optical fibers and optical detectors.

Electronics also suffers from external electromagnetic interference (EMI) effects. For example, a surge of current induced due to a lightning bolt can have a disastrous effect on an electronic system. Once again, an optics-based system would not be affected by EMI.

Finally, we cannot see electrons! And we need to see information. This is, by definition, a "captured market" domain of optoelectronics. Liquid crystal and light emitting diode based displays are already making tremendous impacts in today's technology.

## I.5   THE PROMISE OF OPTICAL INFORMATION PROCESSING

Light has many properties that make it very attractive for information processing. Some of these properties shown in Fig. I.2 are:

i) *Immunity to electromagnetic interference*: Since light particles carry no charge, electromagnetic activities, such as lightning and other potential discharges which can play havoc with electrical signals, have essentially no effect on optical signals;

ii) *Non-interference of crossing light signals*: Two unrelated light beams can cross one another and emerge with little effect on each other—a property that could be exploited in very high density information processing. In electronic signals, two crossing signals will have serious effects and cause loss of information;

iii) *Promise of high parallelism*: The benefits of optics, as far as parallelism is concerned, is obvious to us when we see an image and are able to process it in parallel to make real time decisions, like crossing a busy road. Of course, we do not know how exactly the human brain exploits the parallelism, but this is one of the great challenges for computer scientists;

iv) *High speed/high bandwidth*: Optical pulses have been produced with widths of only a few femtoseconds! In principle, such short pulses could be exploited for a variety of high speed applications;

v) *Signal (beam) steering*: Optical beams can be steered quite easily by the use of lenses or holograms. This is difficult or impossible to do for electron beams in a reasonable manner. The beam steering phenomenon can, in principle, allow one to reconfigure interconnections in very short times and thus, generate circuits which can be flexible (or functional) in real time;

vi) *Special function devices*: This is an exciting property of optical devices which has great potential in high speed information processing. An important example in this case is the lens which, when used with a proper object to image relationship, can produce a Fourier transform of the object image. This property is exploited in numerous recognition-based systems. Another example is the use of optics in spectrum analyzers, which exploits the special diffraction properties of light;

vii) *Ease of coupling with electronics*: This is one of the most important features of optics and one that has paid most dividend so far. Optical and electronic interactions can easily be merged in semiconductor devices. This has led to the most important optoelectronic devices vis., the laser, the detector, and the modulator.

Of course, as noted earlier, we can see by using light, a feature that no other technology offers.

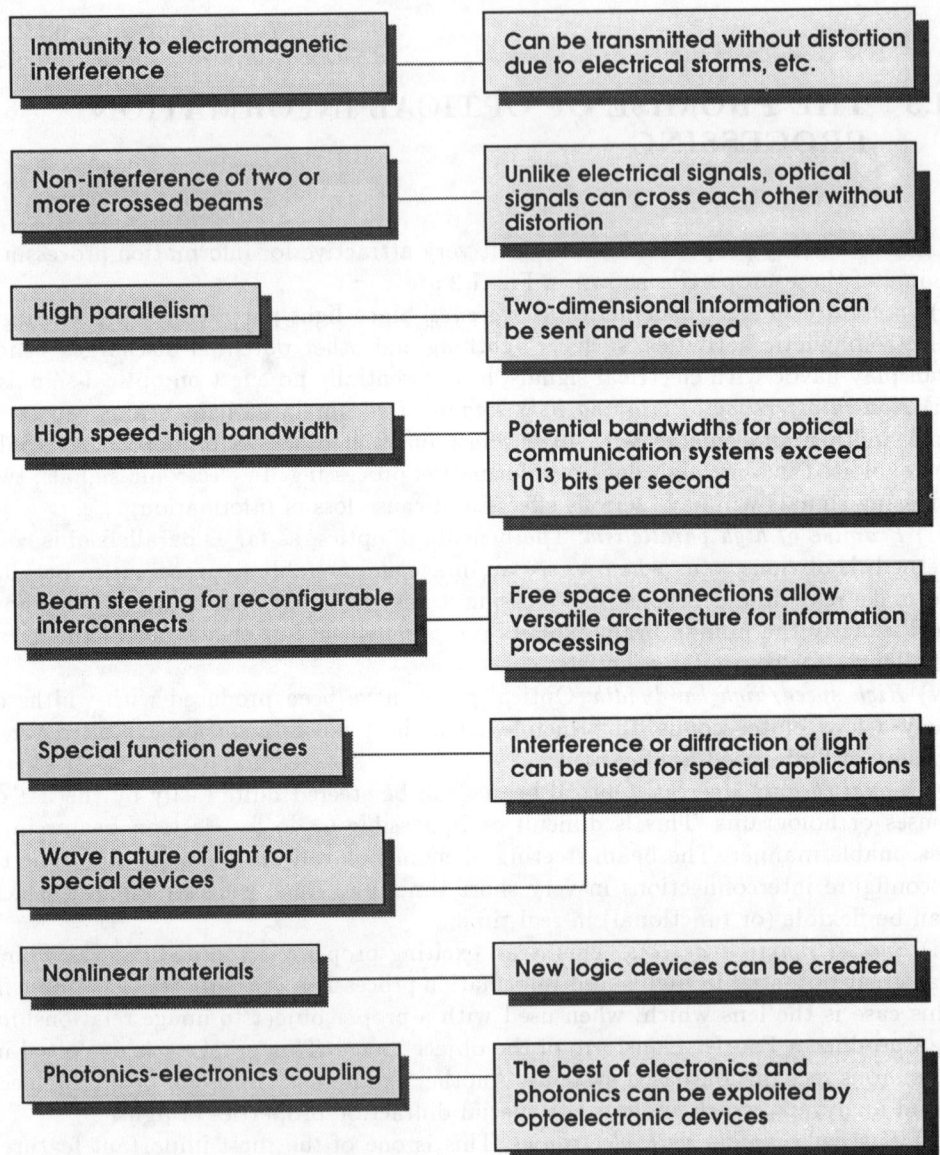

Figure I.2: Special features of light and optical devices which make optics an attractive medium for information processing.

So far, optics has played a very important role in a number of areas shown in Fig. I.3. These areas include:

i) *Memory*: Information is stored digitally on optical discs (compact discs or CDs) as tiny "bumps" which can be read by a solid state laser. This has greatly revolutionized the music industry, as well as the general information storage industry. However, the laser in the CD player still has to be backed up by an electronic chip which does all the signal processing and controls the audio output;

ii) *Optical communication*: This has been the most important area where optoelectronic devices have made inroads into modern technology. This is also an area which has given impetus to compound semiconductor research and development. Optical fibers are rapidly replacing the traditional copper cables for carrying telephone conversations and television programs;

iii) *Local area networks (LANs)*: This is another area where optical interconnects between local computers, telephones, etc., are making office buildings and factories more efficient and capable of handling high volume information;

iv) *Printing and desktop publishing*: This area has received a great boost with the availability of laser printers;

v) *Guidance and control*: Laser guided weapons and unmanned flying crafts have become important components of modern armies;

vi) *Photonic switching and interconnects*: The use of optical devices in chip-to-chip interconnects is becoming increasingly feasible;

vii) *Displays*: The area of displays is, of course, a fate of optoelectronic devices. Portions of this text were typed on a laptop computer in very unacademic surroundings!

## I.6  ROLE OF THIS BOOK

This book has been written to be used in a one semester course on optoelectronics. It is expected that the students have had a senior level course on semiconductor devices and are familiar with basic electronic devices and important semiconductor physics concepts, such as bandgaps, conduction band, etc. However, a brief review of these concepts is given in the text and the instructor can go over these concepts, depending the level of the students.

It would be extremely useful if the students have had a course on electromagnetics and are exposed to an undergraduate course on quantum mechanics. In most electrical engineering schools, students interested in solid state electronics do have such a background by the time they are in the final semester of their senior year. Thus, these students, as well as graduate students, should be comfortable in a course taught from this book.

The book could also be used by electrical engineers and other applied sci-

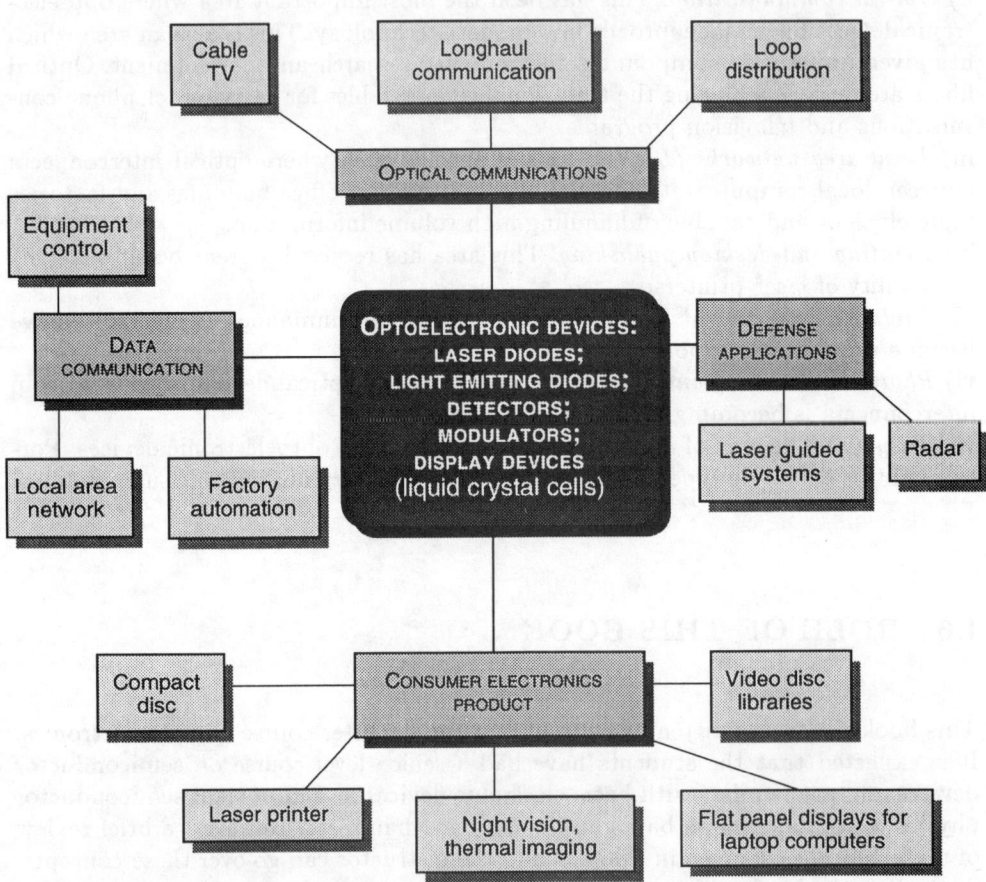

Figure I.3: Some areas where optoelectronics has made impacts in modern technology.

entists to get a good understanding of the field of optoelectronics.

The reader will notice that through the liberal use of pedagogical tools such as tables, charts, as well as figures and solved examples, attempts have been made to make difficult concepts as clear as possible. A few features of this text are outlined below.

**Solved Examples:** The textbook has about one hundred worked examples. This feature should be useful not only in a classroom setting, but also for practicing engineers and applied scientists who may want to learn more about optoelectronic devices.

**Difficulty Levels of the Topics:** The various topics in this text are placed in three categories and are identified by three symbols discussed below:

$\longrightarrow$ : These topics are relatively easy and appropriate for a lower level one semester course on optoelectronics for well prepared seniors and beginning graduate students.

$\rightsquigarrow$ : The sections are somewhat difficult and/or appropriate for students who have had some introduction to semiconductor optoelectronic devices. Depending upon the level of the students taking the course, the instructor may decide to skip some of these sections entirely or just summarize the results without going through the mathematical derivations.

$\mathcal{R}$ : The sections are presented for review or informative reading. The instructors may choose to assign these sections as reading assignments.

**Units:** The book uses SI units throughout. Many worked out examples provide the units at each step. The student may notice the use of centimeters at some places and meters at others. Also, the energy unit is Joules in some places and electron volts at others. These are to conform to standard practices and should not cause the student any difficulty.

# CHAPTER
# 1

# MATERIALS FOR OPTOELECTRONICS: STRUCTURAL PROPERTIES

## 1.1   INTRODUCTION

There are two reasons we like to use photons in information processing: i) As noted in the Introduction to this text, photons are able to accomplish certain functions better than electrons by virtue of their special properties. An important example is the use of photons in optical communication. ii) Another important reason for using photons is the way we humans are "configured!" Nature has decided that sight be an important vehicle used by us to interact with our surroundings. Also, we respond to a specific band of photons with wavelengths between 4000 Å to 8000 Å. We have been endowed with special "hardware and software" to process optical information at extremely fast speeds. We also find certain colors attractive, a feature exploited greatly by the cosmetic industry and the automobile industry!

The first reason outlined above can be called an objective reason for using photons. The choice of the photon wavelength depends upon optimizing the system for an application. The second reason is quite subjective, since the choice of the photon wavelength is intimately tied to the human sight. The materials that are used to generate, manipulate and process the photons for information processing are, accordingly, tremendously varied, ranging from crystalline semiconductors to liquid crystals.

In this chapter we will examine the structure of the optoelectronic materials. This structure, which defines how atoms are arranged in the material, is intimately related to the optical and electronic properties. For most optoelectronic materials, great care has to be taken in the preparation of the material so that the structural quality is excellent. We will discuss some of the techniques used for the material preparation in Chapter 11.

## 1.2   STATES OF MATTER: ORDER

$\mathcal{R}$  The physical matter we observe in our universe can be characterized as solid, liquid and gaseous as shown in Fig. 1.1. One way to characterize these states of matter is by the density and the hardness of the material and the ability of the matter to flow. Thus a solid is usually dense with atomic spacing of a few Angstroms ($10^{-8}$ cm). A solid is also hard and has its own shape, in contrast to a liquid or a gas which takes the shape of the container it is in.

For our purpose, a more useful way of characterizing materials is by the presence of order in the collection of atoms. While order can be defined precisely through mathematical expressions for the correlation functions for the arrangements of atoms, we will simply use a physical and intuitive definition of order. Let us

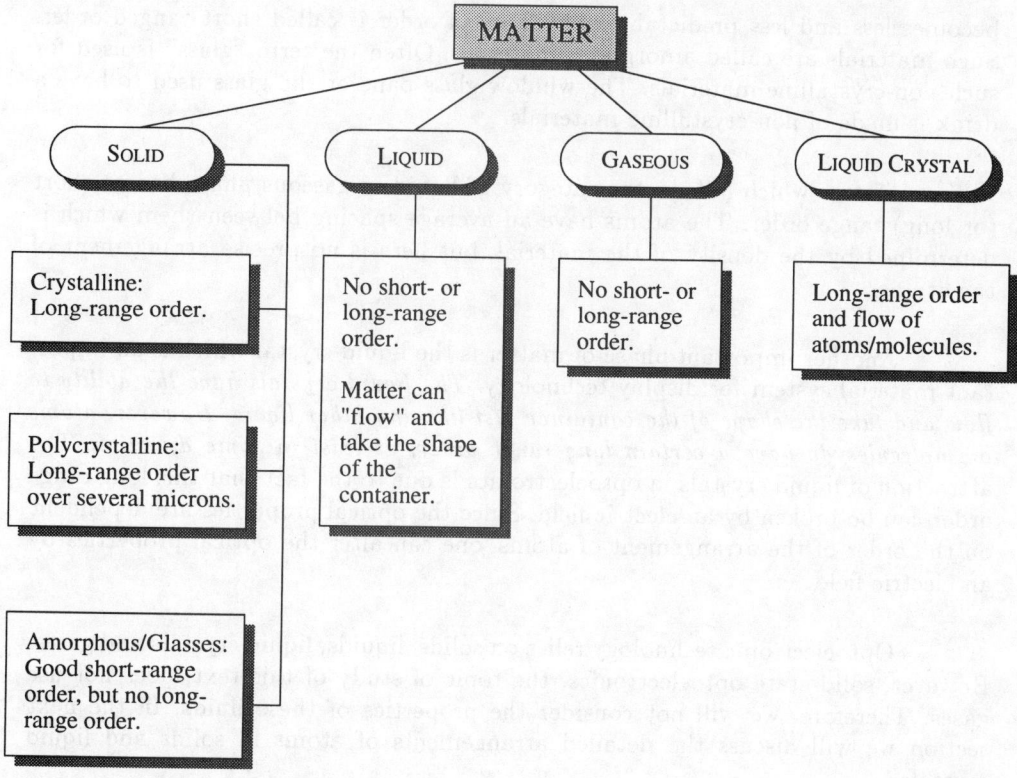

Figure 1.1: The different states of matter and the classification of materials according to the ordered nature of the arrangement of atoms.

imagine that we can actually look at the atoms inside a material. We may find that in some cases *the atoms are arranged in complete precision so that by knowing the position and species of a few atoms, we can predict the position and chemical nature of all the atoms in the sample. When this occurs, we say that the structure has long range order.* Such a long range order is only found in solids and such solids are called crystalline.

In some materials we may find that the precise arrangement of atoms exists over thousands of atoms, but these ordered regions are separated by a boundary across which there is no order. Such solids are called polycrystalline materials. The size of the region across which order persists is called the grain size and, typically, the grain size is a few microns.

There is yet another class of solids in which, if we examine a particular atom, we find that the neighboring atoms (nearest neighbors or second nearest neighbors) are precisely arranged, but as one moves further out, the arrangement becomes less and less predictable. This kind of order is called short ranged order. Such materials are called amorphous materials. Often the term "glass" is used for such non-crystalline materials. The window glass pane or the glass used to have a drink is made of non-crystalline materials.

Matter which falls in the category of liquid or gaseous phase has no short (or long) range order. The atoms have an average spacing between them which is determined by the density of the material, but here is no precise arrangement of the atoms.

Another important phase of matter is the liquid crystal which is an important material system for display technology. *The liquid crystals have the ability to flow and take the shape of the container just like any other liquid. However, atoms or molecules do have a certain long range order, at least in some direction.* The attraction of liquid crystals in optoelectronics is due to the fact that this long range order can be broken by an electric field. Since the optical properties are dependent on the order of the arrangement of atoms, one can alter the optical properties by an electric field.

Optoelectronic technology relies on solids, liquids, liquid crystals and gases. However, solid state optoelectronics, the topic of study of this text, does not use gases. Therefore, we will not consider the properties of these fluids. In the next section we will discuss the detailed arrangements of atoms in solids and liquid crystals.

## 1.3  CRYSTALLINE MATERIALS

$\longrightarrow$  Most of the high performance optoelectronic and electronic devices are made from crystalline materials, where long range order is present among the atoms. Semiconductors form the most important class of materials used in optoelectronics and, as shown in Fig. 1.2, essentially all types of devices can be fabricated from semiconductors. We will discuss the important electronic and optical properties of these materials in later chapters. Certain kinds of materials called ferroelectrics and some dielectrics are also used in modern optoelectronics. We will now discuss some general properties of crystalline materials and will then focus on the specific structural properties of important optoelectronic crystals.

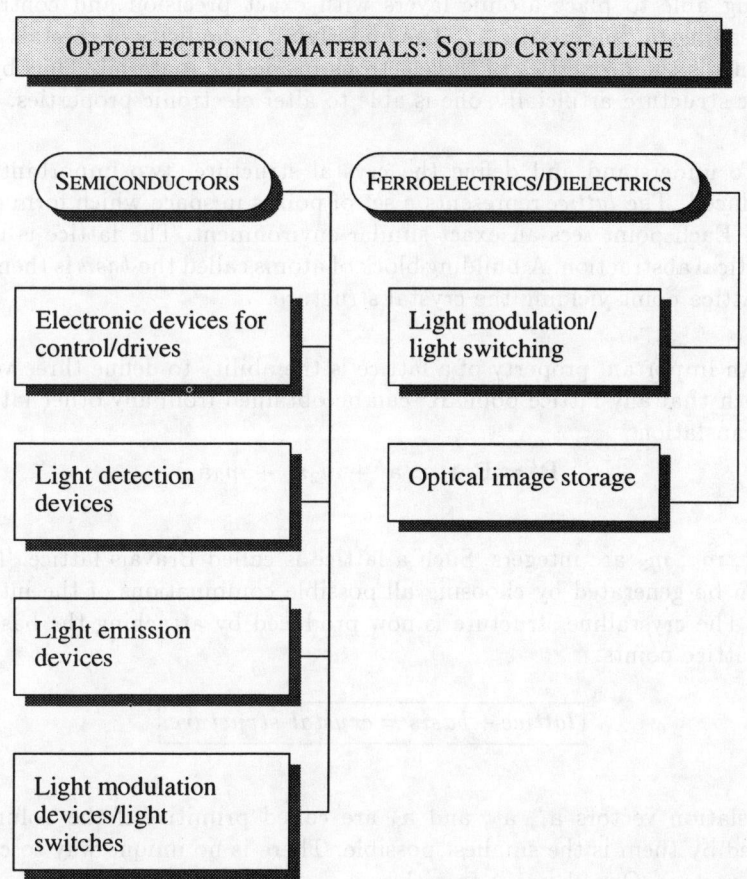

Figure 1.2: In solid state optoelectronics crystalline semiconductors, ferroelectrics and dielectrics are widely used. The various devices based on these materials are outlined.

## 1.3.1   Periodicity of a Crystal

$\longrightarrow$  Crystals are made up of identical building blocks, the block being an atom or a group of atoms. While in "natural" crystals the crystalline symmetry is fixed by nature, new advances in crystal growth techniques are allowing scientists to produce artificial crystals with modified crystalline structure. These advances depend upon being able to place atomic layers with exact precision and control during growth, leading to "superlattices." The underlying periodicity of crystals is the key which controls the properties of the electrons inside the material. Thus by altering crystalline structure artificially, one is able to alter electronic properties.

To understand and define the crystal structure, two important concepts are introduced. The *lattice* represents a set of points in space which form a periodic structure. Each point sees an exact similar environment. The lattice is by itself a mathematical abstraction. A building block of atoms called the *basis* is then attached to each lattice point yielding the crystal structure.

An important property of a lattice is the ability to define three vectors $\mathbf{a}_1$, $\mathbf{a}_2$, $\mathbf{a}_3$, such that any lattice point $\mathbf{R}'$ can be obtained from any other lattice point $\mathbf{R}$ by a translation

$$\mathbf{R}' = \mathbf{R} + m_1\mathbf{a}_1 + m_2\mathbf{a}_2 + m_3\mathbf{a}_3 \qquad (1.1)$$

where $m_1$, $m_2$, $m_3$ are integers. Such a lattice is called Bravais lattice. The entire lattice can be generated by choosing all possible combinations of the integers $m_1$, $m_2$, $m_3$ . The crystalline structure is now produced by attaching the basis to each of these lattice points.

$$\boxed{lattice + basis = crystal\ structure} \qquad (1.2)$$

The translation vectors $\mathbf{a}_1$, $\mathbf{a}_2$, and $\mathbf{a}_3$ are called primitive if the volume of the cell formed by them is the smallest possible. There is no unique way to choose the primitive vectors. One choice is to pick

> $\mathbf{a}_1$ to be the shortest period of the lattice
> $\mathbf{a}_2$ to be the shortest period not parallel to $\mathbf{a}_1$
> $\mathbf{a}_3$ to be the shortest period not coplanar with $\mathbf{a}_1$ and $\mathbf{a}_2$.

It is possible to define more than one set of primitive vectors for a given lattice, and often the choice depends upon convenience. The volume cell enclosed by the primitive vectors is called the *primitive unit cell.*

| System | Number of lattices | Restrictions on conventional cell axes and singles |
|---|---|---|
| Triclinic | 1 | $a_1 \neq a_2 \neq a_3$ $\alpha \neq \beta \neq \gamma$ |
| Monoclinic | 2 | $a_1 \neq a_2 \neq a_3$ $\alpha = \gamma = 90^o \neq \beta$ |
| Orthorhombic | 4 | $a_1 \neq a_2 \neq a_3$ $\alpha = \beta = \gamma = 90^o$ |
| Tetragonal | 2 | $a_1 = a_2 \neq a_3$ $\alpha = \beta = \gamma = 90^o$ |
| Cubic | 3 | $a_1 = a_2 = a_3$ $\alpha = \beta = \gamma = 90^o$ |
| Trigonal | 1 | $a_1 = a_2 = a_3$ $\alpha = \beta = \gamma < 120^o, \neq 90^o$ |
| Hexagonal | 1 | $a_1 = a_2 \neq a_3$ $\alpha = \beta = 90^o$ $\gamma = 120^o$ |

Table 1.1: The 14 Bravais lattices in 3-dimensional systems and their properties

## 1.3.2 Basic Lattice Types

$\longrightarrow$ The various kinds of lattice structures possible in nature are described by the symmetry group that describes their properties. Rotation is one of the important symmetry groups. Lattices can be found which have a rotation symmetry of $2\pi, \frac{2\pi}{2}, \frac{2\pi}{3}, \frac{2\pi}{4}, \frac{2\pi}{6}$. The rotation symmetries are denoted by 1, 2, 3, 4, and 6. No other rotation axes exist, e.g. $\frac{2\pi}{5}$ or $\frac{2\pi}{7}$ are not allowed because such a structure could not fill up an infinite space.

There are 14 types of lattices in 3-dimensions. These lattice classes are defined by the relationships between the primitive vectors $a_1$, $a_2$, and $a_3$, and the angles $\alpha$, $\beta$, and $\gamma$ between them. The general lattice is triclinic ($\alpha \neq \beta \neq \gamma, a_1 \neq a_2 \neq a_3$) and there are 13 special lattices. Table 1.1 provides the basic properties of these three dimensional lattices. We will focus on the cubic lattice which is the structure taken by all semiconductors.

There are 3 kinds of cubic lattices: simple cubic, body-centered cubic, and face-centered cubic.

**Simple Cubic**: The simple cubic lattice shown in Fig. 1.3 is generated by the

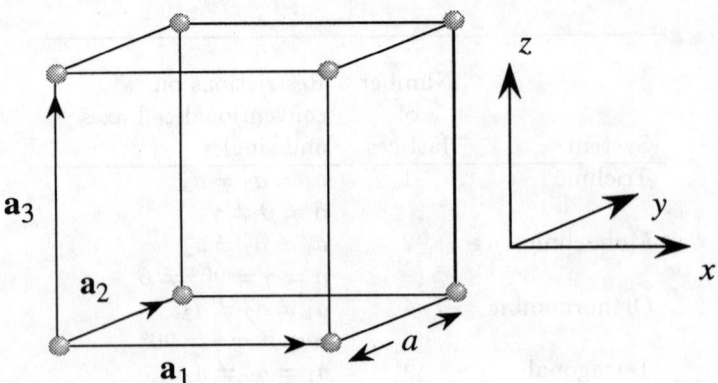

Figure 1.3: A simple cubic lattice showing the primitive vectors. The crystal is produced by repeating the cubic cell through space.

primitive vectors

$$a\mathbf{x}, a\mathbf{y}, a\mathbf{z} \tag{1.3}$$

where the $\mathbf{x}$, $\mathbf{y}$, $\mathbf{z}$ are unit vectors.

**Body-Centered Cubic**: The bcc lattice shown in Fig. 1.4 can be generated from the simple cubic structure by placing a lattice point at the center of the cube. If $\hat{\mathbf{x}}, \hat{\mathbf{y}}$, and $\hat{\mathbf{z}}$ are three orthogonal unit vectors, then a set of primitive vectors for the body-centered cubic lattice could be

$$\mathbf{a}_1 = a\hat{\mathbf{x}}, \mathbf{a}_2 = a\hat{\mathbf{y}}, \mathbf{a}_3 = \frac{a}{2}(\hat{\mathbf{x}} + \hat{\mathbf{y}} + \hat{\mathbf{z}}) \tag{1.4}$$

A more symmetric set for the bcc lattice is

$$\mathbf{a}_1 = \frac{a}{2}(\hat{\mathbf{y}} + \hat{\mathbf{z}} - \hat{\mathbf{x}}), \mathbf{a}_2 = \frac{a}{2}(\hat{\mathbf{z}} + \hat{\mathbf{x}} - \hat{\mathbf{y}}), \mathbf{a}_3 = \frac{a}{2}(\hat{\mathbf{x}} + \hat{\mathbf{y}} - \hat{\mathbf{z}}) \tag{1.5}$$

**Face-Centered Cubic**: A most important lattice for semiconductors is the *face-centered cubic* (fcc) Bravais lattice. To construct the face-centered cubic Bravais lattice add to the simple cubic lattice an additional point in the center of each square face (Fig. 1.5).

A symmetric set of primitive vectors for the face-centered cubic lattice (see Fig. 1.5) is

$$\mathbf{a}_1 = \frac{a}{2}(\hat{\mathbf{y}} + \hat{\mathbf{z}}), \mathbf{a}_2 = \frac{a}{2}(\hat{\mathbf{z}} + \hat{\mathbf{x}}), \mathbf{a}_3 = \frac{a}{2}(\hat{\mathbf{x}} + \hat{\mathbf{y}}) \tag{1.6}$$

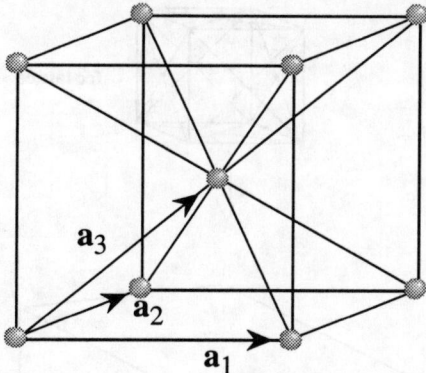

Figure 1.4: The body-centered cubic lattice along with a choice of primitive vectors.

The face-centered cubic and body-centered cubic Bravais lattices are of great importance, since an enormous variety of solids crystallize in these forms with an atom (or ion) at each lattice site. Essentially all semiconductors of interest for electronics and optoelectronics have fcc structure.

## 1.3.3  The Diamond and Zinc Blende Structures

$\longrightarrow$ Most semiconductors of interest for electronics and optoelectronics have an underlying fcc lattice. However, they have two atoms per basis. The coordinates of the two basis atoms are

$$(000) \; and \; (\frac{a}{4}, \frac{a}{4}, \frac{a}{4}) \tag{1.7}$$

Since each atom lies on its own fcc lattice, such a two atom basis structure may be thought of as two inter-penetrating fcc lattices, one displaced from the other by a translation along a body diagonal direction $(\frac{a}{4} \frac{a}{4} \frac{a}{4})$.

Fig. 1.6 gives details of this important structure. If the two atoms of the basis are identical, the structure is called diamond. Semiconductors such as Si, Ge, C, etc., fall in this category. If the two atoms are different, the structure is called the zinc blende structure. Semiconductors such as GaAs, AlAs, CdS, etc., fall in this category. Semiconductors with diamond structure are often called elemental semiconductors, while the zinc blende semiconductors are called compound semi-conductors. The compound semiconductors are also denoted by the position of the atoms in the periodic chart, e.g., GaAs, AlAs, InP are called III-V (three-five) semi-conductors while CdS, HgTe, CdTe, etc., are called II-VI (two-six) semiconductors.

Some semiconductors crystallize in the wurtzite structure which is discussed

Peter Anderson

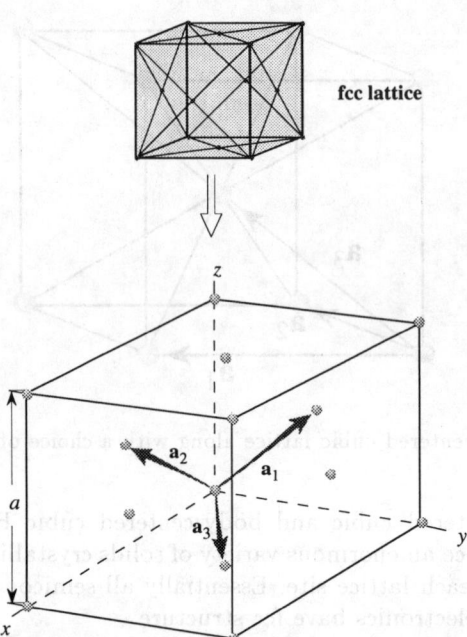

Figure 1.5: Primitive basis vectors for the face-centered cubic lattice.

in problem 10 of this chapter. The structure is shown in Fig. 1.21. The technologies of these semiconductors is not yet well developed, although they have important potential for high power electronic devices and short wavelength optical devices.

## 1.3.4　Ferroelectric Crystals

$\longrightarrow$　An important class of materials used in solid state optoelectronics is called ferroelectric crystals. *A ferroelectric material is characterized by the presence of an electric dipole moment even in the absence of an external electric field.* Thus in ferroelectric crystals, the center of the positive charge of the crystal does not coincide with the center of the negative charge. If we examine a zinc blende material like GaAs, an important optoelectronic material, we note that due to charge transfer, the Ga atom (from the group III of the periodic chart) carries a slightly negative charge (and is called the cation) while the As atom (from the group V of the periodic chart) has a slightly positive charge (and is called the anion). However, there is no net electric dipole in the crystal in spite of the local charge transfer because in the zinc blende structure the cations and anions are placed so that the center of the positive and negative charges coincides when all the atoms are considered. A net dipole could be present if, for example, the entire Ga sublattice moved with respect to the As sublattice. Such a displacement does not occur in GaAs but does occur

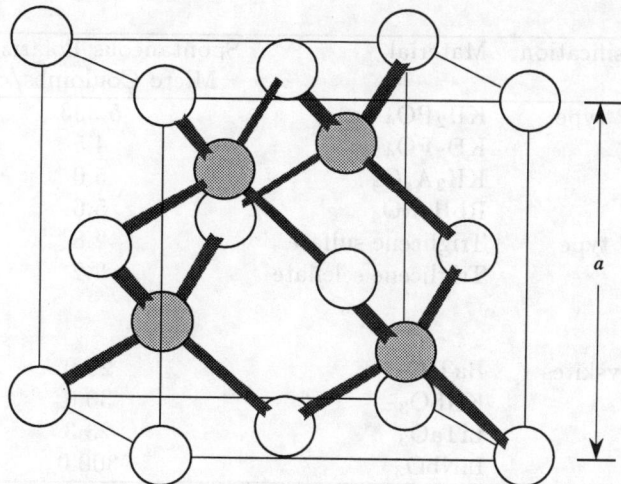

Figure 1.6: The zinc blende crystal structure. The structure consists of the interpenetrating fcc lattices, one displaced from the other by a distance $(\frac{a}{4}\frac{a}{4}\frac{a}{4})$ along the body diagonal. The underlying Bravais lattice is fcc with a two atom basis. The positions of the two atoms is (000) and $(\frac{a}{4}\frac{a}{4}\frac{a}{4})$.

in certain materials.

In Table 1.2 we show some important ferroelectric materials along with the value of the spontaneous polarization (at zero electric field). Ferroelectric crystals are produced by two main effects: i) ordering of ions in such a way as to produce a net electric dipole; ii) displacement of one sublattice with respect to another to produce a net dipole. Some ferroelectrics involve the presence of hydrogen or deuterium (the KDP type shown in Table 1.2). The motion of the proton in the hydrogen or deuterium is the cause of the ferroelectric effect.

An important class of ferroelectric crystals falls in the category of perovskite structure shown in Figs. 1.7a and 1.7b. We can understand the basic perovskite crystal by focusing on a material like barium titanate (BaTiO$_3$). The structure is cubic with Ba$^{2+}$ ions at the cube corners. The O$^{2-}$ ions are at the six face centers of the cube while a $Ti^{4+}$ ion is at the body center. In the absence of any deformation, the material does not have any net electric dipole at zero field, but if there is a displacement of ions, a dipole can develop. In Fig. 1.7b we show how such a displacement can arise. In most materials the net displacement is of the order of $\sim 0.1$ Å to 1 Å.

**EXAMPLE 1.1** The crystal barium titanate has a unit cell volume of $(4 \times 10^{-8}$ cm$)^3$ and a spontaneous polarization of 26.67 $\mu C$ cm$^{-2}$. Calculate the dipole moment of a unit cell and estimate the shift of the positive ions with respect to the negative ions.

| Classification | Material | Spontaneous Polarization Micro Coulombs/cm$^2$ |
|---|---|---|
| KDP type | $KH_2PO_4$ | 5.333 |
| | $KD_2PO_4$ | 4.5 |
| | $KH_2A_sO_4$ | 5.0 |
| | $RbH_2PO_4$ | 5.6 |
| TGS type | Triglicene sulfate | 2.8 |
| | Triglicene selenate | 3.2 |
| | | |
| Perovskites | $BaTiO_3$ | 26.0 |
| | $KNbO_3$ | 30.0 |
| | $LiTaO_3$ | 23.3 |
| | $LiNbO_3$ | 300.0 |

Table 1.2: Some important ferroelectric materials along with the spontaneous polarization values (After C. Kittel, *Introduction to Solid State Physics*, J. Wiley and Sons, New York (1971).)

The dipole moment of the unit cell is

$$p = (26.67 \times 10^{-6} \ C \ cm^{-2})(64 \times 10^{-24} \ cm^3)$$
$$= 1.71 \times 10^{-27} \ C \ cm$$

If the displacement of the positive ions, with respect to the negative ions, is $\delta$, the dipole moment of the unit cell will be $6e\delta$. Thus we have

$$\delta = \frac{1.71 \times 10^{-27} \ C \ cm}{6 \times (1.6 \times 10^{-19} \ C)} = 1.78 \times 10^{-9} \ cm$$
$$= 0.178 \ \text{Å}$$

## 1.3.5   Notation to Denote Planes and Points in a Lattice: Miller Indices

$\longrightarrow$   A simple scheme is used to describe lattice planes, directions and points. For a plane, we use the following procedure:

(1) Define the x, y, z axes (primitive vectors)
(2) Take the intercepts of the plane along the axes in units of lattice constants.

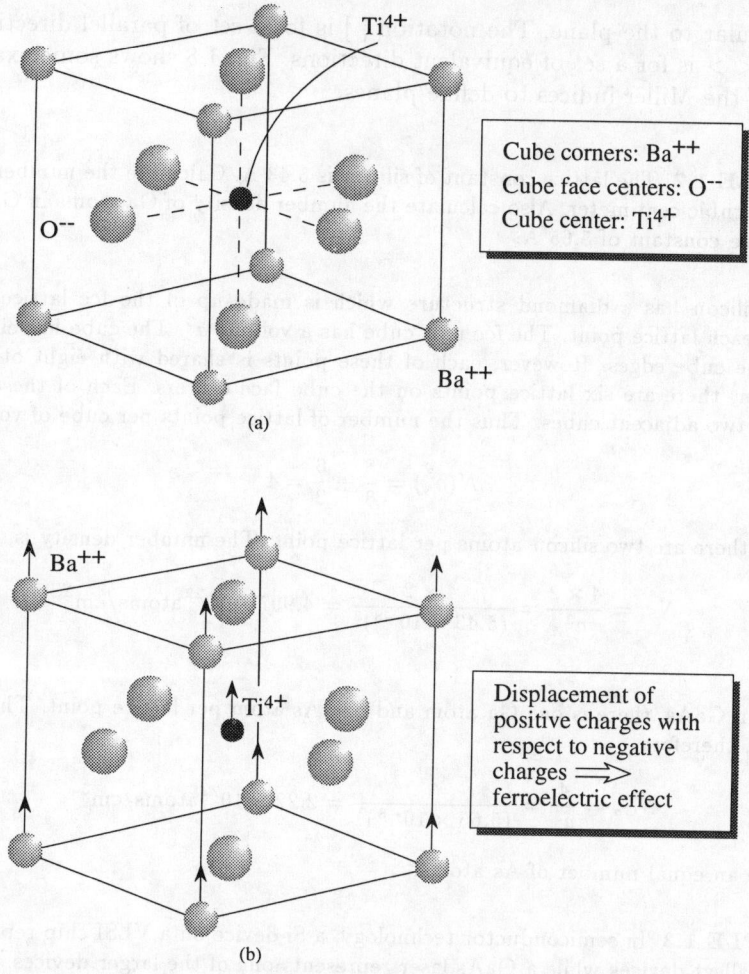

Cube corners: $Ba^{++}$
Cube face centers: $O^{--}$
Cube center: $Ti^{4+}$

(a)

$Ba^{++}$

$Ti^{4+}$

Displacement of positive charges with respect to negative charges $\Longrightarrow$ ferroelectric effect

(b)

Figure 1.7: (a) The structure of a typical perovskite crystal illustrated by examining barium titanate. (b) The ferroelectric effect is produced by a net displacement of the positive ions with respect to the negative ions.

(3) Take the reciprocal of the intercepts and reduce them to the smallest integers.

The notation (hkl) denotes a family of parallel planes.

The notation {hkl} denotes a family of equivalent planes.

To denote directions, we use the smallest set of integers having the same ratio as the direction cosines of the direction.

In a cubic system the Miller indices of a plane are the same as the direction perpendicular to the plane. The notation [ ] is for a set of parallel directions. The notation < > is for a set of equivalent directions. Fig. 1.8 shows some examples of the use of the Miller indices to define planes.

**EXAMPLE 1.2** The lattice constant of silicon is 5.43 Å. Calculate the number of silicon atoms in a cubic centimeter. Also calculate the number density of Ga atoms in GaAs which has a lattice constant of 5.65 Å.

Silicon has a diamond structure which is made up of the fcc lattice with two atoms on each lattice point. The fcc unit cube has a volume $a^3$. The cube has eight lattice sites at the cube edges. However, each of these points is shared with eight other cubes. In addition, there are six lattice points on the cube face centers. Each of these points is shared by two adjacent cubes. Thus the number of lattice points per cube of volume $a^3$ is

$$N(a^3) = \frac{8}{8} + \frac{6}{2} = 4$$

In silicon there are two silicon atoms per lattice point. The number density is, therefore,

$$N_{Si} = \frac{4 \times 2}{a^3} = \frac{4 \times 2}{(5.43 \times 10^{-8})^3} = 4.997 \times 10^{22} \, \text{atoms/cm}^3$$

In GaAs, there is one Ga atom and one As atom per lattice point. The Ga atom density is, therefore,

$$N_{Ga} = \frac{4}{a^3} = \frac{4}{(5.65 \times 10^{-8})^3} = 2.22 \times 10^{22} \, \text{atoms/cm}^3$$

There are an equal number of As atoms.

**EXAMPLE 1.3** In semiconductor technology, a Si device on a VLSI chip represents one of the smallest devices while a GaAs laser represents one of the larger devices. Consider a Si device with dimensions $(5 \times 2 \times 1)$ $\mu m^3$ and a GaAs semiconductor laser with dimensions $(200 \times 10 \times 5)$ $\mu m^3$. Calculate the number of atoms in each device.

From Example 1.2 the number of Si atoms in the Si transistor is

$$N_{Si} = (5 \times 10^{22} \, \text{atoms/cm}^3)(10 \times 10^{-12} \text{cm}^3) = 5 \times 10^{11} \, \text{atoms}$$

The number of Ga atoms in the GaAs laser is

$$N_{Ga} = (2.22 \times 10^{22})(10^4 \times 10^{-12}) = 2.22 \times 10^{14} \, \text{atoms}$$

An equal number of As atoms are also present in the laser.

**EXAMPLE 1.4** Calculate the surface density of Ga atoms on a Ga terminated (001) GaAs surface.

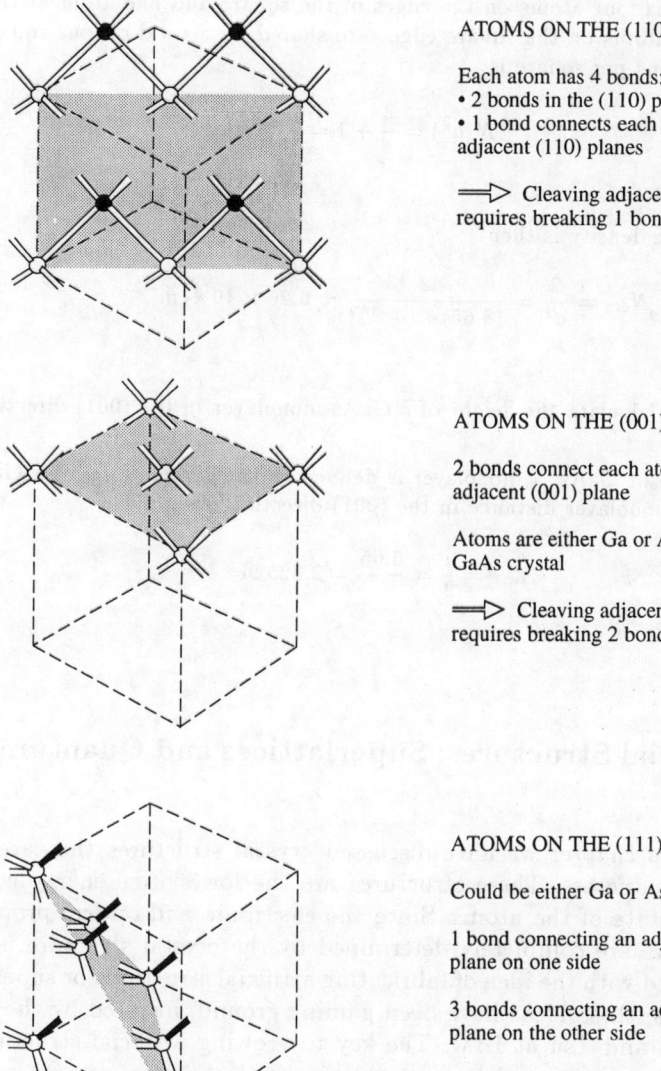

ATOMS ON THE (110) PLANE

Each atom has 4 bonds:
• 2 bonds in the (110) plane
• 1 bond connects each atom to
adjacent (110) planes

⟹ Cleaving adjacent planes
requires breaking 1 bond per atom

ATOMS ON THE (001) PLANE

2 bonds connect each atom to
adjacent (001) plane

Atoms are either Ga or As in a
GaAs crystal

⟹ Cleaving adjacent planes
requires breaking 2 bonds per atom

ATOMS ON THE (111) PLANE

Could be either Ga or As

1 bond connecting an adjacent
plane on one side

3 bonds connecting an adjacent
plane on the other side

Figure 1.8: Some important planes in the cubic system along with their Miller indices. This figure also shows how many bonds connect adjacent planes. This number determines how easy or difficult it is to cleave the crystal along these planes by cutting the bonds joining the adjacent planes.

In the (001) surfaces, the top atoms are either Ga or As leading to the terminology Ga terminated (or Ga stabilized) and As terminated (or As stabilized), respectively. A square of area $a^2$ has four atoms on the edges of the square and one atom at the center of the square. The atoms on the square edges are shared by a total of four squares. The total number of atoms per square is

$$N(a^2) = \frac{4}{4} + 1 = 2$$

The surface density is then

$$N_{Ga} = \frac{2}{a^2} = \frac{2}{(5.65 \times 10^{-8})^2} = 6.26 \times 10^{14} \, \text{cm}^{-2}$$

**EXAMPLE 1.5** Calculate the height of a GaAs monolayer in the (001) direction.

In the case of GaAs, a monolayer is defined as the combination of a Ga and As atomic layer. The monolayer distance in the (001) direction is simply

$$A_{m\ell} = \frac{a}{2} = \frac{5.65}{2} = 2.825 \, \text{Å}$$

## 1.3.6   Artificial Structures: Superlattices and Quantum Wells

$\mathcal{R}$  So far in this chapter we have discussed crystal structures that are present in natural semiconductors. These structures are the lowest free energy configuration of the solid state of the atoms. Since the electronic and optical properties of the semiconductors are completely determined by the crystal structure, scientists have been intrigued with the idea of fabricating artificial structures or superlattices. Since the mid-70's, these ideas have been gaining ground, inspired by the pioneering work of Esaki and Tsu at IBM. The key to growing artificial structures with tailorable crystal structure and hence tailorable optical and electronic properties has been the progress in hetero-epitaxy, i.e., the growth of one material on another. Heteroepitaxial crystal growth techniques such as molecular beam epitaxy (MBE) and metal-organic chemical vapor deposition (MOCVD), disucssed in Chapter 11, have made a tremendous impact on semiconductor physics and technology. From very high speeds, low-noise electronic devices used for satellite communications to low-threshold lasers for communication, semiconductor devices are being made by these techniques. Although, so far, only compound semiconductors have benefitted from these growth techniques, it appears that silicon technology is on the threshold of using hetero-epitaxy for faster devices, by combining Si with Si-Ge alloys.

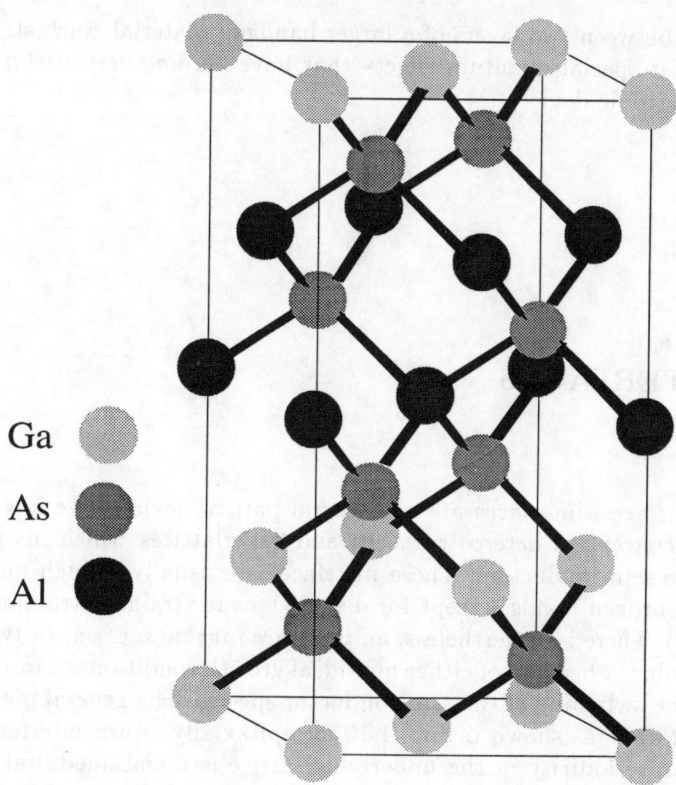

Figure 1.9: Arrangement of atoms in a $(GaAs)_2(AlAs)_2$ superlattice grown along (001) direction.

Ga

As

Al

MBE or MOCVD are techniques which allow monolayer ($\sim 3$ Å) control in the chemical composition of the growing crystal. These techniques will be examined in brief in Chapter 11. Nearly every semiconductor extending from zero bandgap ($\alpha$-Sn,HgCdTe) to large bandgap materials such as ZnSe,CdS, etc., has been grown by epitaxial techniques such as MBE and MOCVD.

Since the heteroepitaxial techniques allow one to grow heterostructures with atomic control, one can change the periodicity of the crystal in the growth direction. This leads to the concept of superlattices where two (or more) semiconductors A and B are grown alternately with thicknesses $d_A$ and $d_B$ respectively. The periodicity of the lattice in the growth direction is then $d_A + d_B$. A $(GaAs)_2$ $(AlAs)_2$ superlattice is illustrated in Fig. 1.9. It is a great testimony to the precision of the new growth techniques that values of $d_A$ and $d_B$ as low as one monolayer have been grown.

It is important to point out that the most widely used heterostructures are not superlattices but quantum wells, in which a single layer of one semiconductor is sandwiched between two layers of a larger bandgap material. Such structures allow one to exploit special quantum effects that have become very useful in electronic and optoelectronic devices.

## 1.4   INTERFACES

$\mathcal{R}$   Like surfaces, interfaces are an integral part of devices. We have already discussed the concept of heterostructures and superlattices which involve interfaces between two semiconductors. These interfaces are usually of high quality with essentially no broken bonds, except for dislocations in strained structures (to be discussed later). There is, nevertheless, an *interface roughness* of one or two monolayers which is produced because of either non-ideal growth conditions or imprecise shutter control in the switching of the semiconductor species. The general picture of such a rough interface is as shown in Fig. 1.10 for epitaxially grown interfaces. The crystallinity and periodicity in the underlying lattice is maintained, but the chemical species have some disorder on interfacial planes. Such a disorder is quite important in many electronic and optoelectronic devices.

One of the most important interfaces in electronics is the $Si/SiO_2$ interface. This interface and its quality are responsible for essentially all of the modern consumer electronic revolution. This interface represents a situation where two materials with very different lattice constants and crystal structures are brought together. However, in spite of these large differences the interface quality is quite good. It appears that the interface has a region of a few monolayers of amorphous or disordered $Si/SiO_2$ region creating fluctuations in the chemical species (and consequently in potential energy) across the interface. This interface roughness is responsible for reducing mobility of electrons and holes in MOS devices. It can also lead to "trap" states, which can seriously deteriorate device performance if the interface quality is poor.

Finally, we have the interfaces formed between metals and semiconductors. Structurally, these important interfaces are hardest to characterize. These interfaces are usually produced in the presence of high temperatures and involve diffusion of metal elements along with complex chemical reactions. The "interfacial region" usually extends over several hundred Angstroms and is a complex non-crystalline region.

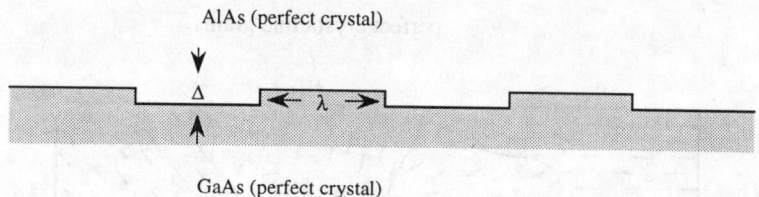

Figure 1.10: A schematic picture of the interfaces between materials with similar lattice constants such as GaAs/AlAs. No loss of crystalline lattice and long range order is suffered in such interfaces. The interface is characterized by islands of height $\Delta$ and lateral extent $\lambda$.

## 1.5 POLYCRYSTALLINE MATERIALS

$\mathcal{R}$ Polycrystalline materials are widely used in electronic and optoelectronic technologies. Polycrystalline structures are produced when a material is deposited on a substrate which does not have a similar crystal structure. For example, if a metal film is deposited on a semiconductor, the film grows in a polycrystalline form. Also, if silicon is deposited on a glass substrate, it grows in a polycrystalline form.

Polycrystalline films are described by their average grain size as shown in Fig. 1.11. Within a grain the atoms are arranged as in crystal, i.e., with perfect order. However, each grain is surrounded by a grain boundary which is a region with a high density of defects. The defects arise due to broken or unfulfilled bonds between atoms. In some cases, chemical impurities may also gather at these grain boundaries. Different grains in the polycrystal have essentially no order between their constituent atoms.

Depending upon the growth process and the differences between the substrate and the deposited film, the grain size of a polycrystal can range from 0.1 $\mu$m to 10 $\mu$m or more. If the grain size exceeds 10 $\mu$m, for some device applications the material can be considered to be crystalline. The presence of grain boundaries has serious consequences on the electrical and optical properties of the material. Indeed, certain devices such as light emitting diodes (LEDs) or laser diodes (LDs) cannot be made from polycrystalline materials. However, some electronic devices used as control transistors for displays can be made from polycrystalline materials. The key advantages of polycrystalline materials is the low cost of the film deposition and the large area of the film possible. Thus polycrystalline technology is an important technology for displays (high density TV, portable and personal computers, etc.). In Fig. 1.12 the important uses of non-crystalline materials are outlined.

It is essential to mention $PbZrO_3$–$PbTiO_3$–$La_2O_3$ (PLZT), an extremely important polycrystalline material that is finding extensive use in optoelectronics.

perfect crystalline grains

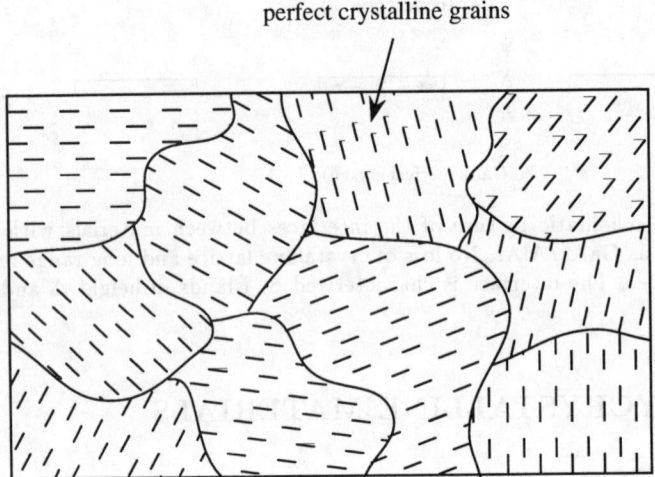

Figure 1.11: A schematic description of a polycrystalline material. Atoms are arranged periodically in a grain, but there is no order between the various grains. The grain boundaries represent regions where defects produced by broken or unfilled bonds are present.

This material is a ceramic oxide with ferroelectric properties. While single crystal electro-optic materials like potassium dihydrogen phosphate (KDP), $BaTiO_3$ and $Gd(M_0O_4)_3$ are important materials, their applications are limited by cost, size and susceptibility to moisture (especially for KDP). In contrast, polycrystalline ceramics are not subject to these limitations. The fabrication technology of PLZT is now highly developed and this ceramic is used for a variety of electro-optic devices.

## 1.6  AMORPHOUS AND GLASSY MATERIALS

$\mathcal{R}$  In amorphous materials (sometimes also called glasses) the order among atoms is even lower than that in polycrystalline materials. The most important amorphous materials in optoelectronic technology are glasses based on $SiO_x$ (with different dopants) and amorphous semiconductors such as amorphous silicon (a-Si). These materials find important uses as shown in Fig. 1.12.

The amorphous materials are characterized by good short range order, but poor long range order. Thus the nearest neighbor and even second neighbor coordination is quite good in amorphous materials. However, the arrangement of atoms which are third nearest neighbors (or further away) is unpredictable. The amorphous material may also have a high density of broken bonds.

The most important amorphous material in optics is, of course, "glass"

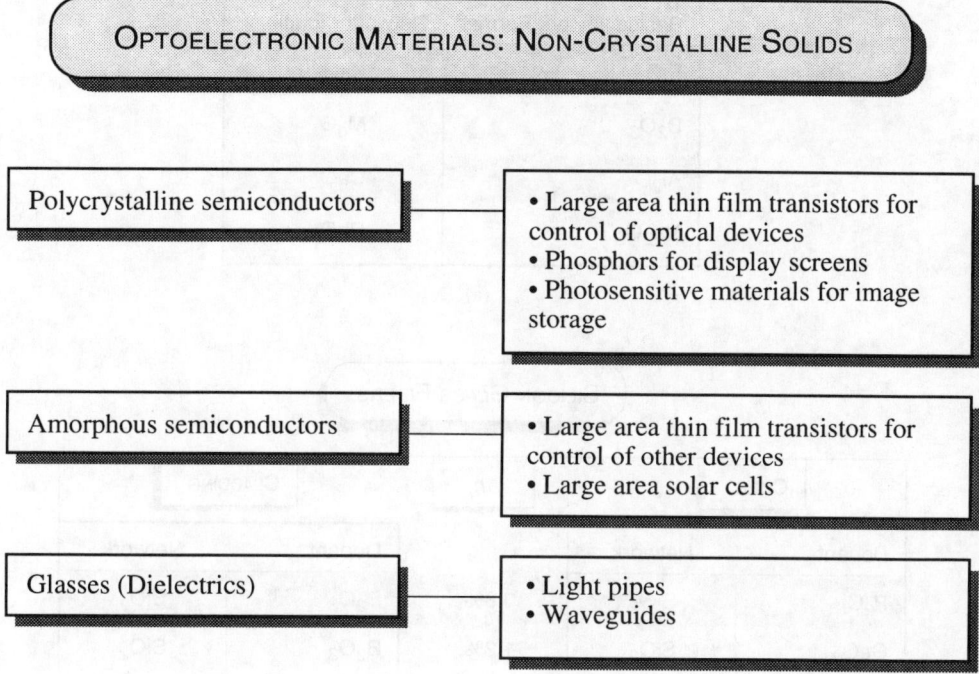

Figure 1.12: Important uses of noncrystalline materials in optoelectronic technology.

which is used in all kinds of optical elements, such as lenses, prisms, etc., as well as in optical fibers. Glass is made from some of the most abundant elements on the earth's crust, vis. oxygen (which forms 62% of earth's crust) and silicon ( which forms 21% of the crust). In silica based glass, silicon and oxygen atoms form a lattice network which is not crystalline, but has a good nearest neighbor ordering. The glass is, in general, made of a component which forms the basic network—the network former—and a component which can introduce subtle changes in the properties of the glass—the network modifier. In Fig. 1.13a we show some important network formers and modifiers. In Fig. 1.13b we show the influence of various dopants on the refractive index for a $SiO_2$-based glass system.

The most widely used glass is silica based as far as optical fibers are concerned due to the high purity level that is possible. Glasses based on $B_2O_3$, $N_2O_3$, etc., are used for other industrial applications. The network modifiers are used to modify the refractive index of the glass and by controlled introduction, they can cause a spatial variation in the index profile.

Next we examine the structure of a-Si which is perhaps the most important amorphous semiconductor material due to its importance in solar cell technology and display technology. In Fig. 1.14, we show a schematic comparison between

| Basic Network Former | Network Modifier |
|---|---|
| $SiO_2$ | $K_2O$ |
| $B_2O_3$ | $M_gO$ |
| $Al_2O_3$ | $CaO$ |
| $Na_2O_3$ | $PbO$ |

(a)

SILICON GLASS FIBERS

| CORE | | $\Delta n_r$ | CLADDING | |
|---|---|---|---|---|
| Dopant | Network | | Dopant | Network |
| $P_2O_5$ | $SiO_2$ | 0.8% | $B_2O_3$ | $SiO_2$ |
| $GeO_2$ | $SiO_2$ | 1.2% | $B_2O_3$ | $SiO_2$ |
| $GeO_2$, $B_2O_3$ | $SiO_2$ | 1.3% | $B_2O_3$ | $SiO_2$ |

(b)

Figure 1.13: Basic materials used to prepare the optical fibers. (a) Some common types of glasses and their compositions. (b) Core and cladding components of silica fibers with borosilicate glass as cladding.

crystalline Si and a-Si. We note that as in crystalline silicon, in a-Si, the Si atoms are 4-fold coordinated, i.e., they have four nearest neighbors. However, some of the atoms have broken or dangling bonds. A high density of dangling bonds can render the material useless electronically as we shall discuss in Chapter 4. Thus in the growth of a-Si, one ensures that a large fraction of H (or F) is incorporated into the film. The H atoms "tie up" the dangling bonds and thus improve the properties of a-Si. Hydrogenated amorphous silicon is usually denoted by a-Si:H.

*An important difference between the crystalline and amorphous materials is in the macroscopic symmetry of the material. The crystals are anisotropic due the precise arrangement of atoms. The amorphous materials, on the other hand, are isotropic.* Some polycrystalline materials such as PLZT discussed above, can be prepared to have anisotropic optoelectronic properties. The anisotropy of crystals results in special cleavage planes along which crystals can be cleaved (split) with ease.

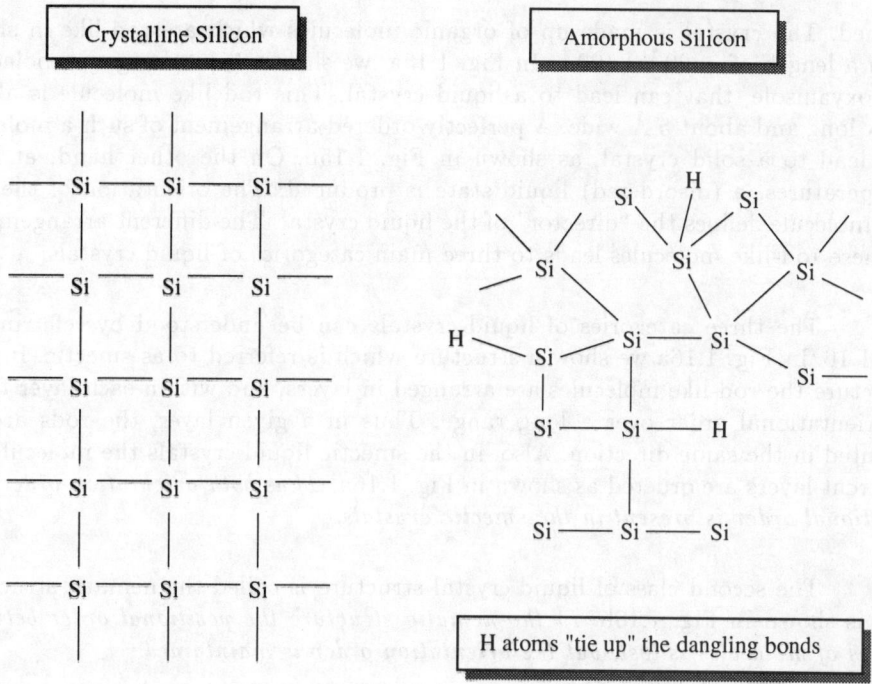

Figure 1.14: A schematic arrangement of the atoms in crystalline and amorphous silicon. The a-Si is hydrogenated to tie up dangling bonds which would otherwise make the material electrically inactive.

# 1.7 LIQUID CRYSTALS

$\longrightarrow$ The liquid crystals are one of the most fascinating material systems in nature, having properties of liquids (such as low viscosity and ability to conform to the shape of a container) as well as of a solid crystal. Their ability to modulate light when an applied electrical signal is used has made them invaluable in flat panel display technology.

As noted in the previous section, crystalline materials have anisotropic properties (they look different from different directions) while noncrystalline materials and liquids are isotropic. Liquid crystals have anisotropic properties similar to solid crystals because of the ordered way in which some of the constituent molecules are arranged. However, the liquid crystals have low viscosity and can flow. The liquid crystals are essentially a stable phase of matter called the mesophase existing between the solid and the liquid.

There are an essentially unlimited number of liquid crystals that can be

formed. The crystal is made up of organic molecules which are rod-like in shape
with a length of ∼ 20Å - 100Å. In Fig. 1.15a, we show a typical organic molecule,
*p*-azoxyanisole, that can lead to a liquid crystal. This rod like molecule is about
20 Å long and about 5 Å wide. A perfectly ordered arrangement of such a molecule
can lead to a solid crystal, as shown in Fig. 1.15b. On the other hand, at high
temperatures, a (disordered) liquid state is produced. The orientation of the rod
like molecule defines the "director" of the liquid crystal. The different arrangements
of these rod-like molecules leads to three main categories of liquid crystals.

The three categories of liquid crystals can be understood by referring to
Fig.1.16. In Fig. 1.16a we show a structure which is referred to as smectic. In this
structure the rod-like molecules are arranged in layers, and within each layer there
is orientational order over a long range. Thus in a given layer, the rods are all
oriented in the same direction. Also, in the smectic liquid crystals the molecules of
different layers are ordered as shown in Fig. 1.16a. *Thus both orientation order and
positional order is present in the smectic crystals.*

The second class of liquid crystal structure is called the nematic structure
and is shown in Fig. 1.16b. *In the nematic structure the positional order between
layers of molecules is lost, but the orientation order is maintained.*

A third class of liquid crystals has the structure shown in Fig. 1.16c and is
called cholesteric. *In these crystals the rod-like molecules in each layer are oriented a
different angle within each layer. Orientation order is maintained within each layer.*
The cholesteric liquid crystal is related to the nematic crystal, with the difference
being the twist of the molecules as one goes from one layer to another.

In addition to the orientational order present in each layer an additional
parameter defining subclasses of a smectic crystal is the chirality (i.e., relative twist)
between molecules. The optical activity of the crystal depends upon the orientation
and the twist present in the molecular layers.

In order that the liquid crystals are useful for optical and optoelectronic applications, it is important that the anisotropy present in the crystal translate into an
optical anisotropy and that a small electrical signal be able to tune the anisotropy.
We will discuss the underlying physics of light propagation in anisotropic media in
Chapter 2. However, here we point out that for low-power optoelectronic devices,
it is important that the rod-like molecules making up the liquid crystal have some
property which can have a strong interaction with light. For this purpose, ferroelectric materials are used in preparation of liquid crystals. As discussed in Section
1.3.4 ferroelectrics are materials which have a net electric dipole even in the absence
of an applied electric field. As a result, these materials have a strong interaction
with light. A very large number of organic molecules can be used to prepare liquid
crystal with good optoelectronic properties. In many cases, these organic molecules

5 Å   CH$_3$—O—⟨⟩—N=N—⟨⟩—O—CH$_3$          p-azoxyanisole (PAA)
                              |                          building block
                              O

←————— ~20 Å —————→

(a)

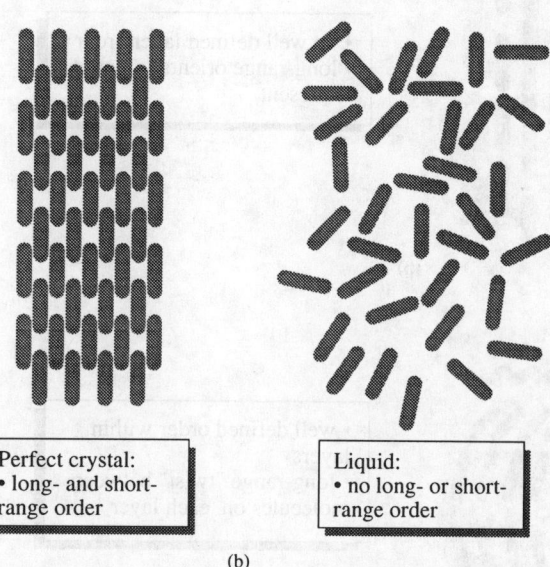

Perfect crystal:                    Liquid:
• long- and short-                  • no long- or short-
  range order                         range order

(b)

Figure 1.15: (a) A typical building block for liquid crystals. (b) A schematic description of a perfect crystal and a liquid. Liquid crystals form a phase of nature in between these extremes.

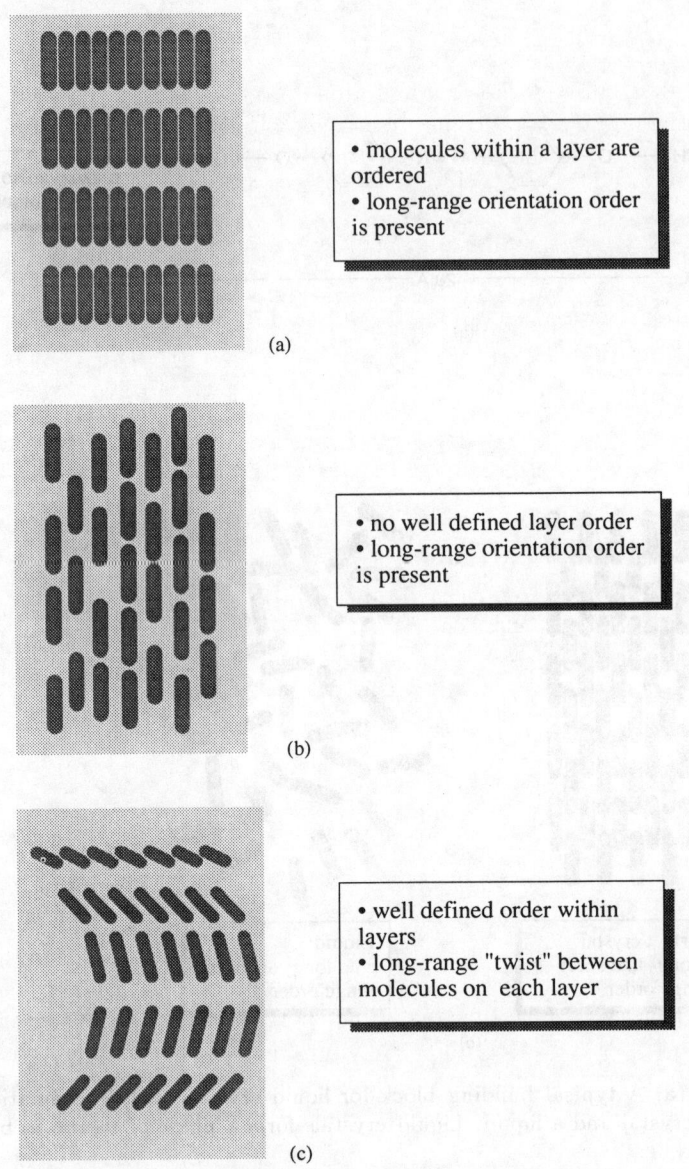

Figure 1.16: A schematic description of the arrangement of molecules in: (a) smectic; (b) nematic; and (c) cholesteric liquid crystals.

are evenly dispersed in polymers to produce what are called *polymer dispersed liquid crystals* (PDLCs).

To fully exploit the potential of liquid crystals, an important feature regarding the interaction of the liquid with surfaces is exploited. It is found that if the surface of a glass plate is rubbed along a certain direction (with, say, a cloth) then when the liquid crystal comes in contact with the surface, the surface molecules align themselves along the rubbed section. Now consider a second rubbed glass plate is placed so that the spacing between the two plates is $\sim$ 5-20 $\mu$m. The orientation of the liquid crystal surface molecules can be prechosen by simply orienting the rubbed direction of the two plates. This produces a twist in the liquid crystal molecules as one goes from one plate to another as shown in Fig. 1.17. Such liquid crystal systems are called *twisted nematic* and a total rotation of 90° can be produced. If the twist angle is increased to enhance the effect, the film becomes unstable if normal nematic films are used. For example, if a twist of 270° is desired, the stable state is one with a −90° twist. *However, if cholesteric liquid crystals are used in which there is already a built-in twist, the* 270° *twist is possible.* Such structures are called *supertwisted.*

The unusual orientation dependence of the rod-like molecules of liquid crystals can be modified by an electric field. This in turn modifies their optical properties resulting in their efficient use as light valves. The physics of these devices will be examined in Chapter 9.

## 1.8  DEFECTS IN MATERIALS

$\mathcal{R}$  In our discussion of crystalline materials we have explored the properties of the perfect crystalline structure. In real materials, crystalline structure invariably has some defects that are introduced due to either thermodynamic considerations (nothing in life is perfect!) or the presence of impurities during the crystal growth process. In general, defects in crystalline semiconductors can be characterized as i) point defects; ii) line defects; iii) planar defects and iv) volume defects. These defects are detrimental to the performance of electronic and optoelectronic devices and are to be avoided as much as possible. We will give a brief overview of the important defects.

### Point Defects
A point defect is a highly localized defect that affects the periodicity of the crystal only in one or a few unit cells. There are a variety of point defects, as shown in Fig. 1.18. An important point in defects is the vacancy that is produced when an atom is missing from a lattice point. The vacancy defects are present in any crystal and

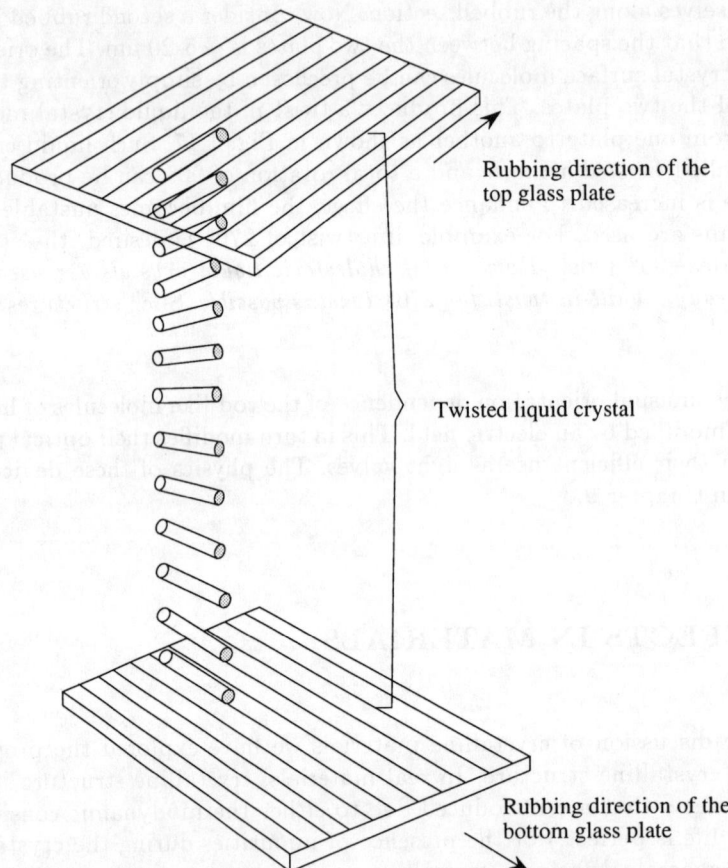

Rubbing direction of the
top glass plate

Twisted liquid crystal

Rubbing direction of the
bottom glass plate

Figure 1.17: A schematic of the twisted nematic liquid crystal produced by using two rubbed glass plates as a container.

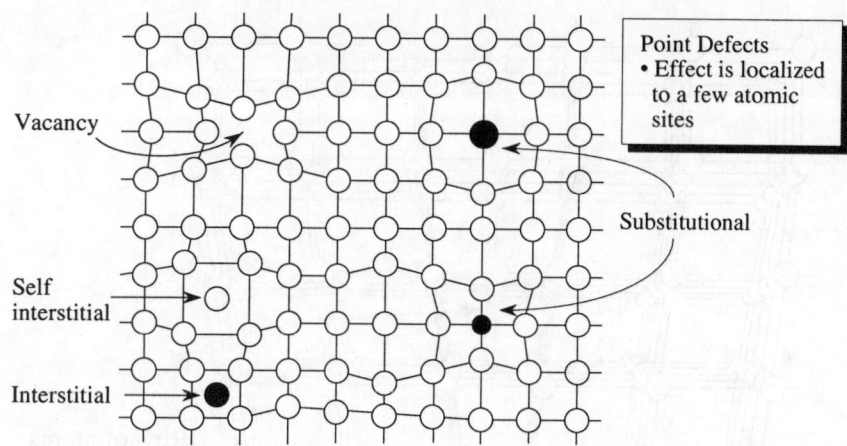

Figure 1.18: A schematic showing some important point defects in a crystal. (Adapted from J.W. Mayer and S.S. Lau, *Electronic Material Science: For Integrated Circuits in Si and GaAs*, MacMillan, New York (1990).)

their concentration is given roughly by the thermodynamics relation

$$\frac{N_{vac}}{N_{Tot}} = exp\left(-\frac{E_{vac}}{k_B T}\right) \tag{1.8}$$

where $N_{vac}$ is the vacancy density, $N_{Tot}$ the total site density in the crystal, $E_{vac}$ the vacancy formation energy and $T$, the crystal growth temperature.

An important point defect in compound semiconductors such as GaAs is the anti-site defect in which one of the atoms, say Ga, sits on the arsenic sublattice instead of the Ga sublattice. Such defects (denoted by $Ga_{As}$) can be a source of reduced device performance.

Other point defects are interstitials in which an atom is sitting in a site that is in between the lattice points as shown in Fig. 1.18, and impurity atoms which involve a wrong chemical species in the lattice. In some cases the defect may involve several sites forming a defect complex.

### Line Defects or Dislocations

In contrast to point defects, line defects (called dislocations) involve a large number of atomic sites that can be connected by a line. Dislocations are produced if, for example, an extra half-plane of atoms is inserted (or taken out) of the crystal as shown in Fig. 1.19. Such dislocations are called *edge dislocations*. Dislocations can also be created if there is a slip in the crystal so that part of the crystal bonds are broken and reconnected with atoms after the slip.

Dislocations can be a serious problem, especially in the growth of strained

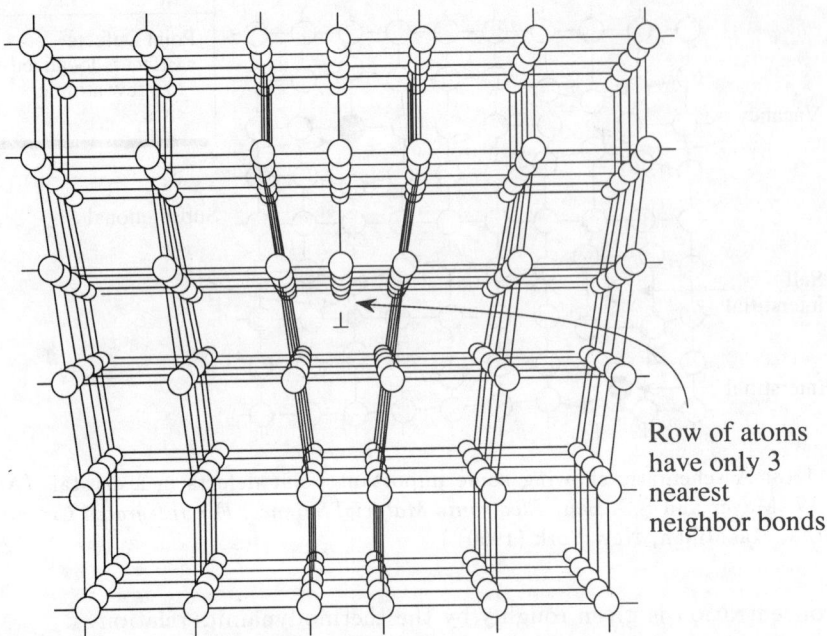

Row of atoms
have only 3
nearest
neighbor bonds

Figure 1.19: A schematic showing the presence of a dislocation. This line defect is produced by adding an extra half-plane of atoms. At the edge of the extra plane, the atoms have a missing bond.

heterostructures, i.e., structures where the lattice constant between the overlayer and the substrate is different. In optoelectronic devices, dislocations can ruin the device performance and render the device useless. Thus the control of dislocations is of great importance.

**Planar Defects and Volume Defects**

Planar defects and volume defects are not important in single crystalline materials, but can be of importance in polycrystalline materials. If, for example, silicon is grown on a glass substrate, it is likely that polycrystalline silicon will be produced. In the polycrystalline material, small regions of Si ($\sim$ a few microns in diameter) are perfectly crystalline, but are next to microcrystallites with different orientations. The interface between these microcrystallites are called grain boundaries. Grain boundaries may be viewed as an array of dislocations.

Volume defects can be produced if the crystal growth process is poor. The crystal may contain regions that are amorphous or may contain voids. In most epitaxial techniques used in modern optoelectronics, these defects are not a problem. However, the developments of new material systems such as diamond (C) or SiC are hampered by such defects.

**EXAMPLE 1.6** Consider an equilibrium growth of a semiconductor at a temperature of 1000 K. The vacancy formation energy is 2.0 eV. Calculate the vacancy density produced if the site density for the semiconductor is $2.5 \times 10^{22}$ cm$^{-3}$.

The vacancy density is

$$
\begin{aligned}
N_{vac} &= N_{Tot} exp\left(-\frac{E_{vac}}{k_B T}\right) \\
&= (2.5 \times 10^{22}\ cm^{-3})\, exp\left(-\frac{2.0\ eV}{0.0867\ eV}\right) \\
&= 2.37 \times 10^{12}\ cm^{-3}
\end{aligned}
$$

This is an extremely low density and will have little effect on the properties of the semiconductor. The defect density would be in mid $10^{15}$ cm$^{-3}$ range if the growth temperature was 1500 K. At such values, the defects can significantly affect device performance.

## 1.9   SCIENCE AND TECHNOLOGY CHALLENGES

In this section we will briefly describe the important challenges that are being faced by scientists in the areas discussed in this chapter.

In Fig. 1.20 we show an outline of the challenges in the area of material synthesis. Let us discuss these challenges and the motivations behind them.

**Crystalline Semiconductors**
In the area of crystalline semiconductors, materials like Si and GaAs have been studied for several decades and as a result, their technologies are quite advanced. As shown in Fig. 1.20, the challenge in semiconductors lies in materials such as GaN, SiC, and C (diamond) which have a very large bandgap. Such materials can be used for high power applications (currently most radar, radio, and satellite transmissions still occur by vacuum tubes) where semiconductor devices are still lagging vacuum tubes. Such materials can also produce light with extremely short wavelength which could be used to read very small feature sizes in compact discs. The challenges in these materials arises due to lack of good substrates (only Si, GaAs and InP substrates are of high quality) which are lattice matched to them, and due to difficulties in doping and high-defect densities.

In addition to high-bandgap materials, there are many challenges in the fabrication of small bandgap materials such as InAs, InSb, and HgCdTe, etc., alloys. These materials can be used for detection of far infrared light (of wavelengths $\sim 14$ $\mu$m) for "nightvision" and other thermal imaging applications. These low-bandgap materials are very "soft" making it difficult to fabricate them with low defects and

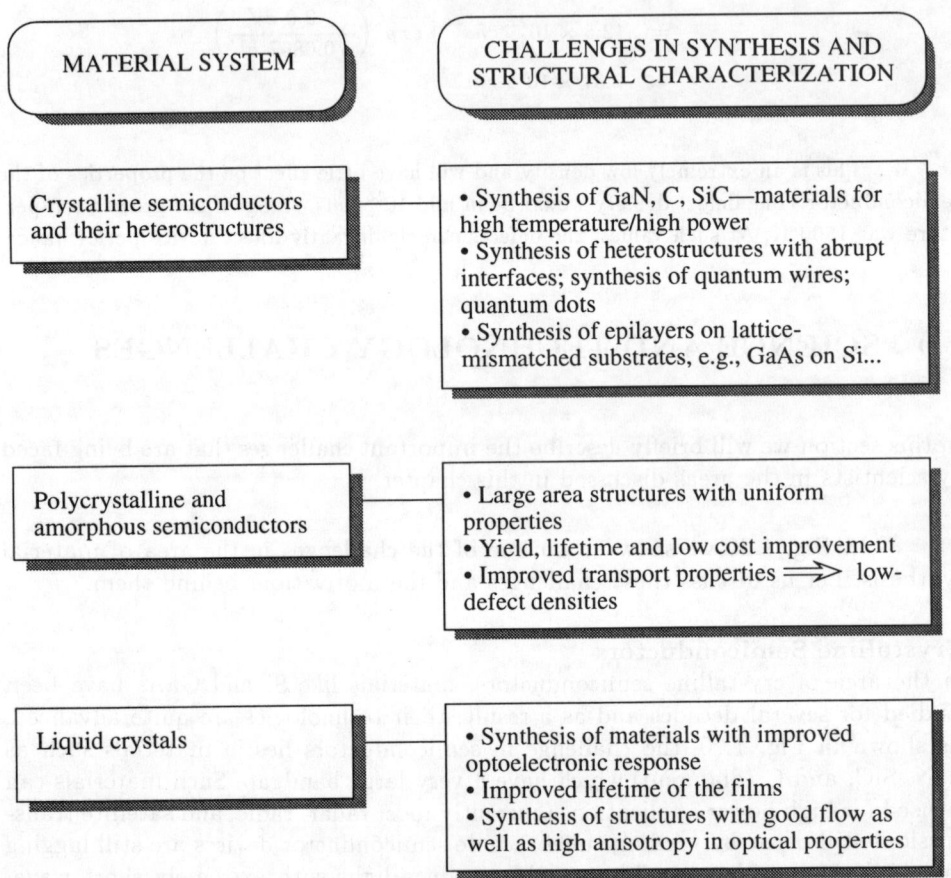

Figure 1.20: Challenges in the area of synthesis and characterization of optoelectronic materials.

high yields. It is also difficult to dope them reliably because of the high background defect density.

The area of heteroepitaxy has led to numerous new devices in both electronic and optoelectronic arena. In these materials, the challenge is to produce high-quality interfaces between two different materials. There is considerable interest in producing structures where a slab of material $\sim$ 100 Å thick is surrounded by another material (the quantum wells). Such structures have been shown to produce excellent lasers. There is also interest in surrounding a thin ($\sim$ 100 Å $\times$ 100 Å) wire of one material by another material (the quantum wire). Such structures have superior predicted optical properties, although these predicted benefits have not yet been realized. There is also some interest in producing quantum dots where one semiconductor of dimensions $\sim$ 100 Å $\times$ 100 Å $\times$ 100 Å is completely surrounded by another material. At present, only quantum wells can be fabricated with ease. The other structures are still in their infancy.

Another important challenge in semiconductor technology is the growth of high quality materials on a substrate with different crystal properties. An important example is the growth of GaAs on Si or InP on Si. Silicon substrates are cheap and can be produced in large sizes (up to 12 inches). However, semiconductor lasers and high-speed detectors cannot be made from silicon. If materials such as GaAs can be deposited on silicon substrates, the cost of optoelectronic devices can be greatly reduced. Unfortunately, it is extremely difficult to grow one material on another one with a different lattice constant without generating a very high density of dislocations.

### Polycrystalline and Amorphous Materials

Polycrystalline and amorphous materials are used in applications where low cost, and not device performance, is of ultimate importance. These applications involve large area solar cell arrays as will as thin film transistors for flat panel displays. Even though device performance can be sacrificed to some extent, one still needs a certain minimum performance. The challenges in the large area growth techniques are the reproducibility of growth yield of low-defect material for improved electrical performance, and compatibility of growth with substrates which are usually glass or quartz.

## 1.10 CHAPTER SUMMARY

In this chapter we have discussed the important structural properties of semiconductors and their heterostructures. The semiconductors we will be dealing with in most optoelectronic devices have a zinc blende or diamond structure. We have discussed the important growth techniques used in producing the semiconductors. We

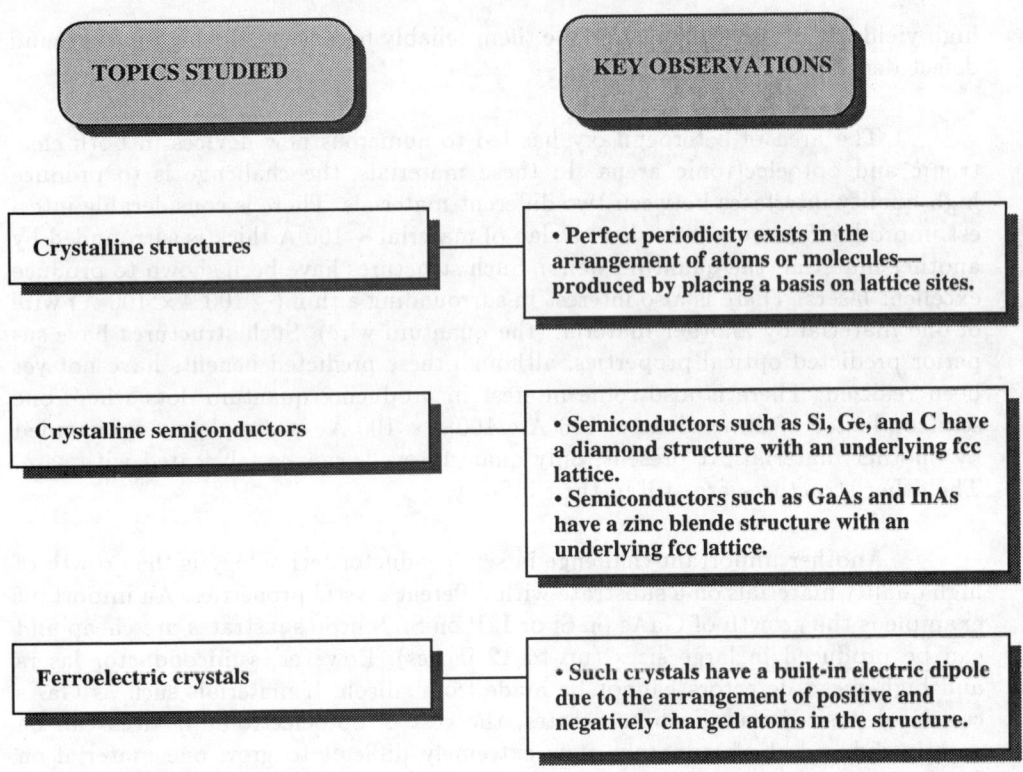

Table 1.3: Summary Table

have also identified the important techniques used to fabricate devices. Tables 1.3 to 1.5 give an overview of the issues that have emerged from this chapter.

## 1.11   PROBLEMS

### Section 1.3
**1.1** a) Find the angles between the tetrahedral bonds of a diamond lattice.

b) What are the direction cosines of the (111) oriented nearest neighbor bond along the x,y,z axes?

**1.2** Consider a semiconductor with the zinc blende structure (such as GaAs).

a) Show that the (100) plane is made up of either cation or anion type atoms.

b) Draw the positions of the atoms on a (110) plane assuming no surface reconstruction.

c) Show that there are two types of (111) surfaces: one where the surface atoms are bonded to three other atoms in the crystal, and another where the surface atoms are bonded to only one. These two types of surfaces area called the A and B surfaces,

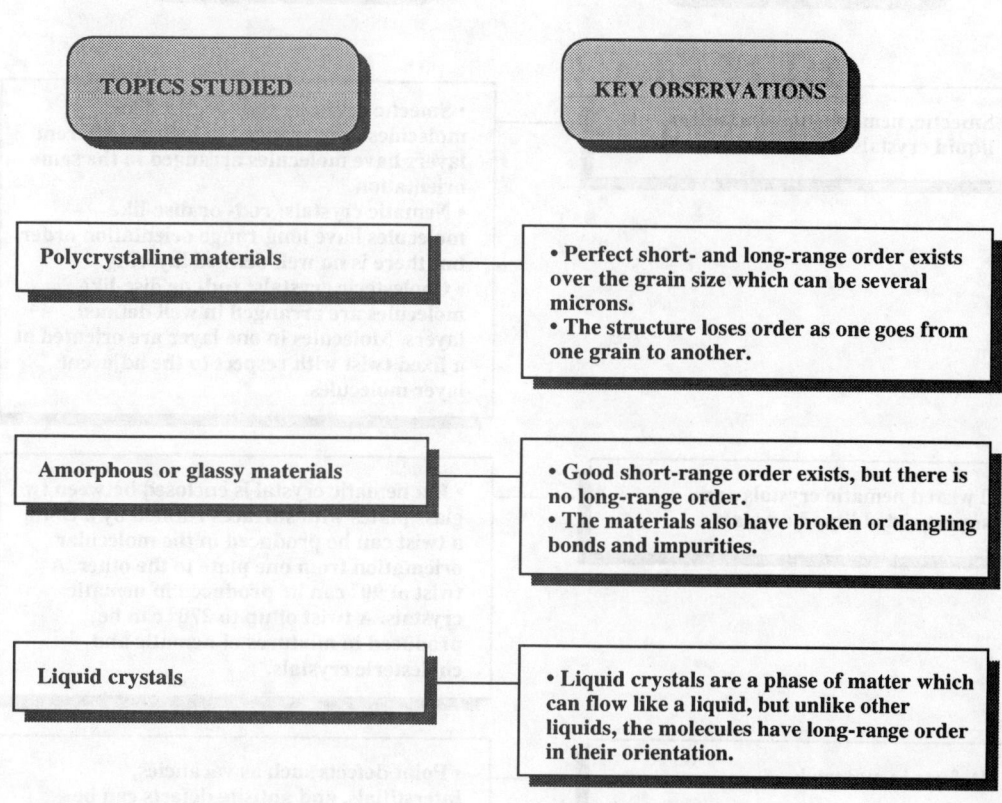

| TOPICS STUDIED | KEY OBSERVATIONS |
|---|---|
| Polycrystalline materials | • Perfect short- and long-range order exists over the grain size which can be several microns.<br>• The structure loses order as one goes from one grain to another. |
| Amorphous or glassy materials | • Good short-range order exists, but there is no long-range order.<br>• The materials also have broken or dangling bonds and impurities. |
| Liquid crystals | • Liquid crystals are a phase of matter which can flow like a liquid, but unlike other liquids, the molecules have long-range order in their orientation. |

Table 1.4: Summary Table

| TOPICS STUDIED | KEY OBSERVATIONS |
|---|---|
| Smectic, nematic, and cholesteric liquid crystals | • Smectic crystals: rod- or disc-like molecules are arranged in layers. Different layers have molecules arranged in the same orientation.<br>• Nematic crystals: rod- or disc-like molecules have long-range orientation order, but there is no well defined layering.<br>• Cholesteric crystals: rod- or disc-like molecules are arranged in well defined layers. Molecules in one layer are oriented at a fixed twist with respect to the adjacent layer molecules. |
| Twisted nematic crystals and supertwisted liquid crystals | • If a nematic crystal is enclosed between two glass plates with surfaces rubbed by a cloth, a twist can be produced in the molecular orientation from one plate to the other. A twist of 90° can be produced in nematic crystals. A twist of up to 270° can be produced in mixtures of nematic and cholesteric crystals. |
| Defects in materials | • Point defects such as vacancies, interstitials, and antisite defects can be generated in materials.<br><br>• Line defects such as dislocations can be a serious problem in heterostructures. |

Table 1.5: Summary Table

respectively.

**1.3** Suppose that identical solid spheres are placed in space so that their centers lie on the atomic points of a crystal and the spheres on the neighboring sites touch each other. Assuming that the spheres have unit density, show that density of such spheres is the following for the various crystal structures:

$$
\begin{aligned}
fcc &: \quad \sqrt{2}\pi/6 = 0.74 \\
bcc &: \quad \sqrt{3}\pi/8 = 0.68 \\
sc &: \quad \pi/6 = 0.52 \\
diamond &: \quad \sqrt{3}\pi/16 = 0.34
\end{aligned}
$$

**1.4** Calculate the number of cells per unit volume in GaAs (a = 5.65 Å). Si has a 4% larger lattice constant. What is the unit cell density for Si? What is the number of atoms per unit volume in each case?

**1.5** A Si wafer is nominally oriented along the (001) direction, but is found to be cut 2° off, towards the (110) axis. This off axis cut produces "steps" on the surface which are 2 monolayers high. What is the lateral spacing between the steps of the 2° off-axis wafer?

**1.6** Conduct a literature search to find out what the lattice mismatch is between GaAs and AlAs at 300 K and 800 K. Calculate the mismatch between GaAs and Si at the same temperatures.

**1.7** In high purity Si crystals, defect densities can be reduced to levels of $10^{13}$ cm$^{-3}$. On an average, what is the spacing between defects in such crystals? In heavily doped Si, the dopant density can approach $10^{19}$ cm$^{-3}$. What is the spacing between defects for such heavily doped semiconductors?

**1.8** A GaAs crystal which is nominally along (001) direction is cut $\theta$ off towards (110) axis. This produces one monolayer high step. If the step size is to be no more than 100 Å, calculate $\theta$.

**1.9** Assume that a Ga-As bond in GaAs has a bond energy of 1.0 eV. Calculate the energy needed to cleave GaAs in the (001) and (110) planes.

**1.10** In this chapter, we have considered only the fcc Bravias lattice, since most semiconductors have this underlying structure. However, some semiconductors have the hexagonal close-packed structure. Semiconductors such as BN, AlN, GaN, InN, SiC, etc., crystallize in the structure.

The hcp structure is formed as shown in Fig. 1.21a. A close-packed layer of spheres is formed with centers at points $A$. A second layer of spheres is placed on top of this with centers at points $B$. The third layer can be placed on points $A$ (giving rise to the hcp structure), or points $C$ (giving rise to the fcc structure). The hcp has the primitive cell of the hexagonal lattice with two atoms on the basis as shown in Fig. 1.21b. The primitive cell has primitive vectors $a_1 = a_2$ and the $c$-axis (vector $a_3$ is parallel to $c$) normal to the $a_1, a_2$ plane. One atom is at the origin and the other, at the point

$$
r = \frac{2}{3}a_1 + \frac{1}{3}a_2 + \frac{1}{2}a_3
$$

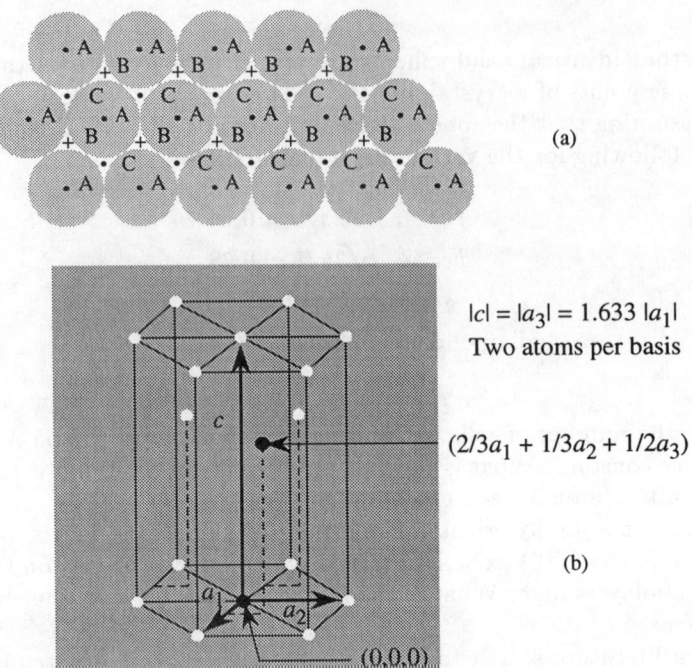

$|c| = |a_3| = 1.633\ |a_1|$
Two atoms per basis

$(2/3a_1 + 1/3a_2 + 1/2a_3)$

(a)

(b)

Figure 1.21: Hexagonal close-packed structure.

Show that the ratio $c/a, (a = a_1, a_2)$ is given by $\sqrt{8/3} = 1.633$. The values of these lattice constants for several semiconductors are given below. These semiconductors are often said to have the Wurtzite structure.

## Section 1.6
**1.11** A silicon polycrystalline film has a grain size of 2.0 $\mu$m. How many atoms are in the grain, assuming that the grain is a cube?

**1.12** A GaAs film is grown on a glass substrate and heat treated to produce a polycrystalline film of grain size 10.0 $\mu$m. If a 2 $\mu$m $\times$2 $\mu$m diode is fabricated on the film, what is the probability that the device has high performance? Assume that the grain size is square on the surface.

**1.13** When a polycrystalline film is heat treated (annealed), the grain size usually increases. Comment on why this occurs.

## Section 1.8
**1.14** Using symmetry arguments and energy minimization arguments, discuss why in a nematic crystal, it is not possible to achieve a 270° twist.

**1.15** Some liquid crystals are used as temperature detectors. Using simple thermodynamic arguments, discuss why the long range order of the liquid crystal is destroyed as temperature increases.

**1.16** A typical thickness of a liquid crystal cell used in laptop computers is 5 Å.

If the thickness of the nematic crystal molecule is 5 $\mu$m and the average spacing between the molecules is 10 Å, how many molecules are stacked in a typical display cell?

**Section 1.9**

**1.17** A serious problem in the growth of a heterostructure made from two semiconductors is due to the difficulty in finding a temperature at which both semiconductors can grow with high quality. Consider the growth of HgTe and CdTe which is usually grown at $\sim 600$ K. Assume that the defect formation energy in HgTe is 1.0 eV and in CdTe is 2.0 eV. Calculate the density of defects in the heterostructure with equal HgTe and CdTe.

**1.18** Calculate the defect density in GaAs grown by LPE at 1000 K. The defect formation energy is 2.0 eV.

**1.19** Why are entropy considerations unimportant in dislocation generation?

## 1.12   REFERENCES

- **Crystal Structures**

  - J. M. Buerger, *Introduction to Crystal Geometry*, McGraw-Hill (1971).

  - M. Lax, *Symmetry Principles in Solid State and Molecular Physics*, Wiley (1974). Has a good description of the Brillouin zones of several structures in Appendix E.

  - J. F. Nye, *Physical Properties of Crystals*, Oxford (1985).

  - F. C. Phillips, *An Introduction of Crystallography*, Wiley (1971).

- **Defects in Semiconductors**

  - P.K. Bhattacharya and S. Dhar, *Deep Levels in III-V Compound Semiconductors Grown by Molecular Beam Epitaxy*, *Semiconductors and Semimetals*, eds. A.C. Willardson and C. Beer, Academic Press, New York, vol. 26 (1988).

  - E.N. Economou, *Green's Functions in Quantum Physics*, Springer-Verlag, Berlin (1979).

  - G.F. Foster and J.C. Slater, *Phys. Rev.*, **96**, 1208 (1954).

  - H.F. Matare, *Defect Electronics in Semiconductors*, Wiley-Interscience, New York (1971).

  - S. Pantelides, *Rev. Mod. Phys.*, **50**, 797 (1978).

- **Dislocations and Lattice Mismatched Epitaxy**

  - S. Amelinckx, *Dislocations in Solids*, ed. F.R.N. Nabarro, North-Holland, New York (1988).

  - C.A.B Ball and J.H. van der Merwe, *Dislocations in Solids*, ed. F.R.N. Nabarro, North-Holland, New York, vol. 5 (1983).

  - H.F. Matare, *Defect Electronics in Semiconductors*, Wiley-Interscience, New York (1971).

# CHAPTER 2

# LIGHT PROPAGATION IN MEDIA

## 2.1   INTRODUCTION

Information processing by optoelectronics involves the propagation, generation and modulation (switching) of information coded in optical beams. It is therefore important to know how light propagates through various media and how the propagation is affected if the optical properties of the media are altered. We will start this chapter with a discussion of light propagation in uniform media and then discuss the Fresnel formulae that tell us how reflection and refraction occur across media.

In Chapter 1 we have identified several important materials that are used in modern optoelectronic systems. These materials can be broadly characterized as crystalline and non-crystalline. The crystalline materials have anisotropic structural and optical properties due to the ordered arrangement of atoms. On the other hand, non-crystalline materials usually have isotropic properties due to the random arrangement of atoms over a long range. The presence or absence of isotropic properties has a profound effect on light propagation and in this chapter we will examine these differences.

In order to understand optical wave propagation we must start with Maxwell equations that describe the physics of electromagnetic waves. We therefore start this chapter with an examination of Maxwell equations and the resulting wave equations.

## 2.2   MAXWELL EQUATIONS AND THE WAVE EQUATION

$\longrightarrow$   This text deals with optoelectronics, i.e., with the interaction of "light" with various electronic materials. Light waves form a small subset of the broader electromagnetic radiation spectrum. In Fig. 2.1a we show the electromagnetic spectrum ranging from gamma rays to radio waves. The fundamental physics behind this vast spectrum with wavelength ranging from Angstroms to kilometers is described by the Maxwell equations which is the topic of this section. However, it is important to note that while the basic physics is the same for all electromagnetic waves, whether or not the waves are useful for a certain application depends upon the details of the application. Thus, x-rays are appropriate for studying the arrangement of atoms in a crystal, but optical waves are appropriate for sending information across the oceans.

The optical band in the electromagnetic spectrum usually refers to waves with wavelengths ranging roughly from 0.3 $\mu$m to 15 $\mu$m as shown in Fig. 2.1a. This

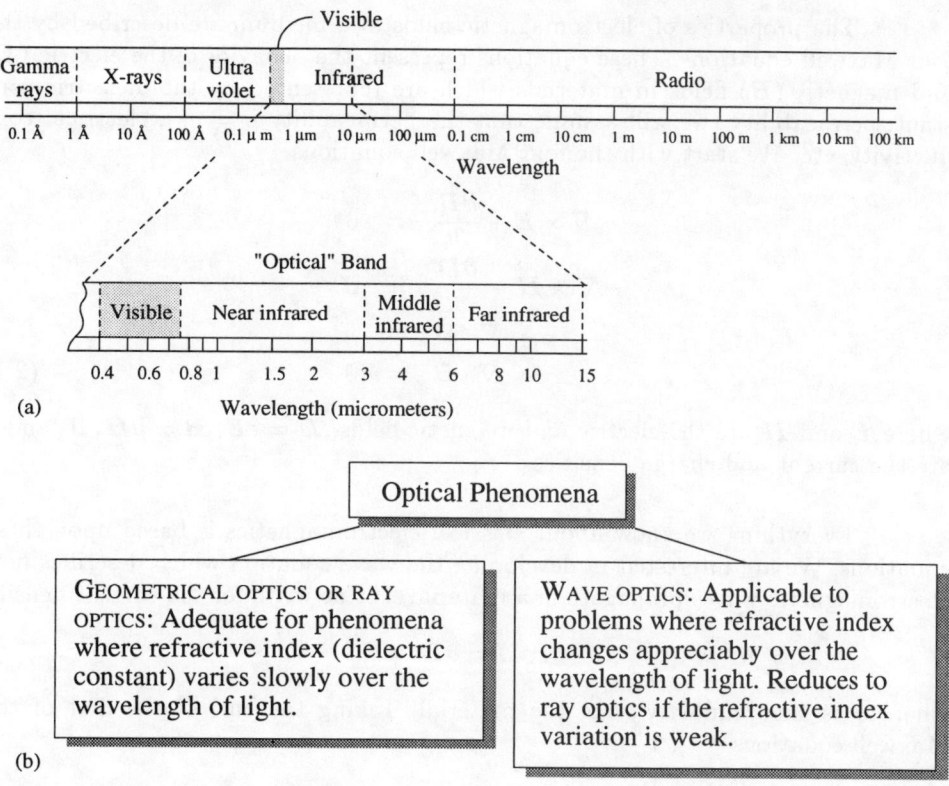

Figure 2.1: (a) The electromagnetic spectrum extends from Gamma rays to radio waves spanning 16 orders of magnitude in frequencies. Optical effects usually involve radiation of wavelength between 0.3 to 15 micrometers. (b) A complete description of optical phenomenon needs the wave description. However, ray optics is adequate for many applications.

includes the "visible" band which is responsible for human sight.

As we start our discussion of the properties of optical waves, it is important to point out two approaches that are taken to describe optical effects. The reader has undoubtedly seen these approaches in describing how a lens works and in describing how diffraction occurs. To describe most of the properties of a lens, one uses a ray approach or "geometrical optics" approach. In this approach, light is treated as traveling in a straight line and all wave effects are suppressed. As noted in Fig. 2.1b, geometric optics is adequate to describe phenomena in materials where the refractive index does not change appreciably over a distance equal to the wavelength of light. However, when we are dealing with optical elements where the refractive index does change significantly over the dimensions of a wavelength, we must use a full wave description to understand optical phenomenon. We will use both geometrical optics and wave optics to study optical effects in various devices.

The properties of electromagnetic fields in a medium are described by the four Maxwell equations. These equations represent the behavior of the electric ($E$) and magnetic ($B$) fields in materials which are represented by the dielectric constant, permeability (we will assume that the permeability $\mu = \mu_0$), electrical conductivity, etc. We start with the four Maxwell equations

$$\nabla \times E + \frac{\partial B}{\partial t} = 0$$

$$\nabla \times H - \frac{\partial D}{\partial t} = J$$

$$\nabla \cdot D = \rho$$

$$\nabla \cdot B = 0 \qquad (2.1)$$

where $E$ and $H$ are the electric and magnetic fields, $D = \epsilon E$, $B = \mu H$, $J$, and $\rho$ are the current and charge densities.

Everything we know about classical electromagnetics is based upon these equations. We are interested in developing the wave equation which describes how electromagnetic waves propagate in a material. Let us write for the current density

$$J = \sigma E$$

where $\sigma$ is the conductivity of the material. Taking the curl of the first of the Maxwell equations we get

$$\nabla \times (\nabla \times E) + \frac{\partial}{\partial t}(\nabla \times B) = 0$$

Using the identity

$$\nabla \times (\nabla \times E) = \nabla(\nabla \cdot E) - \nabla^2 E$$

and the observation that for a neutral material ($\rho = 0$) we have $\nabla \cdot E = 0$, we get, after substituting for $\nabla \times B$ from the second Maxwell equation,

$$\nabla^2 E = \epsilon \mu_0 \frac{\partial^2 E}{\partial t^2} + \sigma \mu_0 \frac{\partial E}{\partial t} \qquad (2.2)$$

This represents a wave equation for a wave propagating with dissipation. The general solution can be chosen to be of the form

$$E = E_0 \exp\left\{i(k \cdot r - \omega t)\right\} \qquad (2.3)$$

so that $k$ is given by

$$- k^2 = -\epsilon \mu_0 \omega^2 - \sigma \mu_0 i \omega \qquad (2.4)$$

or ($c = (\epsilon_0 \mu_0)^{-1/2}$; $\tilde{\epsilon}$ is the relative dielectric constant)

$$k = \frac{\omega}{c}\left(\tilde{\epsilon} + \frac{\sigma \mu_0 i}{\omega}\right)^{1/2} \qquad (2.5)$$

Note that we are using the complex space notation to describe the fields. The fields are, of course, real and are given by the real part of the complex wave, nevertheless, the complex plane notation is very useful. In general, $k$ is a complex number. In free space where $\sigma = 0$ we simply have $(\tilde{\epsilon} = 1)$

$$k = \omega/c \tag{2.6}$$

In a medium, the phase velocity $\omega/k$ is modified by dividing $c$ by a complex refractive index given by

$$n_r = \left( \tilde{\epsilon} + \frac{\sigma \mu_0 i}{\omega} \right)^{1/2}$$

We can write the complex refractive index in terms of its real and imaginary parts

$$n_r = n_r' + i n_r''$$

so that

$$k = \frac{n_r' \omega}{c} + i n_r'' \frac{\omega}{c} \tag{2.7}$$

The electric field given by wave Eqn. 2.3 now becomes (for propagations in the $+z$ direction)

$$E = E_0 \exp \left\{ i\omega \left( \frac{n_r' z}{c} - t \right) \right\} \exp \left( \frac{-n_r'' \omega z}{c} \right) \tag{2.8}$$

The velocity of the wave is reduced by $n_r'$ to $c/n_r'$ and its amplitude is damped exponentially by a fraction $\exp \left( -2\pi n_r''/n_r' \right)$ per wavelength. The damping of the wave is associated with the absorption of the electromagnetic energy. The absorption coefficient $\alpha$ is described by the absorption of the intensity (i.e., square of Eqn. 2.8)

$$\alpha = \frac{2 n_r'' \omega}{c} \tag{2.9}$$

Note that in the absence of absorption, $n_r' = n_r$, and the refraction index will simply be denoted by $n_r$. The absorption coefficient can be measured for any material system and it provides information on $n_r''$.

Repeating the derivation given above for the magnetic fields, we can derive the wave equation for the magnetic field. This equation is

$$\nabla^2 H = \epsilon \mu_0 \frac{\partial^2 H}{\partial t^2} + \sigma \mu_0 \frac{\partial H}{\partial t} \tag{2.10}$$

which has propagating solutions similar to those discussed for the electric field,

$$H(r, t) = H_o exp \left\{ i(k \cdot r - \omega t) \right\} \tag{2.11}$$

We will now summarize the relation between the electric and magnetic fields and the energy and power associated with electromagnetic waves. The energy density associated with the electric and magnetic fields is given by

$$E_{den} = \frac{1}{2} \left( \boldsymbol{E} \cdot \boldsymbol{D} + \boldsymbol{H} \cdot \boldsymbol{B} \right) \qquad (2.12)$$

while the power density associated with radiation is given by the Poynting vector $\boldsymbol{S}$, where

$$\boldsymbol{S} = \boldsymbol{E} \times \boldsymbol{H} \qquad (2.13)$$

The macroscopic material properties of the medium are represented by the real and imaginary parts of the refractive index. In general, as the wave propagates, these parameters may change across some boundary. To solve for the electric and magnetic fields across such boundaries, it is important to use the boundary conditions that are imposed on the solutions. If $z$ is normal to a plane which divides two regions with different material parameters, the normal components of $\boldsymbol{D}$ and $\boldsymbol{B}$ across the boundary are related by (these conditions result from the last two of the equations given by Eqn. 2.1)

$$\begin{aligned} \left( \boldsymbol{D}(z^+) - \boldsymbol{D}(z^-) \right) \cdot \hat{z} &= \sigma_s \\ \left( \boldsymbol{B}(z^+) - \boldsymbol{B}(z^-) \right) \cdot \hat{z} &= 0 \end{aligned} \qquad (2.14)$$

where $\hat{z}$ is a unit vector and $z^+$ and $z^-$ represent points on either side of the boundary. Here $\sigma_s$ is the surface charge density at the boundary. Additionally, the tangential components of the electric and magnetic fields are related by (these conditions result from the first two Maxwell equations)

$$\begin{aligned} \hat{z} \times \left( \boldsymbol{E}(z^+) - \boldsymbol{E}(z^-) \right) &= 0 \\ \hat{z} \times \left( \boldsymbol{H}(z^+) - \boldsymbol{H}(z^-) \right) &= \boldsymbol{J}_s \end{aligned} \qquad (2.15)$$

where $\boldsymbol{J}_s$ is the current density flowing along the surface.

Note that if an electromagnetic plane wave is travelling along the $z$-axis, the $\boldsymbol{E}$ and $\boldsymbol{B}$ fields are in the $x$-$y$ plane and the electric and magnetic field components are related by the relations (using, say, the first Maxwell equation)

$$\begin{aligned} E_y &= \frac{\omega}{k} B_x \\ E_x &= -\frac{\omega}{k} B_y \end{aligned} \qquad (2.16)$$

**EXAMPLE 2.1** The solar constant gives the power density incident on the earth from the sun. Assuming this to be 2.0 calories/minute/cm$^2$, calculate the following:
$i$) The magnitude of the Poynting vector in sunlight in units of $W/m^2$.
$ii$) The rms electric field in sunlight.

*iii)* The rms magnetic field in sunlight.

The Poynting vector magnitude is simply (1 calorie = 4.186 J)

$$S = \frac{2 \times 4.186 \; J}{(60 \; s)(10^{-4} \; m^2)} = 1.395 \; kW/m^2$$

To calculate the electric field we note that

$$B = \frac{kE}{\omega}$$

or

$$H = \frac{E}{c\mu_0}$$

Using $\mu_0 = 4\pi \times 10^{-7}(kg \; m \; A^{-2} \; s^{-2})$ and $c = 3 \times 10^8 \; m/s$, we get

$$| \; E_{rms} \; |^2 = c\mu_0 \; S = 5.25 \times 10^5 \; (V/m)^2$$

or

$$E_{rms} = 725 \; V/m$$

In a similar manner, the magnetic field has a value

$$H_{rms} = 1.93 \; A/m$$

**EXAMPLE 2.2** A detector has been designed to detect a minimum rms power density of 10.0 nW/cm$^2$. Calculate the amplitude of the electric field associated with the minimum detectable optical signal.

The electric field amplitude is given by

$$\begin{aligned} |E|^2 &= 2 \; c\mu_0 S \\ &= 2 \times \left(3 \times 10^8 \; m/s\right) \left(4\pi \times 10^{-7} \; kg \; m \; A^{-2} \; s^{-2}\right) \left(10^{-4} \; W/m^2\right) \end{aligned}$$

This gives

$$E = 0.275 \; V/m$$

**EXAMPLE 2.3** Consider two dielectrics with $\epsilon_1 = 3\epsilon_0$ and $\epsilon_2 = 5\epsilon_0$ with interface plane in the $xy$-plane. The electric field vector has a value $E_1 = (3\hat{x} + 5\hat{y} - 4\hat{z})E_0$ in a medium 1 at the interface. Calculate the electric field at the other side (in medium 2) of the interface. Also calculate the electric energy density $\frac{1}{2} E \cdot D$ at the two sides of the interface.

Since the tangential components of the electric field are continuous at a boundary, we have, for the second medium

$$E_{2x} = 3E_0$$
$$E_{2y} = 5E_0$$

To calculate the $z$-component, use the continuity of the $D$-field

$$D_{1z} = D_{2z}$$

i.e.,

$$3\epsilon_0(-4E_0) = 5\epsilon_0(E_{2z})$$

which gives

$$E_{2z} = \frac{-12}{5}E_0$$

The electric field at the interface in medium 2 is then

$$E_2 = \left(3\hat{x} + 5\hat{y} - \frac{12}{5}\hat{z}\right)E_0$$

The $D$-fields in the two media are

$$
\begin{aligned}
D_1 &= \epsilon_0\left(9\hat{x} + 15\hat{y} - 12\hat{z}\right)E_0 \\
D_2 &= \epsilon_0\left(15\hat{x} + 25\hat{y} - 12\hat{z}\right)E_0
\end{aligned}
$$

The electric energy density changes from 75 $\epsilon_0 E_0^2$ to 99.4 $\epsilon_0 E_0^2$ as one goes from medium 1 to medium 2.

## 2.3   POLARIZATION OF LIGHT

$\longrightarrow$   So far, in our discussion on the solutions of the wave equation, we have only mentioned in passing that the electric and magnetic fields are perpendicular to the propagation direction. The precise orientation of the electric field defines the polarization of the wave. For many optoelectronic devices the polarization of light is of extreme importance since the device operation depends critically on controlling and manipulating the polarization state of the optical wave. We start with the standard nomenclature used in defining the polarization state of the optical wave.

Let us consider a monochromatic plane wave travelling along the $z$-axis. The polarization is defined through the electric field vector which, in the complex-function representation, can be written as

$$E(z,t) = Re[A\,exp\,(i(\omega t - kz))] \tag{2.17}$$

where $A$ is a complex vector in the $xy$ plane. To see the time evolution of the electric field in the complex $xy$ plane, we examine the components $E_x$ and $E_y$,

$$
\begin{aligned}
E_x &= A_x cos(\omega t - kz + \delta_x) \\
E_y &= A_y cos(\omega t - kz + \delta_y)
\end{aligned}
\tag{2.18}
$$

where we have used a general form for $A$

$$A = \hat{x}A_x e^{i\delta_x} + \hat{y}A_y e^{i\delta_y} \tag{2.19}$$

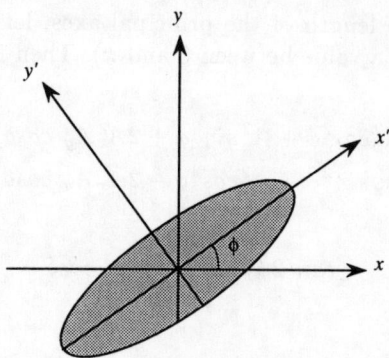

Figure 2.2: A general polarization and ellipse. The major axis of the ellipse are along $x'$ and $y'$.

with $A_x$ and $A_y$ being positive real numbers. Expanding the cosine terms as

$$
\begin{aligned}
cos(\omega t - kz + \delta_x) &= cos(\omega t - kz)cos(\delta_x) - sin(\omega t - kz)sin(\delta_x) \\
cos(\omega t - kz + \delta_y) &= cos(\omega t - kz)cos(\delta_y) - sin(\omega t - kz)sin(\delta_y)
\end{aligned} \quad (2.20)
$$

and eliminating the $(\omega t - kz)$ terms we get a general identity

$$
\begin{aligned}
cos^2(\omega t - kz + \delta_x) + cos^2(\omega t - kz + \delta_y) & \\
-2cos\delta \cdot cos(\omega t - kz + \delta_x) \cdot cos(\omega t - kz + \delta_y) &= sin^2\delta
\end{aligned}
$$

where

$$
\delta = \delta_y - \delta_x \quad (2.21)
$$

Now using Eqn. 2.18 we get

$$
\left(\frac{E_x}{A_x}\right)^2 + \left(\frac{E_y}{A_y}\right)^2 - 2\frac{cos\delta}{A_x A_y}E_x E_y = sin^2\delta \quad (2.22)
$$

Equation 2.22 represents the equation of an ellipse and the optical wave is, in general, said to be elliptically polarized. In general, the principle axes of the ellipse are not in the $x$ and $y$ directions. The general ellipse describing the polarization is shown in Fig. 2.2, where we show the principle axes of the ellipse along $x'$ and $y'$. One can rotate the $xy$ axes to the $x'y'$ axes and represent Eqn. 2.22 as a standard equation for an ellipse.

$$
\left(\frac{E_{x'}}{a}\right)^2 + \left(\frac{E_{y'}}{b}\right)^2 = 1 \quad (2.23)
$$

where $a$ and $b$ are the principal axes of the ellipse and $E_{x'}$ and $E_{y'}$ are the components of the electric field along these axes.

To calculate the length of the principal axes, let us define as $\phi$ the angle between $x$ and $x'$ ($\phi$ has a value between 0 and $\pi$). Then from simple trigonometry we have

$$
\begin{aligned}
a^2 &= A_x^2 cos^2\phi + A_y^2 sin^2\phi + 2A_x A_y \, cos\delta \, cos\phi \, sin\phi \\
b^2 &= A_x^2 sin^2\phi + A_y^2 cos^2\phi - 2A_x A_y \, cos\delta \, cos\phi \, sin\phi
\end{aligned} \tag{2.24}
$$

and

$$
tan(2\phi) = \frac{2A_x A_y}{A_x^2 - A_y^2} \, cos\delta \tag{2.25}
$$

Note that due to the time dependence of $E_x$ and $E_y$, the $E$-field vector rotates in the complex $x$-$y$ plane. We have the following possibilities:

- $\delta = 0$ or $\pi$: In this case the light is linearly polarized and the tip of the $E$-field moves in a straight line as shown in Fig. 2.3.

- $\delta > 0$: The end of the electric field vector moves in a clockwise direction. If the magnitude of $E_x$ and $E_y$ is the same, the light is circularly polarized if $\delta = \pi/2$. Otherwise, it is elliptically polarized. If an observer facing the incoming beam sees the electric field moving clockwise, the light is said to have left-handed polarization.

- $\delta < 0$: The end of the electric field moves in a counter clockwise direction. If an observer sees an incoming beam with the electric field moving counterclockwise, the light is said to have right-handed polarization.

In Fig. 2.3 we show the various polarization states a light beam can have.

*An important class of optoelectronic devices exploits the polarization of light to tailor the device response. Some of these devices use an electric field to introduce a controllable phase difference between $E_x$ and $E_y$. This allows one to alter the polarization state of light after passing through the device. A use of a polarization selection device (a polarizer) then allows one to modulate the output.* This feature can be exploited for design of optical switches and modulators.

We will see later that liquid crystal display devices also use the polarization properties of light to change the transmitted or reflected beam intensities. An electric field alters the properties of the liquid crystal in such a way that light with a particular polarization either transmits through the device or is scattered (or reflected). We will discus this in detail later.

**EXAMPLE 2.4** We will see later that there are materials in which the propagation velocities of light polarized along different directions are different. Let $c/n_{r1}$ and $c/n_{r2}$ be

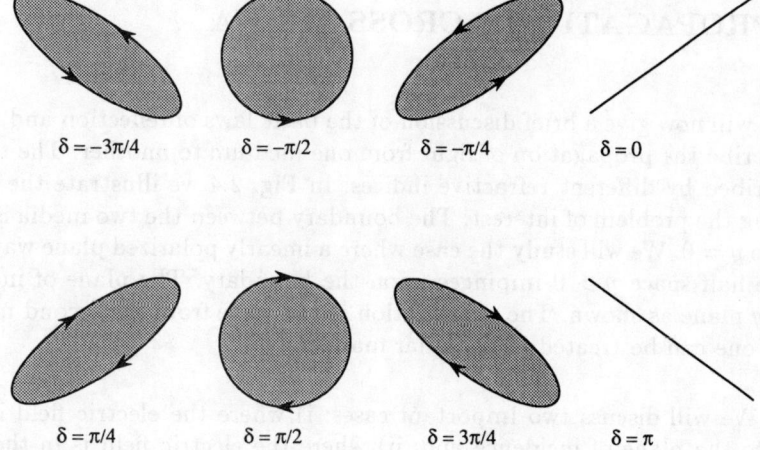

Figure 2.3: The polarization states for a number of different propagating waves. The field amplitudes are $E_x = cos(\omega t - kz)$ and $E_y = cos(\omega t - kz + \delta)$.

the speeds for light polarized along $x$- and $y$-axes. A quarter wave plate is made from such a material so that the phase difference between light initially polarized along the $x$ and $y$ axis is $\pi/2$ after passing through the plate. Calculate the thickness of the plate for the following parameters:

$$n_{r1} = 1.51 \quad n_{r2} = 1.6$$
$$\lambda_0 = \text{free space wavelength} = 0.8 \ \mu m$$

If $d$ is the thickness of the quarter wave plate, we have

$$\left[\left(\frac{2\pi n_{r1}}{\lambda_0} - \frac{2\pi n_{r2}}{\lambda_0}\right) d\right] = \frac{\pi}{2}$$

or

$$d = \frac{\lambda_0}{4|n_{r1} - n_{r2}|} = \frac{(0.8 \ \mu m)}{4 \times 0.09} = 2.22 \ \mu m$$

**EXAMPLE 2.5** Why do car antennas point vertically instead of horizontally? Assume that antenna response is proportional to the electric field component along the direction of the antenna.

It can be expected that usually the car is far from the broadcast stations so that the electromagnetic radiation, though directional, is traveling horizontally. For unpolarized radiation, the component along the vertical antenna is then essentially independent of the propagation direction. On the other hand, if the antenna is horizontal, it's response will depend upon the direction in the horizontal plane, i.e., on the direction in which the car is moving.

## 2.4    FRESNEL FORMULAE: PROPAGATION ACROSS MEDIA

$\longrightarrow$  We will now give a brief discussion of the basic laws of reflection and refraction that describe the propagation of light from one medium to another. The two media are described by different refractive indices. In Fig. 2.4 we illustrate the geometry describing the problem of interest. The boundary between the two media is given by the plane $y = 0$. We will study the case where a linearly polarized plane wave coming from the half space $y > 0$ impinges upon the boundary. The plane of incidence is the $x - y$ plane as shown. The propagation of the wave from the second medium to the first one can be treated in a similar manner.

We will discuss two important cases: i) where the electric field is perpendicular to the plane of incidence and; ii) where the electric field is in the plane of incidence.

### 2.4.1    Electric Vector Perpendicular to the Plane of Incidence

In this case, the electric field is in the $z$-direction. We call this a *transverse electric* *(TE)* wave, since the $E$-field is normal to the plane of incidence. We have an incident and a reflected wave in the $y > 0$ space and a refracted wave in the $y < 0$ space. In the medium 1, we write

$$E_z = A \, e^{ik_1(x\sin\alpha - y\cos\alpha)} + C e^{ik_1(x\sin\alpha' + y\cos\alpha')} \qquad (2.26)$$

and in the medium 2 we have

$$E_z = B \, e^{ik_2(x\sin\beta - y\cos\beta)} \qquad (2.27)$$

where, as shown in Fig. 2.4, the angles of incidence, reflection and refraction are $\alpha, \alpha'$ and $\beta$, respectively. Applying the boundary condition given by Eqn. 2.15 at $y = 0$, we get

$$Ae^{ik_1 x\sin\alpha} + C \, e^{ik_1 x\sin\alpha'} = B \, e^{ik_2 x\sin\beta} \qquad (2.28)$$

Since this equation is true for arbitrary $x$, we require that the phase terms all be equal. This gives

$$\alpha \;\; = \;\; \alpha';$$
$$k_1 \sin\alpha \;\; = \;\; k_2 \sin\beta \qquad (2.29)$$

The first of these equations gives us the law of reflection. The second equation gives the law of refraction and may be written as (Snell's Law)

$$\frac{\sin\alpha}{\sin\beta} = \frac{k_2}{k_1} = \sqrt{\frac{\epsilon_2}{\epsilon_1}} = \frac{n_{r_2}}{n_{r_1}} \qquad (2.30)$$

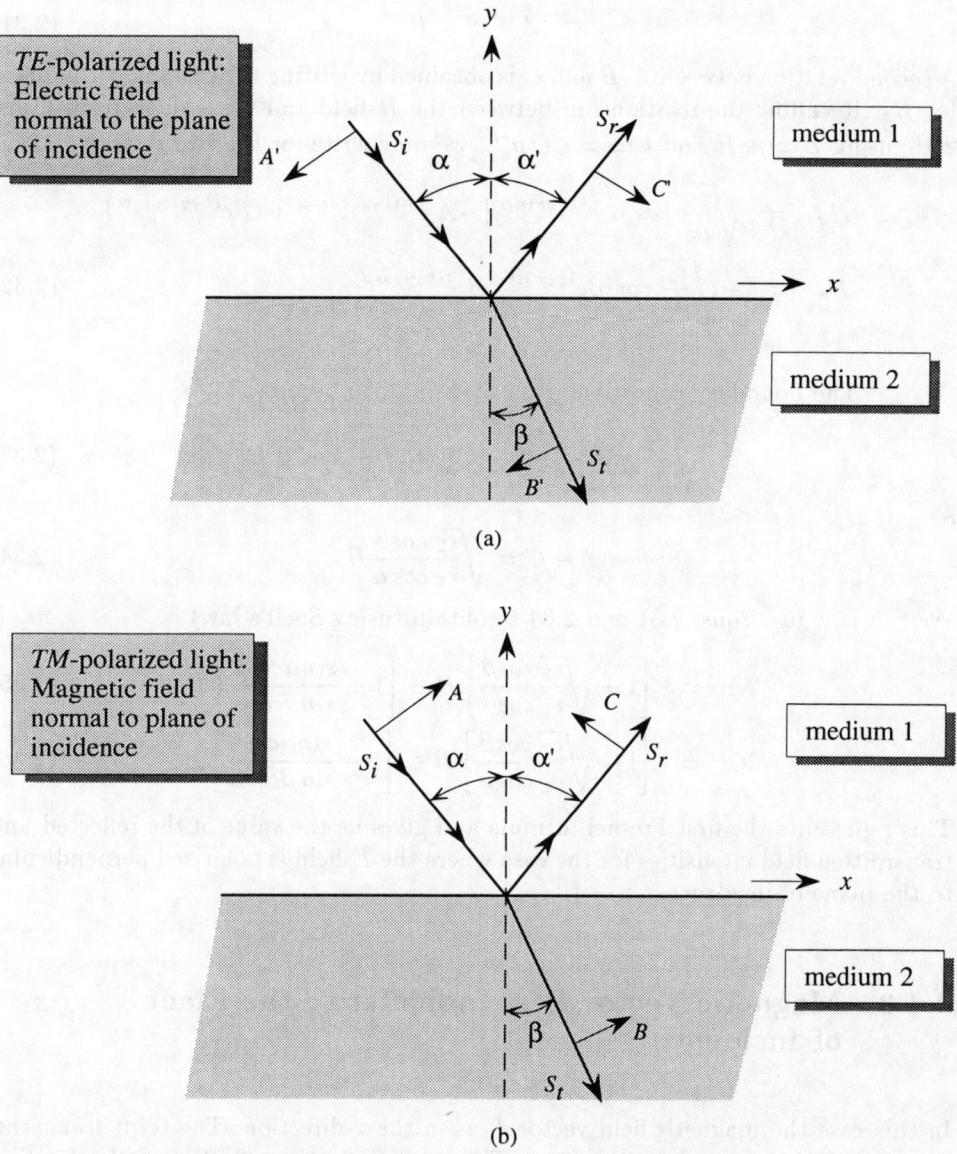

Figure 2.4: The propagation of a light wave from a rarer to a denser medium. (a) The electric field is perpendicular to the plane of the drawing (plane of incidence). The magnetic field amplitudes are $A'$, $B'$ and $C'$ for the incident light, refracted light and reflected light, respectively. The corresponding amplitudes for the electric field (not shown) are $A$, $B$, and $C$. (b) The magnetic field is perpendicular to the plane of incidence. The electric field amplitudes are $A$, $B$, and $C$ for the incident, transmitted and reflected lights, respectively.

Note that we have chosen $\mu_1 = \mu_2 = \mu_0$. We now have from Eqns. 2.28 and 2.29,

$$A + C = B \qquad (2.31)$$

A second relation between $A$, $B$ and $C$ is obtained by writing the boundary condition for $H_x$. Recalling the relationship between the $H$-field and the $E$-field from Eqns. 2.16, using $B = \mu_0 H$ and $k/\omega = \epsilon^{1/2}\mu_0^{1/2}$ we may write for the two regions

$$H_x = \sqrt{\frac{\epsilon_1}{\mu_0}} \cos \alpha \; e^{ik_1 x \sin\alpha} \left\{ -A \, e^{-ik_1 y \cos \alpha} + C \, e^{ik_1 y \cos \alpha} \right\}$$

$$H_x = -\sqrt{\frac{\epsilon_2}{\mu_0}} \cos \beta e^{ik_1 x \sin\beta} B \; e^{ik_2 y \cos\beta} \qquad (2.32)$$

The boundary condition given by Eqn. 2.15 becomes

$$\sqrt{\frac{\epsilon_1}{\mu_0}} \cos \alpha \{ -A + C \} = \sqrt{\frac{\epsilon_2}{\mu_0}} \cos \beta \, B \qquad (2.33)$$

or

$$A - C = \sqrt{\frac{\epsilon_2}{\epsilon_1}} \frac{\cos\beta}{\cos \alpha} B \qquad (2.34)$$

We can now use Eqns. 2.31 and 2.34 to obtain (using Snell's law)

$$2A = \left[ 1 + \sqrt{\frac{\epsilon_2}{\epsilon_1}} \frac{\cos\beta}{\cos\alpha} \right] B \equiv \left[ 1 + \frac{\sin\alpha\cos\beta}{\sin\beta\cos\alpha} \right] B \qquad (2.35)$$

$$2C = \left[ 1 - \sqrt{\frac{\epsilon_2}{\epsilon_1}} \frac{\cos\beta}{\cos \alpha} \right] B \equiv \left[ 1 - \frac{\sin\alpha\cos\beta}{\sin \beta\cos\alpha} \right] B \qquad (2.36)$$

This represents the first Fresnel formula and gives us the value of the reflected and transmitted field intensities for the case where the $E$-field is polarized perpendicular to the plane of incidence.

## 2.4.2  Magnetic Vector Perpendicular to the Plane of Incidence

In this case the magnetic field vector $H$ is in the $z$-direction. The term *transverse magnetic* (*TM*) is used for this case. The solution to the reflection and refraction problem proceeds along the same approach as in the previous case, except that one applies the boundary conditions to $H_z = 0$ at $y = 0$ and to $E_x$ at $y = 0$.

Denoting the $H_z$ amplitudes of the incident, reflected and refracted waves by $A'$, $C'$ and $B'$, respectively, we get, after imposing the boundary condition at $H_z = 0$

$$A' e^{ik_1 x \sin\alpha} + C' e^{ik_1 x \sin\alpha'} = B' e^{ik_2 \sin\beta} \qquad (2.37)$$

Once again, if the phase terms are equated we get the laws of reflection and refraction. Additionally, we then have

$$A' + C' = B' \tag{2.38}$$

For the electric field, after using the relation between the electric field amplitudes $(A, B, C)$ and the magnetic field amplitudes $(A', B', C')$, we get

$$A + C = \sqrt{\frac{\epsilon_2}{\epsilon_1}} B \tag{2.39}$$

A second set of equations results when we impose the boundary condition on the electric field component $E_x$. Keeping in mind that the vectors $\boldsymbol{E}, \boldsymbol{H}$ and $\boldsymbol{S}$ (i.e., the propagation direction ) form a right handed system, the electric field amplitudes have the orientation shown in Fig. 2.4b. Projecting the electric field vector to the x-axis, we get, from the boundary condition

$$(A - C) \cos \alpha = B \cos \beta \tag{2.40}$$

Notice the change in sign between $A$ and $C$. This change in sign corresponds to a phase change of $\pi$ in the reflected wave. From Eqns. 2.39 and 2.40, we get

$$2A = \left( \sqrt{\frac{\epsilon_2}{\epsilon_1}} + \frac{cos\beta}{cos\alpha} \right) B = \left( \frac{sin\alpha}{sin\beta} + \frac{cos\beta}{cos\alpha} \right) B \tag{2.41}$$

$$2C = \left( \sqrt{\frac{\epsilon_2}{\epsilon_1}} - \frac{cos\beta}{cos\alpha} \right) B = \left( \frac{sin\alpha}{sin\beta} - \frac{cos\beta}{cos\alpha} \right) B \tag{2.42}$$

Further simplification gives

$$\frac{4A}{C} = \frac{sin2\alpha + sin2\beta}{sin\beta cos\alpha} = \frac{2sin(\alpha + \beta)cos(\alpha - \beta)}{sin\beta cos\alpha} \tag{2.43}$$

$$\frac{4C}{B} = \frac{sin2\alpha - sin2\beta}{sin\beta cos\alpha} = \frac{2cos(\alpha + \beta)sin(\alpha - \beta)}{sin\beta cos\alpha} \tag{2.44}$$

After simple manipulations of these equations, we get Fresnel's second formula,

$$A : B : C = tan(\alpha + \beta) : \left( \frac{tan(\alpha + \beta)}{cos(\alpha - \beta)} - \frac{tan(\alpha - \beta)}{cos(\alpha + \beta)} \right) : tan(\alpha - \beta) \tag{2.45}$$

We see from the two Fresnel formulae that even though the laws of reflection and refraction that give us the relations between the angles of incidence, reflection and refraction are the same for the two polarizations considered, important differences exist between the amplitude relations. These differences lead to special polarization effects when a beam of light impinges from one medium to another.

## 2.4.3    Polarization Effects: Brewster's Law

Let us consider how the amplitude ratios of the reflected and refracted waves change as a function of the angle of incidence. For focus we assume that a wave is coming from a rarer medium into a denser medium (i.e. $\epsilon_2 > \epsilon_1$). The amplitude ratios of the transmitted and reflected beams are

$$t = \frac{B}{A}, r = -\frac{C}{A} \qquad (2.46)$$

The negative sign is introduced in $R$ since for most range of angle of incidence, the signs of $C$ and $A$ are opposite. Note that a change in sign between $A$ and $C$ means that a phase change of $\pi$ has occurred (since $exp\,(i\pi) = -1$).

Let us first consider the $TE$-polarized light. From Eqns. 2.35 and 2.36, we have

$$\boxed{r_{TE} = \frac{sin(\alpha - \beta)}{sin(\alpha + \beta)}} \qquad (2.47)$$

For small angles of incidence, we get $\left(\text{using } \beta \sim \frac{n_{r_1}\alpha}{n_{r_2}}\right)$

$$r_{TE} = \frac{n_{r_2} - n_{r_1}}{n_{r_2} + n_{r_1}} \qquad (2.48)$$

At the grazing angle where $\alpha \to \frac{\pi}{2}$, we get $r_{TE} \to 1$. The amplitude ratio for the refracted light is simply given by

$$t_{TE} = 1 - r_{TE} \qquad (2.49)$$

In Fig. 2.5, we show a typical plot for the amplitude ratios as a function of the incidence angle.

We now come to the case of $TM$-polarized light where we see that using our definition of $r_{TM}, r_{TM}$ is negative for the cases where $\alpha$ is small. From Eqn. 2.45,

$$\boxed{r_{TM} = -\frac{tan(\alpha - \beta)}{tan(\alpha + \beta)}} \qquad (2.50)$$

For small values of $\alpha$, the value of $r_{TM}$ is the same as that of $r_{TE}$, except for the sign. In Fig. 2.5 we show the behavior of $r_{TE}$ as a function of $\alpha$. As we can see from this figure, $r_{TE}$ starts at a negative value but then reaches a value of unity at $\alpha = \pi/2$. Thus it has a value equal to zero at an angle $\alpha = \alpha_{pol}$ called the angle of polarization.

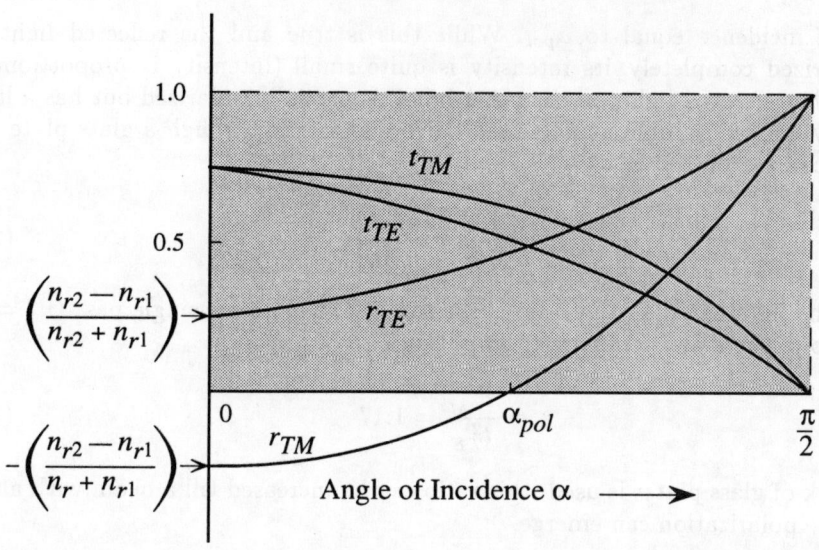

Figure 2.5: A plot of the amplitude ratios of the reflected and transmitted $TE$- and $TE$-polarized light as a function of the angle of incidence. Note that, according to our definition, a positive value of $r$ implies a phase change of $\pi$ upon reflection.

At $\alpha = \alpha_{pol}$, the denominator of Eqn. 2.50 must change signs and, therefore, we have

$$\alpha_{pol} + \beta = \pi/2 \qquad (2.51)$$

which gives us the result

$$sin\beta = cos\alpha_{pol} \qquad (2.52)$$

We also know from the law of refraction that

$$sin\beta = \frac{n_{r_1}}{n_{r_2}} sin\alpha_{pol} \qquad (2.53)$$

This gives us the relation (Brewster's Law)

$$\boxed{tan\alpha_{pol} = \frac{n_{r_2}}{n_{r_1}}} \qquad (2.54)$$

For light going from air to glass we find (using $n_{r_2} = 1.5, n_{r_1} = 1$) $\alpha_{pol} = 57°$. For water ($n_{r_2} = 1.33$) $\alpha_{pol} = 53°$. It is important to note that since $r_{TM}$ vanishes at this angle, the reflected light is polarized completely, with the $E$-field normal to the plane of incidence. *It is also important to note that at $\alpha = \alpha_{pol}$, the reflected wave is perpendicular to the refracted wave.*

The reflection and refraction discussion given here tells us that, in principle, we can generate polarized light from an incident unpolarized light by using an angle of incidence equal to $\alpha_{pol}$. While this is true and the reflected light will be polarized completely, its intensity is quite small (intensity is proportional to $r_{TE}^2$). The refracted light, on the other hand, is partially polarized but has a larger intensity. If, for example, unpolarized light passes once through a glass plate from air, we can see that at $\alpha = \alpha_{pol}$

$$\frac{t_{TM}}{t_{TE}} = \frac{n_r^2 + 1}{2n_r} \tag{2.55}$$

When the light emerges, the ratio is the same. Thus, after a single pass ($n_r = 1.5$) the ratio of the $s$ and $p$ polarized amplitudes is

$$\frac{t_{TM}^2}{t_{TE}^2} = 1.17 \tag{2.56}$$

If a stack of glass plates is used, this ratio can be increased till a beam with almost complete polarization can emerge.

## 2.4.4   Total Internal Reflection and Evanescent Fields

Let us now consider a situation where light is incident from a *denser medium on a boundary with a rarer medium*. For simplicity, let us say the rarer medium is air. We then have ($n_r$ is the refractive index of the denser medium)

$$\frac{sin\alpha}{sin\beta} = \frac{1}{n_r} \tag{2.57}$$

Since $\beta < \alpha$, we see that when

$$n_r sin\alpha > 1 \tag{2.58}$$

$\beta$ is imaginary.

In Fig. 2.6 we show the reflection ratios $r_{TM}$ and $r_{TE}$ (defined as $C/A$ and not $-C/A$ in this discussion) as a function of the angle of incidence. We see that the value of $r_{TE}$ and $r_{TM}$ becomes unity not at $\alpha = \pi/2$ but at $\alpha = \alpha_{tot}$, the critical angle, where

$$sin\alpha_{tot} = n_r \tag{2.59}$$

To proceed beyond $\alpha_{tot}$, we need to allow the refracted angle to have a complex value. When $\alpha$ is less than $\alpha_{tot}$, we have a real value of $\beta$ going from 0 to $\pi/2$. Beyond this point we allow $\beta$ to be in the complex plane having a real and imaginary part. In general we write

$$\beta = \pi/2 \pm i\beta' \tag{2.60}$$

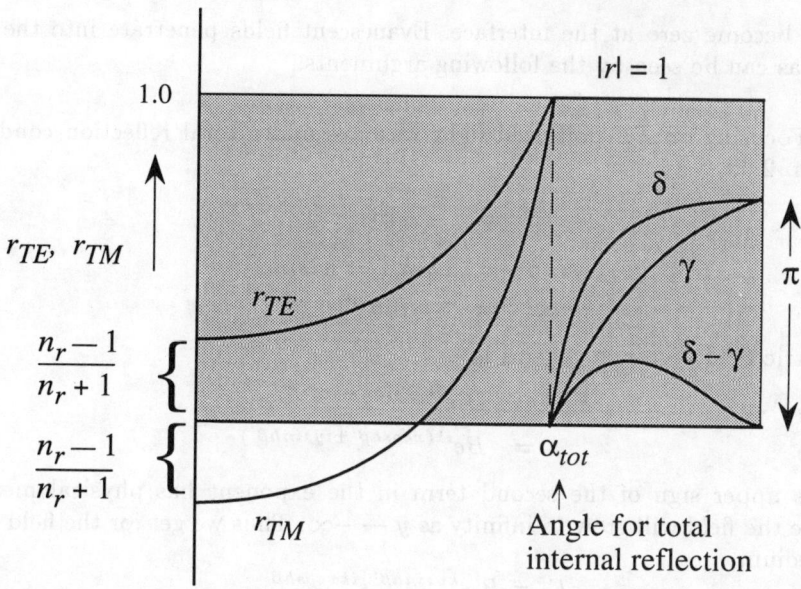

Figure 2.6: A plot of the reflection amplitude ratios $r_{TE}$ and $r_{TM}$ as a function of the angle of incidence. The change of angle produced by total reflection is also shown.

We have for $\alpha > \alpha_{tot}$

$$sin\beta = sin(\pi/2 \pm i\beta') = cos(\pm i\beta') \qquad (2.61)$$
$$= cosh\beta' > 1 \qquad (2.62)$$

This is consistent with the law of refraction where $sin\beta = n_r sin\alpha$ must be larger than 1 when $\alpha$ is greater than $\alpha_{tot}$. We now have, from Eqns. 2.47 and 2.50

$$r_{TE} = \frac{sin\left(\frac{\pi}{2} \pm i\beta' - \alpha\right)}{sin\left(\frac{\pi}{2} \pm i\beta' + \alpha\right)} = \frac{cos(\alpha \mp i\beta')}{cos(\alpha \pm i\beta')} = e^{i\gamma} \qquad (2.63)$$

$$r_{TM} = -\frac{tan\left(\frac{\pi}{2} \pm i\beta' - \alpha\right)}{tan\left(\frac{\pi}{2} \pm i\beta' + \alpha\right)} = \frac{cot(\alpha \mp i\beta')}{cot(\alpha \pm i\beta')} = e^{i\delta} \qquad (2.64)$$

We note that both $r_{TM}$ and $r_{TE}$ have a magnitude of unity. However, a phase change now accompanies the totally reflected wave. In Fig. 2.6 we show a typical plot of the phase changes as the angle of incidence changes. As can be seen from this figure, there is a difference between the phase change produced in $TM$ and $TE$ polarized light. This phase difference can be exploited to produce elliptically polarized light. In Section 3.3 of the next chapter, we will examine the importance of total internal reflection for propagation of guided modes in waveguides.

In total internal reflection there is no refracted beam and the entire electromagnetic energy is reflected back. However, the fields in the rarer medium do not abruptly become zero at the interface. Evanescent fields penetrate into the rarer medium as can be seen by the following arguments.

Focusing on $TE$-polarized light we have under total reflection conditions from Eqn. 2.60,

$$\beta = \frac{\pi}{2} \pm i\beta' \tag{2.65}$$

$$sin\beta = cosh\beta' = n_r sin\alpha \tag{2.66}$$

$$cos\beta = \mp isinh\beta' \tag{2.67}$$

The electric field vector in general is

$$E_z = B \, e^{ik(xsin\beta - ycos\beta)} \tag{2.68}$$

$$= Be^{ik(xcosh\beta' \pm iysinh\beta')} \tag{2.69}$$

Only the upper sign of the second term in the exponent has physical meaning; otherwise the field will grow to infinity as $y \rightarrow -\infty$. Thus we get for the field in the rarer medium,

$$E_z = B \, e^{kysinh\beta'} e^{ikxcosh\beta'} \tag{2.70}$$

This is the evanescent wave which decreases exponentially away from y=0 (y is negative in the rarer medium). Since $ksinh\beta'$ is of the order of $\lambda^{-1}$, the wave decays away in a few wavelengths. Nevertheless, the presence of this wave is of great importance and is exploited in devices such as waveguide couplers, switches, etc. as will be discussed later.

**EXAMPLE 2.6** A light beam emerges from a GaAs crystal. Calculate the fraction of light intensity reflected back into the crystal if the light is normally incident. The refractive index of GaAs is 3.6.

The reflection coefficient for the reflected light intensity is

$$R = \mid r \mid^2 = \frac{(2.6)^2}{(4.6)^2} = 0.31$$

Thus only 69% of the light incident normally escapes to the free space. GaAs is often used to generate optical signals and this example shows that a significant part of the light is unable to escape the device. It is important to note that in semiconductor lasers, the light that is reflected back into the material is used to provide feedback for laser oscillations.

**EXAMPLE 2.7** Consider an interface between two non-magnetic lossless media with $\epsilon_1 = 6.0 \, \epsilon_0$ and $\epsilon_2 = 12.0 \, \epsilon_0$. Calculate the reflection and transmission coefficients for light incident normally from medium 1 to medium 2. Evaluate the coefficients for the electric and magnetic field. The reflection coefficient for the electric field ratios is

$$r(E) = \frac{1 - \sqrt{\frac{\epsilon_2}{\epsilon_1}}}{1 + \sqrt{\frac{\epsilon_2}{\epsilon_1}}} = -0.171$$

The transmission coefficient is
$$t(E) = 0.829$$
The negative sign of the reflection coefficient implies that the direction of the electric field is reversed upon reflection, i.e., a phase change of $\pi$ has occurred.

For the magnetic field, the reflection coefficient is
$$r(H) = 0.171$$
and
$$t(E) = 1 + 0.171 = 1.171$$
Thus the magnetic field of the transmitted beam increases upon passing into the second medium.

**EXAMPLE 2.8** A plane wave is incident at an angle of 38° to the normal from a medium with $\epsilon_1 = 6.0\ \epsilon_0$ on an interface with a medium with $\epsilon_2 = 14.0\ \epsilon_0$. Obtain the reflection and transmission coefficients for $TE$- and $TM$-polarized assuming the two media to be non-magnetic.

From Snell's law, we know that the angle of the refracted wave is
$$\beta = sin^{-1}\left(\sqrt{\frac{\epsilon_2}{\epsilon_1}}\ sin\alpha\right)$$
$$= 27.6°$$
The reflection coefficient is
$$r_{TE} = -\frac{C}{A} = \frac{sin(\alpha - \beta)}{sin(\alpha + \beta)} = 0.198$$
The negative sign in the definition of $R_{TE}$ implies a reversal in the electric field direction at the interface upon reflection. The transmission coefficient is
$$t_{TE} = 1 - r_{TE} = 0.802$$
For the $TM$-polarized light we have
$$r_{TM} = -\frac{tan(\alpha - \beta)}{tan(\alpha + \beta)} = 0.803$$
The transmission coefficient is (see Eqn. 2.39)
$$t_{TM} = \sqrt{\frac{\epsilon_1}{\epsilon_2}}\ (1 - r_{TM})$$
$$= 0.6$$

Note that the coefficients for the magnetic fields can be obtained by using the relation between the electric field and the magnetic field in a medium. The example shows that the electric field of the reflected wave is 20% in strength (and opposed in direction) of the incident field for the $TE$-polarized light. However, the electric field of the $TM$ wave is only 8% of the incident strength.

## 2.5    WAVE PROPAGATION IN CRYSTALS

$\longrightarrow$  In Chapter 1 we have discussed the structural properties of crystals and liquid crystals. We have seen that unlike the non-crystalline materials, there is a long range order in the arrangement of atoms in these materials which leads to anisotropic physical properties. Thus, for example, the light propagation along different directions is not described the same refractive index. In fact, light polarized along different directions will propagate with different speed, in general. These anisotropic properties are of great value in designing remarkable optical devices—both passive and active. Among passive devices that use the anisotropy of light propagation in crystals are quarter wave plates to alter polarization, polarizers, birefringent plates, etc. The active devices that use anisotropy of the material are electro-optic devices, liquid crystal devices, acousto-optic devices, etc. Since electro-optic modulators and liquid crystal display devices will be examined in Chapter 9, we will review the relevant physics of light propagation in anisotropic media in this section.

It should be noted that it is not possible to cover all aspects of wave propagation in crystals in one section. Therefore, we will not provide detailed derivations of some of the results given, but simply focus on the important physics issues.

In an isotropic medium, the propagation of light waves is described by a direction independent dielectric constant (or refractive index). However, in crystalline materials, the medium is not isotropic. It is useful to describe the properties of a crystal by choosing principal axes determined by the crystal symmetry. The displacement $D$ and the electric field $E$ of the light waves, in general, have a relation given by the dielectric tensor

$$
\begin{aligned}
D_1 &= \epsilon_{11}E_1 + \epsilon_{12}E_2 + \epsilon_{13}E_3 \\
D_2 &= \epsilon_{21}E_1 + \epsilon_{22}E_2 + \epsilon_{23}E_3 \\
D_3 &= \epsilon_{31}E_1 + \epsilon_{32}E_2 + \epsilon_{33}E_3
\end{aligned}
\tag{2.71}
$$

with $\epsilon_{ij} = \epsilon_{ji}$. In general, the refractive indices $n_{ri}$ will be different along different directions. A useful concept to describe wave propagation is the concept of Fresnel ellipsoid or the equivalent concept of the index ellipsoid. The Fresnel ellipsoid is given by the relation

$$
\sum \epsilon_{ij} x_i x_j = \text{constant}
\tag{2.72}
$$

A more useful description of the constraints placed on wave propagation is by an equivalent index ellipsoid (or indicatrix)

$$
\frac{x_i^2}{n_{r1}^2} + \frac{x_2^2}{n_{r2}^2} + \frac{x_3^2}{n_{r3}^2} = \text{constant}
\tag{2.73}
$$

This equation describes the surface of an ellipsoid.

We will briefly describe how the index ellipsoid is used to describe the polarization of light propagating in a crystalline material. In Fig. 2.7a, we show the index ellipsoid of a crystal and a light wave propagating along a direction $k$. In an isotropic medium the wave can have an arbitrary polarization in the plane perpendicular to $k$. However, in an anisotropic medium the wave has either of two linear polarizations and the velocity of the light with each polarization is, in general, different.

To calculate the polarization, we use the construction outlined in Fig. 2.7b. A plane is drawn perpendicular to the $k$-vector and the intersection of this plane with the ellipsoid produces an ellipse. The ellipse produced has principal axes $a$ and $b$ as shown in Fig. 2.7b. The directions of polarization allowed for the wave are now given by $D_a$ and $D_b$, i.e., parallel to the principal axes. *The velocities of the light with the two polarizations are inversely proportional to the length of the principal axes.* In particular, if the light is propagating along the axis $i = 3$, the light is polarized along $i = 1$ and $i = 2$ with velocities $c/n_{r1}$ and $c/n_{r2}$, respectively.

In the analysis discussed above, the indices $n_{r1}, n_{r2}, n_{r3}$ are, in general, different. Their values and their differences depend upon the details of the material structure. We can have the following cases:

Isotropic:

$$n_{r1} = n_{r2} = n_{r3} \tag{2.74}$$

Uniaxial:

$$n_{r1} = n_{r2} \neq n_{r3} \tag{2.75}$$

Biaxial:

$$
\begin{aligned}
n_{r1} &\neq n_{r2} \\
n_{r2} &\neq n_{r3} \\
n_{r1} &\neq n_{r3}
\end{aligned}
\tag{2.76}
$$

The case of most interest to us is the uniaxial medium which describes most electro-optic devices used for light modulation and which describes the liquid crystals.

Focusing on the uniaxial crystals, let us denote the axis-3 by $z$, the 1 and 2 being $x$ and $y$. If light is propagating along the $z$-axis it could have any polarization in the $x$-$y$ plane, and all these polarizations will have the same propagation velocities. The $z$-axis (also called the $c$-axis in optics) is then called the optic axis, and $n_{r3}(= n_{rz})$ is denoted by the $n_{re}$, the extraordinary refractive index while $n_{rx}$

| Material | $n_{ro}$ | $n_{re}$ |
|----------|----------|----------|
| Quartz | 1.544 | 1.553 |
| ZnS | 2.354 | 2.358 |
| KDP | 1.507 | 1.467 |
| Calcite | 1.658 | 1.486 |
| $LiNbO_3$ | 2.300 | 2.208 |
| $BaTiO_3$ | 2.416 | 2.364 |

Table 2.1: Refractive indices of some uniaxial crystals. The refractive indices are wavelength dependent and are given for a wavelength of 0.63 $\mu$m.

and $n_{ry}$ are denoted by $n_{ro}$, the ordinary refractive index. If $n_{ro} < n_{re}$, the crystal is said to be positive while if $n_{ro} > n_{re}$, the crystal is said to be negative. Values of some important uniaxial crystals are given in Table 2.1.

If, in a uniaxial crystal, light is propagating in a direction other than the optic axis, a phase delay will develop between the two polarizations of light due to their different propagation velocities. This phase delay is exploited for designing devices that can alter the polarization of light. If somehow the refractive index can be altered, the device can become active and can be used to modulate a light signal as will be discussed later.

It is useful to provide here a brief discussion of the polarization and refractive index of a wave propagating along a general direction $\hat{s}$ making an angle $\theta$ with the optic axis (the $z$-axis). The displacement fields are polarized as shown in Fig. 2.8, and as discussed earlier. For the uniaxial crystal, the wave polarized along the $x$-axis (choosing the $\hat{s}$ direction to be in the $y$-$z$ plane) is the ordinary wave with index $n_{ro}$. The wave polarized along the orthogonal direction along the other semi-major axis of the ellipse (see Fig. 2.8) is the extraordinary wave with a refractive index given by

$$\frac{1}{n_{re}^2(\theta)} = \frac{cos^2\theta}{n_{ro}^2} + \frac{sin^2\theta}{n_{re}^2} \qquad (2.77)$$

If the wave is propagating along the $z$-axis ($\theta = 0$), i.e., the optic axis, the value of $n_{re}(\theta)$ is simply $n_{ro}$ as expected. If the wave is propagating along the $y$-axis, $n_{re}(\theta) = n_{re}$.

**EXAMPLE 2.9** Consider a quarter-wave plate on which light initially polarized along a direction 45° to the $x$-axis impinges. What is the polarization of the emerging light if

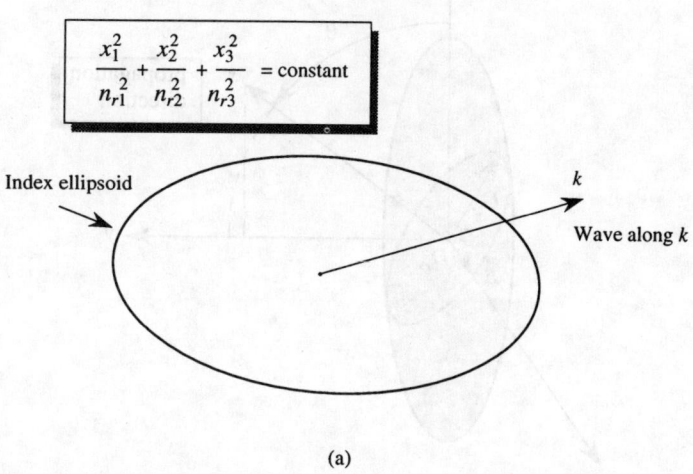

$$\frac{x_1^2}{n_{r1}^2} + \frac{x_2^2}{n_{r2}^2} + \frac{x_3^2}{n_{r3}^2} = \text{constant}$$

(a)

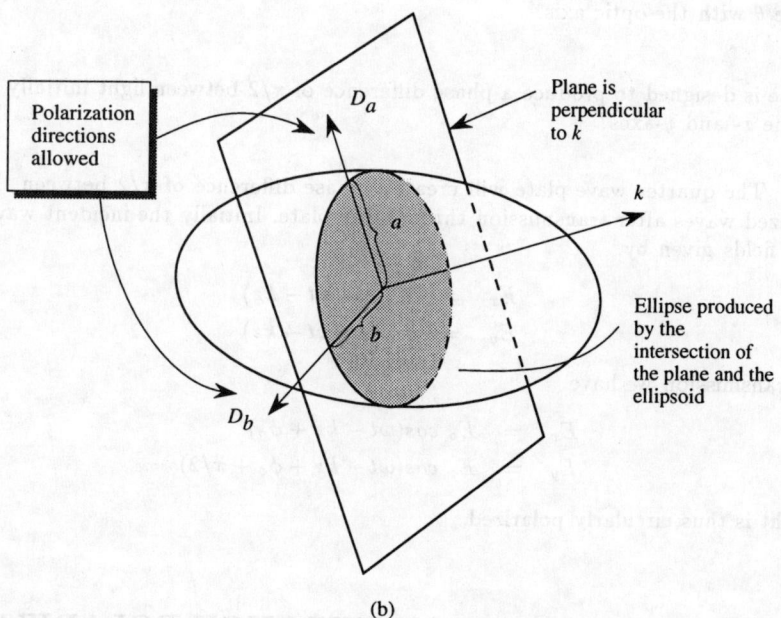

(b)

Figure 2.7: (a) An index ellipsoid for a crystal. Shown is a wave along the direction $k$. (b) The construction used to obtain the polarization of the wave.

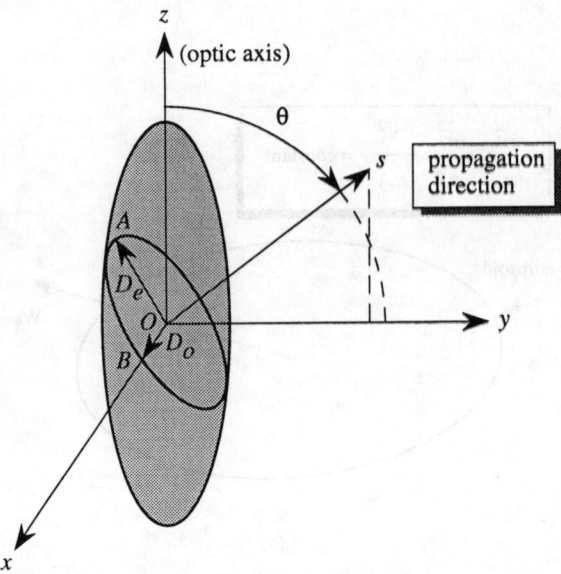

Figure 2.8: The polarization of an optical beam propagating along a direction $S$, making an angle $\theta$ with the optic axis.

the plate is designed to produce a phase difference of $\pi/2$ between light initially polarized along the $x$- and $y$-axes?

The quarter wave plate will create a phase difference of $\pi/2$ between the $x$- and $y$-polarized waves after transmission through the plate. Initially the incident wave has the electric fields given by

$$
\begin{aligned}
E_x &= E_o\,cos(\omega t - kz) \\
E_y &= E_o\,cos(\omega t - kz)
\end{aligned}
$$

After transmission we have

$$
\begin{aligned}
E_x &= E_o\,cos(\omega t - kz + \phi_o) \\
E_y &= E_o\,cos(\omega t - kz + \phi_o + \pi/2)
\end{aligned}
$$

The light is thus circularly polarized.

## 2.6   LIGHT MODULATION THROUGH POLARIZATION CONTROL

$\longrightarrow$   An extremely important need in optoelectronics is the control of the light intensity. We are all familiar with the transparency on which we can write with a special pen. Our writing can be displayed by a simple optical system using lenses and a light source. The places where we have written act to modulate (or modify) the light intensity passing through the transparency. The need for "programmable transparencies" is felt in the area of information display. The liquid crystal displays used in pocket calculators, personal computer notebooks, airplane cockpits, etc., are perhaps the most important programmable transparencies. Light modulators are also used in optical communication and in optical correlators for image comparison. Many other uses of such devices have been suggested in literature for advanced mathematical operations such as matrix-vector multiplication, matrix-matrix multiplication, etc.

A most useful technique to modulate an optical signal is through the use of polarizers and an active device that can change the polarization of light. The general approach is illustrated in Fig. 2.9. In this particular geometry (other geometries are also possible) two polarizers aligned in the cross-polarized configuration are placed on each side of the device. The device consists of a crystal (or liquid crystal) in which the two refractive indices $n_{re}$ and $n_{ro}$ are different. Also, it is possible to alter the difference between $n_{re}$ and $n_{ro}$ by using an external perturbation. This alteration can be done by applying an electric field and utilizing an effect called the electro-optic effect which will be discussed briefly in the next section.

Let us first consider the case of an electro-optic modulator based on crystals such as lithium niobate. Later we will consider the case of a liquid crystal such as a twisted nematic. Let us assume that a linearly polarized light enters the device. As shown in Fig. 2.9, let us assume that a linearly polarized light is incident on the crystal and the $x$-axis and the $y$-axis represent the two polarization axes for the crystal. In general, the two directions have different refractive indices and, as the wave propagates, a phase difference develops between the two polarizations. Consider an input signal that is linearly polarized and given by

$$E_x = \frac{E_o}{\sqrt{2}} \, exp \, (i\omega t) \tag{2.78}$$

$$E_y = \frac{E_o}{\sqrt{2}} \, exp \, (i\omega t) \tag{2.79}$$

After transmission through the modulator, the wave emerges with a general polarization given by

$$E_x = \frac{E_o}{\sqrt{2}} \, exp \, (i\omega t + i\theta_1) \tag{2.80}$$

$$E_y = \frac{E_o}{\sqrt{2}} \, exp \, (i\omega t + i\theta_2) \tag{2.81}$$

with the phase difference given by $\phi = \theta_2 - \theta_1$. If $\phi$ is $\pi/2$, the output beam is circularly polarized and if it is $\pi$, it is linearly polarized with polarization 90° with respect to the input beam. If the output beam passes through a polarizer at 90°

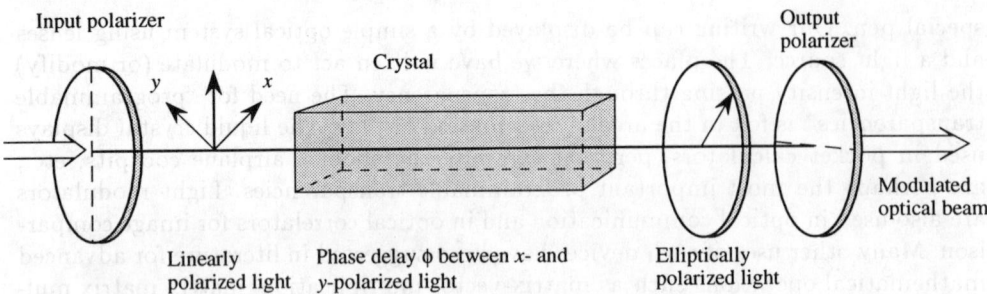

Figure 2.9: A schematic of how a polarization charge produced by a crystal device can alter the intensity of an optical beam.

with respect to the input beam polarizer as shown in Fig. 2.9, the modulation ratio is given by (assuming no absorption losses)

$$\frac{I_{out}}{I_{in}} = sin^2 \frac{\phi}{2} \qquad (2.82)$$

Thus if $\phi$ can be controlled by an electric field, the intensity can be modulated.

**Polarization and Modulation Properties of a Twisted Nematic**

As we have noted in Chapter 1, the liquid crystal is a material which has good long range order along some direction. Also, since the material is made up of rod like or disc like molecules, it has a very strong anisotropy between $n_{re}$ and $n_{ro}$. As noted in Chapter 1, in a nematic liquid crystal, one can introduce a twist in the order in which molecules are arranged by using two glass plates that have been rubbed in a particular orientation. As a result, the optic axis of the liquid crystal changes from point to point, consequently, the direction of the polarizations corresponding to the ordinary and extraordinary rays changes from point to point (as shown in Fig. 2.10).

An important and accurate approximation that is used to describe how light propagates (i.e., how the polarization changes) through a twisted nematic crystal is called the adiabatic approximation. *The adiabatic approximation depends upon the fact that the twist in the crystal is "slowly varying." This is a good approximation for liquid crystals, since a twist of $\pi/2$ is produced over several microns (say $\sim 10$-$20$ $\mu m$). As a result, the light responds according to the local refractive indices and the local polarization axes. Thus, if light enters the crystal along the "slow polarization" direction, it remains along this polarization as it travels down the liquid crystal.*

From the adiabatic approximation discussed above, we can see that there are two sources for the polarization change in a twisted nematic liquid crystal: i) As a result of the difference between $n_{re}$ and $n_{ro}$, the phase difference between the two rays, states changes, thus the polarization changes. This is the effect discussed

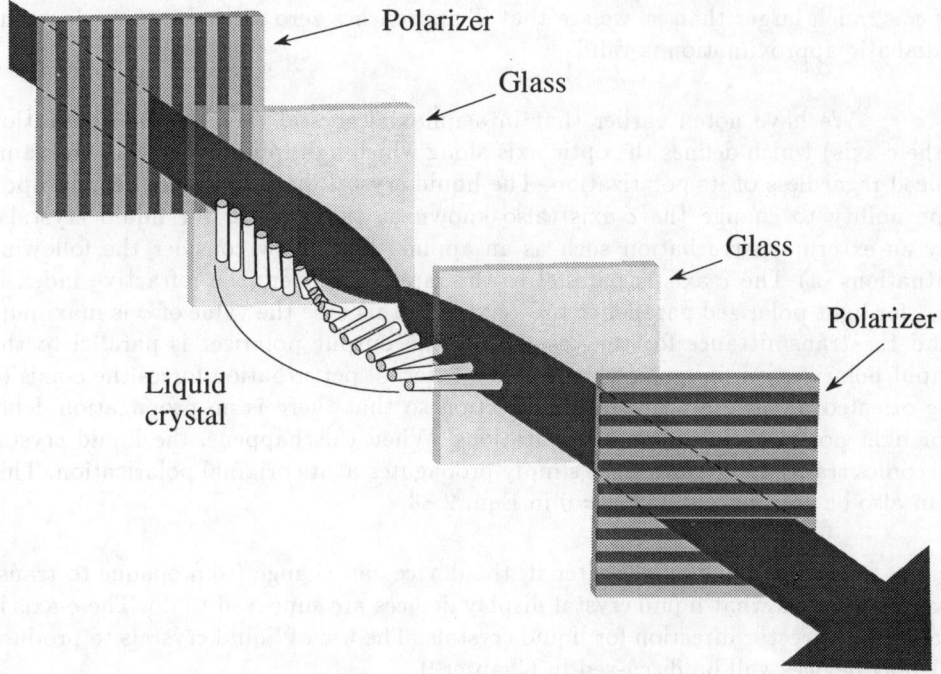

Figure 2.10: A schematic of a twisted nematic crystal. In the adiabatic approximation, if the twist is "slow," the polarization of light simply follows the twist as shown.

above and produces a modulation of light as given by Eqn. 2.82. ii) Additionally, due to the twist in the crystal, the polarization is rotated. This effect is exploited in most liquid crystal displays.

According to the adiabatic approximation, if the twist angle is 90° (or 270°) from the top plate to the lower one, and light polarized as one of the ordinary or extraordinary waves is sent in, we have the following possibilities: i) If the output polarizer is oriented along the input polarizer, the transmitted intensity is zero; ii) if the output and output polarizers are cross-polarized, the light passes through. A more accurate treatment of the problem shows that the transmittance in the first case (i.e., the polarizers having the same orientation), is given by the following relation for a $\pi/2$ twist

$$T = \frac{sin^2\left(\frac{\pi}{2}\sqrt{1 + (\phi/\pi)^2}\right)}{1 + (\phi/\pi)^2} \tag{2.83}$$

where $\phi$ is the phase difference produced due to the difference in the values of $n_{re}$ and $n_{ro}$ and is for a device of thickness $d$,

$$\phi = \frac{2\pi}{\lambda}(n_{re} - n_{ro})d \tag{2.84}$$

If $\phi$ is much larger than $\pi$, we see that $T$ approaches zero as is the case where the adiabatic approximation is valid.

We have noted earlier that in a uniaxial crystal there is one orientation (the $c$-axis) which defines the optic axis along which light propagates with the same speed regardless of its polarization. The liquid crystal display devices depend upon the ability to change the $c$-axis (also known as the director for liquid crystals) by an external perturbation such as an applied field. Now consider the following situations: a) The $c$-axis is parallel to the input polarizer (the refractive index is $n_{re}$ for light polarized parallel to the $c$-axis). In this case the value of $\phi$ is maximum and the transmittance for the case where the output polarizer is parallel to the input polarizer is minimum; b) an applied external perturbation forces the $c$-axis to be oriented along the propagation direction so that there is no propagation delay for light polarized in different orientations. When this happens, the liquid crystal becomes transparent since light simply propagates at its original polarization. This can also be seen by putting $\phi = 0$ in Eqn. 2.83.

Thus, if the $c$-axis is altered, the device can change from opaque to transparent which is what liquid crystal display devices are supposed to do. The $c$-axis is called the director direction for liquid crystals. The use of liquid crystals to produce display devices will be discussed in Chapter 9.

## Electro-Optic Effect in Crystals

In Fig. 2.9, we have seen how a pair of polarizers can be used to transmit or block an optical signal. This concept becomes particularly useful for "intelligent" optoelectronic devices if the optical properties of a device can be modified by some external stimulus. In principle, a variety of external stimulii such as electric field, magnetic field, pressure, etc., can modify the optical properties such as the refractive index. However, for optoelectronic applications we are interested in using electric fields to change the optical properties.

The idea behind optoelectronic devices such as switches and modulators (often called light valves) is quite simple and is illustrated in Fig. 2.11. Cross-polarized or parallel polarized polarizers can be used in the device. An electric field is used to alter the refractive index difference $(n_{re} - n_{ro})$ so that a change in the polarization occurs. This leads to a change in the output intensity. The effect of the electric field on the optical properties depends upon the symmetry of the crystal and whether or not the molecules can be easily rotated as in liquid crystals.

In solid crystals the atoms on the lattice are fixed and do not physically move or rotate as a result of an electric field. However, the "electron cloud" in the outermost shell of the atoms does distort under the influence of an applied field. This distortion manifests itself in small changes in the refractive indices of the material. In Chapter 5 we will discuss the details of how the refractive index is modified by

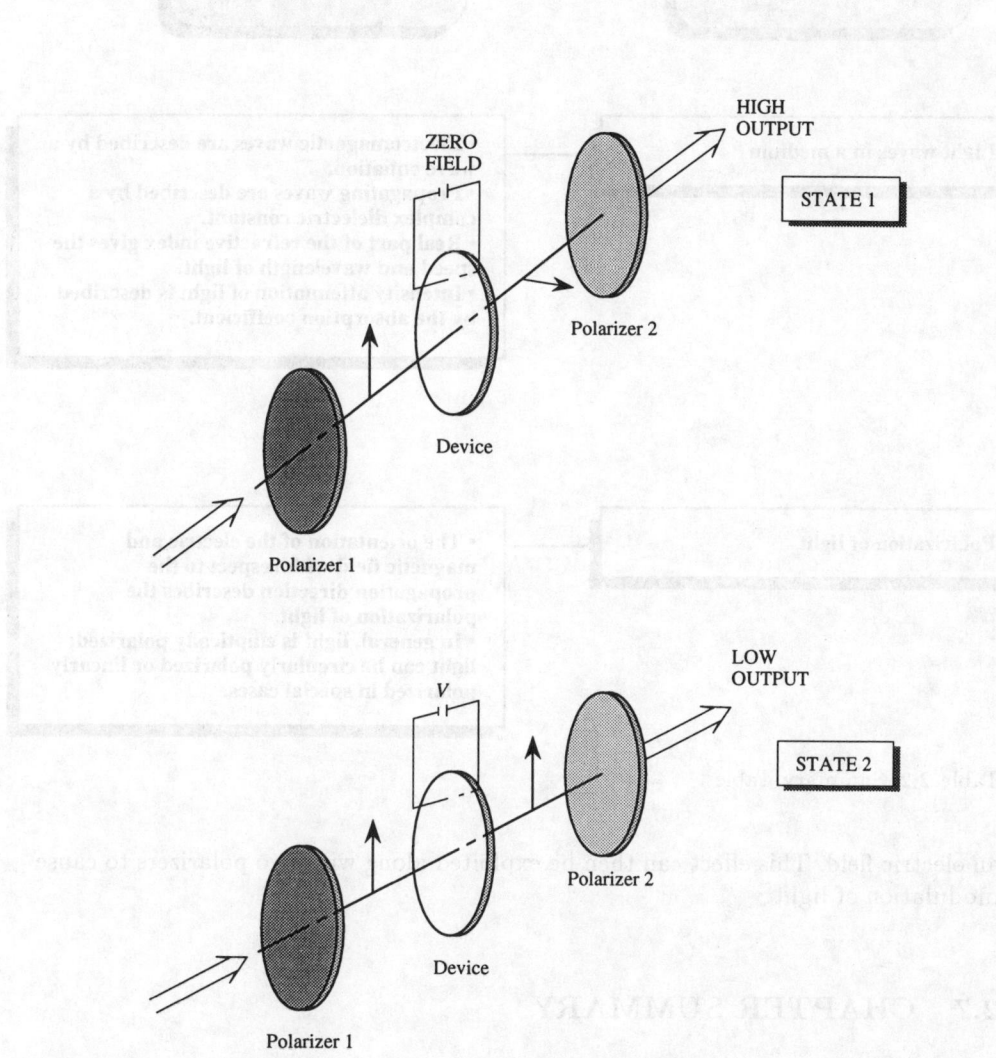

Figure 2.11: A schematic of one of many arrangements of how two polarizers can be used to change the light transmitted through a device. In state 1, the device with no applied field causes a 90° polarization change and the transmittance is high. A field alters the polarization back to the original value and the transmittance is low.

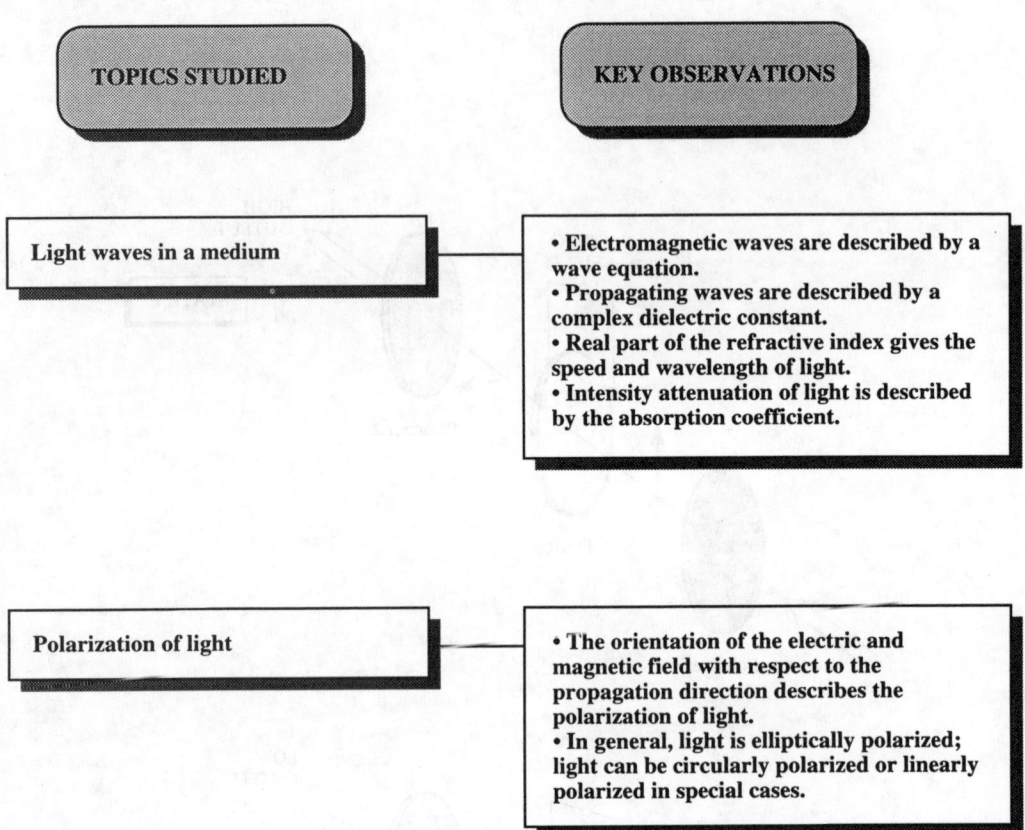

Table 2.2: Summary Table

an electric field. This effect can then be exploited along with two polarizers to cause modulation of light.

## 2.7  CHAPTER SUMMARY

The areas discussed in this chapter are summarized in Tables 2.2 through 2.4.

| TOPICS STUDIED | KEY OBSERVATIONS |
|---|---|
| Fresnel formulae for reflection and refraction of light | These formulae give us the ratio of the amplitudes of the reflected and transmitted lights. The ratios are different for *TE*- and *TM*-polarized waves. |
| Polarization of unpolarized incident light due to reflection and refraction | Since the relative ratios of the reflected and refracted amplitudes differs for *TE*- and *TM*-polarized light, unpolarized incident light is polarized after reflection (refraction) from an interface. At Brewster's angle, the reflected light is only *TE*-polarized. |
| Total internal reflection | When light travels from a denser medium to a rarer medium, it suffers total internal reflection if the angle of incidence is greater than the critical angle given by $\sin^{-1}\left(n_{r2}/n_{r1}\right)$. An evanescent wave penetrates the rarer medium even under total internal reflection conditions. |

Table 2.3: Summary Table

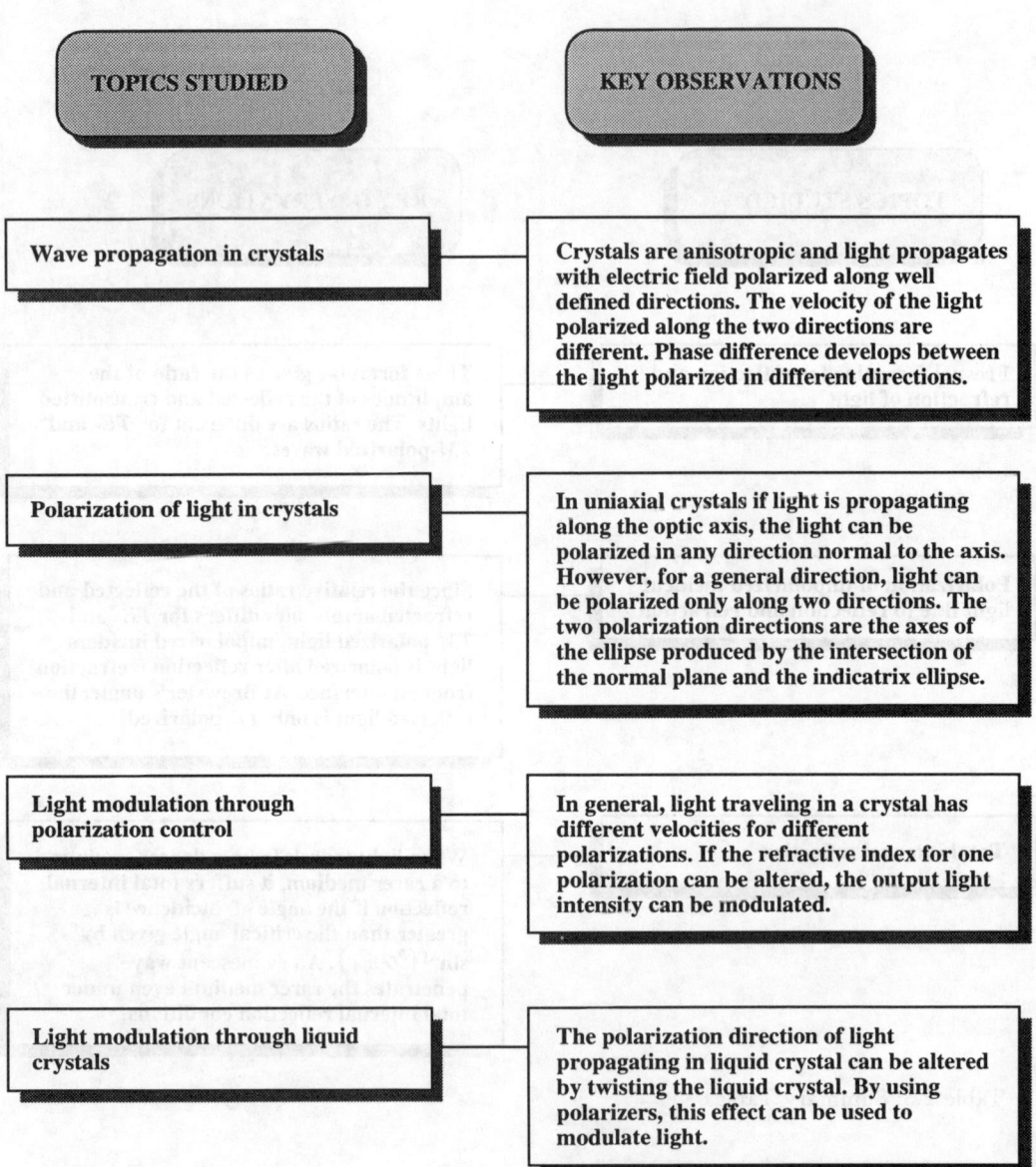

Table 2.4: Summary Table

## 2.8  PROBLEMS

**Section 2.2**

**2.1**  The optical power impinging upon a detector has a value of $10^{-6}$ mW/cm$^2$. Calculate the electric field amplitude associated with this power.

**2.2**  A detector is designed to detect a minimum electric field of 2.0 mV/cm. Calculate the minimum power level this detector can detect.

**2.3**  A microwave oven is designed to produce a maximum electric field (rms) of 1 kV/cm. Calculate the maximum electromagnetic power produced by this oven.

**2.4**  A typical silicon MOSFET "breaks down" when the electric field reaches $2 \times 10^5$ V/cm. Calculate the optical power needed to cause breakdown. Assume that the breakdown is only due to the high electric field. In an actual experiment, other effects such as photogenerated carriers may dominate.

**2.5**  The interface between two media with dielectric constants $\epsilon_1 = \epsilon_0$ and $\epsilon_2 = 12\epsilon_0$ coincides with the $xy$-plane. The electric field at some point at the interface in region 1 is known to have the value $\boldsymbol{E}_1 = (3\hat{x} + 2\hat{y} - 6\hat{z})\,E_0$. Calculate the electric field $\boldsymbol{E}_2$ at the same point in region 2. Also calculate the $D$-fields and the electric energy densities on the two sides of the plane.

**2.6**  The absorption coefficient in a material is 20 cm$^{-1}$ for light of wavelength 0.8 $\mu$m. Calculate the imaginary part of the refractive index.

**2.7**  The absorption coefficient of a light beam in silicon is found to be 600 cm$^{-1}$. Calculate the imaginary part of the refractive index.

**2.8**  The real part of the refractive index in a GaAs sample is found to be 3.6 for light of a free space wavelength of 0.8 $\mu$m. The absorption coefficient is found to be 40 cm$^{-1}$. Calculate the real and imaginary part of the dielectric constant.

**Section 2.3**

**2.9**  The electric fields of an optical beam are represented by the following:

$$E_x = \frac{1}{2}\cos(\omega t - kz)$$
$$E_y = \cos(\omega t - kz + \delta)$$

Sketch the polarization ellipses for the values given by $0, \pi/4, \pi/2$, and $3\pi/4$.

**2.10**  Derive the equation (Eqn. 2.22)

$$\left(\frac{E_x}{A_x}\right)^2 + \left(\frac{E_y}{A_y}\right)^2 - 2\frac{\cos\delta}{A_x A_y}E_x E_y = \sin^2\delta$$

**2.11**  An optical beam is traveling along the $z$-axis, and its fields are given by $E_x = 0.1\cos(\omega t - kz)$; $E_y = \cos(\omega t - kz + \pi/2)$. Calculate the major and minor axis of the polarization ellipse.

**2.12**  Sketch a diagram similar to Fig. 2.3 of the text for various $\delta$ values when the magnitude of the electric field in the $x$-direction is twice the magnitude of that in the $y$-direction.

**2.13** Consider an antenna whose response is proportional to the peak electric field along the antenna direction. The antenna is rotated in a plane perpendicular to the propagation direction of an elliptically polarized light. Plot the antenna response for

$$E_x = \frac{1}{2} \cos(\omega t - kz)$$
$$E_y = \cos(\omega t - kz + \delta)$$

with $\delta = \pi/4$.

## Section 2.4

**2.14** A plane wave is incident from air upon a non magnetic dielectric. It is found that at a certain angle, the power transmission of the $TE$ wave is 80% and that of the $TM$ wave is 100%. Calculate the refractive index of the dielectric and the angle of incidence.

**2.15** Calculate the transmission and reflection coefficients for the amplitudes of the electric field of a $TE$ wave incident at an angle of 30° form a medium with light velocity 0.5 $c$ to a medium with light velocity 0.75 $c$. Assume that both media are non magnetic and lossless. Here $c$ is the velocity of light in a vacuum.

**2.16** A plane wave is incident at an angle of 54° on an interface of two lossless, non magnetic media. No reflection occurs at this angle. Calculate the angle at which the refracted wave travels.

## Section 2.5

**2.17** A light beam enters one side of a prism which is made from an anisotropic material with $n_{ro} = 1.658$ and $n_{re} = 1.486$. The $c$-axis of the prism is normal to the plane of incidence. Calculate the range of internal angle of incidence for which the ordinary wave is totally reflected. The transmitted wave is thus completely polarized which allows one to use such prisms as polarizers.

**2.18** Consider a quartz plate cut along the $yz$ plane with the $c$-axis along the $z$-direction. The ordinary and extraordinary refractive indices are 1.544 and 1.553, respectively. Calculate the thickness of the plate if it is to serve as a quarter-wave plate, i.e., it produces a phase difference of $\pi/2$ between $y$ and $z$ polarized light. The wavelength in free space is 6330 Å.

**2.19** A light beam impinges from air on a uniaxial crystal with the $c$-axis normal to the plane of incidence. The angle of incidence is 45°. If $n_{re}$ and $n_{ro}$ are 1.55 and 1.54, calculate the angle of the refracted beams.

## 2.9 REFERENCES

- **General**

  - M. Born and E. Wolf, *Principles of Optics*, Pergamon, Oxford (1985).
  - P. Diament, *Wave Transmission and Fiber Optics*, Macmillan Publishing Company, New York (1990).
  - A. Yariv and P. Yeh, *Optical Waves in Crystals*, John Wiley and Sons, New York (1984).

# CHAPTER
# 3

# LIGHT PROPAGATION
# IN WAVEGUIDES

## 3.1  INTRODUCTION

In the previous chapter we have discussed the properties of *plane waves* in optical media. In the plane waves, the amplitude of the wave in the plane normal to the propagation direction is uniform. In most applications in optoelectronics, the optical beams are confined laterally to a finite region in space. Special optical elements are used to confine and allow propagation of such optical modes. An important structure used in optical systems is the layered structure or the waveguide structure. As the name waveguide implies, these structures are used to confine the optical waves in a well defined region and guide their propagation. The layered structures can be made from non-crystalline materials or from crystalline materials. For example, glass is used to produce optical fibers used in optical communication while semiconductor waveguides are used in semiconductor lasers, switches and a variety of other devices. In this chapter we will study how light propagates in planar and cylindrical waveguides. We will also study devices that are used to couple light into waveguides and from one waveguide to another.

## 3.2  WAVEGUIDES: SOME PHYSICAL PROPERTIES

$\longrightarrow$  In Chapter 2, in our discussion of the wave equation and wave propagation, we have only considered a single interface between two media. We will now discuss structures with multiple interfaces between materials with different refractive indices. By carefully tailoring the spatial dependence of the refractive indices, devices known as waveguides can be produced. As the name waveguide implies, these devices are designed so that the electromagnetic energy is confined to a narrow region in space and is guided along a channel. By confining the light wave in a narrow region (of typical dimensions of the order of the wavelength) several benefits arise:

- The optical wave may have to be sent from one point in space to another for applications in optical communications. The waveguide in this case is the "light pipe," or optical fiber. By confining the optical energy the light can be sent over thousands of kilometers.

- The wave can be affected by a physical phenomenon such as an altered refractive index due to an applied field. The change occurs over a small volume in space and the entire optical wave can sense the change. This feature is used in the design of waveguide light modulators.

- A particular region in space may have the property that light passing through it is amplified. Since this region usually has a small dimension, waveguiding is used to allow the optical wave to grow. Semiconductor lasers and amplifiers need waveguides for this reason.

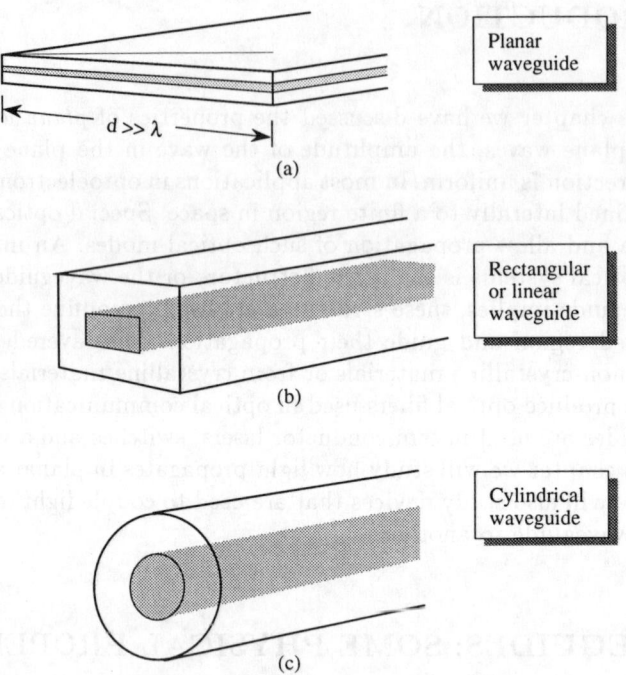

Figure 3.1: Different kinds of waveguide structures.

The confinement of the optical wave in space is done by a variation in the dielectric constant of the guide in space. The variation could be in a single dimension to produce planar waveguides as shown in Fig. 3.1a. One can also have a dielectric constant variation in two dimensions to produce a linear or rectangular waveguide as shown in Fig. 3.1b. Note that air can also be used as an effective confining medium. Finally, one can have a cylindrical waveguide as shown in Fig. 3.1c.

Waveguides produced from dielectrics and semiconductors grown epitaxially are usually of the planar or rectangular form. The optical fibers used in communication applications are pulled from a melt and are of the cylindrical form. Dielectric and semiconductor waveguides are widely used in integrated optoelectronics where electronic signals are superimposed on optical signals.

Techniques like ion implantation, etching, epitaxial growth and regrowth—techniques discussed in Chapter 11 are used to create a spatial variation in the material composition needed for waveguiding.

### 3.2.1  Some Properties of Optical Fibers

$\longrightarrow$ The superior properties of the fiber when compared to the metallic cables have been the most important driving forces for the use of optoelectronics in telecommunications, computer links, industrial automation, medical technology, and military applications. The output of the optical sources such as laser diodes (LDs) and light emitting diodes (LEDs), must be coupled to the optical fiber for most applications. It is thus essential to understand some of the basic principles governing the optical fiber to appreciate the demands placed on optical sources and detectors.

Optical fibers come in three main categories depending upon the ingredients used—silica, glass, and plastic. The silica fibers are made from $SiO_2$ with appropriate addition of metal oxides to fine-tune the refractive index. Important dopants that are used to adjust the refractive index are $TiO_2$, $Al_2O_3$, $GeO_2$ and $P_2O_5$. The glass fibers are made from a variety of glasses which have very high chemical stability. Plastics are also used for many applications. At present the plastics have a higher attenuation than the silica or glass fibers, but continuous improvements in the field may change this. In Chapter 11 we will discuss the fabrication issues of optical fibers.

The optical fiber is a cylindrical waveguide system through which the optical waves can propagate. The basic structure consists of a core at the center and a cladding layer on the outside as shown in Fig. 3.2. The core is a cylinder of transparent dielectric wire with a refractive index $n_{r1}$, and the cladding is a dielectric sheath with a lower index $n_{r2}$. The fibers have several classifications based on the index profile and the core size as shown in Fig. 3.2. The core size determines how many modes of the optical wave can propagate in the fiber. An optical cable consists of several of such optical fibers.

## 3.3  PLANAR WAVEGUIDES: A GEOMETRICAL OPTICS STUDY

$\longrightarrow$ We start our study of waveguides by examining the light propagation using simple ray optics. We have noted in Chapter 2, Section 2.2, that ray optics is adequate when the refractive index of the media does not change appreciably over the wavelength of light. This is often not true for waveguides. Nevertheless, ray optics or geometrical optics techniques have been used to understand some of the basic concepts in waveguides such as waveguide cutoff, propagation modes, etc. However, it must be kept in mind that while this picture is simple and intuitively appealing, it is not as complete as the wave theory discussed later.

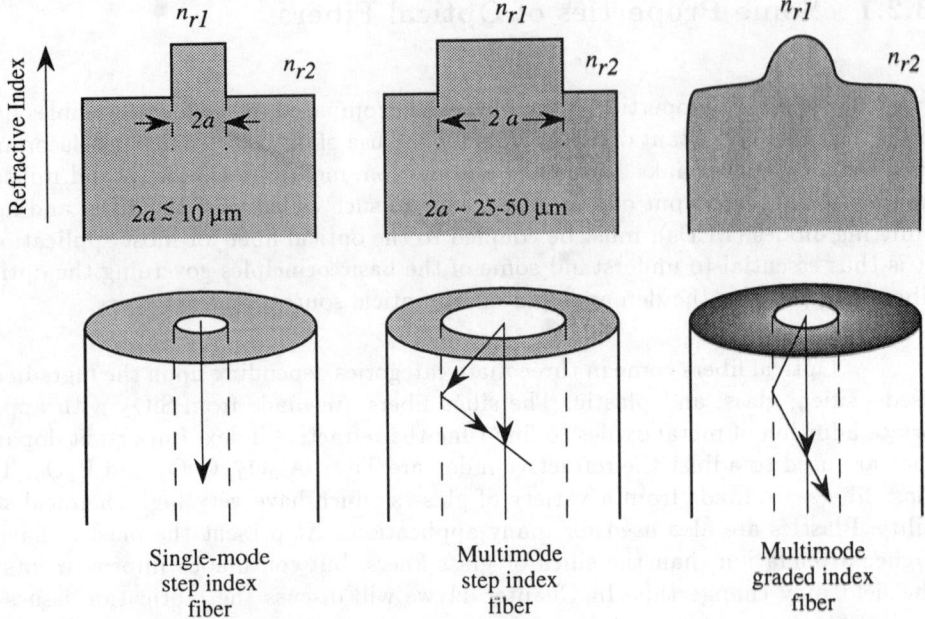

Figure 3.2: Structures of various kinds of optical fibers. Depending upon the size of the inner core, one or more modes of optical waves can propagate in the fiber. The single mode fibers are used for long haul high performance networks while the multimode fibers (which are cheaper) are used for LANs.

In the ray description of light propagation in a waveguide, we imagine the light to trace a zig-zag path in the guide with total internal reflection of light occurring at the interface. We use the Fresnel formulae for reflection and refraction in order to develop our treatment.

In Fig. 3.3 we show an interface separating two media with refractive indices $n_{r1}$ and $n_{r2}$ as shown. In the previous chapter we have discussed the amplitude ratio for the reflected wave and the phase change produced when there is total internal reflection.

Referring to Fig. 3.3a and the discussions in Section 2.4 (Eqns. 2.35 and 2.36), we may write

$$
\begin{aligned}
r_{TE} &= \frac{n_{r1}cos\theta_1 - n_{r2}cos\theta_2}{n_{r1}cos\theta_1 + n_{r2}cos\theta_2} \\
&= \frac{n_{r1}cos\theta_1 - \left(n_{r2}^2 - n_{r1}^2 sin^2\theta_1\right)^{1/2}}{n_{r1}cos\theta_1 + \left(n_{r2}^2 - n_{r1}^2 sin^2\theta_1\right)^{1/2}}
\end{aligned} \tag{3.1}
$$

Also, from Eqns. 2.41 and 2.42, we get

$$
\begin{aligned}
r_{TM} &= \frac{n_{r2}cos\theta_1 - n_{r1}cos\theta_2}{n_{r2}cos\theta_1 + n_{r1}cos\theta_2} \\
&= \frac{n_{r2}^2 cos\theta_1 - n_{r1}\left(n_{r2}^2 - n_{r1}^2 sin^2\theta_1\right)^{1/2}}{n_{r2}^2 cos\theta_1 + n_{r1}\left(n_{r2}^2 - n_{r1}^2 sin^2\theta_1\right)^{1/2}}
\end{aligned} \tag{3.2}
$$

with the critical angle being given by

$$
sin\theta_c = \frac{n_{r2}}{n_{r1}} \tag{3.3}
$$

When the angle of incidence is greater than $\theta_c$, the reflected amplitude ratio is complex. Writing for the complex reflected amplitude ratio,

$$
r = exp\ 2i\phi \tag{3.4}
$$

Upon comparison with Eqns. 3.1 and 3.2, we get, for the phase shift upon total reflection

$$
\boxed{tan\phi_{TE} = \frac{(n_{r_1}^2 sin^2\theta_1 - n_{r_2}^2)^{1/2}}{n_{r1}cos\theta_1}} \tag{3.5}
$$

$$
\boxed{tan\phi_{TM} = n_{r1}\frac{(n_{r_1}^2 sin^2\theta_1 - n_{r_2}^2)^{1/2}}{n_{r_2}^2 cos\theta_1}} \tag{3.6}
$$

These changes in the phase upon total internal reflection are very important in determining the allowed modes in a waveguide as we will see later. In Fig. 3.3b we show a sketch of the values of $\phi_{TE}$ for different refractive index ratios.

Let us now consider an asymmetric slab waveguide as shown in Fig. 3.4 which is produced by placing a film of index $n_{rf} = n_{r1}$ between a substrate of index $n_{rs} = n_{r3}$ and a cover or cladding material of index $n_{rc} = n_{r2}$. Let us assume that we have the relation

$$
n_{r1} > n_{r3} > n_{r2} \tag{3.7}
$$

As a result of the above relation we see that the critical angle for the film-substrate interface $\theta_s$ is larger than the critical angle $\theta_c$ for the film-cover interface. As the angle $\theta$ is changed, we have three distinct possibilities as shown in Fig. 3.4:

- For small angles of incidence, $\theta < \theta_s$, and $\theta_c$, light can escape from the waveguide as shown in Fig. 3.4a. In this case the wave mode is called the radiation mode.

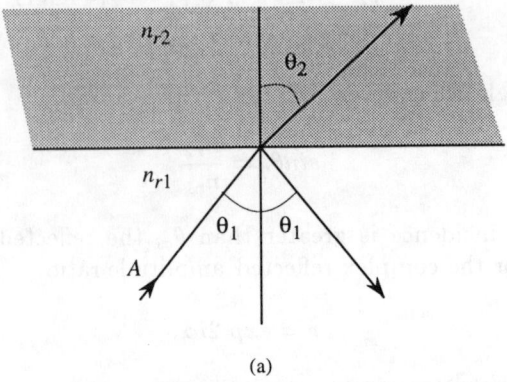

(a)

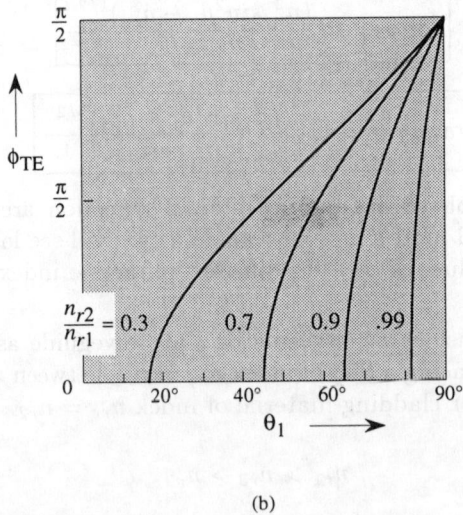

(b)

Figure 3.3: (a) Wave normals of reflected and refracted light at a planar interface between two media; (b) Phase shift $\phi_{TE}$ of the $TE$-polarized light as a function of the angle of incidence. The phase shifts are for angles of incidence greater than the critical angle for total internal reflection.

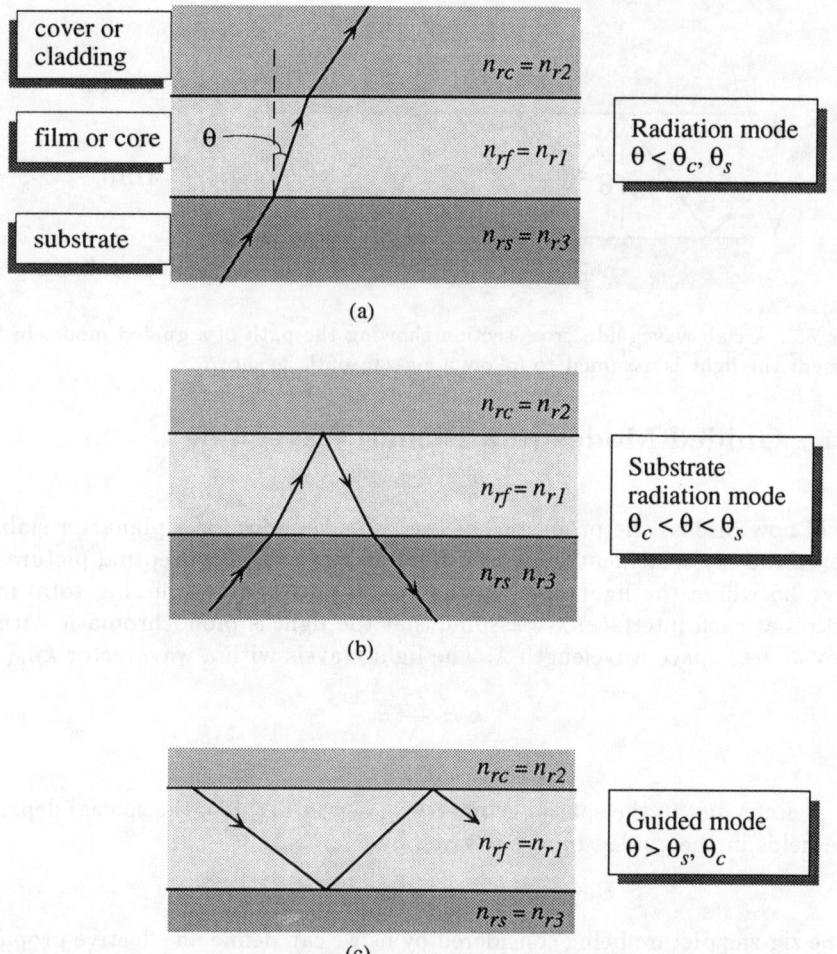

Figure 3.4: A simplified zig-zag description of light propagation in an asymmetric planar waveguide. (a) The radiation mode, (b) the substrate radiation mode, and (c) the guided mode.

- For somewhat larger $\theta$ so that $\theta_c < \theta < \theta_s$, the radiation is totally reflected from the film-cover interface and escapes into the structure as shown in Fig. 3.4b. Such modes are called substrate radiation modes.

- Finally we have the case where $\theta_s, \theta_c < \theta$, i.e., total reflection occurs at both interfaces and the radiation is confined in the guide. Such cases correspond to guided modes of propagation. One uses waveguides for their ability to sustain the guided modes.

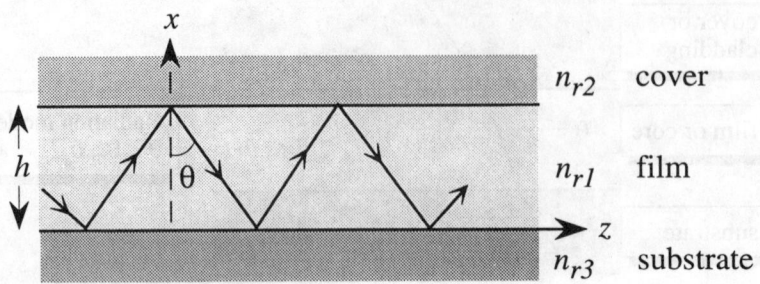

Figure 3.5: A slab waveguide cross-section showing the path of a guided mode. In the ray treatment the light is assumed to follow a zig-zag path as shown.

## 3.3.1  Guided Modes in a Planar Waveguide

We will now discuss the properties of the guided modes for a planar or slab waveguide. In the ray treatment that we will follow, we have a conceptual picture shown in Fig. 3.5 where the light rays zig-zag through the guide suffering total internal reflection at each interface. We assume that the light is monochromatic with a frequency $\omega$, free space wavelength $\lambda$. The light travels with a wavevector $kn_{r1}$ where

$$k = \frac{2\pi}{\lambda} = \frac{\omega}{c} \tag{3.8}$$

Referring to the spatial geometry shown in Fig. 3.5, the spatial dependence of the fields in the guided modes is given by

$$E \sim exp\left[-ikn_{r1}(\pm x cos\theta + z sin\theta)\right] \tag{3.9}$$

For the zig-zag picture being considered by us we can define an effective propagation constant along the $z$-axis and a related phase velocity $v_p$ given by

$$\beta = kn_{r1}sin\theta \equiv \frac{\omega}{v_p} \tag{3.10}$$

This is simply the $z$ component of the wavevector $kn_{r1}$ and represents propagation along the guide axis.

Now let us examine if rays with all possible angles $\theta$ and wavelengths are allowed to propagate or if there are some restrictions on the guided modes. We choose a cross-section of the guide $z = 0$ and sum the phases of the ray as we go from $x = 0$ interface to the opposite boundary $x = h$ and then back to $x = 0$. The total sum must equal $2\pi$ or a multiple of $2\pi$ for a single-valued field. Including the phase shifts produced by total internal reflection ($2\phi_s$ at the film substrate boundary and $2\phi_c$ at the film- cover boundary), we have the condition

$$\boxed{2kn_{r1}hcos\theta - 2\phi_s - 2\phi_c = 2\nu\pi, \nu = 0, 1, 2 \cdots} \tag{3.11}$$

The index $\nu$ defines the mode number. Note that $\phi_s$ and $\phi_c$ are functions of the angle $\theta$ (see Fig. 3.3b) and one needs to solve for the allowed $\theta$ values iteratively or graphically. The equation gives us the allowed $\theta$ values and consequently the propagation constant $\beta$ as a function of $\omega$. This is the dispersion relation for the guide. For the guided modes we also have

$$n_{r1} sin\theta \leq n_{r3} \tag{3.12}$$

and as a result the propagation constant obeys the inequality

$$\boxed{k n_{r3} < \beta < k n_{r1}} \tag{3.13}$$

A simple approach to finding the dispersion relation is the graphical approach illustrated in Fig. 3.6 for the fundamental mode with $\nu = 0$. We plot each of the terms in the resonance condition equation (i.e., Eqn. 3.11) as a function of the angle $\theta$ and look for a graphical solution for the resonance condition. To illustrate some of the important issues, we examine two cases; one of a symmetric guide, and one of an asymmetric guide with phase shifts $\phi_c$ and $\phi_s$. For the symmetric guide $\phi_c = \phi_s$. If we examine the fundamental mode ($\nu = 0$), we see from Fig. 3.6 that the symmetric guide always has a solution, since the dotted curve representing $k n_{r1} h \cos \theta$ intersects the $\phi_c + \phi_s = 2\phi_c$ curve at a point where $\theta > \theta_c$. This implies that there is no cutoff for the fundamental mode. In addition, by examining the curves $k n_{r1} h \cos \theta - 2\pi\nu$, we can see that as the waveguide thickness increases, more and more modes are allowed to propagate in the guide. In general, the higher order modes will have a cutoff frequency below which those modes cannot propagate.

When we examine the asymmetric waveguide, we see that the fundamental mode may not be allowed if the film thickness $h$ is small. For example, in the case shown in Fig. 3.6, only the intersection of the dotted line with the $\phi_s + \phi_c$ curve in the region $\theta > \theta_s$ allows the mode to propagate as a guided mode. Thus, for sufficiently thin waveguides, one will have a cutoff frequency for the fundamental mode for the asymmetric guide.

As one varies the value of $k$ (or $\omega$), one can see how the allowed propagation wavevector $\beta$ evolves. In Fig. 3.7 we show a typical $\omega - \beta$ dispersion relation that arises in the waveguides. We note that the allowed guided modes are enclosed by radiation modes on one side and a forbidden region on the other.

Before finishing up our ray treatment of the planar waveguide and starting with the wave treatment, it is important to note that while in our treatment we have used the structural thickness of the guide $h$ as the region in which the guided mode is confined, in reality the effective thickness is somewhat larger. This is because of the penetration of the evanescent fields into the substrate and the cover regions. In general one may write

$$h_{eff} = h + x_s + x_c \tag{3.14}$$

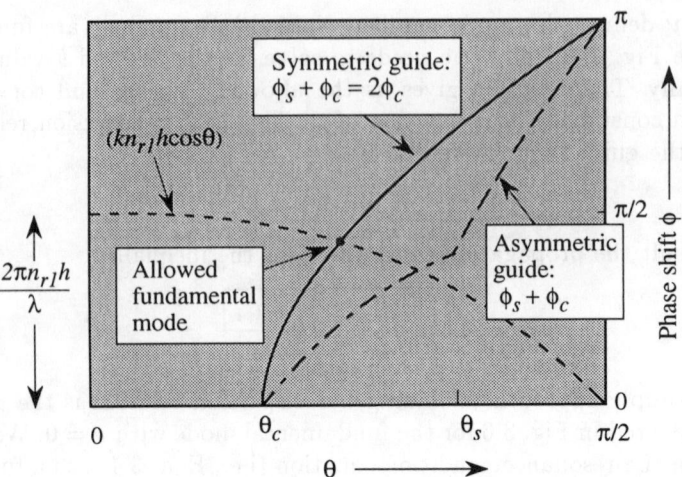

Figure 3.6: A graphical method for solving the resonance condition equation for the allowed guided modes. The results are shown for the fundamental mode where the $kn_{r1}cos\theta = \phi_s + \phi_c$ condition gives the mode. Also $\theta$ must be larger than $\theta_s$ and $\theta_c$.

where $x_s$ and $x_c$ represent the penetration depth of the light. The correction to the thickness is unimportant if $h \gg \lambda$, but if $h$ and $\lambda$ are comparable the correction is quite appreciable and must be used to calculate the dispersion relations for the guided modes.

**EXAMPLE 3.1** A plane wave is incident from a denser medium with $n_{r1} = 3.6$ to a rarer medium with $n_{r2} = 2.2$ at an angle of incidence $\theta_1 = 40°$. Calculate the reflected waves for $TE$ and $TM$ polarizations.

If we attempt to use Snell's law, we see that for the refracted wave

$$sin\theta_2 = \frac{3.6}{2.2} \ sin\theta_1 = 1.052$$

which means that the waves will suffer total internal reflection. For the $TE$ wave, the angle $\phi_{TE}$ giving the phase change is from Eqn. 3.5

$$\phi_{TE} \ = \ 0.254 \ radians$$
$$r_{TE} \ = \ exp(i0.51)$$

For the $TM$ wave we have, from Eqn. 3.6

$$\phi_{TE} \ = \ 0.608 \ radians$$
$$r_{TM} \ = \ exp(i1.217)$$

**EXAMPLE 3.2** A slab symmetric waveguide is made with the core (film) refractive index of 3.5 and cladding and substrate refractive index of 3.0. Calculate the maximum

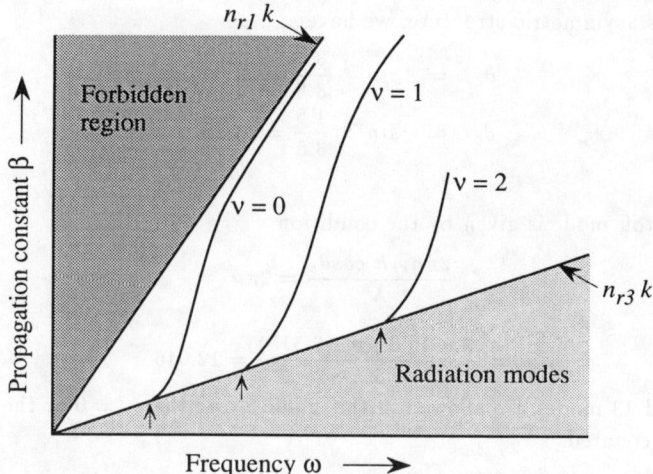

Figure 3.7: A typical dispersion relation for the modes of a slab waveguide. The cutoff frequencies for various modes are indicated by arrows.

thickness of the film that will allow no more than six guided modes to propagate for $\lambda = 1.5\mu m$.

The film thickness should be chosen so that the $\nu = 6$ mode has a cutoff. The critical angle of incidence for the guide is

$$\theta_s = \theta_c = sin^{-1}\frac{3.0}{3.5} = 59°$$

Thus, for the $\nu = 6$ mode we must have the film thickness such that

$$\frac{2\pi n_{r1} h \; cos\theta_c}{\lambda} < 12\pi$$

or

$$h < \frac{12\pi\lambda}{2\pi n_{r1} \; cos\theta_c} = 5.0 \; \mu m$$

Thus, the film (core) thickness for the guide has to be less than 5.0 $\mu m$.

**EXAMPLE 3.3** An asymmetric waveguide is designed with the following parameters:

$$
\begin{aligned}
n_{r1} &= 3.5 \\
n_{r2} &= 1.5 \\
n_{r3} &= 3.0 \\
h &= 10.0 \; \mu m
\end{aligned}
$$

Calculate the number of modes allowed in the waveguide if $\lambda = 1.5\mu m$.

For this asymmetric structure, we have

$$\theta_s = sin^{-1} \frac{3.0}{3.5} = 59°$$

$$\theta_c = sin^{-1} \frac{1.5}{3.5} = 25.38°$$

The cutoff mode is given by the condition

$$\frac{2\pi n_{r1} h \; cos\theta_s}{\lambda} = 2\pi\nu$$

or

$$\nu = \frac{3.5(10.0 \; \mu m)(0.515)}{1.5 \; \mu m} = 12.016$$

Thus $\nu = 12$ and 13 modes are allowed in this guide. Note that $\nu = 0$ or the fundamental mode has to be counted.

## 3.4   OPTICAL FIBERS: GEOMETRICAL OPTICS

$\longrightarrow$   The optical fiber is perhaps the most important component in modern communication system. Without the fiber, the state of semiconductor optoelectronics would not be so advanced. It is, therefore, extremely important to understand how light propagates in fibers. A proper understanding of this requires one to use the wave description of light. However, if the core of the fiber is larger than the wavelength of light, a simple geometric optics treatment provides adequate results. This treatment, given in this section, also provides valuable insight into the performance controlling characteristics of the fiber.

The optical fiber is a carrier of optical information, and one of the important parameters of an optical-fiber-based system is the efficiency with which light can be coupled into a fiber from an optical source. This requires that the incident light arrive with such an angle that it suffers a total internal reflection at the core-cladding layer interface as shown in Fig. 3.8.

The critical angle for total internal reflection can be obtained by applying Snell's law to an optical beam propagating from a medium with index $n_{r1}$ to a medium with index $n_{r2}$. The angles of incidence and refraction are given by

$$\frac{sin\phi_2}{sin\phi_1} = \frac{n_{r1}}{n_{r2}} \tag{3.15}$$

For total reflection $\phi_2 = 90°$ so that the critical angle $\phi_1 = \phi_{1c}$ is

$$\phi_{1c} = sin^{-1}\frac{n_{r2}}{n_{r1}} \tag{3.16}$$

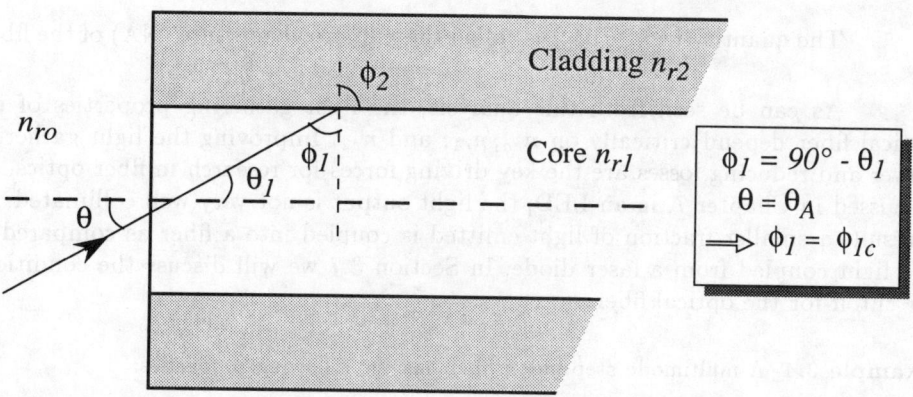

Figure 3.8: The geometry used to calculate the acceptance angle for an optical fiber. When the incident angle exceeds a value $\theta_A$, the ray is not able to propagate within the optical fiber.

Now let us consider the case of an optical fiber in which light is incident from a medium with index $n_{ro}$ into the fiber as shown in Fig. 3.8. We wish to find the maximum angle of acceptance for the fiber. According to Snell's law,

$$n_{ro}sin\theta = n_{r1}sin\theta_1 = n_{r1}cos\phi_1 \tag{3.17}$$

The total internal reflection between the core and the cladding layer interface occurs when

$$sin\phi_1 \geq \frac{n_{r2}}{n_{r1}} \tag{3.18}$$

or

$$cos\phi_1 \leq \left(1 - \frac{n_{r2}^2}{n_{r1}^2}\right)^{1/2} \tag{3.19}$$

Using Eqns. 3.17 and 3.19 we get for the limit on $sin\theta$

$$n_{ro}sin\theta < \left(n_{r1}^2 - n_{r2}^2\right)^{1/2} \tag{3.20}$$

or

$$sin\theta \leq \frac{1}{n_{ro}}\left(n_{r1}^2 - n_{r2}^2\right)^{1/2} \tag{3.21}$$

The maximum angle of acceptance is now

$$\boxed{\theta_A = sin^{-1}\left[\frac{1}{n_{ro}}\left(n_{r1}^2 - n_{r2}^2\right)^{1/2}\right]} \tag{3.22}$$

The quantity $\left(n_{r1}^2 - n_{r2}^2\right)$ is called the numerical aperture (NA) of the fiber.

As can be seen from this analysis, the light gathering properties of the optical fiber depend critically on $n_{ro}, n_{r1}$, and $n_{r2}$. Improving the light gathering power and reducing losses are the key driving forces for research in fiber optics. As discussed in Chapter 7, in an LED, the light output is not very well collimated. As a result, a smaller fraction of light emitted is coupled into a fiber as compared to the light coupled from a laser diode. In Section 3.7 we will discuss the conditions for cutoff for the optical fiber.

**Example 3.4**  A multimode step index fiber has the following properties:

$$\text{core refractive index} \quad n_{r1} = 1.53$$
$$\text{cladding refractive index} \quad n_{r2} = 1.48$$

Calculate the critical angle for total internal reflection and the maximum angle of acceptance when light is incident from air.

The critical angle is

$$\phi_{1c} = sin^{-1}\left(\frac{1.48}{1.53}\right)$$
$$= 75.3°$$

The maximum angle of acceptance is

$$\theta_A = sin^{-1}\left[(1.53)^2 - (1.48)^2\right]^{1/2}$$
$$= 22.8°$$

## 3.5  POLARIZATION CONSTRAINTS IN WAVEGUIDES

$\longrightarrow$ Having examined the geometrical optics treatment of waveguides, let us examine a more rigorous treatment based on wave optics. In the next sections we will consider the optical guided modes in planar and cylindrical waveguides. Before starting the details of the study, let us examine the constraints satisfied by the electric and magnetic field components in the two kinds of waveguides.

The general solutions for the fields in a propagating mode in the planar waveguide with the guide axis along the $z$-direction are of the form

$$\boldsymbol{E} = \boldsymbol{E}(x,y)\, exp\left[i(\omega t - \beta z)\right] \tag{3.23}$$
$$\boldsymbol{H} = \boldsymbol{H}(x,y)\, exp\left[i(\omega t - \beta z)\right] \tag{3.24}$$

In the case of the cylindrical waveguides the solutions have the form

$$\boldsymbol{E} = \boldsymbol{E}(r,\theta) \, exp \, [i(\omega t - \beta z)] \tag{3.25}$$
$$\boldsymbol{H} = \boldsymbol{H}(r,\theta) \, exp \, [i(\omega t - \beta z)] \tag{3.26}$$

Here $\beta$ is the propagation constant for the propagating wave and for a given free space wavelength $\lambda$, its value has to be determined by solving the wavequation with appropriate boundary conditions.

In the case of the planar waveguide where the refractive index variation is along the $x$-axis, since there is no variation in the material parameters in the $y$-direction, we can assume that $\partial/\partial y = 0$. Substituting the values of $\boldsymbol{E}$ and $\boldsymbol{H}$ in the curl equations of the Maxwell equations set, we get a set of relations between $E_y, H_x, H_z$ and an independent set of relations involving $E_x, E_z, H_y$. The former set involves modes in which the electric field is purely transverse to the propagation vector, and the modes are called transverse electric, or $TE$, modes. In the latter set, the magnetic field has only a transverse component. The relationships are:

$TE$ waves (involve $E_y, H_x, H_z$):

$$\beta E_y = -\mu\omega H_x$$
$$-i\beta H_x - \frac{\partial H_z}{\partial x} = i\epsilon\omega E_y$$
$$\frac{\partial E_y}{\partial x} = -i\mu\omega H_z \tag{3.27}$$

$TM$ waves (involve $E_x, E_z, H_y$):

$$\beta H_y = \epsilon\omega E_x$$
$$i\beta E_x + \frac{\partial E_z}{\partial x} = i\omega\mu H_y$$
$$\frac{\partial H_y}{\partial x} = i\epsilon\omega E_z \tag{3.28}$$

In a planar waveguide there is thus a natural separation into $TE$ and $TM$ modes. It is important to point out that if there is no material parameter variation along the $x$-direction, i.e., if there is no guiding action,

$$\frac{\partial E_y}{\partial x} \equiv 0 \Rightarrow H_z = 0 \tag{3.29}$$

and

$$\frac{\partial H_y}{\partial x} \equiv 0 \Rightarrow E_z = 0 \tag{3.30}$$

In this case we have no axial components of the fields and the waves are known as transverse electromagnetic or TEM.

In the case of the cylindrical waveguides, the components do not separate into groups of uncoupled transverse electric or transverse magnetic waves as in the case of the planar guides. In general, using the curl operator in the cylindrical coordinates, i.e.,

$$\nabla \times A = \hat{r}\left(\frac{1}{r}\frac{\partial A_z}{\partial \theta} - \frac{\partial A_\theta}{\partial z}\right) + \hat{\theta}\left(\frac{\partial A_r}{\partial z} - \frac{\partial A_z}{\partial r}\right)$$
$$+ \hat{z}\frac{1}{r}\left(\frac{\partial}{\partial r}(rA_\theta) - \frac{\partial A_r}{\partial \theta}\right) \tag{3.31}$$

we get the following relations from the Maxwell equations:

$$\frac{1}{r}\left(\frac{\partial E_z}{\partial \theta} + i\beta r E_\theta\right) = -i\omega\mu H_r$$

$$i\beta E_r + \frac{\partial E_z}{\partial r} = i\omega\mu H_\theta$$

$$\frac{1}{r}\left(\frac{\partial r E_\theta}{\partial r} - \frac{\partial E_r}{\partial \theta}\right) = -i\omega\mu H_z$$

$$\frac{1}{r}\left(\frac{\partial H_z}{\partial \theta} + i\beta r H_\theta\right) = i\omega\epsilon E_r$$

$$i\beta H_r + \frac{\partial H_z}{\partial r} = -i\omega\epsilon E_\theta$$

$$\frac{1}{r}\left(\frac{\partial r H_\theta}{\partial r} - \frac{\partial H_r}{\partial \theta}\right) = i\omega\epsilon E_z \tag{3.32}$$

Since, in general, we have coupling between $E_z$ and $H_z$, the waves are not simple linearly polarized $TE$ and $TM$ waves. Instead, in general, one can have additional hybrid modes in which neither the electric nor magnetic field is purely transverse.

In Fig. 3.9 we summarize the different polarizations which are possible when waves travel in various structures.

## 3.6 GUIDED OPTICAL MODES IN PLANAR WAVEGUIDES

$\longrightarrow$ In Section 3.3 we discussed a treatment for planar waveguides based upon geometrical optics and the "zig zag" propagation of light. We will now consider the more realistic wave treatment of the planar waveguide. Let us consider a planar waveguide structure as shown in Fig. 3.10. The waveguide has a three-layer structure, as shown, where a high refractive index material is surrounded by a low refractive index material. The thickness of the guide is $d$, as shown. Although in

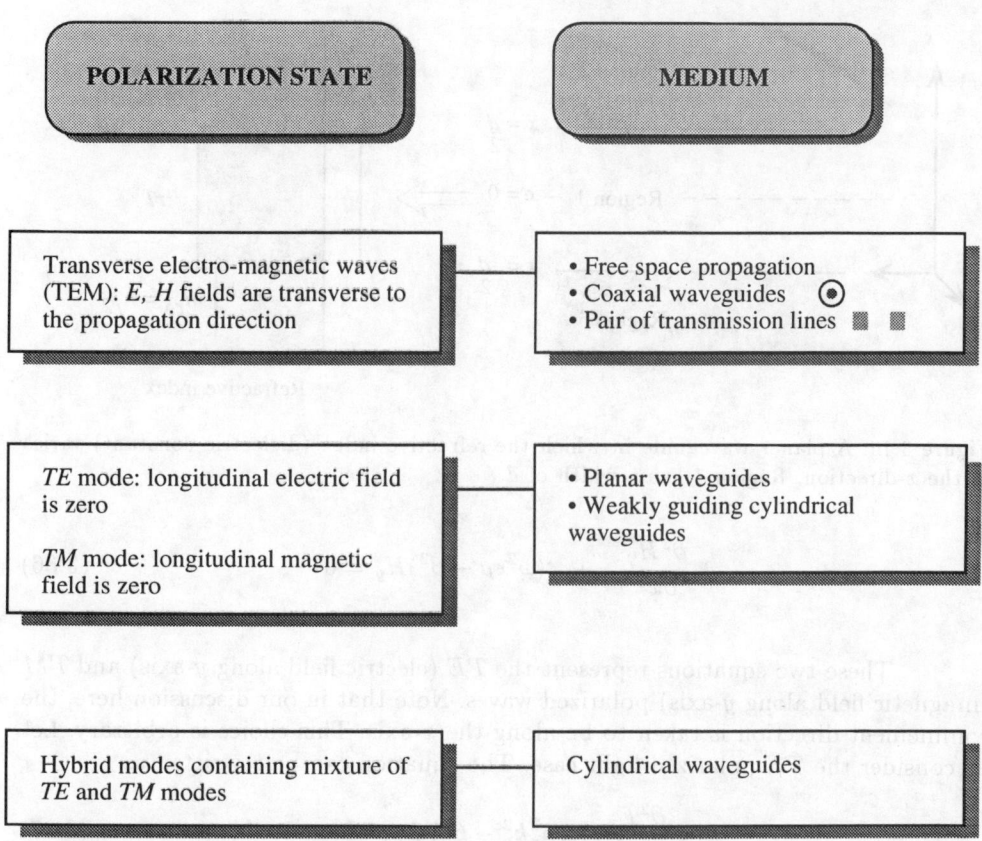

Figure 3.9: Various polarization states and terminology used for wave propagation in free space and in waveguides. The media in which the different polarization states can exist are shown.

general, the refractive indices (dielectric constants) of regions surrounding the guide can be different, we will discuss the case where $n_{r2} = n_{r3}$, as shown in Fig. 3.10.

Let us consider the case of a wave propagating along the $z$-axis. Assuming that there is no spatial variation of the wave in the $y$-direction, the waves for the electric and magnetic fields are, in general,

$$\mathbf{E} = \mathbf{E}(x, y) \, exp \, [i(\omega t - \beta z)] \tag{3.33}$$

$$\mathbf{H} = \mathbf{H}(x, y) \, exp \, [i(\omega t - \beta z)] \tag{3.34}$$

The $x$-dependence of the fields is given from the relations from Eqns. 3.27 (after eliminating $H$-terms) and 3.28 (after eliminating $E$-terms)

$$\frac{\partial^2 E_y}{\partial x^2} + (\omega^2 \epsilon \mu - \beta^2) E_y = 0 \tag{3.35}$$

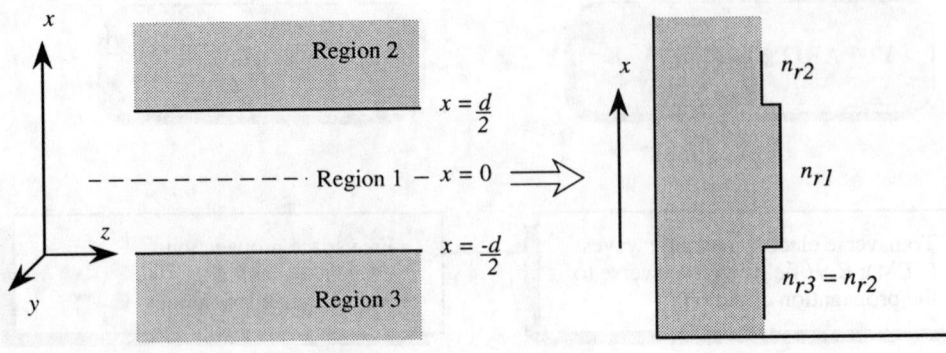

Figure 3.10: A planer waveguide in which the refractive index (dielectric constant) varies in the $x$-direction. Region 1 has a width of $d$.

$$\frac{\partial^2 H_y}{\partial x^2} + (\omega^2 \epsilon \mu - \beta^2) H_y = 0 \qquad (3.36)$$

These two equations represent the $TE$ (electric field along $y$-axis) and $TM$ (magnetic field along $y$-axis) polarized waves. Note that in our discussion here, the confinement direction is taken to be along the $x$-axis. This choice is arbitrary. Let us consider the $TE$ polarized light case. The equation can be rewritten as

$$\frac{\partial^2 E_y}{\partial x^2} + \left(n_r^2 k_o^2 - \beta^2\right) E_y = 0 \qquad (3.37)$$

where we have used the fact that

$$\omega\sqrt{\epsilon\mu} = n_r\, k_o \qquad (3.38)$$

where

$$k_o = \frac{2\pi}{\lambda} \qquad (3.39)$$

is the free space wave vector.

From Eqn. 3.37 we can see that, in general, the guided mode solutions will either have the form

$$E_y \sim exp\,(i\,k_x x) \ \ \text{if} \ \ n_r^2 k_o^2 > \beta^2 \qquad (3.40)$$

where

$$k_x = \sqrt{n_r^2 k_o^2 - \beta^2} \qquad (3.41)$$

or the form

$$E_y \sim exp\,(-\gamma x) \ \ \text{if} \ \ n_r^2 k_o^2 < \beta^2 \qquad (3.42)$$

where

$$\gamma = \sqrt{\beta^2 - n_r^2 k_o^2} \qquad (3.43)$$

Equations 3.40 and 3.42 represent oscillatory and decaying waves. The waveguide is designed so that for some modes (i.e., the guided modes), the field has a general form given by Eqn. 3.40 in the guiding region and the form given by Eqn. 3.42 in the cladding region. Referring to Fig. 3.10, we get the following solutions for the electric field: ($k_x$ and $\gamma$ are given by Eqns. 3.40 and 3.41 with $n_r = n_{r1}$ for the equation for $k_x$ and $n_r = n_{r2}$ for the equation for $\gamma$)

$$E_y = \begin{cases} A\ e^{\gamma x} & x \leq -d/2 \\ B\ cos(k_x x) + C\ sin(k_x x) & -d/2 \leq x \leq d/2 \\ D\ e^{-\gamma x} & x \geq d/2 \end{cases} \tag{3.44}$$

Matching the fields and the derivatives at the boundaries we have the conditions

$$B\ cos\ \frac{k_x d}{2} - C\ sin\ \frac{k_x d}{2} = Ae^{-\gamma d/2}$$

$$k_x B\ sin\ \frac{k_x d}{2} + Ck_x\ cos\ \frac{k_x d}{2} = \gamma Ae^{-\gamma d/2}$$

$$B\ cos\ \frac{k_x d}{2} + C\ sin\ \frac{k_x d}{2} = De^{-\gamma d/2}$$

$$-k_x B\ sin\ \frac{k_x d}{2} + k_x C\ cos\ \frac{k_x d}{2} = -\gamma De^{-\gamma d/2} \tag{3.45}$$

From these equations we get the set of equations

$$2B\ cos\ \frac{k_x d}{2} = (A + D)e^{-\gamma d/2}$$

$$2k_x B\ sin\ \frac{k_x d}{2} = \gamma(A + D)e^{-\gamma d/2} \tag{3.46}$$

and

$$2C\ sin\ \frac{k_x d}{2} = (D - A)e^{-\gamma d/2}$$

$$2k_x C\ cos\ \frac{k_x d}{2} = -\gamma(D - A)e^{-\gamma d/2} \tag{3.47}$$

These pairs give us two separate conditions for the solutions obtained by dividing one equation by the other within each pair. The conditions for the allowed guided modes are the following transcendental equations:

$$\boxed{\frac{k_x d}{2}\ tan\ \frac{k_x d}{2} = \frac{\gamma d}{2}} \tag{3.48}$$

or

$$\boxed{\frac{k_x d}{2}\ cot\ \frac{k_x d}{2} = -\frac{\gamma d}{2}} \tag{3.49}$$

The first of these equations results in modes that have even parity, i.e., with $A = D$ in Eqn. 3.46, while the second equation gives modes with odd parity, with

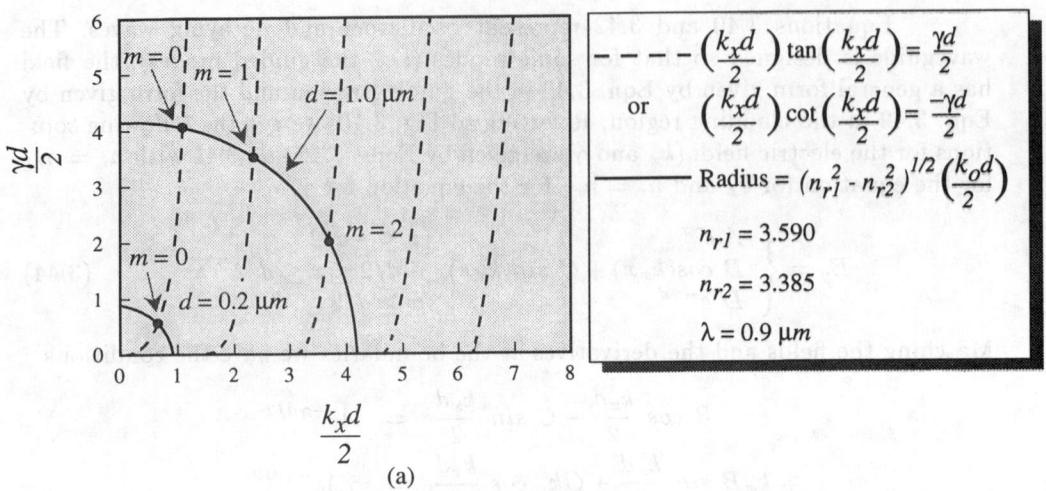

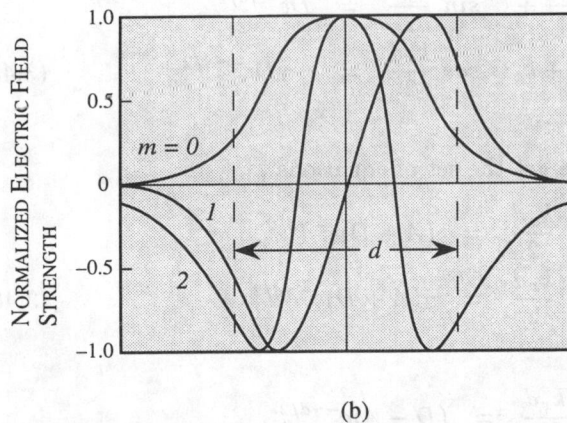

(b)

Figure 3.11: (a) The graphical approach to solving for the allowed modes in a waveguide. (b) Typical solutions for the waveguide modes.

$A = -D$ in Eqn. 3.47. The modes are denoted by $m = 0, 1, 2, \ldots$. The transcendental Eqns. 3.48 and 3.49 can be solved by numerical techniques. One useful approach is a graphical technique shown in Fig. 3.11a. One starts out with plotting curves in the $\gamma d/2$, $k_x d/2$ plane which satisfy Eqns. 3.48 or 3.49. Note that a large number of $k_x$ values satisfy the equations for a given value of $\gamma$. Next, we note that we have the equality (from Eqn. 3.41 with $n_r = n_{r1}$ and Eqn. 3.43 with $n_r = n_{r2}$)

$$\left(\frac{k_x d}{2}\right)^2 + \left(\frac{\gamma d}{2}\right)^2 = (n_{r1}^2 - n_{r2}^2)\left(\frac{k_o d}{2}\right)^2$$

$$\equiv R(d)^2 \tag{3.50}$$

We therefore draw a circle with radius $R(d)$. For a given value of $n_{r1}, n_{r2}$ and $d$, there is one such circle. The intersection of this circle with the first set of curves gives the desired solutions. There may be a number of allowed solutions for a given waveguide thickness. As the waveguide thickness increases, the number of allowed modes also increases, as shown in Fig. 3.11a. To find the cutoff mode for a given thickness, we note that $(k_x d/2) \tan (k_x d/2)$ and $(k_x d/2) \cot (k_x d/2)$ intersect the $(k_x d/2)$ axis at values of $m\pi/2$. Thus the thickness at which a mode $m$ is just allowed is given by

$$R(d_c) = \frac{m\pi}{2}$$

or from Eqn. 3.50

$$\boxed{d_c = \frac{1}{2} \frac{m\lambda}{\left(n_{r1}^2 - n_{r2}^2\right)^{1/2}}} \tag{3.51}$$

Once $k_x$ (or $\gamma$) is known, the propagation constant $\beta$ is known for a given free space $k_o$ from the relation

$$k_x^2 = n_{r1}^2 k_o^2 - \beta^2$$

A typical modal profile for the electric field is shown in Fig. 3.11b.

## 3.6.1 Optical Confinement Factor

An important parameter in optical waveguides is the fraction of the optical energy in the guide region (region 1 of Fig. 3.10). This is known as the *optical confinement factor* $\Gamma$. This factor is known from Eqn. 3.46 or 3.47 once $\gamma$ and $k_x$ are known. The optical confinement factor is

$$\Gamma = \frac{\int_{-d/2}^{d/2} E_y^2 \, dx}{\int_{-\infty}^{\infty} E_y^2 \, dx} \tag{3.52}$$

If the wave is normalized, in terms of $k_x$, the confinement factor is found to have the value

$$\Gamma = \left\{ 1 + \frac{\cos^2(k_x d/2)}{\gamma \left[ d/2 + \left(\frac{1}{k_x}\right) \sin (k_x d/2) \cos (k_x d/2) \right]} \right\}^{-1} \tag{3.53}$$

The value of $\Gamma$ is a strong function of the mode number, the thickness of the guide, and the difference between $n_{r1}$ and $n_{r2}$. Some values for $\Gamma$ are shown in Fig. 3.12 for GaAs/Al$_{03}$Ga$_{0.7}$As waveguides.

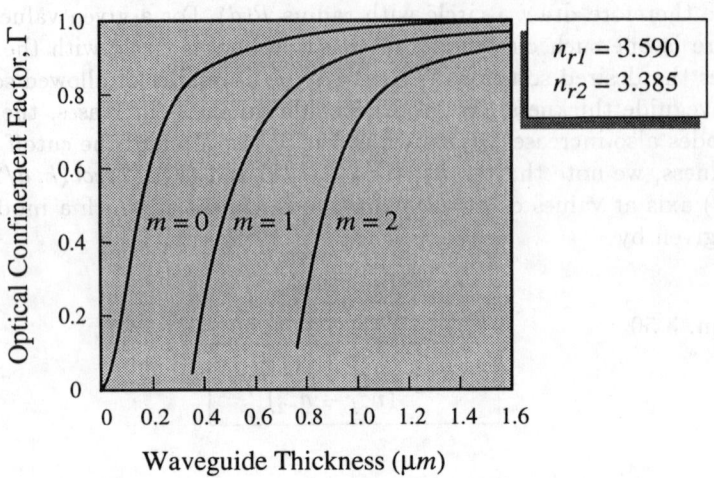

Figure 3.12: Optical confinement factor as a function of waveguide thickness.

**EXAMPLE 3.5**  Consider a symmetric planar wavegide with $n_{r1} = 3.6$ and $n_{r2} = 3.3$. The core region thickness is 2.0 $\mu$m. Calculate the number of $TE$ modes allowed in the guide for a free space wavelength of 0.8 $\mu$m. Compare the results with the geometric optics results derived in Section 3.3.

The number of modes is determined from Eqn. 3.51 by using $d = d_c$

$$
\begin{aligned}
m &= \frac{2d_c \left(n_{r1}^2 - n_{r2}^2\right)^{1/2}}{\lambda} \\
&= \frac{2\,(2.0 \ \mu m)\,(12.96 - 10.89)^{1/2}}{(0.8 \ \mu m)} \\
&= 7.18
\end{aligned}
$$

Thus $m$ has to be 7 and there are eight modes allowed in this guide.

Following the discussion in Section 3.3, we first calculate the critical angle for internal reflection

$$\theta = \theta c = sin^{-1} \frac{3.3}{3.5} = 70.54° = 1.231 \text{ radians}$$

From Eqn. 3.11 we get

$$
\begin{aligned}
\nu &= 1/2\pi \left[2kn_{r1}h \ cos\theta - 2\phi_s - 2\phi_c\right] \\
&= 1/2\pi \left[37.68 - 4.924\right] \\
&= 5.21
\end{aligned}
$$

Thus the geometric optics treatment suggests there are only six allowed modes. However, if we recall that we should use the effective thickness which accounts for the penetration of the waves into the cladding region, we will get a better agreement.

**EXAMPLE 3.6** Consider a symmetric planar wavegide made from two materials with refractive indices 3.4 and 3.2. Calculate the maximum thickness that will sustain a single $TE$ guided mode with $\lambda = 0.8\mu$m. Also calculate the optical confinement factor for this mode.

The maximum thickness we can have is given by Eqn. 3.51 with $m = 1$

$$d < d_c = \frac{0.8}{2(11.56 - 10.24)^{1/2}}$$
$$= 0.348 \ \mu m$$

In order to find the optical confinement factor for this guide with, say $d = 0.345\mu$m, we can use the results in Fig. 3.11a.

We have
$$R(d) = 1.558$$
If a circle with this radius were drawn on Fig. 3.11a, we find that it intersects the first dashed line at approximately

$$\frac{\gamma d}{2} \simeq 1.25; \quad \frac{k_x d}{2} \simeq 0.9$$

From Eqn. 3.53 this gives an optical confinement factor of 0.83.

## 3.7   OPTICAL FIBERS: WAVE OPTICS

↝   In Section 3.4 we have discussed the simpler geometric optics treatment of light propagation in optical fibers. This treatment is adequate if the size of the fiber core is much larger than the wavelength of the light being transmitted. This is the case in multimode fibers which are used for less demanding applications. For many demanding optical communication applications, the fibers used are single mode with core diameters approaching a micrometer. For such cases we need to discuss the appropriate wave picture for light propagation. As discussed in Section 3.6 for the planar guide, the wave description leads to well defined guided modes that describe the form of the optical waves that are allowed to propagate.

In the case of the planar waveguide, we solved the wave equation in the cartesian coordinates $x, y, z$. In the case of the cylindrical waveguide it is appropriate that we work in the cylindrical coordinate system using $r, \theta, z$ as the independent coordinates. We choose a general solution for the electric (or magnetic) field of the form

$$\boldsymbol{E}(r, \theta, \phi) = \boldsymbol{E}(r, \theta) \ exp \ [i(\ell\theta + \beta z)] \tag{3.54}$$

where the integer $\ell$ describes the azimuthal dependence of the field. The propagation wave number is $\beta$. The problem is solved in essentially the same manner as we solved the planar waveguide problem, i.e., we must obtain the equations satisfied by the field in the core region and the cladding region, match the boundary conditions to obtain an eigenvalue problem, and look for the allowed values of $\ell$ and $\beta$.

The details of the cylindrical waveguide are more complicated in general, since the separation of the electric and magnetic fields into $TE$ and $TM$ mode solutions is not possible, in general, as noted in Section 3.5. It is not possible to separate the guided modes into pure $TE$ and $TM$ waves because the boundary conditions on the fields are applied at a curved boundary. Also, the circular symmetry of the guide causes the solutions to be nearly double degenerate. The degeneracy can be appreciated by using the ray picture in which guided modes have helical paths along the fiber axis with opposite helicity. Such helical waves are called *hybrid EH or HE modes*. Additionally, there are modes with no helicity. The general problem of finding the allowed modes is rather involved and, therefore, we will use an approximation of a scalar field, thus avoiding the complications of coupled fields. In this approach the field obeys the scalar wave equation. The results obtained from this approach are accurate when we are describing the cutoff conditions. However, the details of the polarization of the guided modes are lost.

For a step index waveguide, we get, after substituting Eqns. 3.54 in the wave equation written in the cylindrical coordinate system, (core radius is $a$)

$$\frac{d^2 E}{dr^2} + \frac{1}{r}\frac{dE}{dr} + \left( k_o^2 n_{r1}^2 - \beta^2 - \frac{\ell^2}{r^2} \right) E = 0 \qquad r \leq a \qquad (3.55)$$

$$\frac{d^2 E}{dr^2} + \frac{1}{r}\frac{dE}{dr} + \left( k_o^2 n_{r2}^2 - \beta^2 - \frac{\ell^2}{r^2} \right) E = 0 \qquad r > a \qquad (3.56)$$

We define the parameters

$$u^2 = \left( k_o^2 n_{r1}^2 - \beta^2 \right) a^2 \qquad (3.57)$$

$$\gamma^2 = \left( \beta^2 - k_o^2 n_{r2}^2 \right) a^2 \qquad (3.58)$$

and

$$V = \left( u^2 + \gamma^2 \right)^{1/2} = k_o a \left( n_{r1}^2 - n_{r2}^2 \right)^{1.2}$$

$$= \frac{2\pi a}{\lambda}(NA) \qquad (3.59)$$

where $NA$ is the numerical aperture of the fiber (i.e., $(n_{r1}^2 - n_{r2}^2)^{1/2}$).

Note that what we are looking for are the values of $\{\beta, \ell\}$ combinations that satisfy the Eqns. 3.55 and 3.56 when the boundary conditions are applied to the field. In the case of the planar waveguide, we recall that the general solutions involved cosine and sine functions in the core region and exponentially decaying functions in the cladding region. Here the corresponding solutions involve Bessel functions $J_\ell(x)$ in the core region and the second kind of modified Bessel function $K_\ell(x)$ in the cladding region. The general solution is

$$E(r) = A J_\ell\left(\frac{ur}{a}\right) \qquad r < a \tag{3.60}$$

$$E(r) = BK_\ell\left(\frac{\gamma r}{a}\right) \qquad r > a \tag{3.61}$$

The Bessel functions describing the lowest field dependence have the form

$$J_0(x) = 1 - \frac{x^2}{2^2(1!)^2} + \frac{x^4}{2^4(2!)^2} - \frac{x^6}{2^6(3!)^2} \cdots \tag{3.62}$$

$$J_1(x) = \frac{x}{2} - \frac{x^3}{2^3 1! 2!} + \frac{x^5}{2^5 2! 3!} - \frac{x^7}{2^7 3! 4!} + \cdots \tag{3.63}$$

The Bessel functions also have the asymptotic limits

$$J_\ell(x) = \frac{1}{\ell!}\left(\frac{x}{2}\right)^\ell \qquad \text{for } x << 1 \tag{3.64}$$

$$J_\ell(x) = \sqrt{\frac{2}{\pi x}} \cos\left[x - \frac{\pi(2\ell+1)}{4}\right] \text{ for } x >> 1 \tag{3.65}$$

The second kind of modified Bessel functions have the form

$$K_\ell(x) = (\ell-1)! 2^{\ell-1} x^{-\ell} \qquad \text{for } x << 1 \text{ and } \ell \geq 1 \tag{3.66}$$

$$K_\ell(x) = \sqrt{\frac{2}{\pi x}}\, e^{-x}\left(1 + \frac{4\ell^2 - 1}{8x}\right) \qquad \text{for } x >> 1 \tag{3.67}$$

Additionally, the functions $J_\ell(x)$ and $K_\ell(x)$ obey the following recursion relations that are useful for solving for the propagating modes (the prime refers to a spatial derivative):

$$J_{-\ell} = (-1)^\ell J_\ell \tag{3.68}$$

$$J'_\ell = \frac{1}{2}(J_{\ell-1} - J_{\ell+1}) = \pm J_{\ell\mp 1} \mp \frac{\ell J_\ell}{x} \tag{3.69}$$

$$J_{\ell\mp 1} = \frac{2\ell J_\ell}{x} - J_{\ell\pm 1} \tag{3.70}$$

$$J_{\ell\mp 2} = \frac{2(\ell\mp 1)J_{\ell\mp 1}}{x} - J_\ell \tag{3.71}$$

$$K_\ell = K_{-\ell} \tag{3.72}$$

$$K'_\ell = -\frac{1}{2}(K_{\ell-1} + K_{\ell+1}) = \mp \frac{\ell K_\ell}{x} - K_{\ell\mp 1} \tag{3.73}$$

$$K_{\ell\mp1} = \mp\frac{2\ell K_\ell}{x} + K_{\ell\pm1} \tag{3.74}$$

$$K_{\ell\mp2} = \mp\frac{2(\ell\mp1)K_{\ell\mp1}}{x} + K_\ell \tag{3.75}$$

We also have the following asymptotic relations that are satisfied:

For $x \ll 1$

$$\frac{K_0}{K_1} = x\,ln\frac{2}{1.782x} \tag{3.76}$$

$$\frac{K_{\ell-1}}{K_\ell} = \frac{x}{2(\ell-1)} \quad \ell \ge 2 \tag{3.77}$$

$$\frac{K_{\ell+1}}{K_\ell} = \frac{2\ell}{x} \quad \ell \ge 1 \tag{3.78}$$

For $x \gg 1$

$$\frac{K_{\ell\mp1}}{K_\ell} = 1 + \frac{1\mp2\ell}{2x} \tag{3.79}$$

Returning to our solution and equating the field and its derivative across the core-cladding layer boundary, we get

$$\frac{uJ_\ell'(u)}{J_\ell(u)} = \frac{\gamma K_\ell'(\gamma)}{K_\ell(\gamma)} \tag{3.80}$$

Using the recursion relations given by Eqns. 3.69 and 3.73, we get

$$\boxed{\frac{uJ_{\ell\pm1}(u)}{J_\ell(u)} = \pm\frac{\gamma K_{\ell\pm1}(\gamma)}{K_\ell(\gamma)}} \tag{3.81}$$

The solutions are now obtained by finding the value of $u$ or $\gamma$ for a fixed value of $\ell$. The Bessel functions have an oscillatory behavior with a number of nodes as shown for $J_0$ and $J_1$ in Fig. 3.13. Thus for each value of $\ell$, there are a number of possible solutions that satisfy Eqn. 3.81. A given solution is, therefore, characterized by the value of $\ell$ which gives the $\theta$ dependence of the solution and the value $m$ which provides information on the $r$ dependence of the solution.

Two interesting and important cases can be discussed analytically and provide insight into the problem. The first case is that of the fiber in which the core radius is small and only one or two modes are supported. In other words, we are

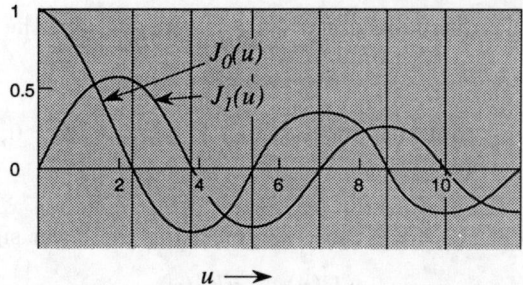

Figure 3.13: A plot of the spatial dependence of the Bessel functions $J_0$ and $J_1$.

near the cutoff condition for the modes. The second case occurs when the mode is very far from cutoff and where we can assume that $r \rightarrow \infty$.

Let us consider the first limiting case where the mode is near cutoff. For $\ell = 0$ we have from Eqn. 3.81

$$\frac{u J_1(u)}{J_0(u)} = \frac{\gamma K_1(\gamma)}{K_0(\gamma)}$$

Inverting this equation and using Eqn. 3.76 for the limiting case near cutoff where $\gamma$ approaches zero, we have

$$\frac{J_0(V)}{V J_1(V)} = \ell n \frac{1.782\gamma}{2} \qquad (3.82)$$

where we have used the approximation $u \sim V$. This give the solution

$$\gamma = 1.122 exp \left[ -\frac{J_0(V)}{V J_1(V)} \right]$$

| zeros of $J_0$ and $J_1$ | | | |
|---|---|---|---|

| | $m$ | | | |
|---|---|---|---|---|
| $l$ | 1 | 2 | 3 | 4 |
| 0 | 2.405 | 5.520 | 8.654 | 11.790 |
| 1 | 0 | 3.832 | 7.016 | 10.173 |

Table 3.1: The first four zeros of the Bessel functions $J_0$ and $J_1$.

The cutoff condition occurs when $\gamma$ is zero, i.e., for the condition,

$$J_1(V) = 0 \tag{3.83}$$

From Table 3.1 we see that this occurs when $V = 0$, i.e., $a = 0$, so that *there is no cutoff for the $\ell = 0$ mode.*

If we consider the $\ell = 1$ case, we get, using the lower signs in Eqn. 3.81

$$\frac{uJ_0(u)}{J_1(u)} = \frac{\gamma K_0(\gamma)}{K_1(\gamma)} \tag{3.84}$$

Once again, in the limit of $\gamma$ approaching zero, we get

$$\frac{VJ_0(V)}{J_1(V)} = \gamma^2 \, \ell n \left(\frac{2}{1.782\gamma}\right)$$

which gives the condition for cutoff

$$J_0(V) = 0 \tag{3.85}$$

or, from Table 3.1,

$$V = 2.405 \tag{3.86}$$

Using the value of $V$ from Eqn. 3.59, we get

$$\boxed{a = \frac{2.405\lambda}{2\pi(NA)}} \tag{3.87}$$

*This is the maximum core radius allowed for a single mode optical fiber for a given wavelength.*

Examining the other extreme where the mode is far from cutoff, we can assume that $\gamma \to \infty$ and we have from Eqn. 3.79

$$\lim_{\gamma \to \infty} \frac{K_{\ell\mp1}(\gamma)}{K_\ell(\gamma)} \to 1 \tag{3.88}$$

This gives us from Eqn. 3.81, the solution

$$\gamma J_\ell(u) = uJ_{\ell+1}(u) \tag{3.89}$$

This can be solved iteratively using a computer. For an optical fiber with a large core diameter, the number of allowed modes is approximately given by

$$\boxed{N \cong \frac{a^2}{2} \left(\frac{2\pi}{\lambda}\right)^2 (NA)^2} \tag{3.90}$$

For single-mode fibers the core radius has to be rather small and demands on high precision manufacturing and system installation are extreme. Earlier, it was believed that such demands could not be met in a commercial environment and only multimode fibers would find use outside the laboratory. However, this expectation has now proven to be unrealistically pessimistic as single-mode fibers are beginning to dominate the optical communication systems. The benefits of single-mode fibers arise from the suppression of optical pulse broadening due to multipath delays, an issue that will be discussed later in Chapter 10.

**EXAMPLE 3.7** An optical fiber has the following properties:

$$\text{core radius} \qquad a = 4\ \mu m$$
$$\text{core refractive index} \qquad n_{r1} = 1.5$$
$$\text{cladding refractive index} \qquad n_{r2} = 1.47$$

Calculate the cutoff wavelength for the fiber.

The cutoff wavelength is given by

$$
\begin{aligned}
\lambda_c &= \frac{2\pi a (n_{r1}^2 - n_{r2}^2)^{1/2}}{2.405} \\
&= \frac{2\pi (4.0\ \mu m)(0.0891)^{1/2}}{2.405} \\
&= 3.12\ \mu m
\end{aligned}
$$

**EXAMPLE 3.8** An optical fiber is to be designed to be single mode for a 1.55 $\mu m$ communication system. If $n_{r1} = 1.55$ and $n_{r2} = 1.48$, calculate the maximum diameter of the core that can be allowed.

The core thickness is given by the radius

$$
\begin{aligned}
a &= \frac{2.405(1.55\ \mu m)}{2\pi(2.4025 - 2.1904)^{1/2}} \\
&= 1.29\ \mu m \\
2a &= \text{diameter} = 2.58\ \mu m
\end{aligned}
$$

If one recognizes that the optical fibers have to be hundreds or even thousands of kilometers in length, one can see the challenge involved in producing single-mode fibers.

## 3.8  WAVEPACKET PROPAGATION: DISPERSION AND GROUP VELOCITY

$\longrightarrow$  In our discussion of optical phenomenon, we have considered the optical waves to have the waveform given by ($\psi$ stands for the electric or magnetic field)

$$\psi \sim exp(ikx)\ exp(-i\omega t) \tag{3.91}$$

where $k$ is the wavevector and, for simplicity, we assume that the wave is propagating along the $x$-direction. If we examine this waveform, we see that the energy density associated with the wave ($\psi^* \psi$) is uniform at all points along the $x$-axis and is constant in time. As a result, this is not a very physical way of describing a propagating optical pulse. In physical situations we are interested in describing optical pulses that are localized in space and propagate from one point to another. Such physical situations are described by constructing *wavepackets*. In this section we will examine some simple wavepackets and their properties. In particular, we will examine how material properties of the propagation medium affects the wavepacket propagation.

Going back to a one-dimensional wave with a wave vector $k_0$,

$$\psi_{k_0}(x) = e^{ik_0 x} \tag{3.92}$$

we note that if a state was constructed not from a single $k_0$ component, but from a spread $\pm \Delta k$, then the function

$$
\begin{aligned}
\psi(x, x_0) &= \int_{k_0-\Delta k}^{k_0+\Delta k} dk\ e^{ik(x-x_0)} \\
&= \frac{2\sin(\Delta k\ (x-x_0))}{(x-x_0)} e^{ik_0(x-x_0)}
\end{aligned} \tag{3.93}
$$

is centered around the point $x_0$ and the probability ($|\psi|^2$) decays from its maximum value at $x_0$ to a very small value within a distance $\pi/\Delta k$ as shown in Fig. 3.14.

If $\Delta k$ is small, this new "wavepacket" has essentially the same properties as $\psi$ at $k_0$, but is localized in space and is thus very useful to describe motion of the pulse. A more useful wavepacket is constructed by multiplying the integrand in the wavepacket by a Gaussian weighting factor

$$f(k - k_0) = \exp\left[-\frac{(k-k_0)^2}{2(\Delta k)^2}\right] \tag{3.94}$$

$$
\begin{aligned}
\psi(x, x_0) &= \int_{-\infty}^{\infty} \exp\left[-\frac{(k-k_0)^2}{2(\Delta k)^2} + ik(x-x_0)\right] dk \\
&= \exp\left[ik_0(x-x_0) - \frac{(x-x_0)^2}{2}(\Delta k)^2\right] \\
&\times \int_{-\infty}^{\infty} \exp\left[-\frac{(k-k_0)^2}{2(\Delta k)^2} + i(k-k_0)(x-x_0) + \frac{(x-x_0)^2}{2}(\Delta k)^2\right] \\
&= \sqrt{2\pi\Delta k}\ \exp\left[ik_0(x-x_0) - \frac{1}{2}(x-x_0)^2(\Delta k)^2\right]
\end{aligned} \tag{3.95}
$$

$\psi(x, x_0)$ represents a Gaussian wavepacket in space which decays rapidly away from $x_0$. We note that when we considered the original state $\exp(ik_0x)$, the wave was spread infinitely in space, but has a precise $k$-value. By constructing a wavepacket, we have sacrificed its precision in $k$-space by $\Delta k$ and gained a precision $\Delta x$ in real space. In general, the width of the wavepacket in real and $k$-space can be seen to have the relation

$$\boxed{\Delta k\, \Delta x \approx 1} \tag{3.96}$$

We can repeat this procedure for a wave of the form

$$\psi \sim e^{i\omega t} \tag{3.97}$$

and also obtain a wavepacket which is localized in time and frequency, the widths again being related by

$$\boxed{\Delta \omega\, \Delta t \approx 1} \tag{3.98}$$

## 3.8.1  Motion of a Wavepacket

Let us now consider how a wavepacket moves through space and time. For this we need to bring in the time dependence of the wavefunction, i.e., the term $\exp(-i\omega t)$

$$\psi(x,t) = \int_{-\infty}^{\infty} f(k - k_0) \, \exp\left\{i[k(x - x_0) - \omega t]\right\} \, dk \tag{3.99}$$

If $\omega$ has a dependence on $k$ similar to the one in free space

$$\omega = ck \tag{3.100}$$

we can write

$$\psi(x,t) = \int_{-\infty}^{\infty} f(k - k_0) \, \exp\left[ik(x - x_0 - ct)\right] \, dk \tag{3.101}$$

which means that the wavepacket simply moves with its center at

$$x - x_0 = ct \tag{3.102}$$

and its shape is unchanged with time. If, however, we have a dispersive media and the $\omega$ vs. $k$ relation is more complex, we can, in general, write

$$\omega(k) = \omega(k_0) + \left.\frac{\partial \omega}{\partial k}\right|_{k=k_0} \cdot (k - k_0) + \frac{1}{2} \left.\frac{\partial^2 \omega}{\partial k^2}\right|_{k=k_0} (k - k_0)^2 + \cdots \tag{3.103}$$

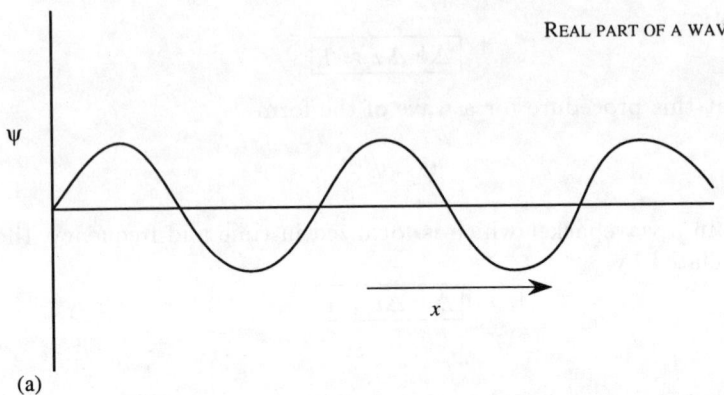

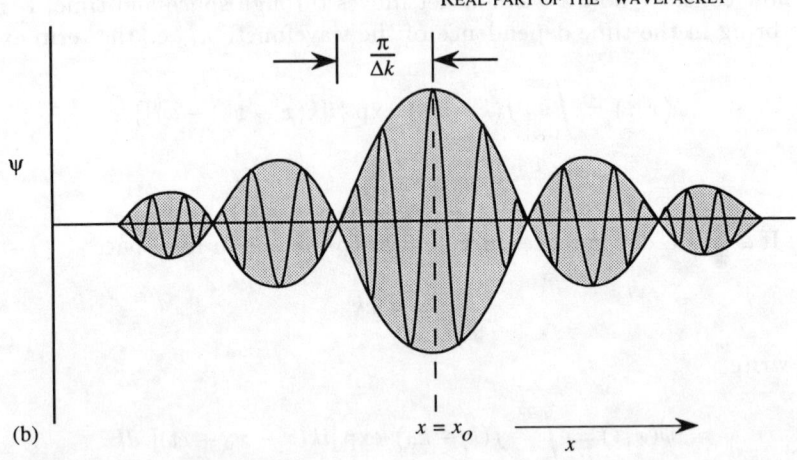

Figure 3.14: (a) A schematic of a 1-dimesional wave $e^{ik.x}$ which is extended over all space. (b) By combining several waves we get a wavepacket that is localized in space with a finite spread. The packet is centered at $x_0$ and has a spread $\Delta x$. The spread is such that $\Delta x . \Delta k \sim 1$.

Setting

$$\omega(k_0) = \omega_0$$

$$\frac{\partial \omega}{\partial k}\bigg|_{k=k_0} = v_g$$

$$\frac{\partial^2 \omega}{\partial k^2}\bigg|_{k=k_0} = \alpha, \tag{3.104}$$

we get

$$\psi(x,t) = \exp\left[i(k_0(x-x_0) - \omega_0 t)\right] \int_{-\infty}^{\infty} f(k-k_0)$$

$$\times \ \exp\left[i(k-k_0)(x-x_0-v_g t) - \frac{i\alpha}{2}(k-k_0)^2 t\right] dk \tag{3.105}$$

If $\alpha$ were zero, the wavepacket would move with its peak centered at

$$x - x_0 = v_g t \tag{3.106}$$

i.e., with a velocity known as the group velocity given by

$$v_g = \frac{\partial \omega}{\partial k}\bigg|_{k=k_0} \tag{3.107}$$

However, for nonzero $\alpha$, we show that the shape of the wavepacket also changes. To see this, let us again assume that we have a Gaussian wavepacket with

$$f(k-k_0) = f(k')$$

$$= \exp\left(\frac{-k'^2}{2\Delta k^2}\right) \tag{3.108}$$

Then

$$\psi(x,t) = \exp\left\{i\left[k_0(x-x_0) - \omega_0 t\right]\right\}$$

$$\times \ \int_{-\infty}^{\infty} \exp\left[ik'(x-x_0-v_g t)\right.$$

$$\left. - \frac{k'^2}{2}\left(i\alpha t + \frac{1}{(\Delta k)^2}\right)\right] dk' \tag{3.109}$$

To evaluate this integral we complete the square in the integrand by adding and subtracting terms

$$\psi(x,t) = \exp\left\{i\left[k_0(x-x_0) - \omega_0 t\right] - \frac{(x-x_0-v_g t)^2(\Delta k)^2}{2\left[1 + i\alpha t(\Delta k)^2\right]}\right\}$$

$$\times \quad \int_{-\infty}^{\infty} \exp\left\{\frac{-1}{2}\left[\frac{1 + i\alpha t\,(\Delta k)^2}{(\Delta k)^2}\right]\right.$$

$$\times \quad \left.\left[k' - i\frac{(x - x_0 - v_g t)\,(\Delta k)^2}{1 + i\alpha t\,(\Delta k)^2}\right]^2\right\}\,dk' \qquad (3.110)$$

The integral has a value

$$\sqrt{\frac{2\pi\,(\Delta k)^2}{1 + i\alpha t\,(\Delta k)^2}}$$

Further multiplying and dividing the right-hand side exponent by $(1 - i\alpha t\,(\Delta k)^2)$ we get

$$\psi(x,t) \quad = \quad \exp\left\{i\left[k_0(x - x_0) - \omega_0 t\right]\right\}\sqrt{\frac{2\pi\,(\Delta k)^2}{1 + i\alpha t\,(\Delta k)^2}}$$

$$\times \quad \exp\left[-\frac{(\Delta k)^2}{2}\,\frac{(x - x_0 - v_g t)^2}{1 + t^2\,(\Delta k)^4 \alpha^2}\right]$$

$$\times \quad \exp\left[\frac{i\alpha t\,(\Delta k)^4}{2}\,\frac{(x - x_0 - v_g t)^2}{1 + t^2 \alpha^2\,(\Delta k)^4}\right] \qquad (3.111)$$

The probability $|\psi|^2$ has the dependence on space and time given by

$$|\psi(x,t)|^2 = \exp\left[-\frac{(\Delta k)^2\,(x - x_0 - v_g t)^2}{1 + \alpha^2 t^2\,(\Delta k)^4}\right] \qquad (3.112)$$

This is a Gaussian distribution centered around $x = x_0 + v_g t$ and the mean width in real space is given by

$$\delta x(t) \quad = \quad \frac{1}{\Delta k}\sqrt{1 + \alpha^2 t^2\,(\Delta k)^4}$$

$$= \quad \delta x(t = 0)\,\sqrt{1 + \frac{\alpha^2 t^2}{[\delta x(t = 0)]^4}} \qquad (3.113)$$

For short times such that

$$\alpha^2 t^2\,(\Delta k)^4 \ll 1 \qquad (3.114)$$

the width does not change appreciably from its starting value, but as time passes, if $\alpha \neq 0$, the wavepacket will start spreading.

In the discussion above, we have introduced the group velocity which represents the velocity at which the wavepacket propagates through a medium. This velocity must be contrasted from the phase velocity

$$v_p = \frac{\omega}{k} \qquad (3.115)$$

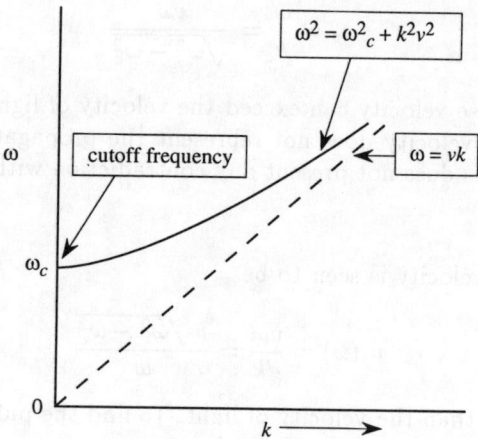

Figure 3.15: A simple approximation for the dispersion relation of a guided mode with cutoff frequency $\omega_c$.

which represents the velocity of a point at which the phase remains constant. The group velocity and phase velocity have the same value in free space, but in a medium, they are different. In fact, *depending upon the dispersion of a medium, the phase velocity can be larger than the velocity of light c.* However, the group velocity which represents the propagation of the wavepacket (or information) is always less than the velocity of light.

## 3.8.2   Waveguide Dispersion

The $\omega$ versus $k$ relation of a medium (i.e., the dispersion relation) describes one of the most important optical properties of the material. In general, this relation need not be a linear relation between $\omega$ and $k$ and, as a result, when pulses propagate in the medium, their shape is altered. When a waveguide is made form a combination of materials, additional dispersion effects come into play since the $\omega$-$k$ relation is modified for the guided modes. The waveguide dispersion relation is quite complicated in general and requires numerical evaluation. However, to see the salient physical features related to waveguide pulse propagation, we use a simple model for the dispersion relation.

If $\omega_c$ is the cutoff frequency of a mode, we assume a dispersion relation of a "shifted" form

$$\omega^2 = \omega_c^2 + k^2 v^2 \qquad (3.116)$$

In this simple form, shown in Fig. 3.15, if the waveguide effect was absent, the group velocity of a pulse would be $v$. The phase velocity for the waveguide is given by

$$v_p = \frac{\omega}{k} = \frac{v\omega}{\sqrt{\omega^2 - \omega_c^2}} \qquad (3.117)$$

We see that the phase velocity can exceed the velocity of light as $\omega$ approaches $\omega_c$. However, the phase velocity does not represent the propagation of any energy, as noted above, and this does not present any contradiction with the special theory of relativity.

The group velocity is seen to be

$$v_g(\omega) = \frac{d\omega}{dk} = \frac{v\sqrt{\omega^2 - \omega_c^2}}{\omega} \qquad (3.118)$$

which is always less than the velocity of light. To find the pulse spreading, we need to examine the second derivative of $\omega$ with respect to $k$

$$\alpha = \frac{d^2\omega}{dk^2} = \frac{v^2 - v_g^2}{\omega} \qquad (3.119)$$

As a result of the waveguide dispersion, a pulse injected into the guide will broaden as it travels. It is important to note that in our simple treatment, we have not assigned a dispersion coming from the material itself. We will discuss this issue in Chapter 10 when we discuss pulse propagation in optical communication systems. The material dispersion can contribute a value of $\alpha$ which can have a positive or negative value. It is thus possible to choose a guide so that the dispersion due to the guiding action and the material dispersion approach zero.

**EXAMPLE 3.9** Consider an optical pulse centered at frequency $1.2\,\omega_c$ propagating in a waveguide with a cutoff frequency of $1.67 \times 10^{14}$ Hertz. The pulse is injected at $t = 0$ with a width of 1.0 cm. Calculate the width of the pulse after traveling a distance of 10 km if the waveguide dispersion is $\omega^2 = \omega_c^2 + v^2 k^2$ with $v = 10^8$ m/s.

The group velocity of the pulse is given by

$$v_g = \frac{10^8 \; m/s(1.44 - 1)^{1/2}}{1.0}$$

$$= 0.66 \times 10^8 \; m/s$$

The time taken to travel a distance of 10 km is

$$t = \frac{10^4 \; m}{(0.66 \times 10^8 \; m/s)} = 1.515 \times 10^{-4} \; s$$

The material dispersion is given by ($\omega = 2\pi\nu$)

$$\alpha = \frac{d^2\omega}{dk^2} = \frac{\left(10^8 \; m/s\right)^2 - \left(0.66 \times 10^8 \; m/s\right)^2}{(2\pi \times 1.67 \times 10^{14} \; rad/s)}$$

$$= 5.378 \; m^2 s^{-1}$$

The pulse width after traveling a distance of 10 km is

$$\Delta x = (10^{-2} \ m)\sqrt{1 + \frac{(28.92 \ m^4 s^{-2})(2.295 \times 10^{-8} s^2)}{(10^{-8} \ m^4)}}$$
$$= 0.082 \ m$$

The original pulse has thus broadened to 8.2 cm.

## 3.9   LIGHT COUPLING DEVICES: WAVEGUIDE TO WAVEGUIDE COUPLERS

$\longrightarrow$   In our discussions so far, we have discussed the nature of various modes in waveguides. In actual devices, as an optical pulse propagates through the device, it often starts out in one mode and gradually transfers to another mode. Coupling of light from one mode to another is an important part of optoelectronics. To understand such coupling, a formalism known as the *coupled mode theory* has been developed. This theory is based on *perturbation theory* according to which the mode transfer is due to a small perturbation in the path of the optical pulse. We will give a simplistic view of this theory in this section.

For many optical information processing applications, it is important to transfer electromagnetic energy from one waveguide to another. It is thus important to have devices that can carry out such coupling. Various couplers of importance include the couplers for rectangular to rectangular guide, planar to planar guide, planar to rectangular guide and waveguide to fiber applications. We will present a discussion of these devices after discussing the coupled mode theory.

### 3.9.1   Coupled Mode Theory and the Directional Coupler

The directional coupler is an important device in which optical energy can be coupled from one guide to another by exploiting the penetration of light in a guide into the cladding region. The directional coupler in its basic form couples two optical inputs $I_1$ and $I_2$ to two outputs $O_1$ and $O_2$ as shown in Fig. 3.16a. In this passive form, the directional coupler couples light entering in one guide to the other guide. However, this device can be designed to operate in an active mode, where an electrical signal can be applied across the structure so that the device can connect an entering optical signal to either $O_1$ or $O_2$. Such a device is extremely important in communication applications, where it can be used to route signals. We will examine this active mode in Chapter 9.

Consider two single-mode waveguides on the same substrate as shown in Fig. 3.16a. The two guides are parallel to each other and are separated by a gap $g$ over a length $L$. Outside the interaction length $L$, the guides separate out as shown. The reason the two waveguides are brought close to each other is to provide a small coupling between the optical modes of the two guides. As the waves progress through the structure (along the $z$-direction) , the optical energy transfers back and forth between the two guides much like the energy transfer between two coupled pendulums.

To understand the operation of the directional coupler, we use a simple coupled mode model to describe how light transfers from one guide to the other guide. For simplicity, we assume that only two modes, one from each waveguide, are involved in pulse propagation. Let us denote these modes by $\psi_1$ and $\psi_2$. Let $A_1$ and $A_2$ be the amplitudes (including the $z$-dependence of the mode) of the optical modes in guides 1 and 2. We assume that the propagation constants in the two guides are the same ($\beta_1 = \beta_2 = \beta$). If there is no interaction between the two guides (i.e., their separation is very large), the propagation equations describing the amplitudes of the two modes are simply

$$\frac{dA_1}{dz} = -i\beta A_1(z)$$
$$\frac{dA_2}{dz} = -i\beta A_2(z) \tag{3.120}$$

with solutions

$$A_1(z) = A_1(0)e^{-i\beta z}$$
$$A_2(z) = A_2(0)e^{-i\beta z}$$

The solutions show that the magnitude of the amplitudes is unchanged as the wave propagates.

In a directional coupler shown in Fig. 3.16 the gap between the two guides is small enough that the evanescent waves of the modes $\psi_1$ and $\psi_2$ have a non-zero overlap with each other. This overlap is the perturbation which couples the two modes. Also let $K$ represent the perturbation for the two modes in the two guides. The coupled mode equations are then written as

$$\frac{dA_1}{dz} = -i\beta A_1(z) - iK A_2(z)$$
$$\frac{dA_2}{dz} = -i\beta A_2(z) - iK A_1(z) \tag{3.121}$$

The propagation constant is, in general

$$\beta = \beta_r - \frac{i\alpha_{\text{loss}}}{2} \tag{3.122}$$

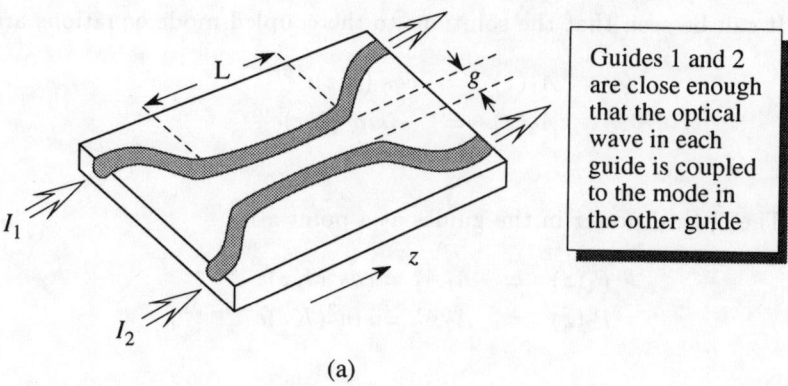

Guides 1 and 2 are close enough that the optical wave in each guide is coupled to the mode in the other guide

(a)

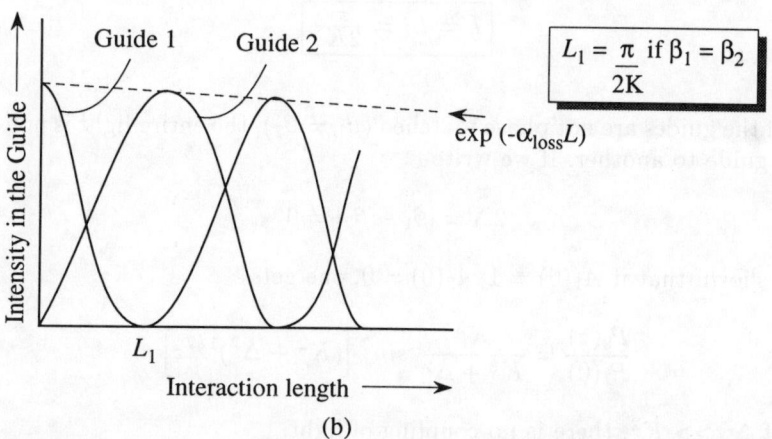

$$L_1 = \frac{\pi}{2K} \text{ if } \beta_1 = \beta_2$$

$\exp(-\alpha_{loss}L)$

(b)

Figure 3.16: (a) A schematic of the directional coupler where two waveguides are separated by a gap $g$ over a length $L$. (b) The transfer of optical energy from one guide to another when initially the light signal enters Guide 1.

where $\alpha_{loss}$ is the loss coefficient in the guide and $\beta_r$ is the real part of the propagation constant.

Let us assume that initially, light is coupled only into the guide 1 at $z = 0$, so that

$$A_1(0) = 1$$
$$A_2(0) = 0 \tag{3.123}$$

It can be seen that the solutions to the coupled mode equations are

$$
\begin{aligned}
A_1(z) &= \cos\,(Kz)e^{i\beta z} \\
A_2(z) &= -i\sin\,(Kz)e^{i\beta z}
\end{aligned}
\tag{3.124}
$$

The optical power in the guides at a point $z$ is

$$
\begin{aligned}
P_1(z) &= A_1 A_1^* = \cos^2(Kz)e^{-\alpha_{loss}z} \\
P_2(z) &= A_2 A_2^* = \sin^2(Kz)e^{-\alpha_{loss}z}
\end{aligned}
\tag{3.125}
$$

As can be seen from these equations and as shown in Fig. 3.16b, the power sloshes back and forth between the two guides as the signal propagates. The shortest distance over which a complete transfer of power occurs when the signal propagates is

$$
\boxed{L = L_1 = \frac{\pi}{2K}}
\tag{3.126}
$$

If the guides are not phase matched ($\beta_1 \neq \beta_2$), the entire light is not coupled from one guide to another. If we write

$$
2\Delta = |\beta_1 - \beta_2| \neq 0
\tag{3.127}
$$

it can be shown that if $A_1(0) = 1; A_2(0) = 0$, one gets

$$
\frac{P_2(z)}{P_1(0)} = \frac{K^2}{K^2 + \Delta^2}\,\sin^2\left[(K^2 + \Delta^2)^{1/2}z\right]
\tag{3.128}
$$

Clearly, if $\Delta^2 \gg K^2$, there is no coupling of light.

## 3.9.2   Planar-to-Planar Guide Couplers

The coupling of a planar guide to another planar structure is another important challenge in optoelectronics. If the two waveguides can be made on the same substrate, two possible coupling approaches are shown in Fig. 3.17a and b. In Fig. 3.17a the core regions of the waveguides have film indices $n_{r1}$ and $n_{r3}$ as shown, and these have to be larger than any other indices in the structure. The coupling occurs because of the evanescent fields of the modes. As discussed in the case of the directional coupler, there is a transfer of energy if the product $Kz = \frac{\pi}{2}$ where $K$ is the coupling constant and $z$ is the distance of overlap. However, for complete transfer the propagation factors must be the same. This is not always possible and, in such cases, the scheme shown in Fig. 3.17b may be used.

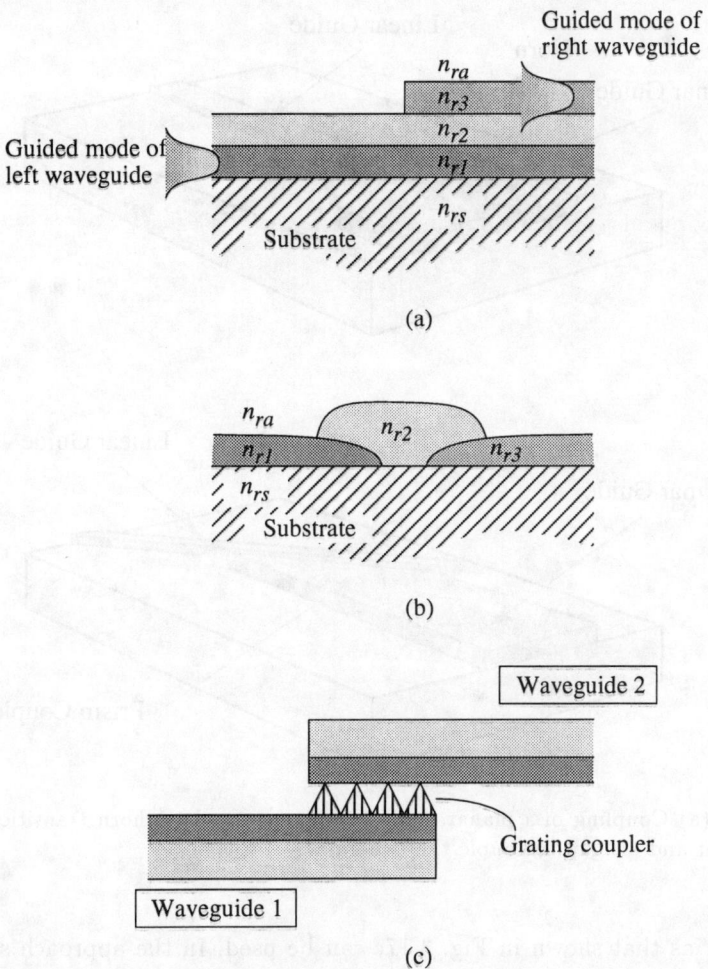

Figure 3.17: Approaches for coupling two planar waveguides. (a) The two guides share the same substrate; (b) Use of tapers allows the mode of one guide to couple into the mode of the next guide; (c) A grating allows coupling of two guides.

In Fig. 3.17b the ends of the two waveguides that are to be coupled are formed into a taper. A transition layer of index $n_{r2}$ (greater than $n_{rs}$) is then used to couple the two guides. When radiation comes from one of the guides, say the left one, it emerges through the taper. The radiation, however, does not go into the substrate but is confined into the intermediate region from where it is coupled into the second guide. Very efficient coupling (up to 100%) can be produced if the taper is gentle.

If the two waveguides to be coupled do not share a common substrate, a

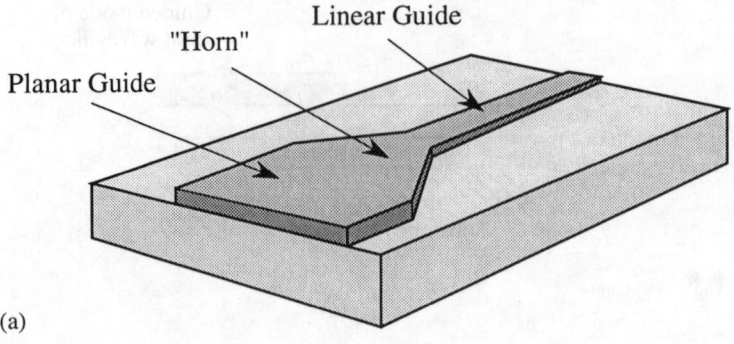

(a)

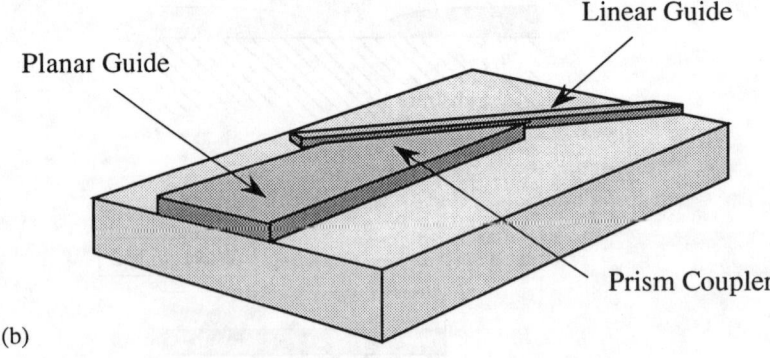

(b)

Figure 3.18: (a) Coupling of a planar guide to a linear one by a horn transition; (b) The use of a prism and air gap to couple the two guides.

scheme such as that shown in Fig. 3.17c can be used. In the approach shown, the two guides are placed on top of each other with a coupling medium. The coupling medium could be a material with a lower refractive index or a grating. The grating provides a high-efficiency coupling provided the grating length is chosen appropriately.

## 3.9.3   Planar-to-Linear Guide Couplers

In coupling a mode of a planar guide to that of a linear (rectangular) guide one has to use a mechanism to reduce the transverse dimension of the planar mode to match that of the linear mode. As shown in Fig. 3.18a, one approach is to use a "horn" that is produced by fabrication techniques to "funnel" the energy from one guide to another.

With advances in integrated optics fabrication techniques, it is possible to taper the planar waveguide into a prism placed next to the linear guide. Light can then be coupled from one guide to another (Fig. 3.18b).

### 3.9.4 Waveguide to Fiber Couplers

The coupling of light from a waveguide to a fiber is of extreme importance since optical fibers are now widely used in long distance communications. The waveguide may be a source of light as in a semiconductor laser or may be a passive element used to route an optical signal. It is important to carry out the coupling, not only for the highest coupling efficiency, but also in a manner that the efficiency does not change with time.

Two popular approaches used with optical sources are called "butt coupling" and "lens coupling". In the butt coupling approach, the fiber is carefully aligned and brought into physical contact with the guide. The fiber is then kept in place by some epoxy and a mechanical holder. In the lens coupling, a microlens is placed between the guide and the fiber for better coupling efficiency.

In addition to the two coupling schemes mentioned, we can also use the coupling of the evanescent fields to allow transfer of light from a waveguide to a fiber. In Chapter 11 we will discuss some of the important issues dealing with fiber to fiber coupling.

**EXAMPLE 3.10** A direction coupler is designed for coupling light between two guides. The coupling constant is found to be $K = 50 \, \text{cm}^{-1}$ and the loss is given by $\alpha_{loss} = 50 \, \text{cm}^{-1}$. Calculate the interaction length of the coupler and the transfer efficiency.

The interaction length is given by

$$
\begin{aligned}
L_1 &= \frac{\pi}{2K} \\
&= \frac{\pi}{100 \, cm^{-1}} = 314.16 \; \mu m
\end{aligned}
$$

The power that is coupled in this transfer is

$$
\begin{aligned}
\frac{P_2(L_1)}{P_1(z=0)} &= e^{-\alpha_{loss}L_1} \\
&= 0.21
\end{aligned}
$$

Thus, in this guide, only 21% of the power is coupled into the second guide and the remainder of the energy is lost.

## 3.10   BEAM-WAVEGUIDE COUPLERS

$\longrightarrow$ The conversion of a free space light beam into the guided modes of the waveguide is a critical task which requires special devices or couplers. In this section we will provide a brief description of these important devices.

Light beams are produced by sources such as lasers and light emitting diodes, in general, not as plane waves with uniform amplitude perpendicular to the propagation, but with a spatially nonuniform amplitude profile. One may associate with the beam a certain width or transverse spread. If this transverse amplitude profile is similar to a particular mode of the waveguide into which the beam is to be launched, it is possible to couple the beam quite efficiently into this mode by "end firing" using transverse couplers. However, such couplers are usually not very efficient and one uses prisms, grating and taper couplers for more efficient coupling.

### 3.10.1   Transverse Couplers

In transverse coupler schemes, the nature of the beam field produced by, say a laser source, is exploited for efficient coupling. A simple lens is used to focus the beam so that the beam mode can match the waveguide mode better. In Fig. 3.19 we show two approaches for this "direct" or "end fire" coupling scheme. In Fig. 3.19a the beam propagating along $z$-axis is directly focused on the waveguide edge. In Fig. 3.19b the waveguide film edge is embedded in the substrate as shown. The structure is cut at an angle and the light is coupled as shown.

While transverse couplers are quite efficient in principle, in real-life situations, slight loss of alignment, mismatch of beam and guiding mode profile, etc., can lead to poor efficiencies.

### 3.10.2   The Prism Coupler

A widely used coupling scheme depends upon using a prism as a coupler. This approach uses the evanescent waves produced in the internal reflection in a prism and coupling these to the "tail" fields of a guided mode of a planar waveguide. When two such waves are coupled, the energy is transferred (as the interaction distance increases) from the starting mode to the second one and back again, as discussed in the previous section for the directional coupler. If the interaction length is chosen carefully, the electromagnetic energy can be transferred entirely from the beam to the guided mode.

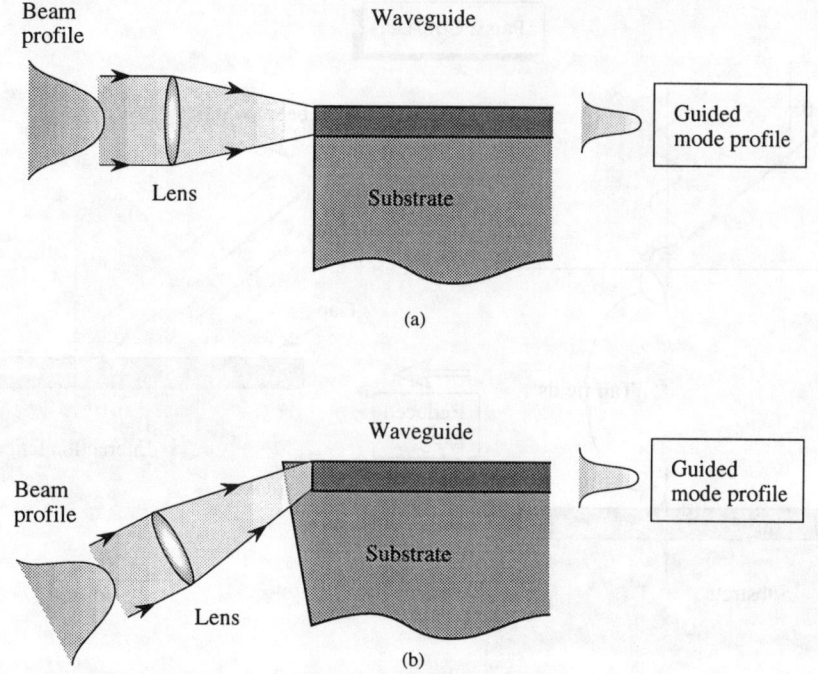

Figure 3.19: (a) A transverse coupling of a beam into a waveguide. Care has to be taken to match the beam mode with the guiding mode. (b) Transverse coupling into a waveguide where the film edge is embedded in the substrate.

We can understand the operation of the prism coupler using Fig. 3.20. In the upper left part of Fig. 3.20 we show the transverse field profile of radiation that is incident with an angle

$$\theta > \theta_c = sin^{-1}\left(\frac{n_{ra}}{n_{rp}}\right) \tag{3.129}$$

where $n_{ra}$ and $n_{rp}$ are the refractive indices of the rarer medium and the prism, respectively. A tail field penetrates into the rarer region as shown.

In the lower left side of Fig. 3.20 we show the field profile of a surface waveguide with the covered medium having a refractive index $n_{ra}$. On the right hand side of Fig. 3.20, the gap of the region "$a$" is reduced so that a small coupling is produced between the prism and the waveguide modes. As discussed for the directional coupler, a transfer of energy can take place if the interaction length $L$ is given by

$$KL = \frac{\pi}{2} \tag{3.130}$$

where $K$ is the coupling coefficient. Also, if the modes are phase matched, the

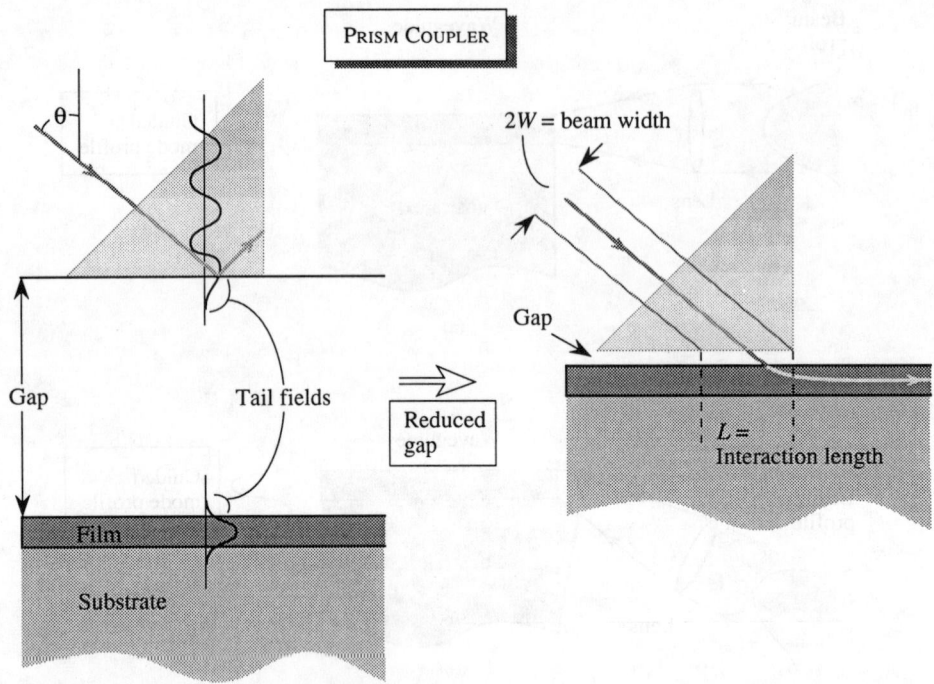

Figure 3.20: The upper left hand side of this figure shows the evanescent fields in a prism of light suffering total internal reflection. The lower left hand side shows the mode profile in a waveguide with air as the cover region. When the gap is reduced, there is coupling between the prism and the waveguide modes. If the interaction length is chosen properly, energy can transfer from the prism to the guide.

transfer of energy is complete. The condition for phase matching is

$$k_p sin\theta = \beta \tag{3.131}$$

where $k_p$ is the propagation constant in the upper medium and $\beta$ is the propagation constant for the guide.

Referring to Fig. 3.20 we see that for a beam with width $2W$, the interaction length is given by

$$L = 2W sec\theta \tag{3.132}$$

so that in a proper design (i.e., for the gap) one has to ensure that the coupling coefficient is such as to optimize the energy transfer.

The prism coupler is quite efficient for transferring light from a beam into a waveguide. It is found that the maximum energy transfer is $\sim 80\%$. Such a transfer efficiency is quite adequate for most applications.

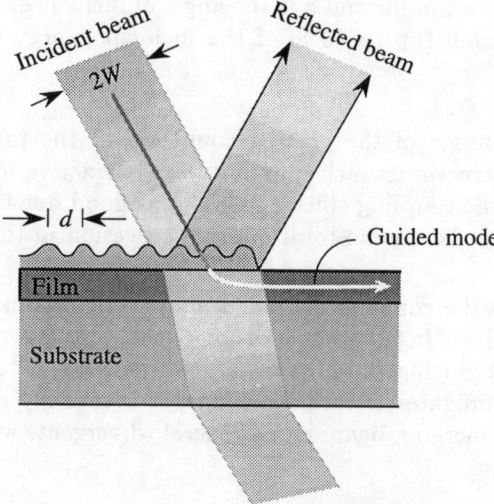

GRATING COUPLER

Figure 3.21: A schematic of a grating coupler which allows the incident beam to be coupled into the surface mode of a waveguide. In a well designed coupler there is only a small loss of the radiation energy into the reflected and transmitted beams. However, compared to the prism coupler, there is a transmission loss which reduces the efficiency of the grating coupler.

### 3.10.3 The Grating Coupler

The grating coupler is another important and efficient coupler of radiation from a beam to a guided mode. The grating couplers are produced by exposing a photoresist (discussed in Chapter 11) to an interference pattern with the desired periodicity. In Fig. 3.21 we show a typical grating coupler profile.

The operation of the grating coupler is understood on the basis of the coupled mode formalism discussed for the prism coupler. However, there is an additional aspect that needs to be considered for the grating coupler. Because of the periodic nature of the grating, the propagation wave vectors in the grating have the form

$$\beta_{gr} = \beta_o + \frac{2\pi\nu}{d} \tag{3.133}$$

where $d$ is the periodicity distance of the grating. If the effect of the grating on the waveguide is small, $\beta_o$ is close to the propagation constant in the guide (without the grating). To create phase matching between the grating-waveguide region and the air, we must have

$$k_a sin\theta = \beta_{gr} \tag{3.134}$$

where $k_a$ is the wavevector in air and $\theta$ is the angle of incidence. With proper phase matching, a large fraction (up to 70%) of the incident energy can be transferred into the guided mode.

The key advantage of the grating coupler over the prism coupler is the planar nature of the structure which makes it very attractive for integrated opto-electronics. However, the coupling efficiency is not as good due to the presence of a transmitted beam since there is no total internal reflection in the grating coupler.

In addition to the couplers discussed above, one can use tapered couplers and holographic couplers. In a tapered coupler, the edge of a waveguide film is simply tapered down reducing the film thickness from, say, $d$ to zero. As a result, as a guided mode proprogates towards the edge, it eventually gets converted to a radiation mode. The emerging beam is, in general, divergent, which limits the use of this taper.

The holographic coupler is another potentially useful coupler where a combination of hologram and a grating can couple a laser beam very efficiently into a waveguide.

## 3.11   CHAPTER SUMMARY

The areas discussed in this chapter are summarized in Tables 3.2 through 3.3.

## 3.12   PROBLEMS

**Section 3.3**
**3.1**  Derive Eqns. 3.1, 3.2, 3.5, and 3.6 for the reflection coefficients and the phase change produced by total internal reflection.
**3.2**  Calculate $\phi_{TE}$ and $\phi_{TM}$ as a function of the angle of incidence for an interface between two materials of refractive index 3.6 and 3.1.
**3.3**  A planar waveguide is made from materials with refractive indices 3.6, 3.1 and 3.0 for the guiding film, substrate and the cover, respectively. Calculate the incidence angles at which the modes become radiation, substrate radiation and guided, respectively.

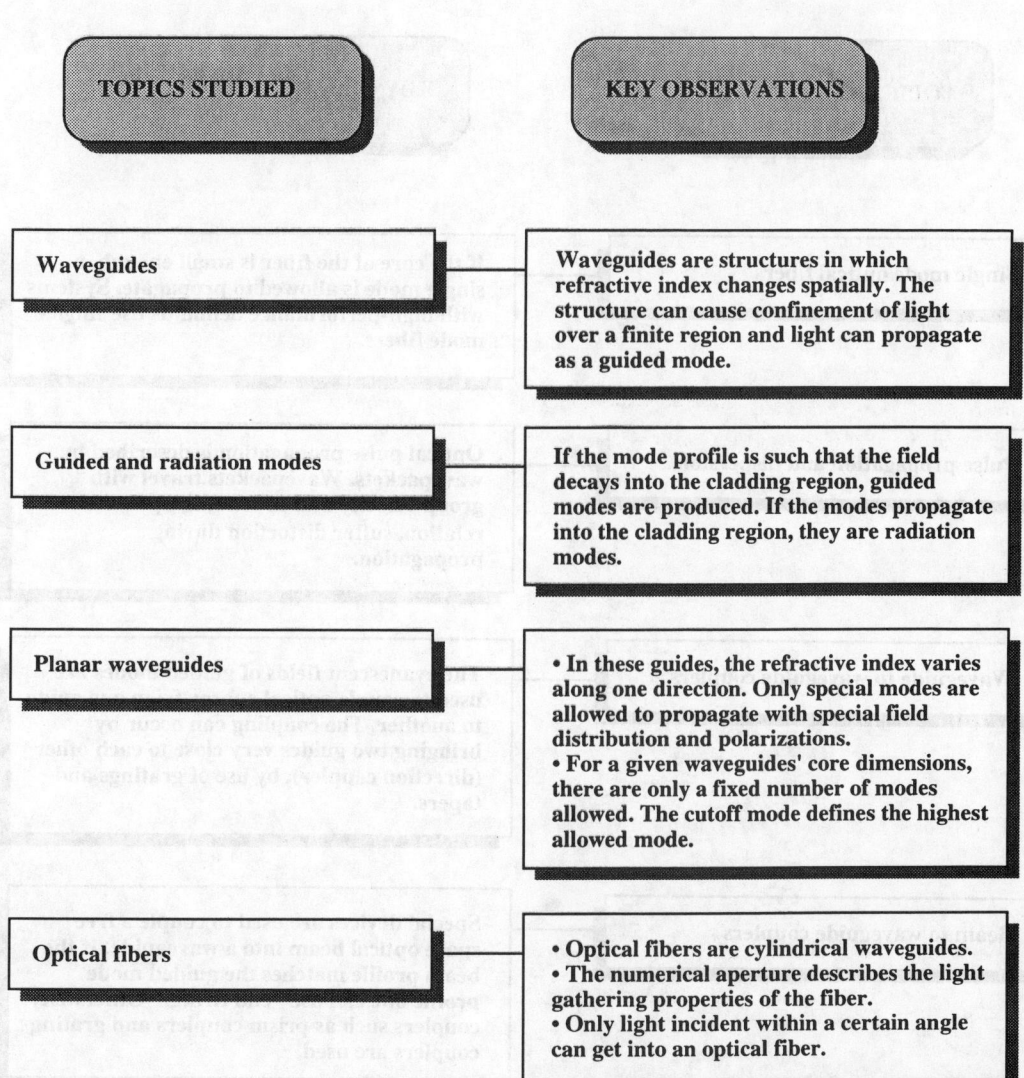

| TOPICS STUDIED | KEY OBSERVATIONS |
| --- | --- |
| Waveguides | Waveguides are structures in which refractive index changes spatially. The structure can cause confinement of light over a finite region and light can propagate as a guided mode. |
| Guided and radiation modes | If the mode profile is such that the field decays into the cladding region, guided modes are produced. If the modes propagate into the cladding region, they are radiation modes. |
| Planar waveguides | • In these guides, the refractive index varies along one direction. Only special modes are allowed to propagate with special field distribution and polarizations.<br>• For a given waveguides' core dimensions, there are only a fixed number of modes allowed. The cutoff mode defines the highest allowed mode. |
| Optical fibers | • Optical fibers are cylindrical waveguides.<br>• The numerical aperture describes the light gathering properties of the fiber.<br>• Only light incident within a certain angle can get into an optical fiber. |

Table 3.2: Summary Table

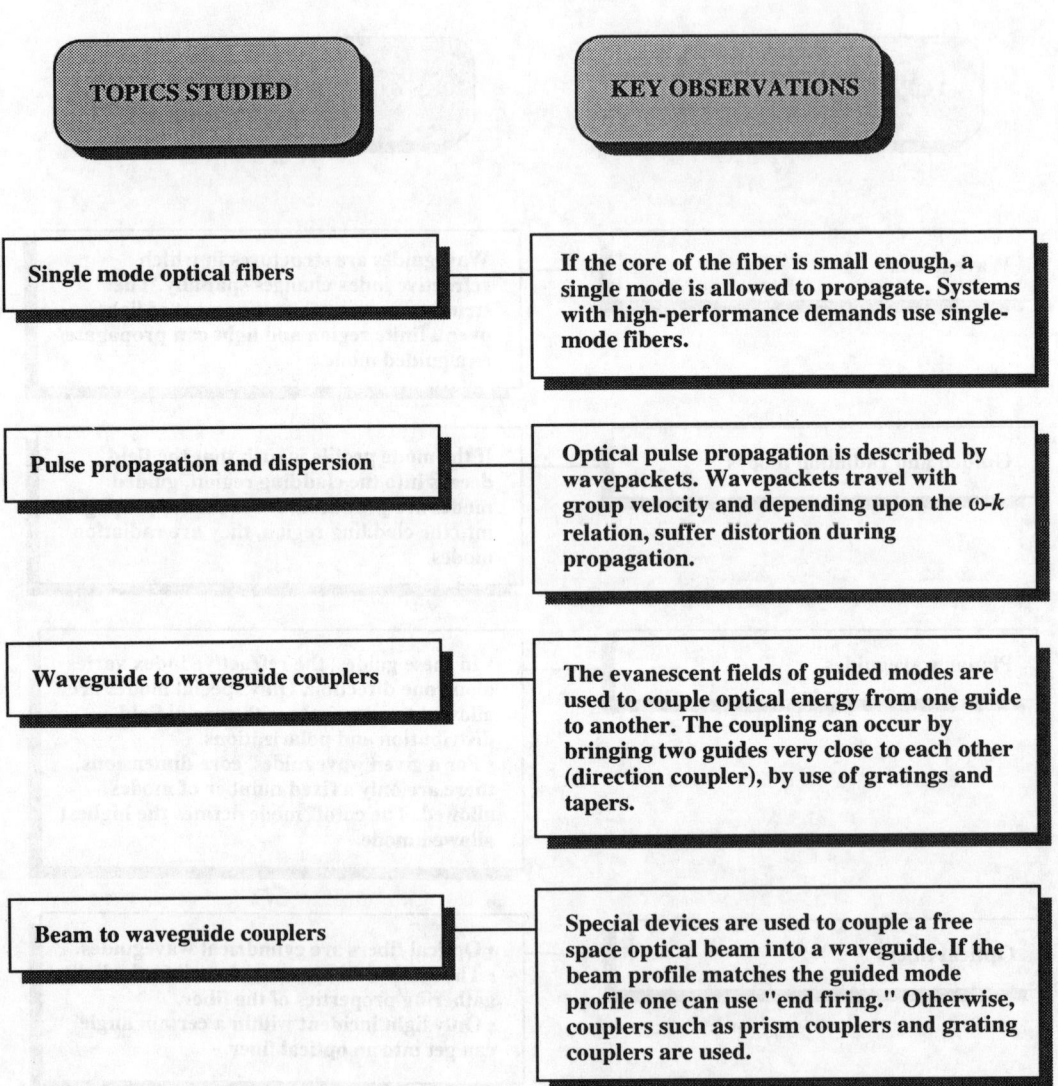

**TOPICS STUDIED**

**KEY OBSERVATIONS**

**Single mode optical fibers**

If the core of the fiber is small enough, a single mode is allowed to propagate. Systems with high-performance demands use single-mode fibers.

**Pulse propagation and dispersion**

Optical pulse propagation is described by wavepackets. Wavepackets travel with group velocity and depending upon the $\omega$-$k$ relation, suffer distortion during propagation.

**Waveguide to waveguide couplers**

The evanescent fields of guided modes are used to couple optical energy from one guide to another. The coupling can occur by bringing two guides very close to each other (direction coupler), by use of gratings and tapers.

**Beam to waveguide couplers**

Special devices are used to couple a free space optical beam into a waveguide. If the beam profile matches the guided mode profile one can use "end firing." Otherwise, couplers such as prism couplers and grating couplers are used.

Table 3.3: Summary Table

**3.4** A waveguide is made with the following parameters:

$$n_{r2} = n_{rc} = 3.1$$
$$n_{r3} = n_{rs} = 3.1$$
$$n_{r1} = n_{rf} = 3.5$$
$$h = 10.0 \ \mu m$$

Calculate the number of allowed modes for the guide if the free space wavelength of light is 1.5 $\mu$m.

**3.5** A waveguide is designed for $\lambda = 1.5$ $\mu$m with the following parameters:

$$n_{r1} = n_{rf} = 3.4$$
$$n_{r2} = n_{rc} = 1.6$$
$$n_{r3} = n_{rs} = 3.2$$
$$h = 3.0 \ \mu m$$

Calculate the number of allowed modes for the guide. If the effective guide thickness due to penetration of the wave into the cover and the substrate is 4.0 $\mu$m, how does the answer change?

**3.6** Design a symmetric slab guide with $n_{rf} = 3.5, n_{rs} = n_{rc} = 3.1$ so that a single mode for $\lambda = 1.5$ $\mu$m is allowed. Assume that the effective film thickness is 0.2 $\mu$m larger than the structural thickness.

## Section 3.4

**3.7** Calculate the maximum acceptance angles $\theta_A$ for the following step index fibers:

i)      $n_{r1} = 1.470; n_{r2} = 1.45; n_{ro} = 1.0$

ii)      $n_{r1} = 1.46; n_{r2} = 1.4; n_{ro} = 1.0$

**3.8** Consider a multimode optical fiber with the following properties:

Core refractive index      $n_{r1} = 1.53$

Cladding refractive index      $n_{r2} = 1.49$

Calculate the critical angle for total internal reflection, the maximum acceptance angle and the numerical aperture.

**3.9** An optical fiber is designed with $n_{r1} = 1.49$ and $n_{r2} = 1.4$. Calculate the numerical aperture for the fiber.

## Section 3.5

**3.10** Derive Eqns. 3.27, 3.28 and 3.32.

## Section 3.6

**3.11** Consider a planar waveguide made from GaAs and AlGaAs with refractive indices of 3.59 and 3.38. If the free space wavelength is 0.9 $\mu$m, calculate the thickness of the guide that will allow only two $TE$ modes.

**3.12** Derive Eqn. 3.53.

**3.13** Using the parameters of the Problem 13.11, calculate the value of $\beta$ for the first mode in a guide with thickness of 1.0 $\mu$m. Also calculate the optical confinement factor.

**3.14** Consider a GaAs/AlGaAs planar waveguide with thickness of 10 $\mu$m. Calculate the number of modes allowed if $n_{r1} = 3.59$ and $n_{r2} = 3.4$.

## Section 3.7

**3.15** A single mode fiber is to be designed for a 0.8 $\mu$m optical system. The refractive indices for the cladding and core regions are 1.49 and 1.53, respectively. Calculate the maximum core radius that can be accepted.

**3.16** A multimode fiber is made with the following properties: $n_{r1} = 1.53$; $n_{r2} = 1.49$; $a = 40$ $\mu$m. Calculate the number of modes this fiber will support in a 1.55 $\mu$m communication system.

**3.17** A fiber is designed with $n_{r1} = 1.5$; $n_{r2} = 1.48$; and $a = 1.0$ $\mu$m. Calculate the cutoff wavelength for this fiber to be a single mode fiber.

**3.18** A single mode optical fiber is designed for a $\lambda = 1.55$ $\mu$m long distance communication system. The core and cladding refractive indices are $n_{r1} = 1.52 \pm 0.01$, $n_{r2} = 1.47 \pm 0.01$ where the error is due to the manufacturing process. Calculate the maximum core diameter for the fiber.

## Section 3.9

**3.19** A directional coupler has a coupling coefficient given by $K = 40$ cm$^{-1}$. The guide loss is given by $\alpha_{loss} = 10$ cm$^{-1}$. Calculate the minimum length for energy transfer from one guide to the other. What fraction of the power is transferred?

**3.20** A particular manufacturing process is used to design directional couplers. It is found that $K = 40 \pm 2$ cm$^{-1}$, $\alpha_{loss} = 15 \pm 2$ cm$^{-1}$. Calculate the minimum length for power transfer between the guides. Also calculate the fluctuations in the power transfer efficiency. At what length would the fluctuation be minimum?

## 3.13   REFERENCES

- General

    - C. Pollack, *Fundamentals of Optoelectronics*, Irwin, Chicago (1995).
    - T. Tamir (editor), *Integrated Optics*, Springer-Verlag, Berlin (1979).

# CHAPTER
# 4

# ELECTRONIC PROPERTIES OF SEMICONDUCTORS

## 4.1    INTRODUCTION

In Chapter 1 we have seen that in crystals, there is a certain periodicity in the way in which atoms are arranged. This periodicity has a profound influence on the properties of electrons inside the crystal. In this chapter we will examine the electronic properties of semiconductors. The electronic properties are represented by what is called the bandstructure which defines the energy levels that an electron can have in the semiconductor. Essentially all the transport and optical properties of a semiconductor are determined by the bandstructure.

We will also examine approaches that can be exploited to modify the bandstructure and thus modify and optimize optoelectronic devices. These approaches, based on alloying (mixing) of semiconductors and upon the use of heterostructures, are widely used in modern optoelectronics.

Semiconductors have little use in their pure form. They become useful when their free carrier density is altered by doping. The concept and use of doping will be discussed in this chapter. The electronic properties of amorphous materials (which, along with polycrystalline materials, are important in optoelectronics) will also be discussed.

## 4.2    ELECTRONS IN A PERIODIC POTENTIAL: BLOCH THEOREM

$\mathcal{R}$ The description of the electron in the semiconductor has to be via the Schrödinger equation

$$\left[\frac{-\hbar^2}{2m_0}\nabla^2 + U(r)\right]\psi(r) = E\psi(r) \tag{4.1}$$

where $U(r)$ is the background potential seen by the electrons. Due to the crystalline nature of the material, the potential $U(r)$ has the same periodicity, $R$, as the lattice

$$U(r) = U(r + R) \tag{4.2}$$

As a result, we expect the *electron probability to be same in all unit cells of the crystal because each cell is identical.* If the potential was random, as in a disordered material, this would not be the case.

The Bloch theorem gives us the form of the electron wavefunction in a periodic structure and states that the eigenfunctions are the product of a plane wave $e^{i\mathbf{k}\cdot\mathbf{r}}$ times a function $u_k(\mathbf{r})$ which has the *same periodicity as the periodic potential*. Thus

$$\boxed{\psi_k(\mathbf{r}) = e^{i\mathbf{k}\cdot\mathbf{r}}u_k(\mathbf{r})} \tag{4.3}$$

is the form of the electronic function. The "cell" periodic part $u_k(r)$ has the same periodicity as the crystal, i.e.,

$$\boxed{u_k(r) = u_k(r + R)} \tag{4.4}$$

The wavefunction has the property

$$
\begin{aligned}
\psi_k(r + R) &= e^{ik\cdot(r+R)}u_k(r + R) = e^{ik\cdot r}u_k(r)e^{ik\cdot R} \\
&= e^{ik\cdot R}\psi_k(r)
\end{aligned} \tag{4.5}
$$

## 4.2.1 From Atomic Levels to Bands

$\mathcal{R}$ The Bloch theorem tells us that the electron wavefunction in a crystal has a particularly simple form: it is the product of a cell periodic part and a plane wave. While this greatly simplifies the Schrödinger equation for the electron in the crystal, the problem is still quite complicated. However, the size of the problem is reduced greatly due to the periodicity. Advanced mathematical models are used to obtain the solution for the electronic levels. These solutions provide the energy of the electron as a function of the quantity $k$ appearing in the Block theorem. This $E$ versus $k$ relation is called the *bandstructure*. In addition, one also obtains the central cell part of the wavefunction which is of great importance in optical transitions.

We will not discuss the derivation of the bandstructure of semiconductors; rather, we will merely provide the reader important facts regarding the bandstructure. To better understand the bandstructure and the makeup of the central cell part of the Bloch state, we start with a brief discussion of an isolated atom. In Fig. 4.1 we show the energy spectrum of a typical isolated atom. The spectrum is made up of two regions: i) bound states in which the electron is bound to the nucleus by Coulombic interactions, and ii) free state where the electron is free. These two regions are separated by the "vacuum energy" $E_{vac}$ as shown, which is chosen as the zero of energy.

In the bound state, the electron energy is negative with respect to the vacuum level. The allowed energy levels are discrete as shown in Fig. 4.1 and are separated from each other by a region of "forbidden gap." The electron wavefunction in general is given by a form $\psi_{n\ell m}$ where the index $n$ is a positive integer $(1, 2, 3 \ldots)$, called the *principle quantum number*, the index $\ell$ is the angular momentum quantum number and gives the value of the orbital angular momentum of the electron's

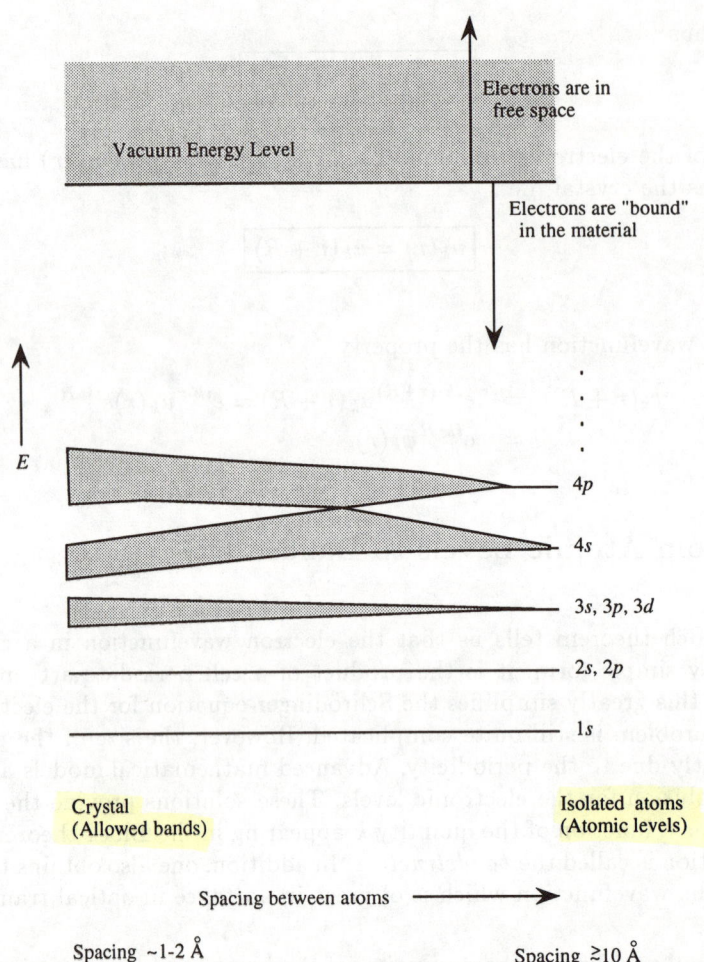

Figure 4.1: A schematic of how allowed and forbidden bands form. When the spacing between atoms is large, the allowed levels are discrete. As the atoms come closer, the electron energy levels interact with each other forming complicated bands. The deep core levels are relatively unaffected, but the higher levels broaden into bands.

motion around the nucleus. Its values are $\hbar, 2\hbar, 3\hbar \ldots$. Finally, the index $m$ gives the value of the projection of the angular momentum along the $z$-axis. It's values can be $0, \pm\hbar, \pm 2\hbar \ldots$.

If the atoms are brought together to form a crystal, the electron on a given atom starts to see the neighboring nuclei and other electrons, and the discrete levels start to broaden until eventually one starts to get bands, as shown schematically in Fig. 4.1.

The allowed energy levels (i.e., the solutions of Eqn. 4.1) are not continuous in general as for the free electron case (i.e., when $V(r)$ is zero), but have *regions where there are no allowed energy values*. This means that the electron cannot exist with energies corresponding to the energies in these "gaps."

## 4.2.2   The Crystal Momentum

$\longrightarrow$   What is the significance of the wavevector $k$ that appears in the electron wavefunction in the Bloch's theorem? For the free-space electron, i.e. the solution of Eqn. 4.1 with $U(r) = 0$, the $k$ that appears in the electron wavefunction is related to the momentum of the electron by $\mathbf{p} = \hbar\mathbf{k}$. For electrons moving in free space there are two very important laws which are used to describe their properties: i) Newton's second law of motion which tells us how the electron's trajectory evolves in the presence of an external force; and ii) the law of conservation of momentum which allows us to determine the electron's trajectory when there is a collision. We are obviously interested in finding out what the analogous laws are when an electron is inside a crystal and not in free space.

*The quantity $\hbar k$ takes on exactly the same role for the electron in the perfect semiconductor as a "momentum."* However, $\hbar k$ only reacts to the external forces such as an electric field *as if* it was the momentum of the electron. It is called crystal momentum because it includes the effect of the atoms in the crystal on the electrons. Let us examine the correspondence of the equations satisfied by the electron crystal momentum and the electron momentum in free space:

**Electron Momentum Relations**:

$$\text{Free space}\quad:\quad E = \frac{\hbar^2 k^2}{2m_0}$$
$$\text{Crystal}\quad:\quad E = E(k) \qquad (4.6)$$

The relation between $E$ and $k$ in the material is the semiconductor bandstructure.

**Equation of Motion (in the absence of any collisions)**: The equation of motion is the same for the free electron $k$ and the crystal momentum $k$

$$\frac{\hbar dk}{dt} = F_{ext} \qquad (4.7)$$

**Collisions of Electrons**: Momentum is conserved in collisions just as in free space. In reality, there are some rare scattering processes in semiconductors where momentum is not conserved. However, for most cases of interest, especially in optical processes, momentum is conserved.

Important high symmetry points

$\Gamma$ point: $k_x = 0 = k_y = k_z$

X point: $k_x = \dfrac{2\pi}{a}$ ; $k_y = k_z = 0$

L point: $k_x = k_y = k_z = \dfrac{\pi}{a}$

$a$ = lattice constant (cube edge)

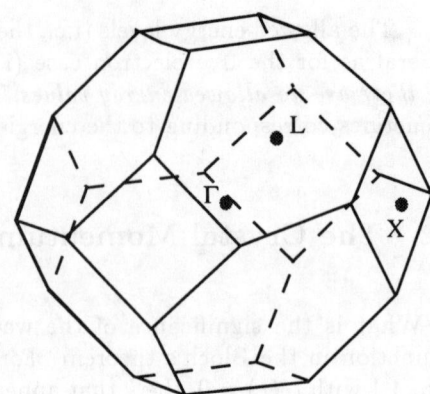

Figure 4.2: The $k$-vector for the electrons in a crystal is limited to a space called the Brillouin zone. The figure shows the Brillouin zone for the fcc lattice relevant for most semiconductors. The values and notations of certain important $k$-points are also shown. Most semiconductors have bandedges of allowed bands at one of these points.

Thus, the only effect of the crystal is to modify the $E$ vs. $k$ relation which can be quite complicated for real semiconductors. The vector $k$ for the electron is related to the wavelength $\lambda$ of the electron wavefunction by

$$k = \frac{2\pi}{\lambda} \qquad (4.8)$$

The shortest wavelength that is allowed in a crystal is governed by spacing between the lattice points in space. Thus the wavelength ranges from infinite to lattice spacing distances. The value of $k$ correspondingly varies from 0 to some finite values determined by the lattice spacing. In any general lattice, the $k$-vector values are thus confined to a volume called the Brillouin zone. For the fcc lattice, which is the lattice for most semiconductors, the Brillouin zone is shown in Fig. 4.2. The origin of the Brillouin zone is $k = (0,0,0)$ and this point is called the $\Gamma$-point. There are other points in the Brillouin zone that occur at high symmetry points. These are all denoted by standard notations and are given in Fig. 4.2. These points are important because the bandedges of the allowed bands generally occur at $k$-values corresponding to these points.

## 4.3  CONDUCTION AND VALENCE BANDEDGES IN SEMICONDUCTORS

$\longrightarrow$ We have discussed in section 4.2 that the solution of the electron problem in a crystal gives an $E$ vs. $k$ relationship which has regions of allowed bands separated by forbidden bandgaps. The question now arises: which of these allowed states are occupied by electrons and which are unoccupied? Two important situations arise when we examine the electron occupation of allowed bands: In one case we have a situation where an allowed band is completely filled with electrons while the next allowed band is separated in energy by a gap $E_g$ and is completely empty at 0 K. In a second case, the highest occupied band is only half full (or partially full).

At this point a very important concept needs to be introduced. *When an allowed band is completely filled with electrons, the electrons in the band cannot conduct any current.* This important concept is central to the special properties of semiconductors.

The electrons are Fermions, particles which have a property that one allowed energy level can contain, at the most, one particle. As a result, in a filled band, electrons cannot carry any current, since the electrons cannot move from one state to another. *Motion of electrons requires that empty states be present.* Because of this, when we have a material in which a band is completely filled, while the next allowed band is separated in energy and empty, the material has, in principle, infinite resistivity and is called an *insulator* or a *semiconductor*. The material in which a band is only half full with electrons has a very low resistivity and is called a *metal*.

*The band that is normally filled with electrons at 0 K in semiconductors is called the valence band while the upper unfilled band is called the conduction band.* The energy difference between the vacuum level and the highest occupied electronic state in a metal is called the metal work function. The energy between the vacuum level and the bottom of the conduction band is called the electron affinity.

We have already discussed that the solution of the Schrödinger equation leads to the bandstructure of the semiconductor. The top of the valence band of most semiconductors occurs at $k=0$, i.e., at effective momentum equal to zero. A typical bandstructure of a semiconductor near the top of the valence band is shown in Fig. 4.3.

The bottom of the conduction band in some semiconductors occurs at $k=0$. Such semiconductors are called direct bandgap materials and, for reasons discussed in Chapter 5, are "optically active." Semiconductors such as GaAs, InP, InGaAs, etc., are direct bandgap semiconductors. In other semiconductors, the bottom of the conduction band does not occur at the $k=0$ point but at certain other points. Such semiconductors are called indirect semiconductors. Examples are Si, Ge, AlAs, etc. These materials have very weak interactions with light and cannot be used for efficient optical devices. The reasons are based on momentum conservation rules in optical transitions which make it difficult to have strong transitions in indirect semiconductors. This will be discussed further in Chapter 5.

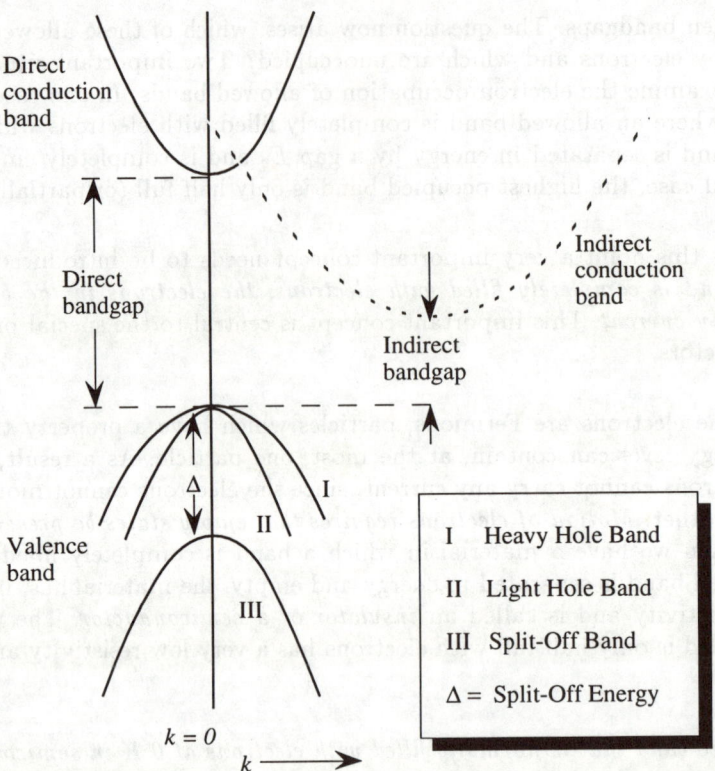

Figure 4.3: Schematic of the valence band, direct bandgap and indirect bandgap conduction bands. The conduction band of the direct gap semiconductor is shown in the solid line while the conduction band of the indirect semiconductor is shown in the dashed line. The curves I, II, III in the valence band are called heavy hole, light hole and split-off hole states, respectively. The reason why valence band states are called holes will be discussed in Section 4.6.

Near the conduction bandedges (occurring at $k = k_o$) it is usually possible to represent the bandstructure by a simple relation of the form

$$E(k, k_o) = E(k_o) + \hbar^2 \sum_{i=x,y,z} \frac{(k_i - k_{oi})^2}{2m_i^*} \qquad (4.9)$$

where the index $i$ represents the $x, y, z$ components of $k$ or $k_o$. For direct bandgap materials $k_o = (0,0,0)$.

If the bandstructure is isotropic as is the case for direct gap semiconductors, the relation becomes

$$E(k) = E_c + \frac{\hbar^2 k^2}{2m^*} \qquad (4.10)$$

where $E_c$ is the conduction bandedge, and the bandstructure is a simple parabola with equal energy surfaces being the surfaces of a sphere.

For indirect materials like Si, the bottom of the conduction band occurs at six equivalent points: $\frac{2\pi}{a}$ (0.85,0,0), $\frac{2\pi}{a}$ (0,0.85,0), $\frac{2\pi}{a}$ (0,0,0.85) and their inverses. The energy momentum relation has the form

$$E(k) = E_c + \frac{\hbar^2 k_l^2}{2m_l^*} + \frac{\hbar^2 k_t^2}{2m_t^*} \tag{4.11}$$

where $k_l$ is the longitudinal part of $k$ (i.e., parallel to the $k$-value at the bandedge) and $k_t$ is the transverse part *measured from the conduction bandedge k-value.* The constant energy surface of the bandstructure is an ellipsoid.

*Near the bandedges, the electrons in semiconductors behave as if they have a mass m\* which is called the effective mass.* For direct gap semiconductors, the effective mass of the electrons at the conduction bandedge is given by the following approximate relation which results from detailed bandstructure calculations

$$\frac{1}{m^*} \cong \frac{1}{m_0} + \frac{2p_{cv}^2}{m_0^2 E_g} \tag{4.12}$$

where, for most semiconductors,

$$\frac{2p_{cv}^2}{m_0} \cong 20.0 \; eV \tag{4.13}$$

where $m_0$ is the free electron mass.

The conduction band effective mass thus decreases rapidly with decreasing bandgap as shown in Fig. 4.4.

Near the top of the valence band (see Fig. 4.3) there are three important curves. The heavier effective mass band is called the heavy hole band (I in Fig. 4.3). The second lighter band is called light hole band (II in Fig. 4.3) and the third band, separated by an energy $\Delta$, is called the split-off band (III in Fig. 4.3). The masses of the valence band electrons are usually much heavier than those in the conduction band and are also negative. *The reason we call the valence band states "holes" will be discussed in Section 4.4.* In general, the effective mass is defined by the relation

$$\boxed{\frac{1}{m^*} = \frac{1}{\hbar^2} \; \frac{d^2 E}{dk^2}} \tag{4.14}$$

**EXAMPLE 4.1** An electron in GaAs is subjected to a 1 ps pulse of electric field of magnitude 10 kV/cm. Assuming that there is no scattering, calculate the energy of the

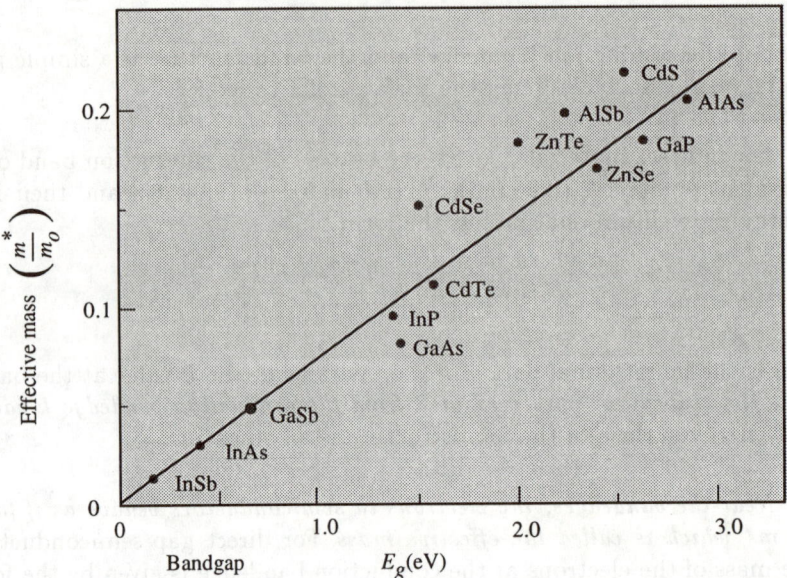

Figure 4.4: Electron effective mass, m* as a function of the lowest-direct gap $E_g$ for various compounds semiconductors. It is interesting to note that the effective mass decreases as the bandgap decreases.

electron at the end of the pulse.

The electrons obey the equation (in the absence of scattering)

$$\frac{\hbar dk}{dt} = eE$$

After a time $t$ the change in the electron momentum is

$$\hbar k_f = eEt$$

and the change in energy is

$$\begin{aligned}
\Delta E &= \frac{(\hbar k_f)^2}{2m^*} = \frac{(eEt)^2}{2m^*} \\
&= \frac{\left[1.6 \times 10^{-19}\ C \times 10^6\ V/m \times 10^{-12}(s)\right]^2}{2 \times 0.067 \times 0.91 \times 10^{-30}\ kg} \\
&= 2.1 \times 10^{-19}\ J \\
&= \frac{2.1 \times 10^{-19}}{1.6 \times 10^{-19}} = 1.31\ eV
\end{aligned}$$

The scattering of electrons does not allow such high energy to be reached in steady state. In fact in steady state conditions, due to scattering, the energy of the electron on an average is $\sim 0.3$ eV. The scattering effects lead to finite velocities (to be discussed later).

## 4.3.1   Density of States

$\longrightarrow$   The electronic states in a semiconductor have a plane wave form modulated by the central cell part. Also, as we have seen near the bandedges, it is possible to write the energy momentum relation in the simple parabolic form

$$E = \frac{\hbar^2 k^2}{2m^*} \qquad (4.15)$$

where the quantity $m^*$ is the effective mass.

We will now introduce the extremely important concept of density of states. The concept of density of states is extremely powerful, and important physical properties such as optical absorption, transport, etc., are intimately dependent upon this concept. Density of states is the number of available electronic states *per unit volume per unit energy* around an energy $E$. If we denote the density of states by $N(E)$, the number of states in a unit volume in an energy interval $dE$ around an energy $E$ is $N(E)dE$. To calculate the density of states, we need to know the dimensionality of the system and the energy vs. $k$ relation that the electrons obey. We assume that we have a parabolic $E$-$k$ relation as given above. In this case, as discussed in Appendix C, the density of states $N(E)$ is given by the following relation for a 3-dimensional system,

$$N(E)dE = \frac{(m^*)^{3/2} E^{1/2} dE}{\sqrt{2}\pi^2 \hbar^3} \qquad (4.16)$$

From quantum mechanics, we know that the electron has an intrinsic angular momentum called *spin*. An electron can have two possible spin states with a given energy. Accounting for spin, the density of states obtained above is simply multiplied by 2:

$$N(E) = \frac{\sqrt{2}(m^*)^{3/2} E^{1/2}}{\pi^2 \hbar^3} \qquad (4.17)$$

In semiconductor heterostructures it is possible to produce a "quantum well" or a "2-dimensional" world for the electron as we will discuss later in this chapter. It can be shown that the density of states for a parabolic band is (including spin)

$$N(E) = \frac{m^*}{\pi \hbar^2} \qquad (4.18)$$

Finally, in a 1D system or a "quantum wire," the density of states is (including spin)

$$N(E) = \frac{\sqrt{2}m^{*1/2}}{\pi\hbar}E^{-1/2} \tag{4.19}$$

We note that as the dimensionality of the system changes, the energy dependence of the density of states also changes. As shown in Fig. 4.5, for a 3-dimensional system we have a $E^{1/2}$ dependence; for a 2-dimensional system we have no energy dependence; and for a 1-dimensional system we have an $E^{-1/2}$ dependence.

One may wonder why we are interested in 2- and 1-dimensional systems since our life is 3-dimensional. *It is possible, indeed, to create semiconductor heterostructures where the electron feels it is in a 2-dimensional space or even in a 1-dimensional space.* This will be discussed in Section 4.7.

For direct bandgap semiconductors, the density of states mass (i.e., the mass used in the expressions for $N(E)$) for the conduction band is the same as the effective mass discussed above

$$m_{dos}^* = m^* \tag{4.20}$$

For materials like silicon in which the effective mass is different in various directions, the density of states mass for one valley is

$$m_{dos}^* = (m_1 m_2 m_3)^{1/3} \tag{4.21}$$

where $m_1, m_2$ and $m_3$ are the effective masses in the three principle directions. For silicon

$$m_\ell^* = m_1 \ ; \ m_t^* = m_2 = m_3 \tag{4.22}$$

Since silicon has 6 conduction bandedge valleys, the density of states calculated for one valley must be multiplied by six to obtain the total density of states.

In the case of valance band masses, there is the heavy hole and light hole mass. One can define the effect of both these bands by an effective density of states mass given by (the mass for holes—to be discussed next—is positive)

$$m_{dos}^{*3/2} = \left(m_{\ell h}^{*3/2} + m_{hh}^{*3/2}\right) \tag{4.23}$$

**EXAMPLE 4.2** Calculate the density of states of a 3D system and a 2D system at an energy of 0.1 eV, if the effective mass is $m_0$.

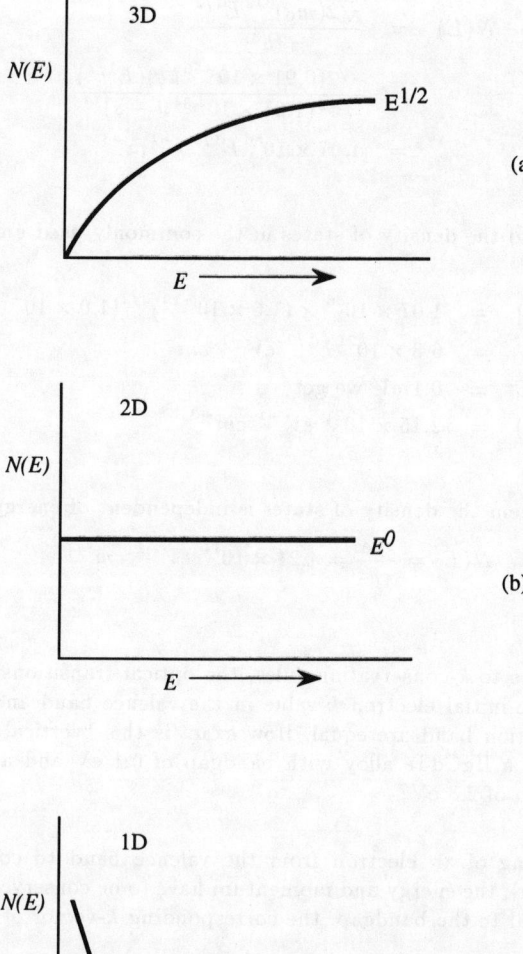

Figure 4.5: Variation in the energy dependence of the density of states in (a) 3-dimensional, (b) 2-dimensional, and (c) 1-dimensional systems. The energy dependence of the density of states is determined by the dimensionality of the system. This feature is exploited in advanced semiconductor devices.

The density of states in a 3D system (including the spin of the electron) is given by ($E$ is the energy in Joules)

$$\begin{aligned} N(E) &= \frac{\sqrt{2}(m_0)^{3/2}E^{1/2}}{\pi^2\hbar^3} \\ &= \frac{\sqrt{2}(0.91\times10^{-30}\,kg)(E^{1/2})}{\pi^2(1.05\times10^{-34}\,J-s)^3} \\ &= 1.07\times10^{56}E^{1/2}\,J^{-1}m^{-3} \end{aligned}$$

Expressing $E$ in eV and the density of states in the commonly used units of $eV^{-1}\ cm^{-3}$, we get

$$\begin{aligned} N(E) &= 1.07\times10^{56}\times(1.6\times10^{-19})^{3/2}(1.0\times10^{-6}) \\ &= 6.8\times10^{21}E^{1/2}\ eV^{-1}\ cm^{-3} \\ \text{at } E &= 0.1\ eV \text{ we get} \\ N(E) &= 2.15\times10^{21}\ eV^{-1}\ cm^{-3} \end{aligned}$$

For a 2D system the density of states is independent of energy and is

$$N(E) = \frac{m_0}{\pi\hbar^2} = 4.21\times10^{14}\ eV^{-1}\ cm^{-2}$$

**EXAMPLE 4.3** Due to $k$-conservation rules, the optical transitions in semiconductors are "vertical", i.e., the initial electron $k$-value in the valence band and the final electron $k$-value in the conduction band are equal. How exact is this "vertical transition rule" in GaAs ($E_g = 1.5\ eV$), a HgCdTe alloy with bandgap of 0.1 eV and a "blue laser" CdSe material with bandgap of 2.8 eV?

In a scattering of an electron from the valence band to conduction band (or absorption of a photon), the energy and momentum have to be conserved. Since the photon energy is roughly equal to the bandgap, the corresponding $k$-vector of the photon is

$$k_{ph} = \frac{2\pi}{\lambda_{ph}} = \frac{(\hbar\omega)}{\hbar v} = \frac{E_g}{\hbar v}$$

where $v$ is the velocity of light in the material. For GaAs the photon $k$-vector is

$$\begin{aligned} k_{ph} &\sim \frac{1.5\times1.6\times10^{-19}\ J}{10^8\ m/s\times1.05\times10^{-34}\ J-sec} \\ &= 2.29\times10^7\ m^{-1} \end{aligned}$$

This is essentially negligible on the scale of $k$ used in the bandstructure which ranges from 0 to $\sim 10^{10}\ m^{-1}$. The transitions thus appear to be "vertical." The $k_{ph}$ values for a HgCdTe alloy ($E_g = 0.1$ eV) is $1.53\times10^6\ m^{-1}$ and for CdSe ($E_g = 2.8$ eV) is $4.27\times10^7\ m^{-1}$.

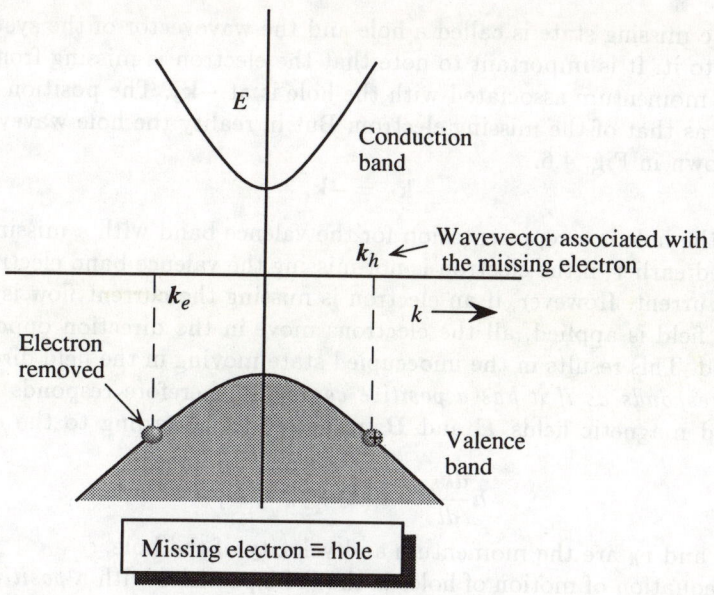

Figure 4.6: Diagram illustrating the wavevector of the missing electron $\mathbf{k}_e$. The wavevector of the system with the missing electron is $-\mathbf{k}_e$, which is associated with the hole.

## 4.4   HOLES IN SEMICONDUCTORS

$\longrightarrow$   As noted in this chapter, semiconductors are defined as materials in which the valence band is full of electrons and the conduction band is empty at 0 K. At finite temperatures some of the electrons leave the valence band and occupy the conduction band. The valence band is then left with some unoccupied states. Let us consider the situation as shown in Fig. 4.6 where an electron with momentum $\mathbf{k}_e$ is missing from the valence band.

When all the valence band states are occupied, the sum over all wavevector states is zero, i.e.,

$$\sum_{\mathbf{k}_i} \mathbf{k}_i = 0 = \sum_{\mathbf{k}_i \neq \mathbf{k}_e} \mathbf{k}_i + \mathbf{k}_e \qquad (4.24)$$

This result is just an indication that there are as many positive $k$ states occupied as negative. Now in the situation where the electron at wavevector $\mathbf{k}_e$ is missing, the total wavevector is

$$\sum_{\mathbf{k}_i \neq \mathbf{k}_e} \mathbf{k}_i = -\mathbf{k}_e \qquad (4.25)$$

The missing state is called a hole and the wavevector of the system $-\mathbf{k}_e$ is attributed to it. It is important to note that the electron is missing from the state $\mathbf{k}_e$ and the momentum associated with the hole is at $-\mathbf{k}_e$. The position of the hole is depicted as that of the missing electron. But in reality the hole wavevector $\mathbf{k}_h$ is $-\mathbf{k}_e$, as shown in Fig. 4.6.

$$\mathbf{k}_h = -\mathbf{k}_e \qquad (4.26)$$

Note that the hole is a representation for the valence band with a missing electron. As discussed earlier, if the electron is not missing the valence band electrons cannot carry any current. However, if an electron is missing the current flow is allowed. If an electric field is applied, all the electrons move in the direction opposite to the electric field. This results in the unoccupied state moving in the field direction. *The hole thus responds as if it has a positive charge.* It therefore responds to external electric and magnetic fields $\mathbf{E}$ and $\mathbf{B}$, respectively, according to the equation of motion

$$\hbar \frac{dk_h}{dt} = e\left[\mathbf{E} + v_h \times \mathbf{B}\right] \qquad (4.27)$$

where $\hbar k_h$ and $v_h$ are the momentum and velocity of the hole.

Thus, the equation of motion of holes is that of a particles with a *positive* charge $e$. The mass of the hole has a positive value, although the electron mass in its valence band is negative. For most semiconductors, the split-off band energy shown in Fig. 4.3 is quite large, and therefore there is a negligible hole density in this band. As a result it is sufficient to only consider the heavy-hole and the light-hole bands.

*When we discuss the conduction band properties of semiconductors we talk about electrons, but when we discuss the valence band properties, we talk about holes. This is because in the valence band, only the missing electrons or holes lead to charge transport and current flow.*

## 4.5  BANDSTRUCTURES OF SOME SEMICONDUCTORS

$\longrightarrow$  We will now examine special features of some semiconductors. Of particular interest are the bandedge properties since they dominate the transport and optical properties. We will examine some important semiconductors in this section. The optoelectronic technology relies on both elemental and compound semiconductors. Thus while silicon is the material of choice for electronic applications, materials like GaAs, AlAs, InAs, InP, GaSb, HgTe, CdTe etc. all play an important role in optoelectronics.

### Silicon

Silicon is the unchallenged material of choice for electronic products. The bandstructure of Si is shown in Fig. 4.7a. The bandgap is 1.1 eV with the bottom of the conduction band occurring at $k = (0.85\frac{2\pi}{a}, 0, 0)$ and the five other equivalent

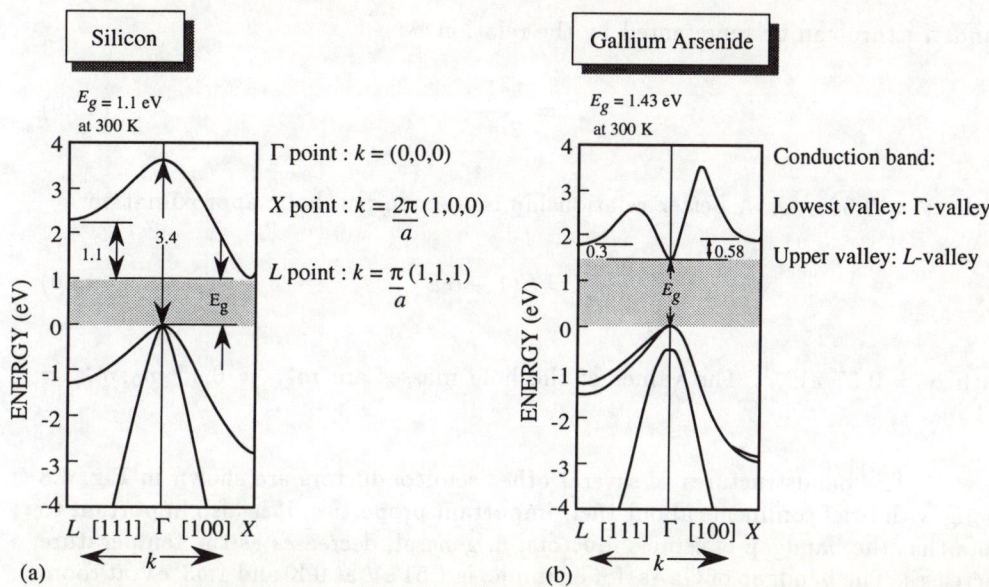

Figure 4.7: (a) Bandstructure of Si. Although the bandstructure of Si is far from ideal, having an indirect bandgap, high hole masses and small spin-orbit splitting, processing-related advantages make Si the premier semiconductor for consumer electronics. (b) Bandstructure of GaAs

points, where $a$ is the lattice constant (5.43 Å). The bandstructure near the conduction band minima is ($k$ is measured from the bandedge value)

$$E(k) = \frac{\hbar^2 k_l^2}{2m_l^*} + \frac{\hbar^2 k_t^2}{2m_t^*} \qquad (4.28)$$

where $m_l^* = 0.98\, m_0$ and $m_t^* = 0.19\, m_0$, and gives ellipsoid constant energy surfaces.

Being an indirect material, Si has very poor optical properties and cannot be used to make lasers. The reason for this is the need for "vertical $k$" transitions in optical processes due to momentum conservation (see Example 4.3). The hole transport properties of Si are also quite poor, since the hole masses are quite large. However, for electronic devices, silicon is the material of choice because of its highly reliable processing technology.

## GaAs

The bandstructure of GaAs is shown in Fig. 4.7b. The bandgap is direct which is the chief attraction of GaAs. The direct bandgap ensures excellent optical properties of GaAs as well as superior electron transport in the conduction band. The bandedge $E$ vs. $k$ relation is quite isotropic leading to spherical equal energy surfaces. The bandstructure can be represented by the relation

$$E = \frac{\hbar^2 k^2}{2m^*}$$

(4.29)

with $m^* = 0.067\ m_0$. A better relationship is the non-parabolic approximation

$$\frac{\hbar^2 k^2}{2m^*} = E(1 + \alpha E)$$

(4.30)

with $\alpha = 0.67\ eV^{-1}$. The values of the hole masses are $m_{hh}^* = 0.45\ m_0; m_{lh}^* = 0.1\ m_0$.

The bandstructures of several other semiconductors are shown in Fig. 4.8 along with brief comments about their important properties. It is also important to note that the bandgap of semiconductors, in general, decreases as the temperature increases. The bandgap of GaAs, for example, is 1.51 eV at 0 K and 1.43 eV at room temperature. These changes have very important consequences for both electronic and optoelectronic devices. The temperature variation alters the laser frequency in solid state lasers, and alters the response of modulators and detectors.

## 4.6    MODIFICATION OF BANDSTRUCTURE BY ALLOYING

$\longrightarrow$ Since essentially all the electronic and optical properties of semiconductor devices are dependent upon the bandstructure, an obvious question that arises is—can the bandstructure of a material be changed? The ability to tailor the bandstructure can obviously become a powerful tool in the hands of an insightful engineer. Novel devices can be conceived and designed for superior and tailorable performance. The answer to the question above is an emphatic yes. The bandstructure of semiconductors can, indeed, be changed and since the mid-1970s, this has become one of the driving forces in semiconductor physics.

In principle, many physical phenomena can modify the electronic bandstructures but we will focus on two important ones, since these are widely used for band tailoring. These involve: i) alloying of two or more semiconductors; ii) use of heterostructures to cause quantum confinement or formation of "superlattices." These two concepts are increasingly being used for improved-performance electronic and optical devices, and their importance is expected to become greater with each passing year. In this section we will briefly examine the effect of alloying on the bandstructure.

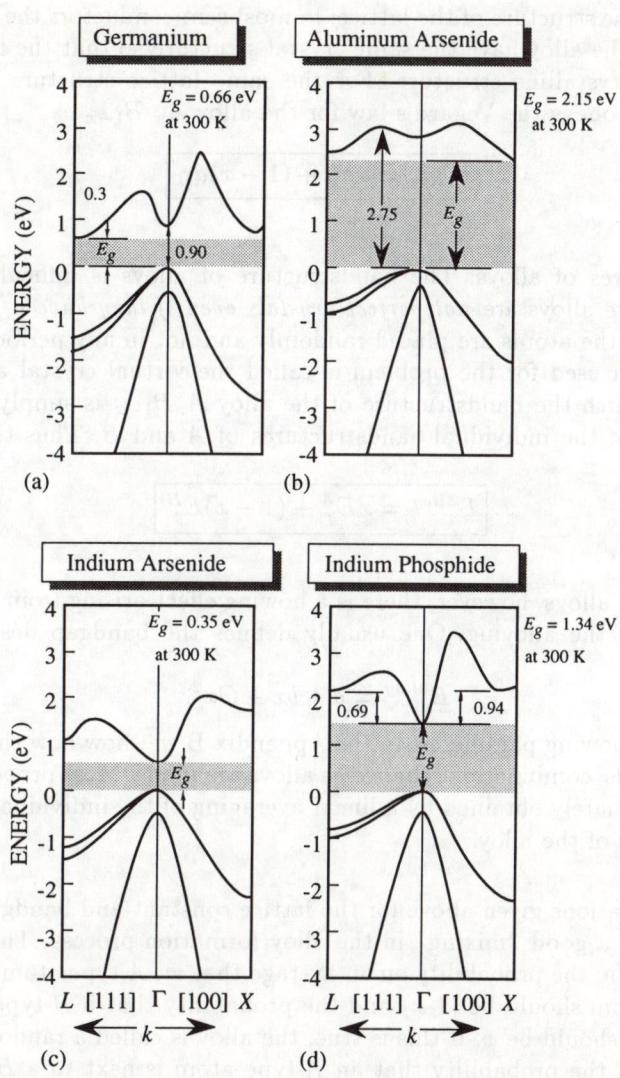

Figure 4.8: (a) Bandstructure of Ge. (b) Bandstructure of AlAs. (c) Bandstructure of InAs. Since no good substrate matches InAs directly, it is often used as an alloy (InGaAs, InAlAs, etc.) for devices. (d) Bandstructure of InP. InP is a very important material for high speed devices as well as a substrate and barrier layer material for semiconductor lasers.

When two semiconductors $A$ and $B$ are mixed via an appropriate growth technique, one has the following properties of the alloy:

a) The crystalline structure of the lattice: In most semiconductors the two (or more) components of the alloy have the same crystal structure so that the final alloy also has the same crystalline structure. For the same lattice structure materials the lattice constant obeys the Vegard's law for the alloy $A_x B_{1-x}$

$$\boxed{a_{alloy} = xa_A + (1 - x)a_B} \tag{4.31}$$

b) Bandstructures of alloys: The bandstructure of alloys is difficult to calculate in principle since alloys are *not perfect crystals even if they have a perfect lattice*. This is because the atoms are placed randomly and not in any periodic manner. A simple approach used for the problem is called the virtual crystal approximation according to which the bandstructure of the alloy $A_x B_{1-x}$ is simply the weighted bandstructure of the individual bandstructures of $A$ and $B$. Thus the bandgap is given by

$$\boxed{E_g^{alloy} = xE_g^A + (1 - x)E_g^B} \tag{4.32}$$

In most alloys, however, there is a bowing effect arising from the increasing disorder due to the alloying. One usually defines the bandgap described by the relation

$$E_g^{alloy} = a + bx + Cx^2 \tag{4.33}$$

where $C$ is the bowing parameter. In the Appendix B we show how the bandgaps of various materials' combinations change as alloys are made. Most properties of alloys can be approximately obtained by a linear averaging of the individual properties of the components of the alloy.

The relations given above for the lattice constant and bandgap are strictly valid if there is a good "mixing" in the alloy formation process. Thus, if an alloy $A_x B_{1-x}$ is grown, the probability on an average that an $A$-type atom is surrounded by a $B$-type atom should be $(1-x)$ and the probability that a $B$-type atom has an $A$-type register should be $x$. If this is true, the alloy is called a random alloy. If, on the other hand, the probability that an $A$-type atom is next to a $B$-type atom is smaller than $x$, the alloy is clustered or phase separated.

It is essential that an alloy not be phase separated, otherwise the material is not very useful for optoelectronic devices and cannot be used in any reliable device process. Certain semiconductor combinations are not allowed by thermodynamics to produce a miscible (or random) alloy at some compositions. The key reason for this is that the sum of the bond energies between $A$-type atoms and $B$-type atoms is larger than twice the bond energies between $A$- and $B$-type bonds. Thus, in the lowest free energy the system prefers to be segregated. The growth of such alloys is quite difficult if a miscible state is to be reached and non-equilibrium growth approaches are used to overcome the "natural dislike" of such materials for each other. Water and oil "alloy" is a typical example of an immiscible system.

**EXAMPLE 4.4** Calculate the bandgap of $Al_{0.3}Ga_{0.7}As$. Use the virtual crystal approximation. In the virtual crystal approximation one calculates the position of an energy point in an alloy by simply taking the *weighted average of the same energy points in the alloy components.*

For the $Al_{0.3}Ga_{0.7}As$ alloy, the bandgap energies are

$$E_g = 2.75(0.3) + (1.43)(0.7) = 1.826 \ eV$$

We see that in $Al_{0.3}Ga_{0.7}As$, the bandgap is 1.86 eV.

## 4.7 BANDSTRUCTURE MODIFICATION BY HETEROSTRUCTURES: QUANTUM WELLS

In Chapter 11, we will discuss some important epitaxial crystal growth techniques, such as MBE and MOCVD. Using these epitaxial growth techniques, it is possible to grow a sequence of semiconductor layers so that a narrow bandgap material is surrounded by a larger bandgap material. In some semiconductors making up such a heterostructure, the bandgap of the narrow gap material is completely enclosed by the bandgap of the larger gap material. An important example is the GaAs and the $Al_xGa_{1-x}As$ system, where the components have a very good match of their lattice constants. In Fig. 4.9 we show schematically the change in the potential created when a heterostructure is formed. Quantum wells are formed in both the conduction band and the valence bands . As a result, the electrons and holes are unable to move freely in the crystal growth (confinement) direction. They can still move freely in the plane perpendicular to the growth direction.

An extremely important parameter in the quantum well problem is the bandedge discontinuity produced when two semiconductors are brought together. As shown in Fig. 4.9, a part of the bandgap discontinuity $(Eg^A - Eg^B)$ of two semiconductors $A$ and $B$ would appear in the conduction band and a part would appear in the valence band.

Note that the important semiconductor system of GaAs and AlAs has a discontinuity ratio given by

$$\frac{\Delta E_c}{\Delta E_v} \cong \frac{60}{40} \text{to} \frac{65}{35} \tag{4.34}$$

i.e., 60 to 65% of the direct bandgap difference between AlGaAs and GaAs is in the conduction band.

Once the band discontinuity is known, the bandstructure of a quantum well structure can be calculated. The simplest way to do this is to use the effective mass theory in which the electron in each region of the structure is represented by its effective mass. The effect of the quantum well is to impose a background confining potential.

The simple quantum well structure such as shown in Fig. 4.9 is one of the most studied heterostructures. Its simple square well shape allows for easy solutions which are quite accurate and can be used to do first order comparisons with many experiments. The Schrödinger equation for the electron states in the quantum well can be written in a simple approximation as

$$\left[-\frac{\hbar^2\nabla^2}{2m^*} + V(z)\right]\psi = E\psi \tag{4.35}$$

where $m^*$ is the effective mass of the electron. The equation written above is strictly valid only for the conduction band of direct bandgap materials where the effective mass has a simple isotropic form. For the valence band to a first approximation, one can use the heavy-hole mass $m^*_{hh}$ and the light- hole mass $m^*_{\ell h}$ to represent the problem.

The Schrödinger equation for the quantum well in which the confinement is along the $z$-axis can be separated into equations in the $x$- and $y$-direction and in the $z$-direction as follows:

$$\Psi = \psi_x\psi_y\psi_z \tag{4.36}$$

$$\frac{-\hbar^2}{2m^*}\frac{\partial^2\psi_x}{\partial x^2} = E_x\psi_x$$

$$\frac{-\hbar^2}{2m^*}\frac{\partial^2\psi_y}{\partial y^2} = E_y\psi_y$$

$$\left[\frac{-\hbar^2}{2m^*}\frac{\partial^2\psi}{\partial z^2} + V(z)\right]\psi_z = E_z\psi_z \tag{4.37}$$

The solutions along the $x,y$ plane in the quantum well are very simple and can be written as

$$\psi_x = \frac{1}{\sqrt{L_x}}e^{ik_x\cdot x}; E_x = \frac{\hbar^2k_x^2}{2m^*} \tag{4.38}$$

$$\psi_y = \frac{1}{\sqrt{L_y}}e^{ik_y y}; E_y = \frac{\hbar^2k_y^2}{2m^*} \tag{4.39}$$

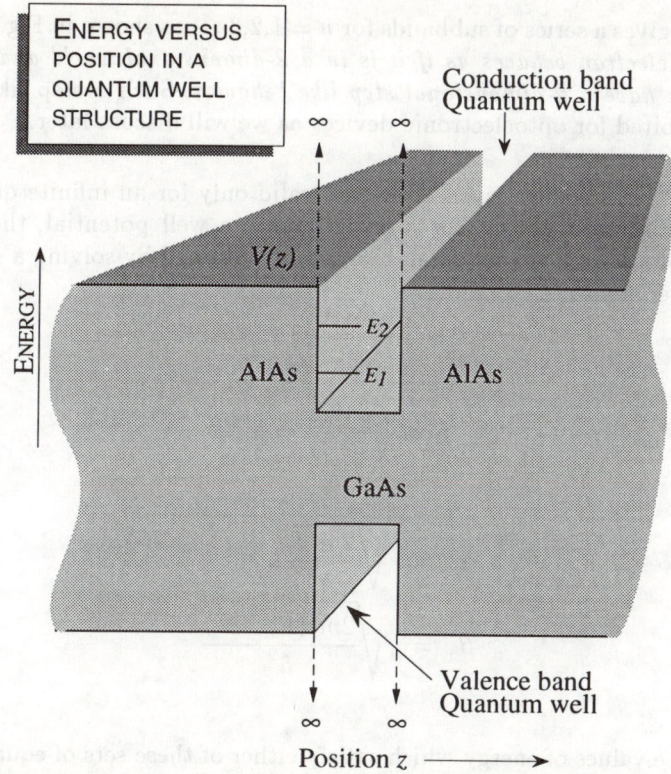

Figure 4.9: Schematic of a quantum well and the subband levels. Note that in a semiconductor quantum well, one has a quantum well for the conduction band and one for the valance band. Subbands are produced in the conduction band and the valence band. In the infinite barrier model, the barriers are chosen to have an infinite potential, as shown.

where $L_x, L_y$ represent the dimensions for normalization along the $x$ and $y$ direction. If the potential $V(z)$ is assumed to be zero in the quantum well and infinite outside the well of dimension $W$, the $z$-direction solutions are quite simple and are given by:

$$\psi_z = \frac{\sqrt{2}}{\sqrt{W}} \cos \frac{n\pi z}{W}; \quad n \text{ is odd}$$

$$= \frac{\sqrt{2}}{\sqrt{W}} \sin \frac{n\pi z}{W}; \quad n \text{ is even} \tag{4.40}$$

$$E_z = \frac{\pi^2 \hbar^2 n^2}{2m^* W^2} \tag{4.41}$$

The total energy of the electron is then (measured from the bulk material bandedge)

$$E_n(k_x, k_y) = \frac{\pi^2 \hbar^2 n^2}{2m^* W^2} + \frac{\hbar^2 k_x^2}{2m*} + \frac{\hbar^2 k_y^2}{2m*} \tag{4.42}$$

This gives a series of subbands for $n = 1,2,3 \ldots$ as shown in Fig. 4.9. *In each subband, the electron behaves as if it is in a 2-dimensional world and the density of states thus have a 2-dimensional step like behavior.* Such a step like density of states is exploited for optoelectronic devices as we will discuss later.

The energy values given above are valid only for an infinite quantum well potential as discussed above. For a finite quantum well potential, the results for the energy values are not analytical, but can be obtained by solving a simple set of equations given below:

$$\alpha tan \left( \frac{\alpha W}{2} \right) = \beta \tag{4.43}$$

or

$$\alpha cot \left( \frac{\alpha W}{2} \right) = -\beta \tag{4.44}$$

where

$$\alpha = \sqrt{\frac{2m^* E}{\hbar^2}}$$

$$\beta = \sqrt{\frac{2m^*(V_o - E)}{\hbar^2}}$$

Those values of energy which satisfy either of these sets of equations are allowed. The energy values one obtains by this more accurate treatment are somewhat smaller than the values given by Eqn. 4.41 and the difference is more pronounced for the higher subband energies. The reader may note the similarity of this problem and the problem of the planar waveguide discussed in Section 3.6.

**EXAMPLE 4.5** In the GaAs/AlGaAs heterostructure, 60% of the bandgap discontinuity is in the conduction band of the narrow gap material. Calculate the conduction band and valence band quantum well potentials for GaAs/$Al_{0.3}Ga_{0.7}As$.

The bandgap difference between GaAs and $Al_{0.3}Ga_{0.7}As$ is (from Appendix B)

$$\Delta E_g = 1.247 \times 0.3 = 0.374 \ eV$$

Since 60% of this difference is in the conduction band, the discontinuity of the quantum well formed in the conduction band is (this is the barrier height)

$$\Delta E_c = 0.374 \times 0.6 = 0.224 \ eV$$

The barrier height for the valence band is

$$\Delta E_v = 0.15 \ eV$$

These barrier heights are not infinity and so our simple model is only an approximation to the real problem. However, the use of the infinite barrier problem is reasonable once the well size is $\sim 150$ Å for this system.

**EXAMPLE 4.6** Using a simple infinite barrier approximation, calculate the "effective bandgap" of a 100 Å GaAs/AlAs quantum well. If there is a one-monolayer fluctuation in the well size, how much will the effective bandgap change? This example gives an idea of how stringent the control has to be to exploit heterostructures.

The confinement of the electron ($m^* = 0.067\ m_0$) pushes the effective conduction band up, and the confinement of electrons at the valence band push the effective edge down ($m_{hh}^* = 0.4\ m_0$). The change in the electron ground state is (using $n = 1$) 55.77 meV. The shift (downwards) in the valence band is (why do we only need to worry about the shift of the HH band to find the effective gap?) 9.34 meV. The net shift is 65.11 meV.

The effective bandgap is thus 65.11 meV larger than the bulk GaAs bandgap.

If the well size changes by one monolayer (e.g., goes from 100 Å to 102.86 Å), the change in the electron level is

$$\Delta E_e \;=\; E_e \left[ 1 - \frac{(100)^2}{(102.86)^2} \right] = E_e \times 0.055$$
$$=\; 3.06\ meV$$

The hole energy changes by

$$\Delta E_{hh} = E_{hh} \times 0.055 = 0.51\ meV.$$

Thus, the bandgap changes by 3.56 meV for a one monolayer variation. In optical frequencies this represents a change of 0.86 Terrahertz, which is too large a shift for many optoelectronic applications.

## 4.8 INTRINSIC CARRIER CONCENTRATION

$\longrightarrow$ We have discussed in the previous sections that a semiconductor is characterized by the fact that at zero Kelvin, the valence band is completely occupied while the conduction band is completely empty. We also saw that a completely full band does not conduct charge. Thus, at low temperatures the pure semiconductor offers an extremely high resistance to current transport. At finite temperatures, the occupation of electrons and holes is described by the Fermi distribution function given by

$$f(E) = \frac{1}{1 + exp\frac{E - E_F}{k_B T}} \tag{4.45}$$

As the temperature is raised, the Fermi distribution function smears and some electrons are emitted from the valence band into the conduction band. Now there are electrons in the conduction band and holes in the valence band which can carry current. However, such current carrying electrons produced by raising the temperature, known as intrinsic carriers, are not useful in semiconductor devices and often are a nuisance. The intrinsic carriers often are a source of limitation for high temperature operation of devices, since they cannot be controlled effectively by electric fields.

The intrinsic carrier concentration depends upon the bandgap and temperature as well as the details of the bandedge masses. We will assume that the bandedge density of states for electrons and holes originate from parabolic $E$-$k$ relationships. The conduction and valence band density of states are shown in Fig. 4.10 along with the position of a Fermi level.

The concentration of electrons in the conduction band is

$$n = \int_{E_c}^{\infty} N_e(E)f(E)dE \tag{4.46}$$

where $N_e(E)$ is the electron density of states near the conduction bandedge and $f(E)$ is the Fermi function. Using the appropriate expressions for $N_e$ and $f$ we get, for a 3D system, (the conduction band density of states starts at $E = E_c$ as shown in Fig. 4.10)

$$n = \frac{1}{2\pi^2}\left(\frac{2m_e^*}{\hbar^2}\right)^{3/2} \int_{E_c}^{\infty} \frac{(E - E_c)^{1/2}dE}{exp(\frac{E-E_F}{k_BT}) + 1} \tag{4.47}$$

If the Fermi level is far from the bandedge, then the unity in the denominator can be neglected. *This approximation leads to the Boltzmann statistics and is valid when n is small ($\stackrel{<}{\sim} 10^{17}$ cm$^{-3}$ for most semiconductors), and is usually valid* for intrinsic concentrations. Then we get

$$n = \frac{1}{2\pi^2}\left(\frac{2m_e^*}{\hbar^2}\right)^{3/2} exp\left(\frac{E_F}{k_BT}\right) \int_{E_c}^{\infty} (E - E_c)^{1/2} \, exp(-E/k_BT)dE$$

$$= 2\left(\frac{m_e^* k_B T}{2\pi\hbar^2}\right)^{3/2} exp[(E_F - E_c)/k_BT] = N_c \, exp[(E_F - E_c)/k_BT]$$

where

$$N_c = 2\left(\frac{m_e^* k_B T}{2\pi\hbar^2}\right)^{3/2} \tag{4.48}$$

$N_c$ is known as the effective density of states at the conduction bandedge. Note that the units of the density of states $N_e$ are eV$^{-1}$ cm$^{-3}$ while those of the effective density of states $N_c$ are cm$^{-3}$.

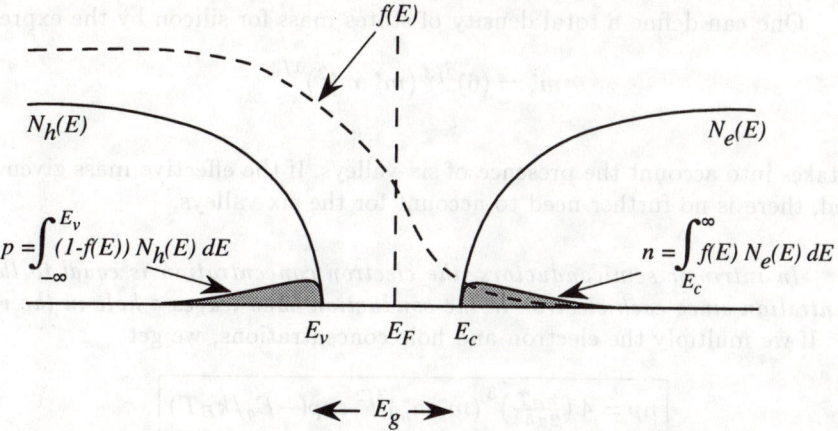

Figure 4.10: A schematic of the density of states of the conduction and valence band. $N_e$ and $N_h$ are the electron and hole density of states. Also shown is the Fermi function giving the occupation probability for the electrons. The resulting electron and hole concentrations are shown. For an intrinsic semiconductor $n = p$, each electron produced in the conduction band leaves behind a hole in the valence band.

The carrier concentration is known *when $E_F$ is calculated*. To find the intrinsic carrier concentration requires finding the hole concentration $p$ as well.

The hole distribution function $f_h$ is given by (remember the hole is the absence of an electron)

$$f_h = 1 - f_e = 1 - \frac{1}{exp(\frac{E - E_F}{k_B T}) + 1} = \frac{1}{exp\left[\frac{(E_F - E)}{k_B T}\right] + 1}$$

$$\cong exp - \left[\frac{(E_F - E)}{k_B T}\right] \qquad (4.49)$$

The approximation is again based on our assumption that $E_F - E \gg k_B T$, which is a good approximation for pure semiconductors. Carrying out the mathematics similar to that for electrons, we find that

$$p = 2 \left(\frac{m_h^* k_B T}{2 \pi \hbar^2}\right)^{3/2} exp \ [(E_v - E_F)/k_B T] \qquad (4.50)$$

$$= N_v exp \ [(E_v - E_F)/k_B T] \qquad (4.51)$$

where $N_v$ is the effective density of states for the valence bandedge.

In the expressions above it is important to note that the relevant masses are the density of states masses. These were given in Eqns. 4.20 to 4.23.

One can define a total density of states mass for silicon by the expression

$$m_e^* = (6)^{2/3} \left(m_\ell^* \, m_t^{*2}\right)^{1/3} \tag{4.52}$$

This takes into account the presence of six valleys. If the effective mass given above is used, there is no further need to account for the six valleys.

*In intrinsic semiconductors, the electron concentration is equal to the hole concentration since each electron in the conduction band leaves a hole in the valence band.* If we multiply the electron and hole concentrations, we get

$$np = 4 \left(\tfrac{k_B T}{2\pi\hbar^2}\right)^3 (m_e^* m_h^*)^{3/2} \, exp(-E_g/k_B T) \tag{4.53}$$

and since for the intrinsic case $n = n_i = p = p_i$, we have from the square root of the equation above

$$n_i = p_i = 2 \left(\tfrac{k_B T}{2\pi\hbar^2}\right)^{3/2} (m_e^* m_h^*)^{3/4} \, exp\left(-E_g/2k_B T\right) \tag{4.54}$$

If we set $n = p$, we also obtain the Fermi level position measured from the valence bandedge using Eqns. 4.48 and 4.50. We denote the intrinsic Fermi level by $E_{Fi}$

$$exp\left(2E_{Fi}/k_B T\right) = (m_h^*/m_e^*)^{3/2} exp\left[(E_c + E_v)/k_B T\right] \tag{4.55}$$

or

$$E_{Fi} = \tfrac{E_c + E_v}{2} + \tfrac{3}{4}k_B T \ell n \left(m_h^*/m_e^*\right) \tag{4.56}$$

Thus the Fermi level of an intrinsic material lies close to the midgap.

We note that the carrier concentration increases exponentially as the bandgap decreases and has a strong temperature dependence. It is quite clear from the discussion above that pure semiconductors have a very low concentration of carriers that can conduct current. One must compare the room temperature concentrations of $\sim 10^{11}$ cm$^{-3}$ to the carrier concentrations of $\sim 10^{21}$ cm$^{-3}$ for metals.

## 4.9   DEFECT LEVELS IN SEMICONDUCTORS

$\mathcal{R}$   We have seen that in a perfectly periodic structure, the electronic spectrum consists of regions of allowed energy band and regions where there is a bandgap.

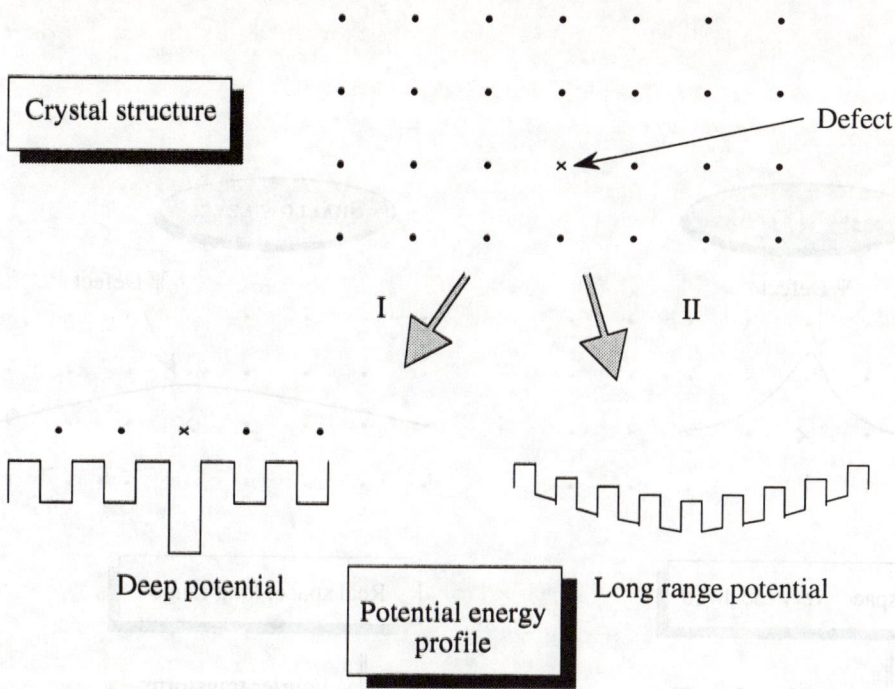

**Figure 4.11:** A schematic of a point defect in a crystal. The effect of the defect on the crystal background potential can be such that a "deep potential" variation is produced in a narrow spatial region, or a long-range disturbance is created.

In a real crystal, as discussed in Chapter 1, Section 1.8, there are a number of defects that are present. These defects may be native defects (vacancies, anti-site defects, etc.) or extrinsic defects (chemical impurities) which are either intentionally or unintentionally introduced in the crystal.

The defects of key interest in semiconductors are the point defects which create a local disturbance in the crystal structure as discussed in Chapter 1. The effect of this crystal disturbance can be of two kinds as shown in Fig. 4.11:

i) The disturbance may create a potential profile which differs from the periodic potential only over one or a few unit cells. This potential is deep and localized and the defect is called a deep-level defect.

ii) The disturbance may create a long range potential disturbance which may extend over tens or more unit cells. Such defects are called shallow-level defects.

Associated with the defects are new electronic states that are called defect levels. These new electronic states can be produced in the regions of allowed bands

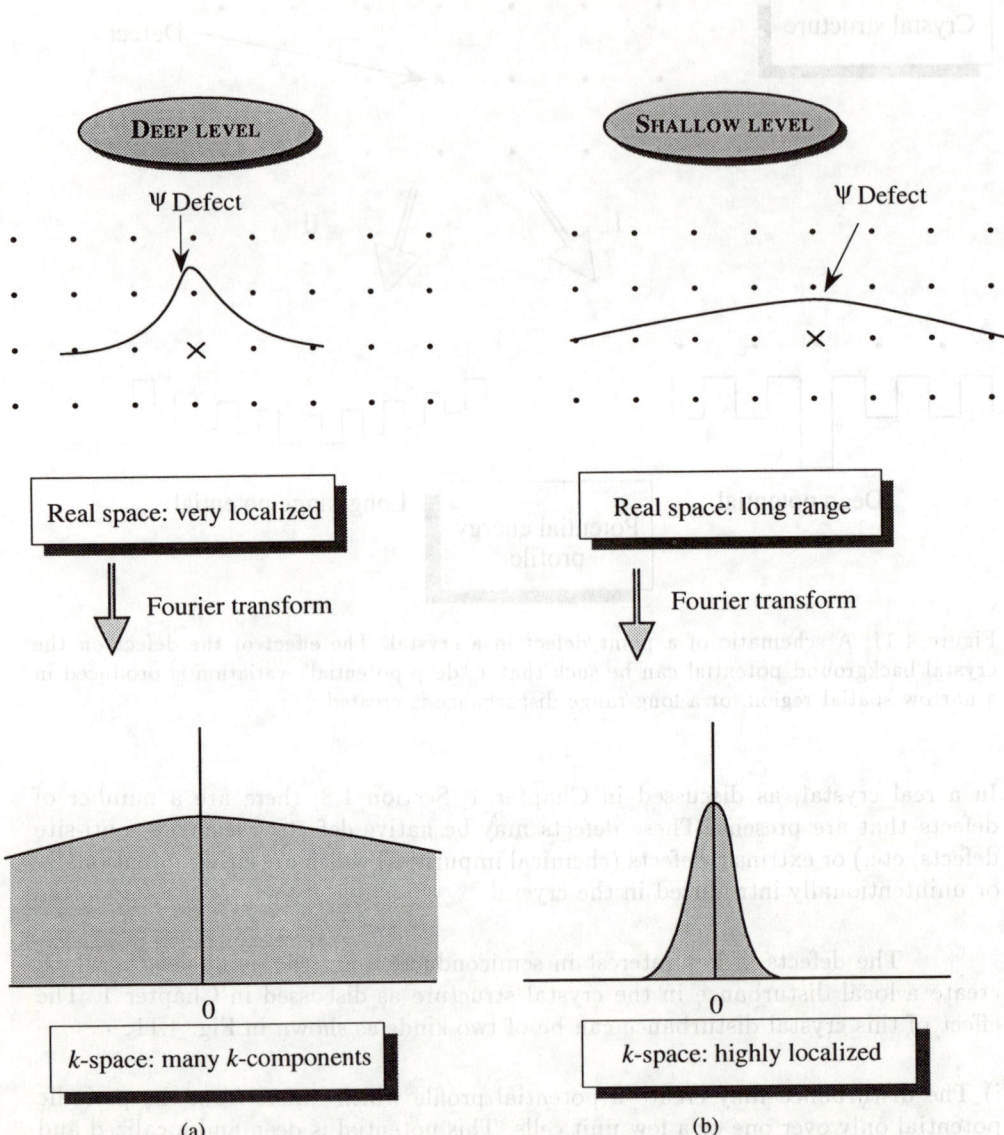

Figure 4.12: A schematic of the wavefunctions associated with (a) a deep short-range potential defect and (b) a shallow long-range potential defect.

(i.e., the conduction or valence band) in which case their effects are minimal. However, the new levels could be produced in the bandgap region in which case they could greatly alter the electronic and optical properties of the semiconductor.

The bandgap levels have an associated wavefunction which no longer has a Bloch form (i.e., a form having a plane wave part $\exp(ik \cdot r)$). However, in general, the defect level can be expressed in terms of the perfect crystal states $\psi_k(\nu)$ . The deep defect has associated with it a wavefunction that is highly localized in space as shown in Fig. 4.12a, and as a consequence is made up of a large number of $k$-states. *Thus the Fourier transform of the deep level function has essentially all k-values in more or less equal proportion.*

The wavefunction associated with a shallow long range impurity is extended in real space as shown in Fig. 4.12b and is made up of only a few $k$-value functions arising from the bandedge states. The most important shallow defect levels are dopants that will be discussed in the next section.

Normally, deep-level defects are to be avoided as much as possible in semiconductors. The key reason for this is that an electron in the defect level is "trapped" near the defect since the wavefunction of the electron is localized in space. Such electrons cannot participate in current flow very easily. *However, in the case of indirect gap materials, deep-level defects are often purposely introduced to increase the optical response of the material.*

## 4.10 DOPING IN SEMICONDUCTORS

$\longrightarrow$ We have seen that the free-carriers that can carry current in pure semiconductors have a very low density. To increase the free carrier density, impurities known as dopants are introduced. The dopants are chosen from the periodic table so that they either have an extra electron in their outer shell compared to the host semiconductor atom they replace, or they have one less electron. The resulting dopant is called a donor or acceptor.

A donor has an extra electron in its valence shell compared to the atom of the crystal it replaces. This electron is bound very weakly to the donor, and at sufficiently high temperatures (say $\sim 200$ K ) it "ionizes" and goes to the conduction band where it can carry current. Similarly an acceptor has one electron less than the host atom it replaces. An electron from the valence band can be "accepted" by the acceptor producing a hole in the valence band.

Most optoelectronic devices operate at room temperature where it is reasonable to assume that all dopants are ionized. Thus in an *n*-type material we assume that the electron density is $\sim N_d$, the donor density, and in a *p*-type material the hole density is $\sim N_a$, the acceptor density. In an *n*-type material the electrons are called the majority carriers while the holes are the minority carriers and vice versa. Let us now consider the position of the Fermi level in a doped material.

Using the approximation that the unity in the denominator of the distribution function can be neglected, we get $f(E) = exp(-(E - E_F)/k_BT)$. This exponential form is often termed the *Boltzmann approximation*. We get (using Eqn. 4.48 for $n = n$ and $n = n_i$)

$$\frac{n}{n_i} = e^{(E_F - E_{Fi})/k_BT} \tag{4.57}$$

and similarly from Eqn. 4.51 ($n_i = p_i$)

$$\frac{p}{n_i} = e^{-(E_F - E_{Fi})/k_BT} \tag{4.58}$$

where $E_F$ is the Fermi level corresponding to the doped semiconductor. When the semiconductor is doped *n*-type (*p*-type), the Fermi level moves towards the conduction (valence) bandedge. In this case, the Boltzmann approximation is not very good, and the simple expressions relating the carrier concentration and the Fermi level are not very accurate. The carrier density is, in general,

$$n = \frac{1}{2\pi^2} \left( \frac{2m^*}{\hbar^2} \right)^{3/2} \int_{E_c}^{\infty} \frac{(E - E_c)^{1/2} dE}{exp\left( \frac{E - E_F}{k_BT} \right) + 1} \tag{4.59}$$

A useful expression for the relation between the carrier concentration and the Fermi level is the Joyce-Dixon approximation. According to this relation we have

$$E_F = E_c + k_BT \left[ \ell n \, \frac{n}{N_c} + \frac{1}{\sqrt{8}} \, \frac{n}{N_c} \right] = E_v - k_BT \left[ \ell n \, \frac{p}{N_v} + \frac{1}{\sqrt{8}} \, \frac{p}{N_v} \right] \tag{4.60}$$

This relation can be used to obtain the Fermi level if $n$ is specified. Or else, it can be used to obtain $n$ if $E_F$ is known by solving for $n$ iteratively. *If the term* $(n/\sqrt{8} \, N_c)$ *is ignored the result corresponds to the Boltzmann approximation.*

As far as the temperature dependence of the ionization of dopants is concerned, three regions can be identified. At low temperatures, the electrons are localized at the donors so that the free carrier density is very low. This is the freezeout regime. At higher temperatures ($k_BT \gtrsim E_c - E_d$, where $E_d$ is the dopant energy), the donors are ionized and the free carrier density (electrons) is essentially equal to the donor density. This is the saturation regime. Finally, at very high temperatures, the intrinsic carrier density dominates. We will assume in all of our subsequent device analyses, that we are in the saturation region where the free electron density is equal to the donor density for n-type materials and hole density is equal to acceptor density for p-type materials.

**EXAMPLE 4.7** A sample of GaAs has a free electron density of $10^{17}$ cm$^{-3}$. Calculate the position of the Fermi level using the Boltzmann statistics and the Joyce-Dixon approximation at 300 K.

In the Boltzmann statistics, the carrier concentration and the Fermi level are related by the following equation ($E_F$ is measured from the bandedge):

$$
\begin{aligned}
E_F &= k_B T \left[ \ell n \frac{n}{N_c} \right] \\
&= 0.026 \left[ \ell n \left( \frac{10^{17}}{4.45 \times 10^{17}} \right) \right] = -0.039 \; eV
\end{aligned}
$$

The Fermi level is 39 meV below the conduction band. In the Joyce-Dixon approximation we have

$$
\begin{aligned}
E_F &= k_B T \left[ \ell n \left( \frac{n}{N_c} \right) + \frac{1}{\sqrt{8}} \frac{n}{N_c} \right] \\
&= 0.026 \left[ \ell n \left( \frac{10^{17}}{4.45 \times 10^{17}} \right) + \frac{10^{17}}{\sqrt{8}(4.45 \times 10^{17})} \right] \\
&= -0.039 + 0.002 = -0.037 \; eV
\end{aligned}
$$

The error produced by using the Boltzmann statistics (compared to the more accurate Joyce-Dixon approximation) is 2 meV.

**EXAMPLE 4.8** Assume that the Fermi level in silicon coincides with the conduction bandedge at 300 K. Calculate the electron carrier concentration using the Boltzmann statistics, the Joyce-Dixon approximation, and the Fermi-Dirac integral.

In the Boltzmann statistics, the carrier density is simply

$$
n = N_c = 2.78 \times 10^{19} \; cm^{-3}
$$

According to the Joyce-Dixon approximation, the carrier density is obtained from the solution of the equation

$$
E_F = 0 = k_B T \left[ \ell n \frac{n}{N_c} + \frac{n}{\sqrt{8 N_c}} \right]
$$

This gives
$$\frac{n}{N_c} = 0.76 \text{ or } n = 2.11 \times 10^{19} cm^{-3}$$

We see that the Boltzmann statistics gives a higher charge density.

**EXAMPLE 4.9** A semiconductor is said to be $n$-type degenerate when the probability that the conduction bandedge electronic levels are occupied by electrons is close to unity. Similarly one can define a $p$-type degenerate semiconductor. Assume a criterion that the Fermi level has to be $\sim 3k_BT$ into the band before the material can be called degenerate. What are the free electron densities in Si and GaAs when the semiconductors just become $n$-type degenerate?

We will use the Joyce-Dixon approximation

$$E_F - E_c = k_B T \left[ \ell n \frac{n}{N_c} + \frac{1}{\sqrt{8}} \frac{n}{N_c} \right]$$

Using $E_F - E_c = 3k_BT$ we get

$$\frac{n}{N_c} = 4.4$$

For Si this leads to a carrier density of

$$n = 4.4 \times 2.78 \times 10^{19} = 1.22 \times 10^{20} cm^{-3}$$

For GaAs the density for degeneracy is

$$n = 4.4 \times 4.45 \times 10^{17} = 1.96 \times 10^{18} cm^{-3}$$

It must be noted, however, that the condition for degeneracy used in this example is somewhat arbitrary.

## 4.10.1   Heavily Doped Semiconductors

$\mathcal{R}$ In our discussions on doping thus far, we have made several important assumptions which are valid only when the doping levels are low: i) we have assumed that the bandstructure of the host crystal is not seriously perturbed and the bandedge states are still described by simple parabolic bands; ii) the dopants are assumed to be independent of each other and their potential is thus a simple Coulombic potential. These assumptions become invalid as the doping levels become higher. When the spacing of the dopant atoms is of the order of 100 Å the potential seen by the impurity electron is influenced by the neighboring impurities. In a sense this is like the problem of electrons in atoms. When the atoms are far apart, we get discrete atomic levels. However, when the atomic separation reaches a few angstroms, as in a crystal, we get electronic bands. At high doping levels we get impurity bands. Several other important effects occur at high doping levels. All these effects require us to abandon our simple picture that works well for low doping levels. The many body effects which self-consistently include the effects of other free electrons present in the system at high doping levels are an important area of research.

An important effect of heavy doping is the narrowing of the bandgap. This can have serious effects on performance of devices such as bipolar transistors. In silicon, if $N_d$ is the donor density (cm$^{-3}$), the bandgap narrowing is given by a simple expression:

$$\Delta E_g \cong -22.5 \left( \frac{N_d}{10^{18}} \frac{300}{T(K)} \right)^{1/2} meV \qquad (4.61)$$

This expression gives reasonable agreement with experiments at low doping levels. At high doping levels it over estimates the bandgap narrowing. However, due to its simplicity, we can use it to get an estimate of bandgap narrowing.

## 4.10.2 Modulation Doping

$\mathcal{R}$ We have already discussed how doping is necessary for all semiconductor devices. The purpose behind doping of semiconductors is to controllably change the free-carrier density in the semiconductor. This requires that the dopant be ionized. When the donor is ionized, a positively charged ion remains in the crystal. This fixed charged center causes scattering for the free electron, as will be discussed in the next chapter. An obvious question that arises is whether one can have a controllable free electron density *without* scattering. The answer to this question is yes and this is realized through the concept of modulation doping. Before addressing this concept, it is worth remarking that the modulation doping also overcomes another problem with doping—the carrier freeze-out problem. At low temperatures, the electrons are localized at the donor sites, thus reducing the free carriers available for conduction. This effect can negate some of the benefits of operating devices at low temperatures. The concept of modulation doping is able to overcome this problem as well.

For modulation doping, a heterostructure is grown (say, GaAs/AlGaAs) and the *high bandgap material* is doped. In equilibrium the *electrons associated with the donors see lower lying energy states in the narrow bandgap material and thus transfer to the GaAs region.* This spatial separation of the positively charged donors and negatively charged electrons produces an electric field profile governed by the

Poisson equation which causes a band-bending. Usually the dopants are placed some distance away from the heterointerface by including an undoped "spacer" region. The ionized impurity scattering is essentially eliminated by this physical separation between the mobile electrons and the fixed ionized scattering centers. *Also, since the electrons are at energy positions lower than the localized ground state of the donor atoms, the electrons remain mobile even at the lowest temperatures provided the material quality is pure.* Extremely high sheet charge density of electrons ($10^{13}$ cm$^{-2}$) can thus be maintained at low temperatures. Transistors based on such concepts (modulation doped field effect transistors—MODFETs) can operate at low temperatures and are often used for detection of very weak signals from space and in other applications where low noise devices are required.

## 4.11  BANDSTRUCTURE OF AMORPHOUS SEMICONDUCTORS

$\longrightarrow$  So far in this chapter we have dealt with the properties of the perfect crystalline structure. In the previous section we have discussed an important defect, the dopant atom which is a necessary defect for semiconductor technology. The effect of dopant is to produce a new electronic level in the bandgap of the semiconductor. In an amorphous material one has a situation where there is some "defect" at almost every site of the crystal. These semiconductors are usually grown under highly non-equilibrium growth conditions so that the growing structure is unable to reach the equilibrium crystalline structure. The advantage, of course, is that the structure is grown at a very low cost and, perhaps, can be grown over large areas. Hydrogenated amorphous silicon is a prime example of such semiconductors and offers the possibility of a cheap material for solar energy conversion and thin film transistors for display devices. As discussed in Chapter 1, Section 1.6, in such amorphous semiconductors, the nearest neighbor coordination is maintained, but long-range order is lost. Since often such semiconductors have broken bonds, certain additional defects are introduced into the structure to passivate the defects. For example, in amorphous silicon, hydrogen serves this purpose. An important question that arises is the following: can one define a bandstructure of an amorphous material? We cannot define an $E$ vs. $k$ relation as we did for a crystal, but a general description of the electronic levels can be given.

### 4.11.1  Extended and Localized States: Mobility Edges

To physically see the effect disorder in an amorphous material has on the electronic states, consider a rough bottom of a river with stones and dirt arranged randomly as shown in Fig. 4.13a. If a little water is poured into this pond, small puddles

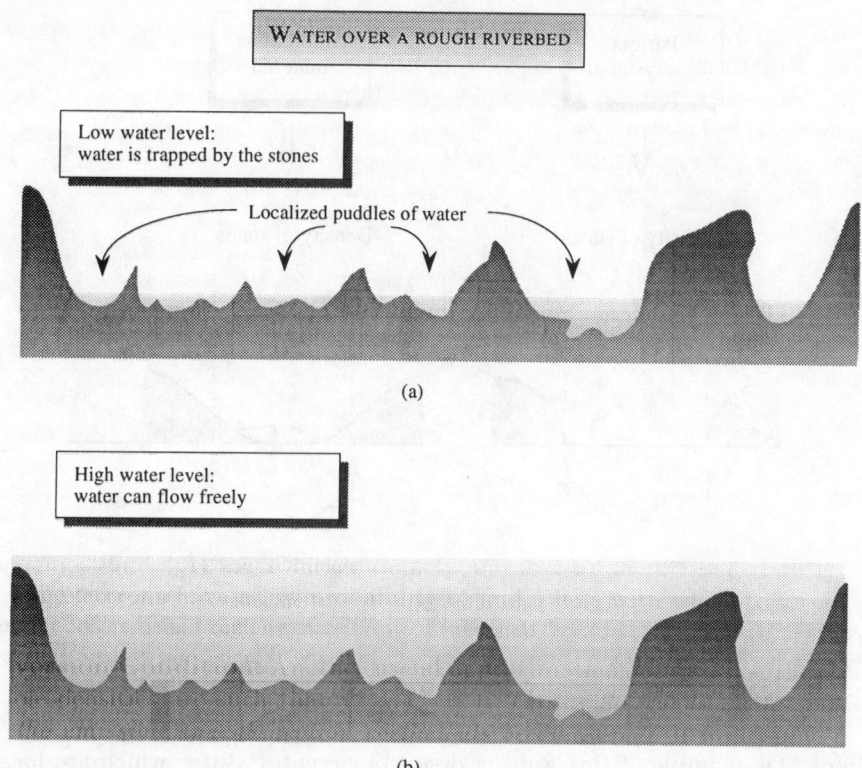

Figure 4.13: A schematic of water over a rough "disordered" river bed. (a) When the water level is low, the water is localized and cannot flow freely. (b) When the water level is high, the water can flow freely, in spite of the disorder.

of water will be trapped at low lying regions. These trapped or "localized" water regions cannot connect from one edge of the river to another. As the water level is raised so that the potential energy of the water is raised, the higher regions of water will see less effect of the underlying disorder and the water can flow freely from one part of the river to another, as shown in Fig. 4.13b. We can say that this water is in an "extended" state.

A situation quite similar to the one described above occurs for electrons which see a disordered potential produced by the random arrangement of atoms. In general, the electron states are no longer plane waves given by the Bloch theorem

$$\psi_{\text{ex}} = \frac{1}{\sqrt{V}} u_{\mathbf{k}}(\boldsymbol{r}) e^{i\mathbf{k}\cdot\boldsymbol{r}} \qquad (4.62)$$

but are given by localized states with a finite extent in space. The defect states have a general form $\psi_{\text{loc}}(\boldsymbol{r}, \boldsymbol{r}_0)$ representing the fact that they are localized around a point $\boldsymbol{r}_0$ in space. Typical localized states may have an exponentially decaying

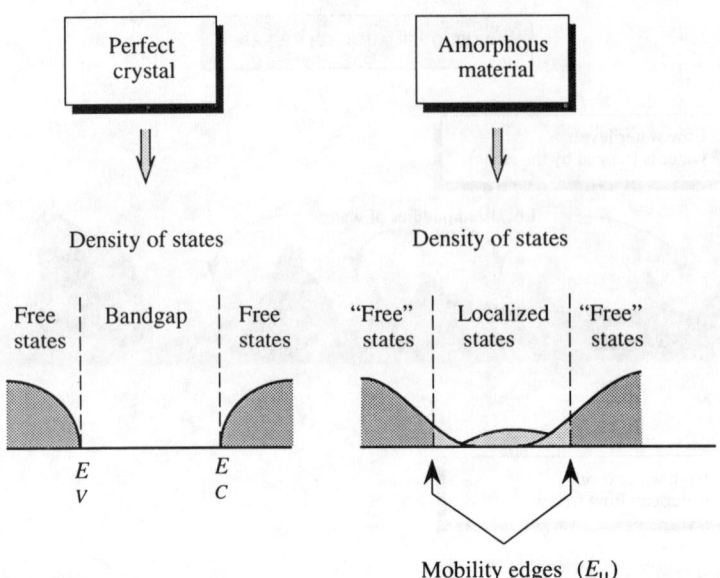

Figure 4.14: Density of states and the influence of disorder. The shaded region represents the region where the electronic states are localized in space. The mobility edges $E_\mu$ separate the region of localized and extended states.

behavior. An example of this are the donor or acceptor states which are localized around the impurity position. This concept of localized and extended states is further generalized as the extent of disorder is increased. For point defects and dopant atoms, the associated localized states are energetically separated. As the level of disorder increases, the extended and localized states merge, forming a continuum in the electronic spectrum.

An important point to note is that if the level of disorder in the amorphous film is not too large, many important properties which make semiconductors useful survive. Thus, one can dope amorphous films, both $n$- and $p$-type. Also, a well-defined "bandgap" can be present. As shown in Fig. 4.14, the "bandgap" is not made up of a region in energy that is forbidden to the electrons. However, the electronic levels in this gap are so highly localized that there is very little current-carrying capability, and for all electrical purposes, there is a bandgap. As noted earlier, the localized states have a wavefunction of the form

$$\psi_{loc}(r, r_0) = g(r, r_0)\, exp\, [-(r - r_0)/l_{loc}] \tag{4.63}$$

where $g(r, r_0)$ is an oscillatory function. The wavefunction decays over a length $l_{loc}$ called the *localization length*. As one moves from the center of the bandgap toward the valence or conduction band, the degree of localization decreases, i.e., $l_{loc}$ increases, eventually becoming as large as the sample dimension. When this occurs, the electron wavefunction is extended over the entire sample and such states are called *extended states*. However, it is important to note that *these states are not Bloch states*. But having an extended nature, electrons in these states can contribute significantly to carrying current.

The energies (on the conduction and valence band sides) where the nature of the electron states changes, as discussed above, so that conductivity changes from very low to a reasonable value are called *mobility edges*. These are shown in Fig. 4.14.

The success of an amorphous film technology depends upon having as small a localized state density as possible. This involves having as small a disorder as possible and, more importantly, a low density of defects such as broken bonds. Focusing on amorphous silicon, perhaps the most important amorphous semiconductor from the technology standpoint, it should be noted that normally a large density of broken bonds are produced in the amorphous film. This density could be as high as $10^{20}$ cm$^{-3}$, creating such a high density of gap states that the material is electrically "dead." However, it is seen that if, during the growth process (discussed in Chapter 9), a considerable fraction ($\sim 10\%$ atomic) of $H$ gets incorporated, the broken bond density decreases by up to three orders of magnitude. Under such conditions, the material has acceptable electronic properties.

## 4.12  SUMMARY

$\mathcal{R}$ In this chapter we have discussed the bandstructure of semiconductors and their heterostructures. Essentially all optical and transport properties of semiconductors are controlled by the bandstructure. We summarize the key findings of this chapter in Tables 4.1—4.3.

| TOPICS STUDIED | KEY OBSERVATIONS |
|---|---|
| Electrons in a periodic potential | • The wavefunction has a plane wave-like behavior $\psi \sim u_k e^{i\mathbf{k}\bullet\mathbf{r}}$ where $u_k$ is periodic.<br>• The energy of the electron is not continuous. Allowed energy bands are separated by bandgaps.<br>• A new parameter k is introduced. For most purposes, $\hbar k$ acts as a momentum of the electron inside the crystal. |
| Semiconductors and metals | • If at 0 K, the highest occupied band is filled completely and the next band (separated by bandgap) is completely empty, a semiconductor results.<br>• If the highest band is partially filled, a metal results. |
| Valence and conduction bands | • The highest occupied band in a semiconductor at 0 K is called the *valence band*. The empty band is called the *conduction band*.<br>• A completely filled band cannot carry any current, so the conductivity of a pure semiconductor at 0 K is zero. |

Table 4.1: Summary Table

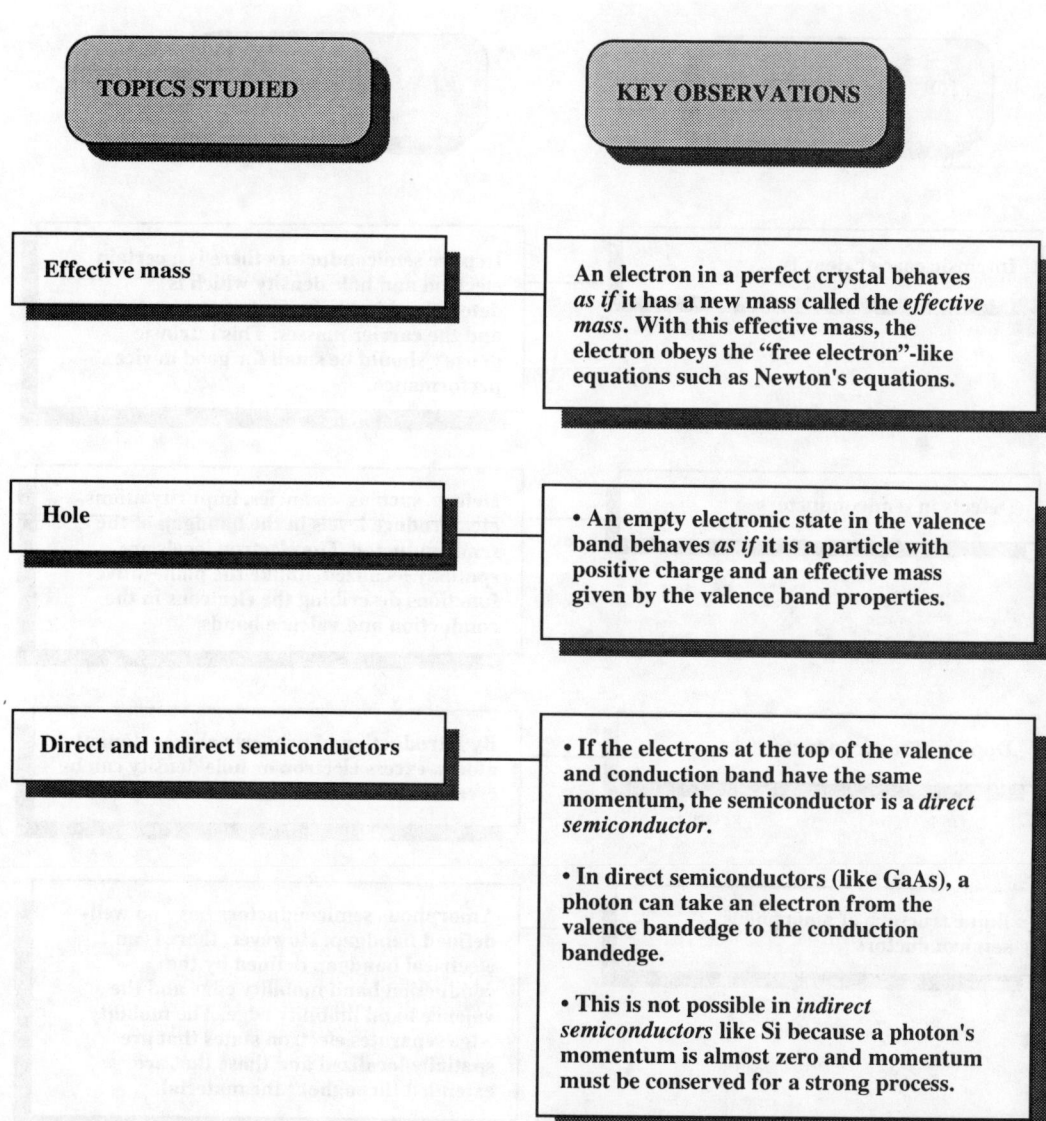

**TOPICS STUDIED**

**KEY OBSERVATIONS**

**Effective mass**

An electron in a perfect crystal behaves *as if* it has a new mass called the *effective mass*. With this effective mass, the electron obeys the "free electron"-like equations such as Newton's equations.

**Hole**

• An empty electronic state in the valence band behaves *as if* it is a particle with positive charge and an effective mass given by the valence band properties.

**Direct and indirect semiconductors**

• If the electrons at the top of the valence and conduction band have the same momentum, the semiconductor is a *direct semiconductor*.

• In direct semiconductors (like GaAs), a photon can take an electron from the valence bandedge to the conduction bandedge.

• This is not possible in *indirect semiconductors* like Si because a photon's momentum is almost zero and momentum must be conserved for a strong process.

Table 4.2: Summary Table

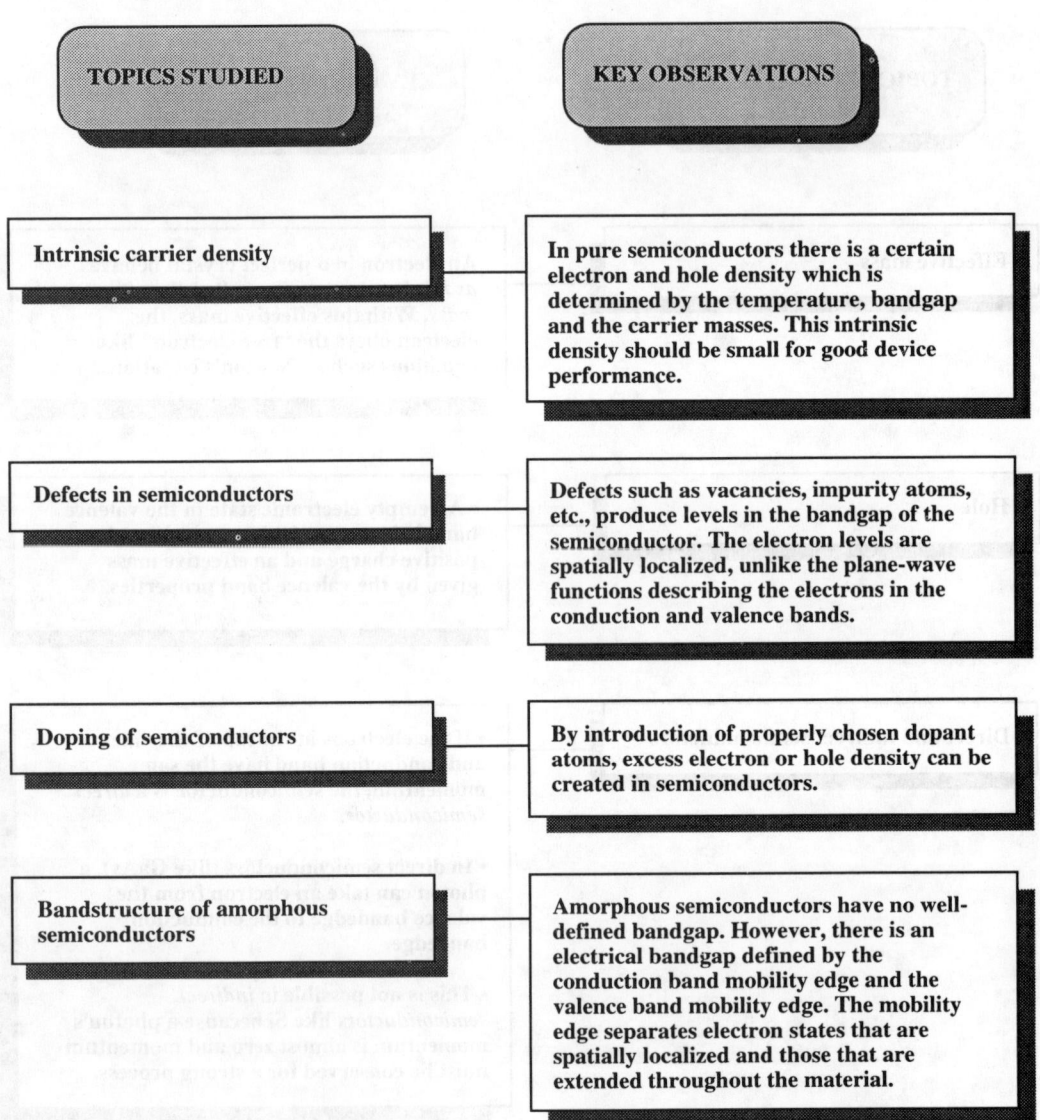

**TOPICS STUDIED**

**KEY OBSERVATIONS**

**Intrinsic carrier density**

In pure semiconductors there is a certain electron and hole density which is determined by the temperature, bandgap and the carrier masses. This intrinsic density should be small for good device performance.

**Defects in semiconductors**

Defects such as vacancies, impurity atoms, etc., produce levels in the bandgap of the semiconductor. The electron levels are spatially localized, unlike the plane-wave functions describing the electrons in the conduction and valence bands.

**Doping of semiconductors**

By introduction of properly chosen dopant atoms, excess electron or hole density can be created in semiconductors.

**Bandstructure of amorphous semiconductors**

Amorphous semiconductors have no well-defined bandgap. However, there is an electrical bandgap defined by the conduction band mobility edge and the valence band mobility edge. The mobility edge separates electron states that are spatially localized and those that are extended throughout the material.

Table 4.3: Summary Table

## 4.13  PROBLEMS

**Sections 4.2-4.5**

**4.1** When a photon impinges upon a semiconductor, it can take an electron from the valence band to the conduction band. The momentum is conserved in such transitions. Calculate the momentum that a 2.0 eV photon carries. The electron in the valence band can go into the conduction band with the momentum change of the photon.

**4.2** Use the data given in Appendix B to identify semiconductors with bandgaps large enough to emit photons with wavelengths less than 0.5 $\mu$m.

**4.3** Plot the conduction band and valence band density of states in Si, Ge and GaAs from the bandedges to 0.5 eV into the bands. Use the units eV$^{-1}$ cm$^{-3}$. Use the following data:

$$Si : m_1^* = m_\ell^* = 0.98\, m_0$$
$$m_2^* = m_3^* = m_t^* = 0.19\, m_0$$
$$m_{hh}^* = 0.49\, m_0$$
$$m_{\ell h}^* = 0.16\, m_0$$

$$Ge : m_\ell^* = 1.64\, m_0$$
$$m_t^* = 0.082\, m_0$$
$$m_{hh}^* = 0.29\, m_0$$
$$m_{\ell h}^* = 0.044\, m_0$$

$$GaAs : m_e^* = m_{dos}^* = 0.067\, m_0$$
$$m_{hh}^* = 0.45\, m_0$$
$$m_{\ell h}^* = 0.08\, m_0$$

**4.4** The wavevector of a conduction band electron in GaAs is $k = (0.1, 0.1, 0.0)$ Å$^{-1}$. Calculate the energy of the electron measured from the conduction bandedge.

**4.5** A conduction band electron in silicon is in the (100) valley and has a $k$-vector of $2\pi/a$ (1.0, 0.1, 0.1). Calculate the energy of the electron measured from the conduction bandedge. Here $a$ is the lattice constant of silicon.

**4.6** An electron in the $\Gamma$-valley of GaAs is to be transferred to the $L$-valley. Using the bandstructure of GaAs estimate the smallest $k$-vector change that is needed for this transition. The electron in the $\Gamma$-valley must have an energy equal to the position of the $L$-valley.

**4.7** Calculate the energies of electrons in GaAs and InAs conduction band with $k$-vectors (0.01, 0.01, 0.01) Å$^{-1}$. Refer the energies to the conduction bandedge values.

**4.8** Consider an electron at the bottom of the conduction band in GaAs. An electric field of $10^4$ V/cm is applied to the materials in the $x$-direction. Calculate the time it takes the electron to reach the Brillouin zone edge. Consult the bandstructure of GaAs to see what the electron energy is at this point. What happens to the electron after it has reached the Brillouin zone edge?

**4.9** An electric field of $10^5$ V/cm is applied for 0.1 ps to holes at the top of the valence band in GaAs. Calculate the hole energies at the end of the pulse. Note that holes will move in both the heavy-hole and light-hole bands.

## Section 4.6

**4.10** According to the virtual crystal approximation for the electronic properties of alloys (discussed in the text), the high symmetry point (i.e., bandedges and other symmetry points like $X$ point, $L$ point, etc.) energy of the alloy is simply given by a composition weighted averaging of the values for the individual semiconductors. Based on this, calculate the bandgap of $Si_x Ge_{1-x}$ alloy as $x$ goes from 1.0 to 0. Note that the conduction bandedge is $X$-like in Si and $L$-like in Ge.

**4.11** Using the Vegard's Law for the lattice constant of an alloy (i.e., lattice constant is the weighted average), find the bandgaps of alloys made in InAs, InP, GaAs, GaP which can be lattice matched to InP.

**4.12** For long haul optical communication, the optical transmission losses in a fiber dictate that the optical beam must have a wavelength of either 1.3 $\mu$m or 1.55 $\mu$m. Which alloy combinations lattice matched to InP have a bandgap corresponding to these wavelengths?

**4.13** Using the virtual crystal approximation, up to what Al composition does the alloy $Al_x Ga_{1-x}As$ remain a direct gap semiconductor? What is the maximum bandgap achievable in the direct alloy?

**4.14** Using the bandstructure of GaAs and AlAs, calculate the conduction band minima at $\Gamma$, $X$, $L$ points in $Al_x Ga_{1-x}As$ alloy, as $x$ varies from 0 to 1.

**4.15** Calculate the composition of $Hg_x Cd_{1-x}Te$ which can be used for a night vision detector with bandgap corresponding to a photon energy of 0.1 eV. Bandgap of CdTe is 1.6 eV *and that of HgTe is −0.3 eV at low tempertures around 4 K.*

## Section 4.7

**4.16** At room temperature the bandgap of GaAs is 1.43 eV. Assuming an infinite barrier approximation, what is the well size of a GaAs/AlAs quantum well which can produce an effective bandgap of 1.50 eV?

**4.17** If a 100 Å cubic dot of GaAs is embedded in AlAs, what is the approximate bandgap of the quantum dot? Assume an infinite barrier model. Use an effective mass of $0.067m_0$ and $0.4m_0$ for the electron and heavy holes, respectively.

**4.18** In the $In_{0.53}Ga_{0.47}As/InP$ system, 40% of the bandgap discontinuity is in the conduction band. Calculate the conduction and valence band discontinuities. Calculate the effective bandgap of a 100 Å quantum well. Use the infinite potential approximation and the finite potential approximation and compare the results.

**4.19** Calculate the first and second subband energy levels for the conduction band in a GaAs/$Al_{0.3}Ga_{0.7}As$ quantum well as a function of well size. Assume an infinite barrier approximation.

**Section 4.8**

**4.20** Calculate the effective density of states at the conduction and valence bands of Si and GaAs at 77 K, 300 K and 500 K.

**4.21** Calculate the intrinsic carrier concentration of Si, Ge and GaAs as a function of temperature from 4 K to 600 K. Assume that the bandgap is given by

$$E_g(T) = E_g(0) - \frac{\alpha T^2}{T + \beta}$$

where $E_g(0)$, $\alpha$, and $\beta$ are given by

$$
\begin{aligned}
Si &: \quad E_g(0) = 1.17 eV; \alpha = 4.37 \times 10^{-4} \, eVK^{-1}; \beta = 636 \, K \\
Ge &: \quad E_g(0) = 0.74 eV; \alpha = 4.77 \times 10^{-4} \, eVK^{-1}; \beta = 235 \, K \\
GaAs &: \quad E_g(0) = 1.519 eV; \alpha = 5.4 \times 10^{-4} \, eVK^{-1}; \beta = 204 \, K
\end{aligned}
$$

**4.22** Estimate the intrinsic carrier concentration of diamond at 700 K (you can assume that the carrier masses are similar to those in Si). Compare the results with those for GaAs and Si. The result illustrates one reason why diamond is useful for high temperature electronics.

**4.23** Estimate the change in intrinsic carrier concentration per K change in temperature for InAs, Si, and GaAs near room temperature.

**4.24** Calculate the position of the intrinsic Fermi level measured from the midgap for InAs.

**Section 4.10**

**4.25** Calculate the density of electrons in silicon if the Fermi level is 0.2 eV below the conduction band at 300 K. Compare the results by using the Boltzmann approximation and the Joyce-Dixon approximation.

**4.26** In a GaAs sample at 300 K, the Fermi level coincides with the valence band-edge. Calculate the hole density using a) the Boltzmann approximation and b) the Joyce-Dixon approximation. Also calculate the electron density using the law of mass action (i.e., $n_i p_i = np$).

**4.27** The electron density in a silicon sample at 300 K is $10^{16}$ cm$^{-3}$. Calculate $E_c - E_F$ and the hole density using the Boltzmann approximation.

**4.28** A GaAs sample is doped $n$-type at $5 \times 10^{17}$ cm$^{-3}$. Assume that all the donors are ionized. What is the position of the Fermi level at 300 K?

## 4.14     REFERENCES

- **General Bandstructure**

    - R. E. Hummel, *Electronic Properties of Materials—An Introduction for Engineers*, Springer, New York (1985).

    - K. Seeger, *Semiconductor Physics: An Introduction*, Springer, Berlin (1985).

    - H. F. Wolf, *Semiconductors*, Wiley-Interscience, New York (1971).

- **Bandstructure Modification**

    - A. G. Milnes and D. L. Feucht, *Heterojunctions and Metal Semiconductor Junctions*, Academic Press, New York (1972).

    - For a simple discussion of electrons in quantum wells any book on basic quantum mechanics is adequate. An example is L. Schiff, *Quantum Mechanics*, McGraw-Hill, New York (1968).

    - J. Singh, *Physics of Semiconductors and Their Heterostructures*, McGraw-Hill, New York (1993).

- **Intrinsic and Extrinsic Carriers**

    - J. S. Blakemore, *Electron. Commun.*, 29, 131 (1952).

    - J. S. Blakemore, *Semiconductor Statistics*, Pergamon Press, New York (1962) reprinted by Dover, New York (1988).

    - K. Seeger, *Semiconductor Physics: An Introduction*, Springer, Berlin (1985).

# CHAPTER
# 5

# TRANSPORT AND OPTICAL PROPERTIES OF SEMICONDUCTORS

## 5.1   INTRODUCTION

Semiconductors are ubiquitous in every aspect of information processing. Almost all active electronic and optoelectronic devices are based on semiconductors. What is so special about semiconductors that has given them such a privilege? The answer is very simple: their electrical properties (e.g., conductivity) and/or optical properties (e.g., light being emitted) can be altered with high speed and with great precision. This cannot be done with metals or insulators.

In the previous chapter we have studied how electrons behave inside a semiconductor. We have seen that, in a perfect material, the electrons have well defined states and energies which give the bandstructure of the material. What happens to these electrons when an electric field is applied? What happens if light shines on a semiconductor? These are the physical responses which determine how semiconductor devices perform. To answer these questions we need to understand something very important—the concept of scattering. The concept of scattering is simple—an electron is in a certain state described by its crystal momentum and suddenly it scatters and moves in a different state. Such scattering events caused by various disturbances form the basis of macroscopic physical properties such as mobility, absorption coefficient, diffusion coefficient, etc.

While the general process of scattering is conceptually simple to understand, the scattering of electrons in a crystal cannot be understood without considerable quantum mechanics. In fact, the full understanding of scattering cannot be acquired without several graduate-level courses. The approach we will take in this text is to provide the reader a sound physical basis for quantum mechanical scattering. *The mathematical rigor will be sacrificed, but important quantum concepts will be motivated.* In a sense, a bit of "leap of faith" will be expected from the reader.

## 5.2   QUANTUM MECHANICS OF SCATTERING PROCESSES

$\mathcal{R}$   In the previous chapters we have discussed the solutions for the electron-allowed energy states (i.e., the bandstructure) in a semiconductor. The electron wavefunction has the general form

$$\psi(k,t) = u_{nk} exp\,(ik \cdot r) exp\,(i\omega t) \tag{5.1}$$

where

$$\hbar\omega = E(k) \tag{5.2}$$

Notice that if an electron occupies an energy level described by energy $E$ and crystal momentum $k$ at a time $t_o$, it remains in the same level at all other times as well.

*There is no scattering in the perfect crystal!* This does not seem to be intuitively correct, but it comes right out of Bloch's theorem. The electron in a perfect crystal suffers no scattering and, apart from its altered $E$ vs. $k$ relation, behaves just as if it were in free space! The electron in the crystal obeys a free-electron-like equation of motion

$$\hbar \frac{dk}{dt} = F_{ext} \tag{5.3}$$

where $F_{ext}$ is the external force. What happens when an electric field is applied to the electron? According to Eqn. 5.3 the $k$-value of the electron will increase and the energy of the electron will follow according to the $E$ vs. $k$ relation. Let us consider the lowest conduction band of a direct gap material like GaAs. If the electron starts at the bottom of the band at time $t = t_o$, it climbs up in energy as shown in Fig. 5.1. Eventually, the $k$-value reaches the value at the Brillouin zone edge. We can describe the progress of the electron by either saying that the electron goes into the neighboring identical Brillouin zone and traces the $E$ vs. $k$ curve losing its energy, or by keeping our focus on the same Brillouin zone and reflecting the electron by a vector $G$ as shown. In either case, the electron will start losing energy as it hits the Brillouin zone edge.

According to the picture discussed above, the electron will gain and lose energy when a dc electric field is applied. Does this really happen? It turns out that the central feature of no scattering is not really true in real semiconductors and the oscillations described above (called Bloch oscillations) do not occur except under some very stringent conditions. In fact, before electrons can travel enough time to reach the zone edge, they scatter. The electrons scatter because the world they live in is not perfect. Even the best of crystals has imperfections and, even if there were no imperfections, the semiconductor is not periodic because of the presence of thermal energy. The thermal energy is due to the vibration of the atoms making up the material, and the electron sees these vibrations as deviations from perfect periodicity. In fact, scattering is an underlying feature determining the response of both electronic and optoelectronic devices. It is, thus, of extreme importance to understand how electrons scatter. In this section we will develop a simplistic description of the scattering process within quantum mechanics. A rigorous description requires advanced quantum mechanics concepts which are beyond the discussion level of this text.

To understand the scattering process we need to identify three components: i) what is the cause of the scattering and how does the electron sense the scatterer? ii) What is the effect of the scattering on the electron's wavefunction and energy? iii) How does the scattering manifest itself in macroscopic properties measured in the laboratory? We will first discuss these issues in a general sense and then consider some specific scattering processes of importance.

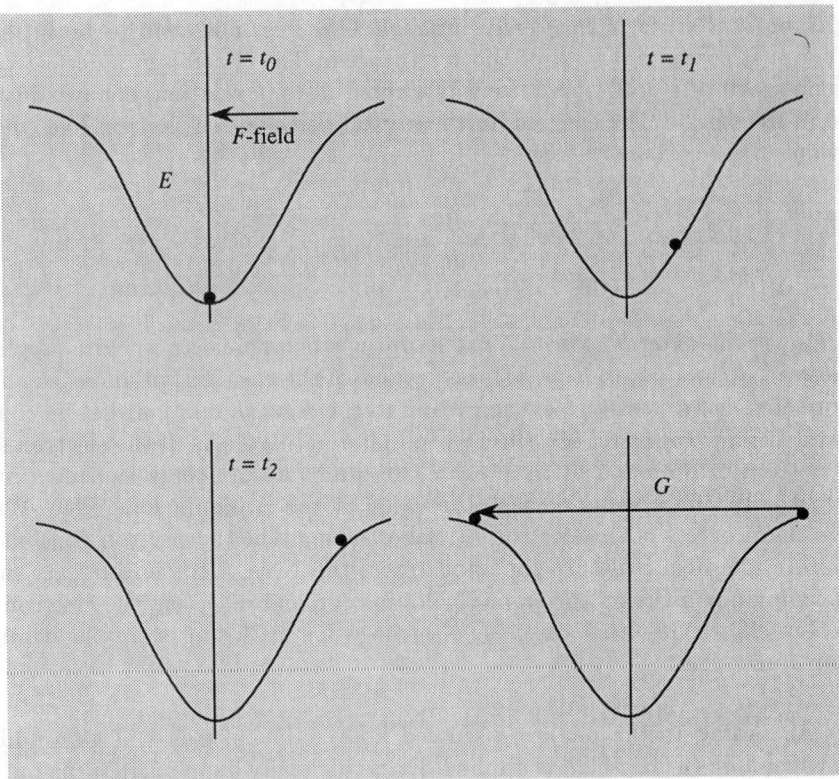

Figure 5.1: The motion of an electron in a band in the absence of any scattering and in the presence of an electric field. The electron oscillates in $k$-space gaining and losing energy from the field.

## 5.2.1   The Perturbation

When the electron is in the perfect crystal, its properties are defined by the quantum mechanical energy operator or the Hamiltonian $H_o$ for the perfect crystal. Formally, we write the Schrödinger equation as

$$H_o \psi_n = E_n \psi_n \tag{5.4}$$

where the set $\{E_n, \psi_n\}$ defines the energy values and wavefunctions allowed by the problem. If $H_o$ has no time dependence, the electron stays in the same state at all times.

Now things are never so simple for the electron and it almost invariably finds itself in a situation where the Hamiltonian is not $H_o$, but $H$ where

$$H = H_o + H' \tag{5.5}$$

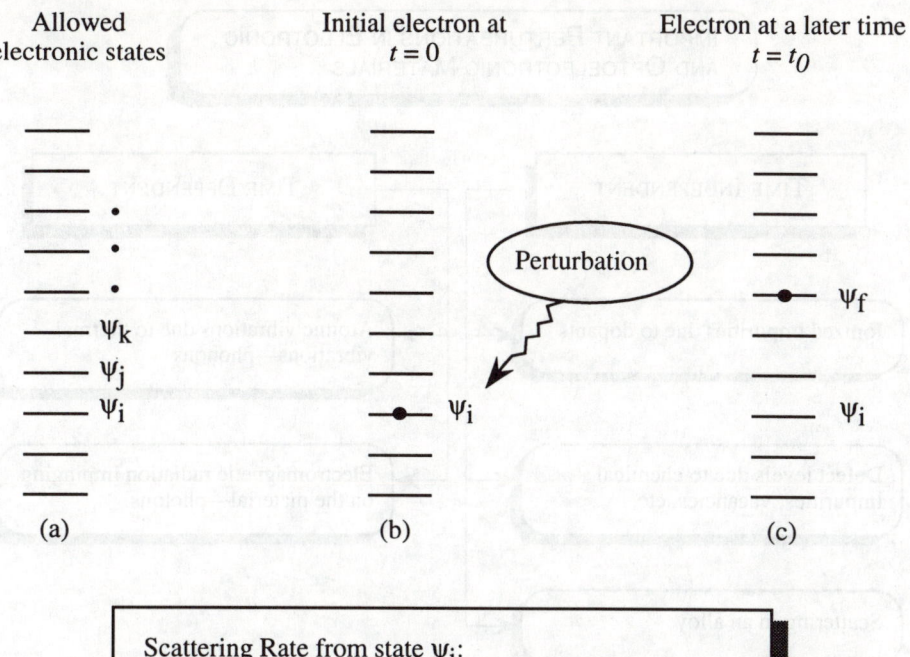

Figure 5.2: (a) The allowed electronic states in a system. (b) The electron is in an initial state $\psi_i$ when a perturbation disturbs the electron. (c) The electron at a later time is in a different state due to the scattering.

The $H'$ part arises from the deviation from the "perfect" world of $H_o$ and is called a disturbance or a perturbation. The perturbation may arise because an atom in the crystal lattice is missing or a host atom (eg., Ga in a GaAs crystal) is replaced by a dopant (eg., a Si atom). It may also arise if all the atoms are vibrating due to the thermal energy of the system or light is shining on the crystal.

The effect of the perturbation can be described by the scattering picture in quantum mechanics according to which, if at a particular instant in time, the electron is occupying the state $\psi_i$, after a certain time it has a finite probability to occupy other states as well, as shown schematically in Fig. 5.2. Of course, in quantum mechanics we can only talk of the probability of the scattering from, say, a state $\psi_i$ to a state $\psi_f$. We will discuss how one obtains this probability of scattering or, equivalently, the rate of scattering from one state to another.

In Fig. 5.3 we show some of the important perturbations in electronic and

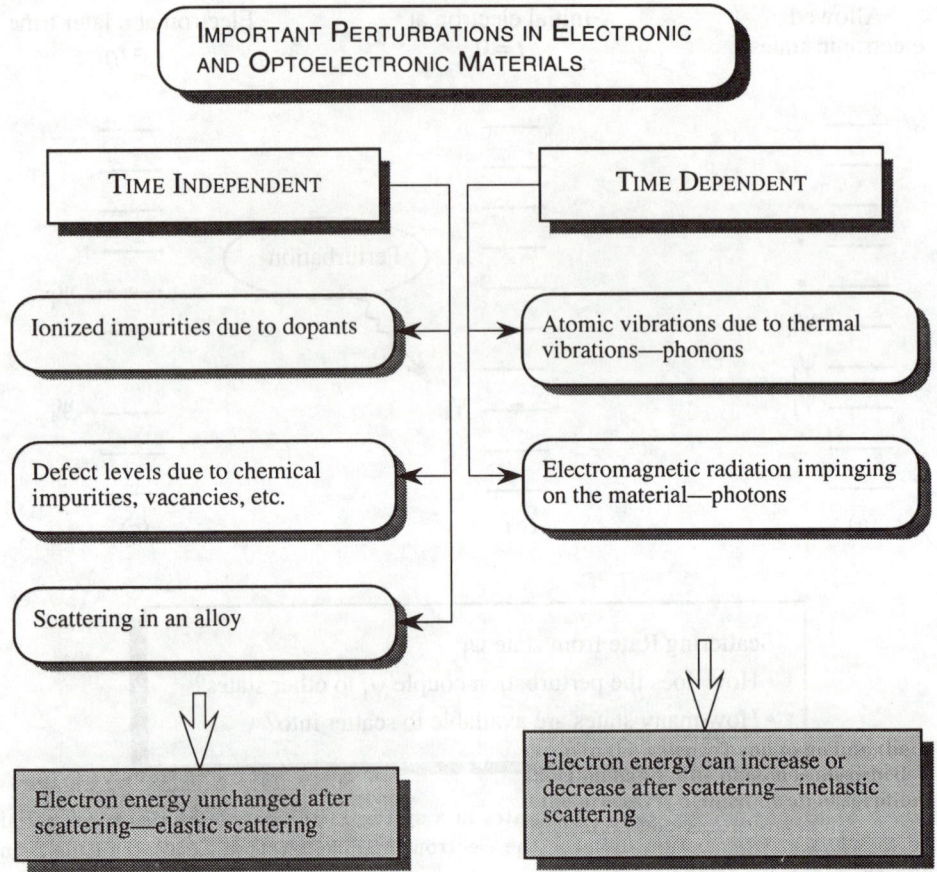

Figure 5.3: Important perturbations that are responsible for scattering of electrons.

optoelectronic materials. We have divided the perturbations into two categories—time-independent and time-dependent. As the terms imply, the time-independent perturbations are those in which $H'$ has no time-dependence. The presence of a defect in a crystal is one such example, since the defect is at a fixed point and does not have any time dependence. Time-dependent perturbations are those where $H'$ has an explicit time-dependence and examples are light shining on a semiconductor or vibrations of the atoms in a crystal. Let us briefly discuss the important differences between these two categories.

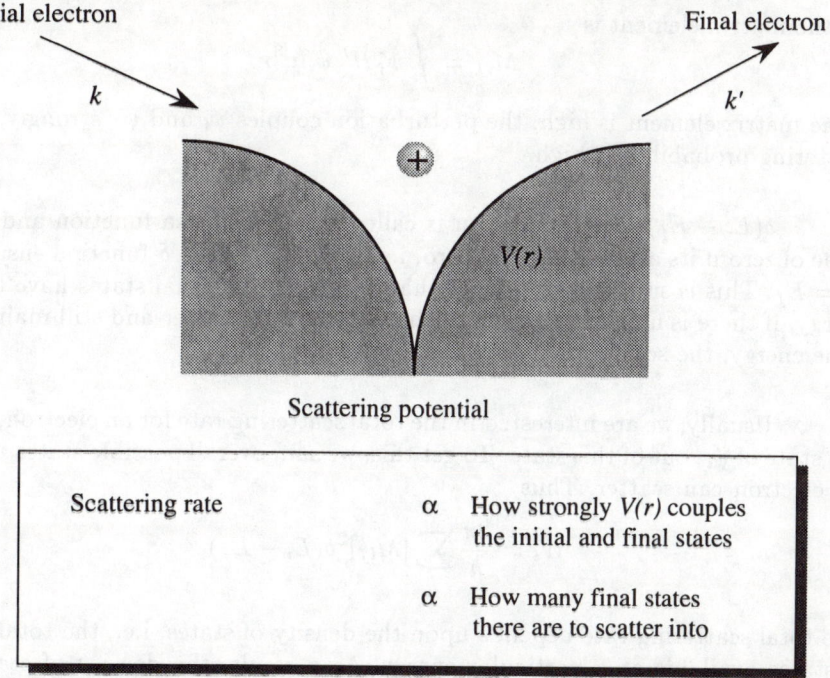

Figure 5.4: The scattering of an electron initially with momentum $\hbar k$ from a scattering potential $V(r)$. The final momentum is $\hbar k'$. The scattering process is assumed to be instantaneous. The scattering depends upon the coupling of the initial state to the final state by scattering potential.

## Time-Independent Perturbations

Any kind of defect causes scattering of an electron much like a pothole in a road causes a car to bump off track. *An important outcome of the time independent nature of the perturbation is that the energy of the electron after scattering is the same as that before scattering.* Thus the scattering is elastic, due to the much smaller mass of the electron compared to that of the defect.

The rate at which the electron scatters from an initial state $\psi_i$ to a final state $\psi_f$ is given by the following expression (see Fig. 5.4)

$$W_{if} = \frac{2\pi}{\hbar} \mid M_{if} \mid^2 \delta\left(E_i - E_f\right) \qquad (5.6)$$

where the various terms have the following significance.

$\frac{2\pi}{\hbar}$ : This comes from the details of the quantum mechanics calculations.

$\mid M_{if} \mid^2$: The term $M_{if}$ is called a matrix element and it determines how strongly the perturbation $H'$ is able to couple the states $\psi_i$ and $\psi_f$. The expression for the matrix element is

$$M_{if} = \int \psi_f^* H' \, \psi_i \, d^3r \qquad (5.7)$$

If the matrix element is high, the perturbation couples $\psi_i$ and $\psi_f$ strongly, and the scattering probability is high.

$\delta(E_i - E_f)$: The $\delta$ function is called the Dirac delta function and it has a value of zero if its argument is not zero, i.e., for our case the $\delta$-function ensures that $E_i = E_f$. This is simply a statement that the initial and final states have the same energy. If there is no state into which the electron can scatter and still maintain the same energy, the scattering rate is zero.

Usually, we are interested in the total scattering rate for an electron, initially in a state of $\psi_i$, out of this state. To get this, we sum overall possible states $\psi_f$ where the electron can scatter. Thus

$$W_i = \frac{2\pi}{\hbar} \sum_f |M_{if}|^2 \delta(E_i - E_f) \qquad (5.8)$$

The total scattering rate depends upon the density of states, i.e., the total number of states available at a particular energy. As a result, the density of states is of great importance in scattering problems. The inverse of $W_i$ defines an important time called the relaxation time. It sort of represents the average time in between collisions or scattering events and is closely related to macroscopic properties such as mobility, diffusion coefficient, and absorption coefficient.

**Time-Dependent Perturbations**

An important class of perturbations which cause scattering is the *time-dependent perturbation* where the perturbation has a general time dependence of the form

$$H'(r, t) = H'(r) \, exp \, (i\omega t) \qquad (5.9)$$

where $\omega$ is the frequency associated with the perturbation. The two most important perturbations in this category are the lattice vibrations produced by the thermal energy of the material and electromagnetic radiation impinging on the material.

The proper treatment of scattering from a time dependent perturbation which appears as a "wave" is outlined in Fig. 5.5. The scattering problem requires a higher level of quantum mechanics understanding—a level often called second quantization.

The reader is probably familiar with the "first" quantization in which the classical particles (an electron, for example) is treated as a wave via the wave equation (the Schrödinger equation). The bandstructure of the semiconductors we have discussed is a solution of this equation.

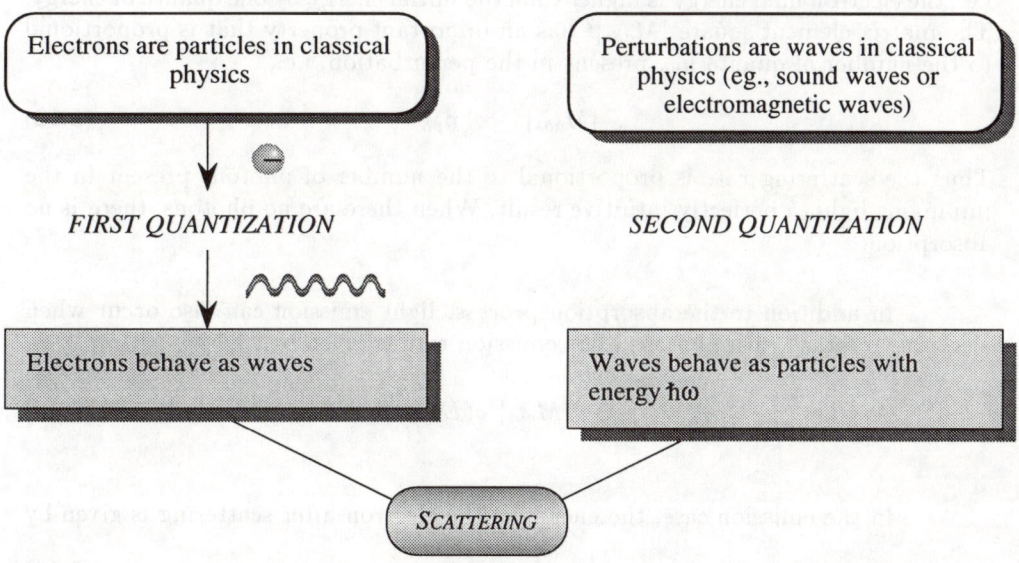

Figure 5.5: A schematic description of how the scattering of electrons from a time-dependent field is treated in quantum mechanics.

In the "second" quantization, physical effects, which in classical physics appear as waves (sound waves, electromagnetic waves, etc.), are given a particle-like nature. This gives a complete dual particle-wave character to all physical objects. *When a wave is second quantized, its description in some domains of physical experimental conditions is that of particles, or quantas. Thus we speak of the quantas of the electromagnetic field— the photons. When the wavefield associated with atomic vibrations of a lattice are second quantized, the particles are called phonons.*

Coming back to our scattering problem, the time-dependent perturbation is described by a collection of particles. Let us say for better focus that we are interested in the problem of light shining on a semiconductor. Light is now described by the "photon occupation number", i.e., how many photons are present in a particular frequency, $\omega$, and polarization mode. This number is defined by the intensity of the light. In the scattering process, the electron can absorb the photon which has an energy $\hbar\omega$. The absorption process has a rate given by

$$W_i = \frac{2\pi}{\hbar} \sum_f |M_{abs}|^2 \delta(E_f - E_i - \hbar\omega) \tag{5.10}$$

This expression is similar to the expression we discussed in the case of the scattering from a time-independent perturbation except that the final energy of the electron after absorption is given by

$$E_f = E_i + \hbar\omega \tag{5.11}$$

i.e., the electron final energy is higher than the initial energy by one quanta of energy. The matrix element square $|M_{abs}|^2$ has an important property that is proportional to the number of quanta $n_{ph}$ present in the perturbation, i.e.,

$$|M_{abs}|^2 \propto n_{ph} \tag{5.12}$$

Thus the scattering rate is proportional to the number of photons present in the impinging light, a perfectly intuitive result. When there are no photons, there is no absorption.

In addition to the absorption process, light emission can also occur when electrons interact with photons. The emission rate is given by

$$W_i = \frac{2\pi}{\hbar} \sum |M_{em}|^2 \delta(E_f - E_i + \hbar\omega) \tag{5.13}$$

In the emission case, the energy of the electron after scattering is given by

$$E_f = E_i - \hbar\omega \tag{5.14}$$

i.e., it is lower than the initial energy by a quanta of energy. The most unexpected part in the emission process is the dependence of the matrix element on the photon number. It is found that the matrix element can be written as

$$|M_{em}|^2 = |M_{st}|^2 + |M_{sp}|^2 \tag{5.15}$$

where $M_{st}$ is the matrix element for a process called stimulated emission and $M_{sp}$ is the matrix element for the process called spontaneous emission. It is found that

$$|M_{st}|^2 \propto n_{ph} \tag{5.16}$$

and

$$|M_{sp}| \propto 1 \tag{5.17}$$

so that

$$|M_{em}|^2 \propto (n_{ph} + 1) \tag{5.18}$$

According to these relations the stimulated emission is proportional to the photon density present, i.e., it is proportional to the perturbation intensity. *On the other hand, the spontaneous emission process does not depend upon the photons present, i.e., even if there are no photons scattering can occur!* This concept is difficult to grasp since we cannot conceive of how scattering can occur when there is no source of scattering.

The conceptual difficulty discussed above is resolved by quantum mechanics by presenting an understanding of the state where no photons are present, i.e., the $n_{ph} = 0$ state. This state is called the *vacuum state*. However, even though no photons are present in this state, this does not mean that there is no electromagnetic field energy in the state. Quantum mechanics says that even in the lowest energy state, the energy of the field is nonzero. Each photon mode of frequency $\omega$ has an energy $\frac{1}{2}\hbar\omega$ in the vacuum state. This energy is called the *zero point energy*. This quantum mechanical observation is related to the uncertainty relation according to which, if we have a particle, the uncertainty in its momentum ($\Delta p$) and its position ($\Delta x$) must satisfy

$$\Delta p \, \Delta x \geq \frac{\hbar}{2} \tag{5.19}$$

so that the system cannot have zero momentum (and zero energy) and be well defined in space.

Due to the presence of the zero point energy, the electron can interact with an electromagnetic field, even if $n_{ph} = 0$ and this leads to spontaneous emission discussed above. The spontaneous emission and stimulated emission of photons play extremely important roles in light emission devices. We also point out that *when spontaneous emission occurs, the emitted photons have no phase coherence with each other since the vacuum energy is sort of like noise. However, when stimulated emission occurs, the emitted photon is in phase with the photon that caused the emission to occur.* This has a profound influence on lasers, as we shall see later.

## 5.3 TRANSPORT PROPERTIES OF SEMICONDUCTORS: DRIFT IN AN ELECTRIC FIELD

$\longrightarrow$ We will now examine how electrons behave in the presence of imperfections when an external field is applied. This involves the understanding of transport of electrons and holes—an area of great importance to electronic and optoelectronic devices. If one considers a beam of electrons moving initially with the same momentum, then due to the scattering processes, the momentum and energy will gradually lose coherence with the initial state values. The average time it takes to lose coherence or memory of the initial state properties is called the "relaxation time" or scattering time. Thus one can define a relaxation time for momentum ($\tau_{sc}$) or velocity as shown in Fig. 5.6.

The scattering process and the relaxation times are related to the microscopic properties of the electrons in the semiconductor and the scattering potential of the imperfection as discussed above. The transport of the electron is schematically shown in Fig. 5.7. The electron suffers collisions which change its momentum.

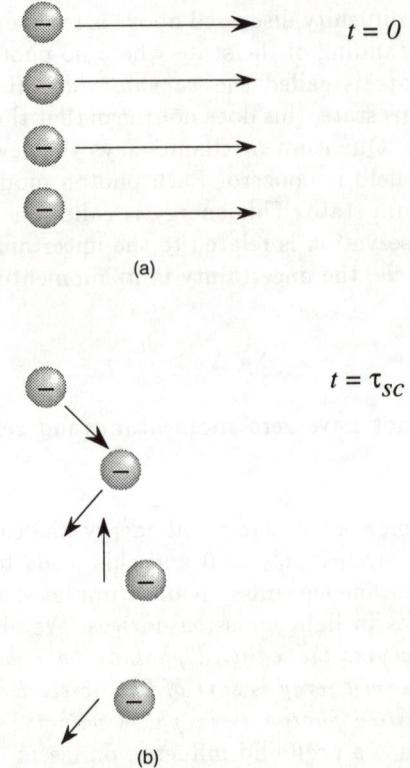

Figure 5.6: A schematic of the effect of scattering on electron velocities. (a) A beam of electrons with constant velocity is considered at time $t = 0$; the arrow is the vector representing the velocity. (b) After a relaxation time (scattering time) $\tau_{sc}$, the electrons have lost memory of their velocities.

*In between collisions, the electron moves under the electric field as a "free particle," obeying Eqn. 5.3.*

## 5.3.1  Velocity-Electric Field Relations in Semiconductors

→  When an electron distribution is subjected to an electric field, the electrons tend to move in the field direction (opposite to the field **E**) and gain velocity from the field. However, because of imperfections, they scatter in random directions. A steady state is established in which the electrons have some net drift velocity in the field direction.

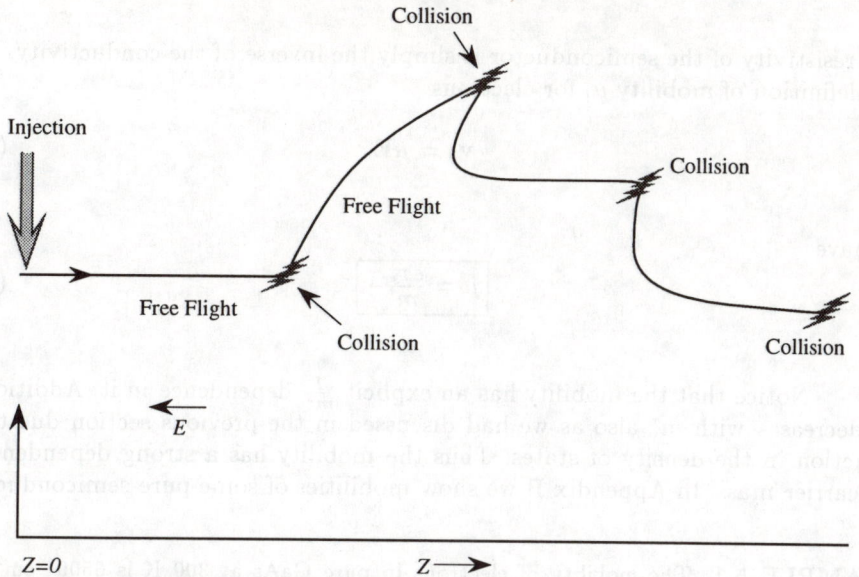

Figure 5.7: A schematic view of an electron moving in a semiconductor under an applied field. The flight of the electron is made up of a series of free flights followed by scattering.

**Low Field Response: Mobility**

At low electric fields, the drift velocity of the electrons which, for simplicity, is called the *velocity*, is proportional to the applied electric field. The proportionality constant is called the mobility $\mu$. In a simple model, the average drift velocity is just the velocity gained by the electron during the time when the electron is moving without scattering. When the electron is moving without scattering, it simply gains velocity according to Newton's laws. This average velocity gain is then that of an electron with mass $m^*$ travelling in a field $\boldsymbol{E}$ for a time $\tau_{sc}$:

$$\mathbf{v}_{avg} = -\frac{e\boldsymbol{E}\tau_{sc}}{m^*} = \mathbf{v}_d \tag{5.20}$$

where $\mathbf{v}_d$ is the drift velocity. The current density is now

$$\mathbf{J} = -ne\mathbf{v}_d = \frac{ne^2\tau_{sc}}{m^*}\mathbf{E} \tag{5.21}$$

Comparing this with the Ohm's law result for conductivity $\sigma$

$$\mathbf{J} = \sigma\mathbf{E} \tag{5.22}$$

we have

$$\sigma = \frac{ne^2\tau_{sc}}{m^*} \tag{5.23}$$

The resistivity of the semiconductor is simply the inverse of the conductivity. From the definition of mobility $\mu$, for electrons

$$\mathbf{v}_d = \mu \mathbf{E} \tag{5.24}$$

we have

$$\boxed{\mu = \frac{e\tau_{sc}}{m^*}} \tag{5.25}$$

Notice that the mobility has an explicit $\frac{1}{m^*}$ dependence in it. Additionally $\tau_{sc}$ decreases with $m^*$ also as we had discussed in the previous section due to the reduction in the density of states. Thus the mobility has a strong dependence on the carrier mass. In Appendix B we show mobilities of some pure semiconductors.

**EXAMPLE 5.1** The mobility of electrons in pure GaAs at 300 K is 8500 cm$^2$/V-s. Calculate the relaxation time. If the GaAs sample is doped at $N_d = 10^{17}$ cm$^{-3}$, the mobility decreases to 5000 cm$^2$/V-s. Calculate the relaxation time due to ionized impurity scattering.

The relaxation time is related to the mobility by

$$\tau_{sc}^{(1)} = \frac{m^* \mu}{e} = \frac{(0.067 \times 0.91 \times 10^{-30}~kg)(8500 \times 10^{-4} m^2/V - s)}{1.6 \times 10^{-19}~C}$$
$$= 3.24 \times 10^{-13}~s$$

If the ionized impurities are present, the time is

$$\tau_{sc}^{(2)} = \frac{m^* \mu}{e} = 1.9 \times 10^{-13}~s$$

The total scattering rate in the presence of ionized impurity scattering is

$$\frac{1}{\tau_{sc}^{(2)}} = \frac{1}{\tau_{sc}^{(1)}} + \frac{1}{\tau_{sc}^{(imp)}}$$

which gives for the ionized impurity scattering time

$$\tau_{sc}^{(imp)} = 4.6 \times 10^{-13}~s$$

**EXAMPLE 5.2** The mobility of electrons in pure silicon is 1500 cm$^2$/V-s. Calculate the time between scattering events using the conductivity effective mass.

The conductivity mass in a material like silicon is given by

$$m_\sigma^* = 3 \left( \frac{2}{m_t^*} + \frac{1}{m_\ell^*} \right)^{-1}$$

$$= 3 \left( \frac{2}{0.19 \ m_0} + \frac{1}{0.98 \ m_0} \right)^{-1} = 0.26 \ m_0$$

The scattering time is then

$$\tau_{sc} = \frac{\mu m_\sigma^*}{e} = \frac{(0.26 \times 0.91 \times 10^{-30})(1500 \times 10^{-4})}{1.6 \times 10^{-19}}$$

$$= 2.2 \times 10^{-13} \ s$$

## High Field Transport

In most electronic devices a significant portion of the electronic transport occurs under strong electric fields. This is especially true of field effect transistors. At such high fields ($F \sim 1$ - $100 \ \mathrm{kV/cm^{-1}}$) the electrons get "hot" and their temperature (electron temperature defined through their average energy) can be much higher than the lattice temperature. The extra energy comes due to the strong electric fields. The drift velocities are also quite high. The description of electrons at such high electric fields is quite complex and requires numerical techniques. The velocity-field relations are no longer linear as can be seen in Fig. 5.8. An important point to note is that the velocity at high fields tends to saturate. Typically the velocity saturates at a value of $\sim 10^7$ cm/s in almost all semiconductors. The field where this occurs is different for different semiconductors.

## Transport in Amorphous Semiconductors

The problem of electron (hole) transport in amorphous semiconductors is considerably more complicated than in crystalline semiconductors. In Chapter 4, Section 4.11, we have discussed the electronic properties of amorphous materials. A key point noted there is that the electronic states do not have a plane-wave-like nature. There are localized states in which electrons are spatially trapped in a local disturbance, and extended states where the electrons are able to connect one region of the material to another. In the localized states, the electrons cannot move in the presence of an electric field, since it is trapped in space. *To be able to move, it needs to scatter from one point to another. Thus, scattering by, say, lattice vibrations helps the electron move* in contrast to crystalline materials where the scattering hinders transport. This transport mechanism applicable to localized states is called the *hopping mechanism* and is shown in Fig. 5.9. The carrier mobility improves with temperature, since as the temperature is raised, there is more scattering and, hence, hopping.

If the electron is in the extended states, it does not need to hop from one point to another. However, there is a great deal of scattering from the disorder in

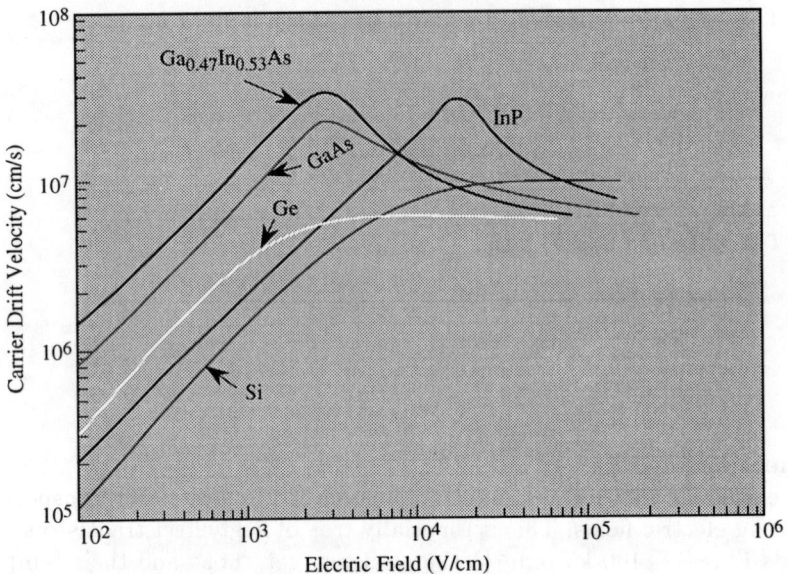

Figure 5.8: Velocity-field relation for electrons in Si, Ge, InP and GaAs.

the amorphous material. The transport properties are, however, much better for the extended electrons.

If the position of the Fermi level is as shown in Fig. 5.9, some of the electrons are in the localized states and some in the extended states. Also, the greater the number of electrons, the more of them will be in the extended states. Thus, mobility is a function of carrier density. In most a-Si structures, electron mobility is $\sim 1$ to $10 \ \text{cm}^2/\text{V-s}$. This number is to be compared with a mobility of $\sim 1,000 \ \text{cm}^2/\text{V-s}$ for crystalline silicon. Improving the mobility in a-Si is a big challenge for scientists.

## 5.3.2   Very High Field Transport: Breakdown Phenomenon

$\longrightarrow$ When the electric field becomes extremely high ($\gtrsim 100 \ \text{kV/cm}^{-1}$), the semiconductor suffers a "breakdown" in which the current has a "runaway" behavior. The breakdown occurs due to carrier multiplication which arises from the two sources discussed below. By carrier multiplication we mean that the number of electrons in the conduction band and holes in the valence band that can participate in current flow increases. Of course, the total number of electrons is always conserved.

**Impact Ionization or Avalanche Breakdown**
In the transport considered in the previous sections, the electron (hole) remains in

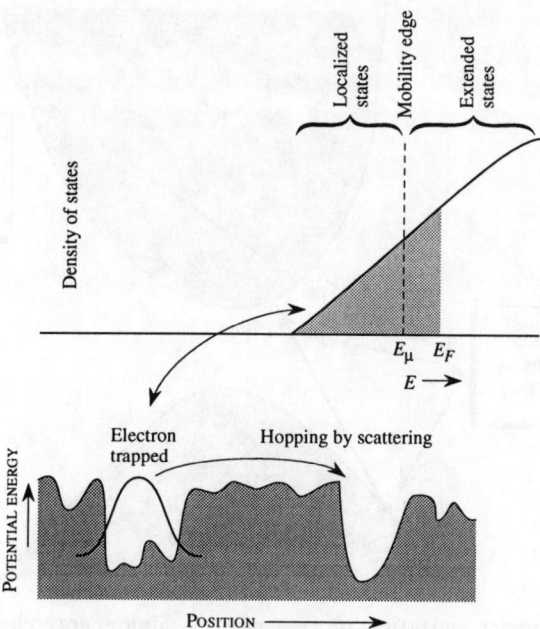

Figure 5.9: A schematic of localized and extended states in a-Si. In the localized states, electrons move by "hopping" from one region to another. Scattering is required to make such a hop.

the same band during the transport. At very high electric fields, this does not hold true. In the impact ionization process shown schematically in Fig. 5.10, an electron which is "very hot" scatters with an electron in the valence band via the Coulombic interaction and knocks it out into the conduction band. The initial electron must provide enough energy to bring the valence band electron up into the conduction band. Thus the *initial electron should have energy slightly larger than the bandgap* (measured from the conduction band minimum). In the final state we now have two electrons in the conduction band and one hole in the valence band. Thus the number of current- carrying charges have multiplied, and the process is often called avalanching. Note that the same could happen to "hot holes" that could then trigger the avalanche.

In order for the impact ionization process to begin, the initiating electron with momentum $\hbar k_1$ must have enough energy to knock an electron from the valence band to the conduction band. Additionally, momentum must be conserved in the scattering. Thus the initial electron must overcome a threshold energy before the carrier multiplication can begin.

For the purpose of device applications, the current in the device, once avalanching

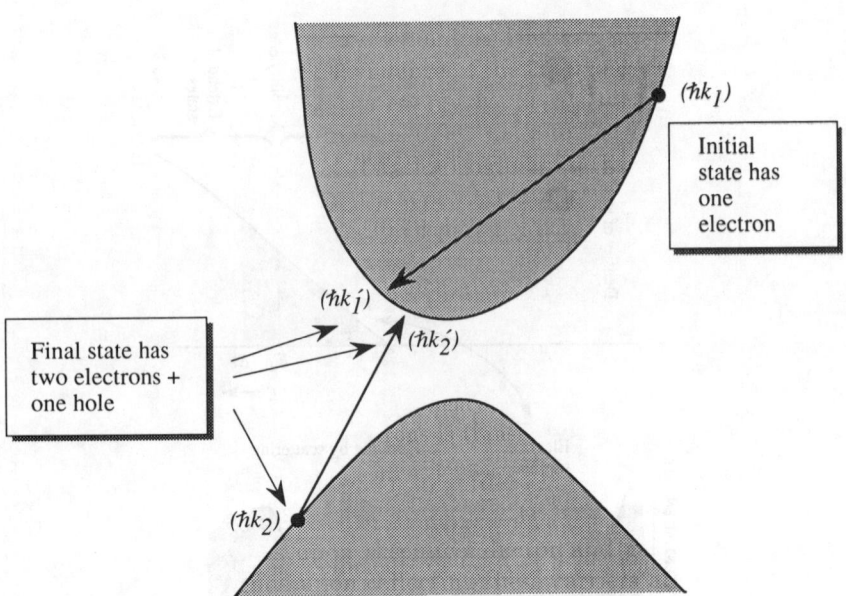

Figure 5.10: The impact ionization process where a high energy electron scatters from a valence band electron producing two conduction band electrons and a hole. Hot holes can undergo a similar process. Both energy and momentum conservation are required in such a process. Thus the initiating electron must have an energy above a minimum threshold.

starts, is given by

$$\frac{dI(t)}{dt} = \alpha_t I \quad \text{or} \quad \frac{dI(z)}{dz} = \alpha_z I \tag{5.26}$$

where $I$ is the current and $\alpha_t$ or $\alpha_z$ represent the average rate of ionization per unit time or distance respectively.

The coefficients $\alpha_z$ (or simply $\alpha_{imp}$) for electrons and $\beta_z$ (or $\beta_{imp}$) for holes strongly depend upon the bandgap of the material. This is because, as discussed above, the process can only start if the initial electron has a kinetic energy equal to a certain threshold (roughly equal to the bandgap). This is achieved for lower electric fields in narrow gap materials. In Fig. 5.11 we show the impact ionization coefficients for some materials.

If the electric field is constant so that $\alpha_{imp}$ is constant, the number of times an initial electron will suffer impact ionization after travelling a distance $x$ is

$$N(x) = \frac{I(x)}{I(0)} = exp\left(\alpha_{imp}x\right) \tag{5.27}$$

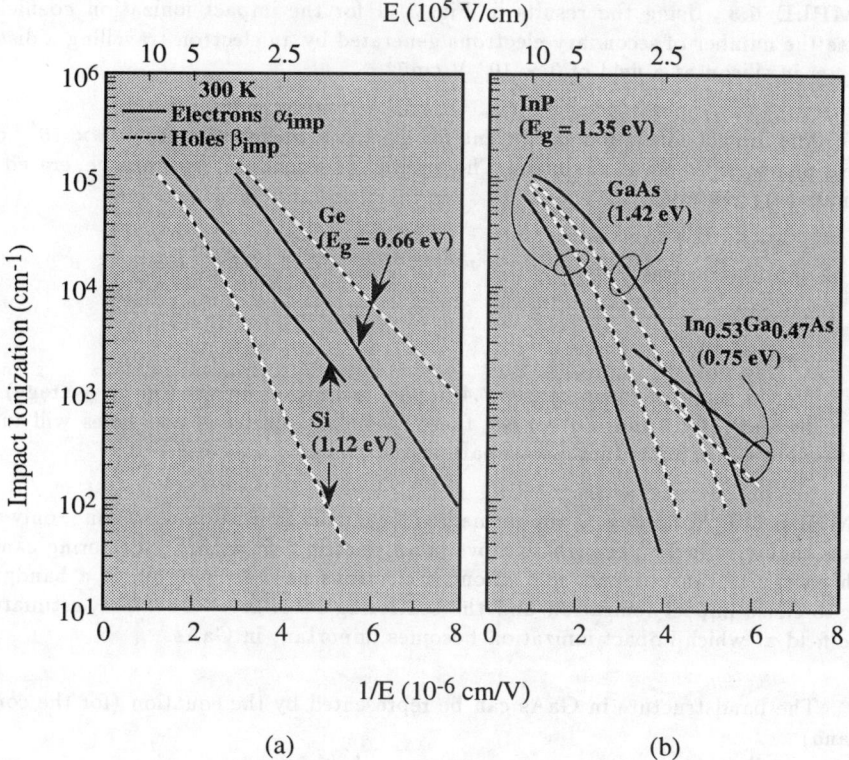

Figure 5.11: Ionization rates for electrons and holes at 300 K versus reciprocal electric field for Ge, Si, GaAs, $In_{0.53}Ga_{0.47}$ and InP. (Si, Ge results are after S.M. Sze, *Physics of Semiconductor Devices*, John Wiley and Sons (1981); InP, GaAs, InGaAs results are after G. Stillman, *Properties of Lattice Matched and Strained Indium Gallium Arsenide*, ed. P. Bhattacharya, INSPEC, London (1993).)

Of course, the "daughter" electrons and holes produced will also ionize further.

A critical breakdown field $F_{crit}$ is defined where $\alpha_{imp}$ or $\beta_{imp}$ approaches $10^4$ cm$^{-1}$. When $\alpha_{imp}$ ($\beta_{imp}$) approaches $10^4$ cm$^{-1}$, there is about one impact ionization when a carrier travels a distance of one micrometer. The avalanche process places an important limitation on the power output of devices. Once the process starts, the current rapidly increases due to carrier multiplication and the control over the device is lost. The push for high power devices is one of the reasons for research in large gap semiconductor devices. It must be noted that in certain devices such as avalanche photodetectors, the process is exploited for high gain detection. The process is also exploited in special microwave devices.

**EXAMPLE 5.3** Using the results in Fig. 5.11 for the impact ionization coefficients, calculate the number of secondary electrons generated by an electron travelling a distance of 1.0 $\mu$m in silicon at a field of $3 \times 10^5$ V cm$^{-1}$.

The impact ionization coefficient for electrons is approximately $2 \times 10^4$ cm$^{-1}$ at a field of $3 \times 10^5$V cm$^{-1}$ in silicon. The number of secondary electrons generated by a single initiating electron is

$$\frac{I(x)}{I(0)} = exp\ \alpha_{imp}W = exp\ (2 \times 10^4 \times 10^{-4})$$

$$= 7.4$$

Thus a single electron causes 7.4 impact ionization events (on an average) as it moves a distance of 1.0 $\mu$m. Of course, these secondary electrons and holes will further generate more $e$-$h$ pairs as they participate in the process.

**EXAMPLE 5.4** According to one formalism for impact ionization scattering, only those electrons that are "lucky" enough to move in an electric field without scattering can gain enough energy to cause impact ionization. If electrons need to gain about a bandgap of energy to cause impact ionization and the scattering rates are $\sim 10^{13}$ s$^{-1}$, estimate the electric field at which impact ionization becomes important in GaAs.

The bandstructure in GaAs can be represented by the equation (for the conduction band)

$$E(1 + \alpha E(k)) = \frac{\hbar^2 k^2}{2m^*}$$

where $m^* = 0.067\ m_0$ and $\alpha\ (= 0.6\ eV^{-1})$ is called the non-parabolicity factor. If $\alpha = 0$, we have the usual parabolic $E$-$k$ relation. The value of $k$ at the point where the electron energy is 1.5 eV (= bandgap of GaAs) is given by

$$k^2 = \frac{2m^* E(1 + \alpha E)}{\hbar^2}$$

$$= \frac{2 \times (0.067 \times 0.91 \times 10^{-30}\ kg)(1.5 \times 1.6 \times 10^{-19}\ J)(1 + 0.6 \times 1.5)}{(1.05 \times 10^{-34}\ J - s)^2}$$

$$= 5.04 \times 10^{18}\ m^{-2}$$

$$k = 2.25 \times 10^9\ m^{-1}$$

If an electron is moving without scattering starting at $k = 0$, after time $\tau_{sc}$, its $k$-value is

$$k = \frac{eE\tau}{\hbar}$$

where $E$ is the applied field. Using $k = 2.25 \times 10^9 m^{-1}$ and $\tau = 10^{-13}$ s, the field is

$$E = \frac{\hbar k}{e\tau} = \frac{(1.05 \times 10^{-34}\ J - s)(2.25 \times 10^9\ m^{-1})}{(1.6 \times 10^{-19}\ C) \times (10^{-13}\ s)}$$

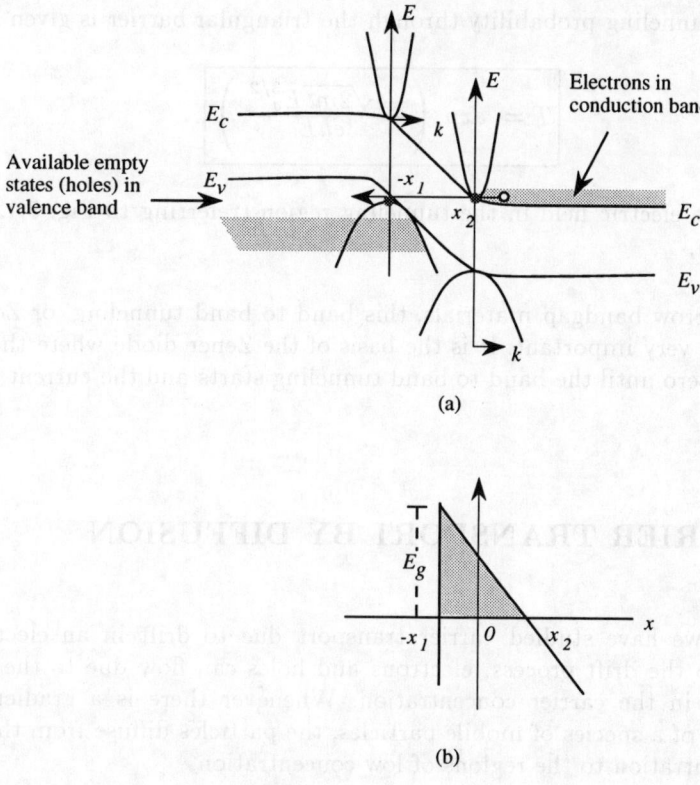

(a)

(b)

Figure 5.12: (a) A schematic showing the $E$-$x$ and $E$-$k$ diagram for a $p$-$n$ junction. An electron in the conduction band can tunnel into an unoccupied state in the valence band or vice-versa. (b) The potential profile seen by the electron during the tunneling process.

$$= 1.47 \times 10^7 \ V/m$$
$$= 1.47 \times 10^5 \ V/cm$$

**Band to Band Tunneling: Zener Tunneling**

In addition to impact ionization, another breakdown mechanism is the band to band or Zener tunneling. Consider a $p$-$n$ junction which is strongly reverse biased (the junction theory is discussed in Appendix D) as shown in Fig. 5.12. At strong electric fields, the electrons in the valence band can tunnel into an unoccupied state in the conduction band. As the electron tunnels, it sees a potential profile shown in Fig. 5.12b.

The tunneling probability through the triangular barrier is given by

$$T = exp\left(\frac{-4\sqrt{2m^*}E_g^{3/2}}{3e\hbar E}\right) \tag{5.28}$$

where $E$ is the electric field in the tunneling region (referring to Fig. 5.12b, $eE = E_g/(x_1 + x_2)$),

In narrow bandgap materials, this band to band tunneling, or Zener tunneling, can be very important. It is the basis of the Zener diode where the current is essentially zero until the band to band tunneling starts and the current increases very sharply.

## 5.4   CARRIER TRANSPORT BY DIFFUSION

$\longrightarrow$   So far we have studied carrier transport due to drift in an electric field. In addition to the drift process, electrons and holes can flow due to the presence of a gradient in the carrier concentration. Whenever there is a gradient in the concentration of a species of mobile particles, the particles diffuse from the regions of high concentration to the regions of low concentration.

In the case of electrons (or holes), as the particles move they suffer random collisions as we had discussed earlier in the previous section. The collision process can be described by the mean free path $\ell$ and the mean collision time $\tau_{sc}$. The mean free path is the average distance the electron (hole) travels in between successive collisions. These collisions are due to the various scattering processes that were discussed for the drift problem. In between the collisions, the electrons move randomly with equal probability of moving in any direction (remember there is no electric field). We are interested in finding out how the electrons move (diffuse) when there is a concentration gradient in space.

The net flow of carriers due to diffusion is given by

$$\phi_n(x,t) = -\frac{\ell^2}{2\tau_{sc}}\frac{dn(x,t)}{dx} = -D_n\frac{dn(x,t)}{dx} \tag{5.29}$$

where $D_n$ is called the diffusion coefficient of the electron system and depends upon the scattering processes which control $\ell$ and the $\tau_{sc}$. *Since the mean free path is essentially $v_{th}\tau_{sc}$, where $v_{th}$ is the mean thermal speed, the diffusion coefficient depends upon the temperature as well.*

In a similar manner, the hole diffusion coefficient gives the hole flux due to a hole density gradient

$$\phi_p(x,t) = -D_p \frac{dp(x,t)}{dx} \tag{5.30}$$

Because of this electron and hole flux, a current can flow in the structure which in the absence of an electric field is given by (current is just charge multiplied by particle flux)

$$
\begin{aligned}
\mathbf{J}_{tot}(diff) &= \mathbf{J}_n(diff) + \mathbf{J}_p(diff) \\
&= eD_n \frac{dn(x,t)}{dx} - eD_p \frac{dp(x,t)}{dx}
\end{aligned}
\tag{5.31}
$$

While both electrons and holes move in the direction of less concentration of electrons and holes respectively, the currents they carry are opposite due to their charge difference.

## 5.4.1   Transport by Drift and Diffusion

$\longrightarrow$  In many electronic devices the charge moves under the combined influence of electric fields and concentration gradients. The current is then given by

$$
\begin{aligned}
\mathbf{J}_n(x) &= -e\mu_n n(x)\mathbf{E}(x) + eD_n \frac{dn(x)}{dx} \\
\mathbf{J}_p(x) &= e\mu_p p(x)\mathbf{E}(x) - eD_p \frac{dp(x)}{dx}
\end{aligned}
\tag{5.32}
$$

While using such a relation, it is important to keep in mind that the electron and hole mobilities are not constant values. They are constant only at low electric fields. At high fields the mobilities decrease and the product $\mu E$ reaches a constant corresponding to the saturation velocity. Thus the drift current saturates and becomes independent of the field.

In our discussion of the diffusion coefficient we indicated that it is controlled by essentially the same scattering mechanisms that control the mobility. There is an important relationship between them given by

$$\frac{D_n}{\mu_n} = \frac{k_B T}{e} \tag{5.33}$$

which is the Einstein relation satisfied for electrons. A similar relation exists for the holes.

**EXAMPLE 5.5**  Use the velocity-field relations for electrons in silicon to obtain the diffusion coefficient at an electric field of 1 kV/cm and 10 kV/cm at 300 K.

According to the $v$-$E$ relations given in Fig. 5.8, the velocity of electrons in silicon is $\sim 1.4 \times 10^6$ cm/s and $\sim 7 \times 10^6$ cm/s at 1 kV/cm and 10 kV/cm. Using the Einstein relation we have for the diffusion coefficient

$$D = \frac{\mu k_B T}{e} = \frac{v k_B T}{eE}$$

This gives

$$
\begin{aligned}
D(1 kV/cm^{-1}) &= \frac{(1.4 \times 10^4 \ m/s)(0.026 \times 1.6 \times 10^{-19} \ J)}{(1.6 \times 10^{-19} \ C)(10^5 \ V m^{-1})} \\
&= 3.64 \times 10^{-3} \ m^2/s = 36.4 \ cm^2/s \\
D(10 \ kV/cm^{-1}) &= \frac{(7 \times 10^4 \ m/s)(0.026 \times 1.6 \times 10^{-19} \ J)}{(1.6 \times 10^{-19} \ C)(10^6 \ V m^{-1})} \\
&= 1.82 \times 10^{-3} \ m^2/s = 18.2 \ cm^2/s
\end{aligned}
$$

The diffusion coefficient decreases with the field because of the higher scattering rate at higher fields.

## 5.5  OPTICAL PROPERTIES OF SEMICONDUCTORS

$\longrightarrow$  In the previous sections we have seen how the transport of electrons and holes is affected by scattering from various imperfections in the crystal. We will now discuss how electrons respond to electromagnetic fields or photons. The interaction of electrons and photons is the basis of all semiconductor optoelectronic devices. We have discussed in Section 5.2 the physical basis of the scattering of electrons from photons. We characterized two classes of scattering: i) absorption of photons where the electron gains energy by absorbing a photon; and ii) emission where the electron emits a photon and loses energy. The emission process itself is characterized as spontaneous emission and stimulated emission. Spontaneous emission occurs even if there are no photons present while stimulated emission occurs because of the presence of photons. In this section we will give quantitative expressions for the various processes characterizing electron-photon interactions. These expressions will not be derived but will be simply introduced to the student. The operation of the various optoelectronic devices is governed by the expressions presented in this and the next few sections.

Light is represented by electromagnetic waves as discussed in Chapter 2. Electromagnetic waves travelling through a medium like a semiconductor are described by Maxwell's equations which show that the waves have a form given by the electric field vector dependence

$$\mathbf{E} = \mathbf{E}_o \, exp \left\{ i\omega \left( \frac{n_r}{c} - t \right) \right\} \, exp \left( -\frac{\alpha z}{2} \right) \tag{5.34}$$

Here $z$ is the propagation direction, $\omega$ the frequency, $n_r$ the refractive index (the real part), and $\alpha$ the absorption coefficient of the medium. If $\alpha$ is zero, the wave propagates without attenuation with a velocity $\frac{c}{n_r}$. However, for nonzero $\alpha$, the photon flux $I \, (\sim E^* E)$ falls as

$$I(z) = I(0) \, exp \left\{ -\alpha z \right\} \tag{5.35}$$

The absorption of light can occur for a variety of reasons including absorption by impurities in the material, intraband absorption where electrons in the conduction band absorb the radiation. *However, the most important optoelectronic interaction in semiconductors as far as devices are concerned is the band to band transition* shown in Fig. 5.13. In the photon absorption process, a photon scatters an electron in the valence band causing the electron to go into the conduction band. In the reverse process the electron in the conduction band recombines with a hole in the valence band to generate a photon. These two processes are of obvious importance for light detection and light emission devices. As has been noted in the beginning of this chapter, the detailed scattering theory is beyond the scope of this text. However, the important expressions which control the performance of detectors and lasers will be examined in this section. The rates for the light emission and absorption processes are determined by quantum mechanics. The scattering involves the following issues:

i) **Conservation of Energy**: In the absorption and emission process we have for the initial and final energies of the electrons $E_i$ and $E_f$

$$\text{absorption}: \quad E_f \; = \; E_i + \hbar\omega \tag{5.36}$$

$$\text{emission}: \quad E_f \; = \; E_i - \hbar\omega \tag{5.37}$$

where $\hbar\omega$ is the photon energy. Since the minimum energy difference between the conduction and valence band states is the bandgap $E_g$, *the photon energy must be larger than the bandgap for absorption to occur.*

ii) **Conservation of Momentum**: In addition to energy conservation, one also needs to conserve the momentum $\hbar k$ for the electrons and the photon system. The

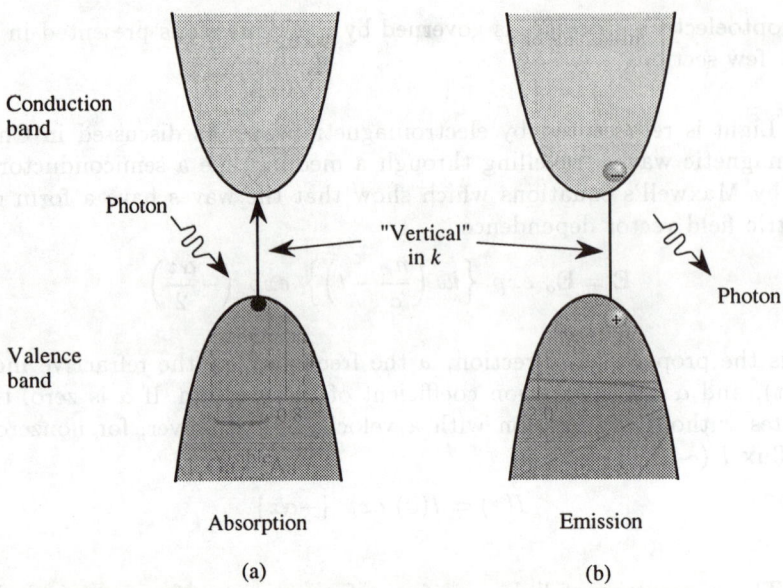

Figure 5.13: Band to band absorption in semiconductors. (a) An electron in the valence band "absorbs" a photon and moves into the conduction band. (b) In the reverse process, the electron recombines with a hole to emit a photon. Momentum conservation ensures that only vertical transitions are allowed.

photon $k_{ph}$ value is given by

$$k_{ph} = \frac{2\pi}{\lambda} \qquad (5.38)$$

Since 1 eV photons correspond to a wavelength 1.24 $\mu$m, the $k$-values of relevance are $\sim 10^{-4}$ Å$^{-1}$ which is essentially zero compared to the $k$-values for electrons. Thus $k$-conservation ensures that the initial and final electrons should have the same $k$-value. Another way to say this is that *only "vertical" k transitions are allowed* in the bandstructure picture as shown in Fig. 5.13.

Because of this constraint of $k$-conservation, in semiconductors where the valence band and conduction bandedges are at the same $k = 0$ value (the direct semiconductors), the optical transitions are quite strong. In indirect materials like Si, Ge, etc., the optical transitions are very weak and require the help of lattice vibrations to satisfy $k$-conservation. This makes a tremendous difference in the optical properties of these two kinds of materials. The transitions are very weak near the bandedges of indirect semiconductors.

Because the $k$-values for the electron and hole are the same in vertical transitions, we have (for photon energies larger than the bandgap)

$$\hbar\omega = E_g + \frac{\hbar^2 k^2}{2}\left(\frac{1}{m_e^*} + \frac{1}{m_h^*}\right)$$

$$= E_g + \frac{\hbar^2 k^2}{2m_r^*} \tag{5.39}$$

where $m_r^*$ is the reduced mass of the electron-hole system. Thus the *relevant density of states function in the scattering process is that where the mass used is the reduced mass*. This density of states is called the *joint density of states*.

Keeping all these issues in mind, one can calculate the absorption coefficient for a semiconductor. For light polarized along direction **a**, the absorption coefficient turns out to be

$$\alpha = \frac{\pi e^2 \hbar}{m_0^2 c n_r \epsilon_o} \frac{1}{\hbar\omega} \mid (\mathbf{a} \cdot \mathbf{p})_{if} \mid^2 N_{cv}(\hbar\omega) \tag{5.40}$$

with $N_{cv}$, the joint density of states given by

$$N_{cv}(\hbar\omega) = \frac{\sqrt{2}(m_r^*)^{3/2}(\hbar\omega - E_g)^{1/2}}{\pi^2 \hbar^3} \tag{5.41}$$

The quantity $\mid (\mathbf{a} \cdot \mathbf{p})_{if} \mid^2$ is called the momentum or dipole matrix element between the conduction and the valence band. The polarization averaged matrix element turns out to be $(2/3)p_{cv}^2$ where it is found that for most semiconductors

$$\frac{2p_{cv}^2}{m_0} \cong 20 \ eV \tag{5.42}$$

If we plug in some numbers for the absorption coefficient of unpolarized light we get for GaAs ($E_g \cong 1.5$ eV) (see Example 5.9)

$$\alpha(\hbar\omega) \cong 5.6 \times 10^4 \frac{(\hbar\omega - E_g)^{1/2}}{\hbar\omega} \ cm^{-1} \tag{5.43}$$

where $\hbar\omega$ and $E_g$ are in units of electron volts. The prefactor for any other direct semiconductor can be obtained by scaling the reduced mass ratio by a power of 3/2 according to the reduced density of states.

In Fig. 5.14 we show the absorption coefficient for several direct and indirect bandgap semiconductors. In the case of indirect bandgap materials, the absorption coefficient is considerably smaller at the bandedges since the process is not allowed unless lattice vibrations are involved. The coefficient for indirect semiconductors increases with temperature since at high temperatures there are more vibrations.

If we have an electron in a conduction band state and a hole in the valence band state with the same $k$-value, the two can recombine and emit a photon. There are two important classes of emission processes. *In spontaneous emission, as discussed in Section 5.2.1, an electron recombines with a hole, even though no photons are present in the region, and emits a photon.* The rate for this emission process is given by

$$W_{em}(\hbar\omega) = \frac{e^2 n_r \hbar\omega}{3\pi\epsilon_0 m_o^2 c^3 \hbar^2} \mid p_{cv} \mid^2 \qquad (5.44)$$

*If photons with frequency $\omega$ are already present in the cavity with the semiconductor, the recombination rate is enhanced by the presence of these photons. If $n_{ph}(\hbar\omega)$ is the photon occupation (i.e., the number of photons in a particular mode), the emission (called stimulated emission) rate is given by*

$$W_{em}^{st}(\hbar\omega) = \frac{e^2 n_r \hbar\omega}{3\pi\epsilon_o m_o^2 c^3 \hbar^2} \mid p_{cv} \mid^2 \cdot n_{ph}(\hbar\omega) \qquad (5.45)$$

Thus the rate is increased in proportion to the photon density already present in the cavity.

*In spontaneous emission, the photons that are emitted have no particular phase relationship with each other and are thus incoherent. Light emission in light emitting diodes (LEDs) is due to spontaneous emission. In stimulated emission, however, the photons that are emitted are in phase with the original photons that are present. The radiation is thus coherent. Laser diodes depend upon stimulated emission.*

The radiative recombination rate of an electron having a momentum $\hbar k$ with a hole (in absence of photons) having the same momentum is

$$\tau_o = \frac{1}{W_{em}} \qquad (5.46)$$

*In the definition of $\tau_o$ it is assumed that the electron can find a hole with which to recombine. If the probability of finding the hole is small, the radiative time can be much longer, as discussed in Section 5.7. For materials like GaAs, the value of $\tau_o$ is about 1 ns (see Example 5.10). However, for indirect materials the recombination time can be as large as 1 $\mu$s. The recombination process is not only* important for optical emission devices, but the rate also plays a key role in the speed of many electronic devices, e.g., bipolar devices, diodes, etc.

*The electron-hole recombination during stimulated emission can be quite a bit smaller than $\tau_o$ depending upon photon intensity present.*

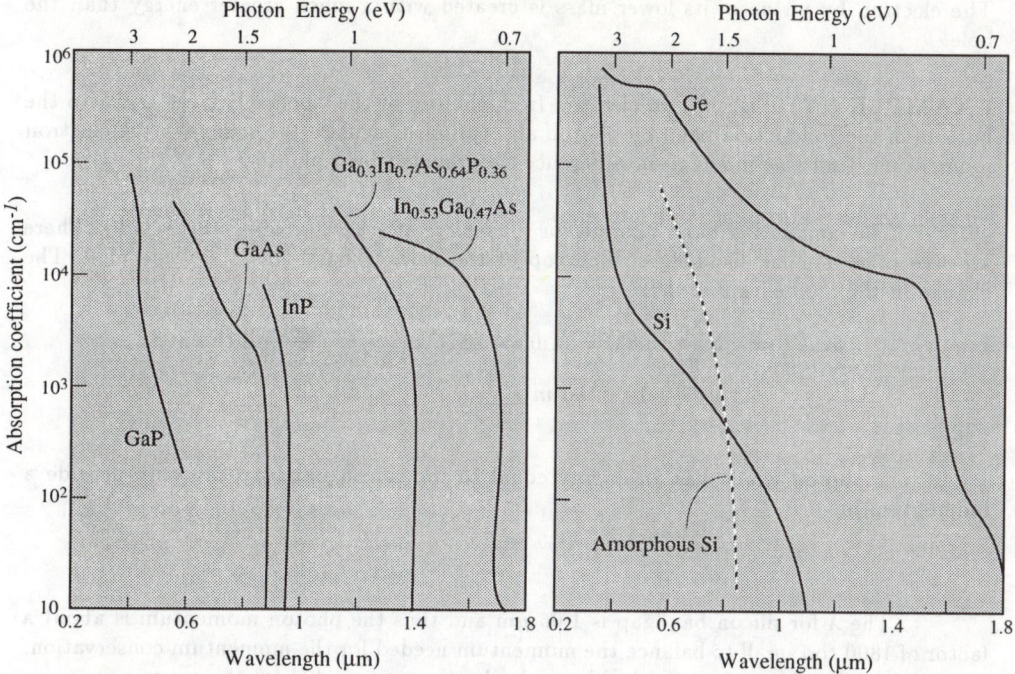

Figure 5.14: Absorption coefficient of several direct and indirect semiconductors. For the direct gap material, the absorption coefficient is very strong once the photon energy exceeds the bandgap. For indirect materials the absorption coefficient rises much more gradually. Once the photon energy is more than the direct gap, the absorption coefficient increases rapidly.

**EXAMPLE 5.6** A 1.6 eV photon is absorbed by a valence band electron in GaAs. If the bandgap of GaAs is 1.41 eV, calculate the energy of the electron and heavy hole produced by the photon absorption.

The electron, heavy-hole, and reduced mass of GaAs are 0.067 $m_0$, 0.45 $m_0$, and 0.058 $m_0$, respectively. The electron and the hole generated by photon absorption have the same momentum. The energy of the electron is

$$E^e = E_c + \frac{m_r^*}{m_e^*}(\hbar\omega - E_g)$$

$$E^e - E_c = \frac{0.058}{0.067}(1.6 - 1.41) = 0.164 \; eV$$

The hole energy is

$$E^h - E_v = -\frac{m_r^*}{m_h^*}(\hbar\omega - E_g) = -\frac{0.058}{0.45}(1.6 - 1.41)$$

$$= -0.025 \ eV$$

The electron by virtue of its lower mass is created with a much greater energy than the hole.

**EXAMPLE 5.7** In silicon, an electron from the top of the valence band is taken to the bottom of the conduction band by photon absorption. Calculate the change in the electron momentum. Can this momentum difference be provided by a photon?

The conduction band minima for silicon are at a $k$-value of $\frac{2\pi}{a}$ (0.85, 0, 0). There are five other similar bandedges. The top of the valence band has a $k$-value of 0. The change in the momentum is thus

$$\hbar\Delta k = \hbar\frac{2\pi}{a}(0.85) = (1.05 \times 10^{-34})\left(\frac{2\pi}{5.43 \times 10^{-10}}\right)(0.85)$$

$$= 1.03 \times 10^{-24} kg \ m \ s^{-1}$$

A photon which has an energy equal to the silicon bandgap can only provide a momentum of

$$\hbar k_{ph} = \hbar \cdot \frac{2\pi}{\lambda}$$

The $\lambda$ for silicon bandgap is 1.06 $\mu$m and thus the photon momentum is about a factor of 1800 too small to balance the momentum needed for the momentum conservation. The lattice vibrations produced by thermal vibration are needed for the process.

**EXAMPLE 5.8** The absorption coefficient near the bandedges of GaAs and Si are $\sim 10^4$ cm$^{-1}$ and $10^3$ cm$^{-1}$ respectively. What is the minimum thickness of a sample in each case which can absorb 90% of the incident light?

The light absorbed in a sample of length $L$ is

$$\frac{I_{abs}}{I_{inc}} = 1 - exp(-\alpha L)$$

$$\text{or} \qquad L = = \frac{1}{\alpha} \ ln\left(1 - \frac{I_{abs}}{I_{inc}}\right)$$

Using $\frac{I_{abs}}{I_{inc}}$ equal to 0.9, we get

$$L(GaAs) = -\frac{1}{10^4} \ ln(0.1) = 2.3 \times 10^{-4} \ cm$$

$$= 2.3 \ \mu m$$

$$L(Si) = -\frac{1}{10^3} \ ln(0.1) = 23 \ \mu m$$

Thus an Si detector requires a very thick active absorption layer to function.

**EXAMPLE 5.9** Calculate the absorption coefficient of GaAs as a function of photon frequency.

The joint density of states for GaAs is (using a reduced mass of $0.065m_0$)

$$
\begin{aligned}
N_{cv}(E) &= \frac{\sqrt{2}(m_r^*)^{3/2}(E - E_g)^{1/2}}{\pi^2 \hbar^3} \\
&= \frac{1.414 \times (0.065 \times 0.91 \times 10^{-30} \ kg)^{3/2}(E - E_g)^{1/2}}{9.87 \times (1.05 \times 10^{-34})^3} \\
&= 1.78 \times 10^{54}(E - \hbar\omega)^{1/2} \ J^{-1} m^{-3}
\end{aligned}
$$

The absorption coefficient is for unpolarized light

$$
\alpha(\hbar\omega) = \frac{\pi e^2 \hbar}{2 n_r c \epsilon_o m_0} \left( \frac{2 p_{cv}^2}{m_0} \right) \frac{N_{cv}(\hbar\omega)}{\hbar\omega} \cdot \frac{2}{3}
$$

The term $\frac{2 p_{cv}^2}{m_0}$ is $\sim 23.0$ eV for GaAs. This gives

$$
\begin{aligned}
\alpha(\hbar\omega) &= \frac{3.1416 \times (1.6 \times 10^{-19} \ C)^2 (1.05 \times 10^{-34} \ J - s)}{2 \times 3.4 \times (3 \times 10^8 \ m/s)(8.84 \times 10^{-12}(F/m)^2)} \\
&\quad \cdot \frac{(23.0 \times 1.6 \times 10^{-19} \ J)}{(0.91 \times 10^{-30} kg)} \frac{(\hbar\omega - E_g)^{1/2}}{\hbar\omega} \times 1.78 \times 10^{54} \ \times \ \frac{2}{3} \\
\alpha(\hbar\omega) &= 2.25 \times 10^{-3} \frac{(\hbar\omega - E_g)^{1/2}}{\hbar\omega} \ m^{-1}
\end{aligned}
$$

Here the energy and $\hbar\omega$ are in units of Joules. It is usual to express the energy in eV, and the absorption coefficient in $cm^{-1}$. This is obtained by multiplying the result by

$$
\left[ \frac{1}{(1.6 \times 10^{-19})^{1/2}} \times \frac{1}{100} \right]
$$

$$
\alpha(\hbar\omega) = 5.6 \times 10^4 \frac{(\hbar\omega - E_g)^{1/2}}{\hbar\omega} \ cm^{-1}
$$

For GaAs the bandgap is 1.5 eV at low temperatures and 1.43 eV at room temperatures. From the value of $\alpha$, we can see that a few microns of GaAs are adequate to absorb a significant fraction of light above the bandgap.

**EXAMPLE 5.10** Calculate the electron-hole recombination time in GaAs.

The recombination rate is given by

$$
W_{em} = \frac{e^2 n_r}{6 \pi \epsilon_o m_0 c^3 \hbar^2} \left( \frac{2 p_{cv}^2}{m_0} \right) \hbar\omega
$$

with $\frac{2p_{cv}^2}{m_0}$ being 23 eV for GaAs.

$$W_{em} = \frac{(1.6 \times 10^{-19}\ C)^2 \times 3.4 \times (23 \times 1.6 \times 10^{-19}\ J)\hbar\omega}{6 \times 3.1416 \times (8.84 \times 10^{-12}\ F/m) \times (0.91 \times 10^{-30}\ kg)}$$

$$\cdot \frac{1}{(3 \times 10^8\ m/s)^3 \times (1.05 \times 10^{-34}\ J - s)^2}$$

$$= 7.1 \times 10^{27}\hbar\omega\ s^{-1}$$

If we require the value of $\hbar\omega$ in eV instead of Joules we get

$$W_{em} = 7.1 \times 10^{27} \times (1.6 \times 10^{-19})\hbar\omega\ s^{-1}$$

$$= 1.14 \times 10^9\ \hbar\omega\ s^{-1}$$

For GaAs, $\hbar\omega \sim 1.5$ eV so that

$$W_{em} = 1.71 \times 10^9\ s^{-1}$$

The corresponding recombination time is

$$\tau_o = \frac{1}{W_{em}} = 0.58\ ns$$

*Remember that this is the recombination time when an electron can find a hole to recombine with.* This happens when there is a high concentration of electrons and holes, i.e., at high injection of electrons and holes or when a minority carrier is injected into a heavily doped majority carrier region.

## 5.6 CHARGE INJECTION AND QUASI-FERMI LEVELS

$\mathcal{R}$ In the discussion of absorption coefficient and recombination time, we have not mentioned whether *electrons and holes are actually present in the states which are involved in the optical processes.* For example, in the absorption of photons we have assumed that the *initial state in the valence band is occupied while the final state is empty.* In general, this may not be true. If electrons and holes are injected into a semiconductor, either by external contacts or by optical excitation, the system may not be in equilibrium, and the question then arises: What kind of distribution function describes the electron and hole occupation? We know that in equilibrium the electron and hole occupation is represented by the Fermi function. A new function is needed to describe the system when the electrons and holes are not in equilibrium.

## 5.6.1 Quasi-Fermi Levels

$\longrightarrow$ We know that in equilibrium the distribution of electrons and holes is given by the Fermi function which is defined once one knows the Fermi level. Also the product of electrons and holes, $np$ is a constant. If excess electrons and holes are injected into the semiconductor, clearly the same function will not describe the occupation of states. Under certain assumptions the electron and hole occupation can be described by the use of *quasi-Fermi levels*. These assumptions are:

i) The electrons are essentially in thermal equilibrium in the conduction band and the holes are in equilibrium in the valence band. This means that the carriers are neither gaining nor losing energy from the crystal lattice atoms.

ii) The electron-hole recombination time is much larger than the time for the electrons and holes to reach equilibrium within the conduction and valence band, respectively.

In most problems of interest, the time to reach equilibrium in the same band is approximately a few picoseconds while the *e-h* recombination time is anywhere from a nanosecond to a microsecond. Thus, the above assumptions are usually met. In this case, the quasi-equilibrium electron and holes can be represented by an electron Fermi function $f^e$ (with electron Fermi level) and a hole Fermi function $f^h$ (with a *different* hole Fermi level). We now have

$$n = \int_{E_c}^{\infty} N_e(E) f^e(E) dE \qquad (5.47)$$

$$p = \int_{-\infty}^{E_v} N_h(E) f^h(E) dE \qquad (5.48)$$

where

$$f^e(E) = \frac{1}{exp\left(\frac{E - E_{Fn}}{k_B T}\right) + 1} \qquad (5.49)$$

If $f^v(E)$ is the electron occupation in the valence band, the hole occupation is

$$f^h(E) = 1 - f^v(E) = 1 - \frac{1}{exp\left(\frac{E - E_{Fp}}{k_B T}\right) + 1}$$

$$= \frac{1}{exp\left(\frac{E_{Fp} - E}{k_B T}\right) + 1} \qquad (5.50)$$

At equilibrium $E_{Fn} = E_{Fp}$. If excess electrons and holes are injected into the semiconductor, the electron quasi-Fermi level $E_{Fn}$ moves towards the conduction band, while the hole quasi-Fermi level $E_{Fp}$ moves towards the valence band.

The ability to define quasi-Fermi levels $E_{Fn}$ and $E_{Fp}$ provides us a very powerful approach to solve non-equilibrium problems which are, of course, of greatest interest in devices.

By defining separate Fermi levels for the electrons and holes, one can study the properties of excess carriers using the same relationship between Fermi level and carrier density as we developed for the equilibrium problem (see Section 4.10). Thus, in the approximation where the Fermi distribution is replaced by an exponential, we have

$$
\begin{aligned}
n &= N_c \, exp\left[\frac{(E_{Fn} - E_c)}{k_B T}\right] \\
p &= N_v \, exp\left[\frac{(E_v - E_{Fp})}{k_B T}\right]
\end{aligned}
\tag{5.51}
$$

In the more accurate Joyce-Dixon approximation we have (compare these with Eqn. 4.60)

$$
\begin{aligned}
(E_{Fn} - E_c) &= k_B T\left[\ell n \, \frac{n}{N_c} + \frac{n}{\sqrt{8}N_c}\right] \\
(E_v - E_{Fp}) &= k_B T\left[\ell n \, \frac{p}{N_v} + \frac{p}{\sqrt{8}N_v}\right]
\end{aligned}
\tag{5.52}
$$

The dependence of the quasi-Fermi levels on the electron and hole densities in GaAs at 300 K are shown in Fig. 5.15. Note that for the same carrier injection ($n = p$), the electron quasi-Fermi level moves a greater amount than the hole quasi-Fermi level. This is because of the smaller electron effective mass.

**EXAMPLE 5.11** Using Boltzmann statistics calculate the position of the electron and hole quasi-Fermi levels when an $e$-$h$ density of $10^{17}$ cm$^{-3}$ is injected into pure (undoped) silicon at 300 K.

At room temperature for Si we have

$$
\begin{aligned}
N_c &= 2.8 \times 10^{19} \; cm^{-3} \\
N_v &= 1.04 \times 10^{19} \; cm^{-3}
\end{aligned}
$$

If $n_c = p_v = 10^{17}$ cm$^{-3}$, we obtain ($k_B T = 0.026$ eV)

$$
\begin{aligned}
E_{Fn} &= k_B T \, \ell n \left[\frac{n}{N_c}\right] + E_c \\
&= E_c - 0.146 \; eV \\
E_{Fp} &= E_v - k_B T \left[\ell n \, \frac{p}{N_v}\right] \\
&= E_v + 0.121 \; eV
\end{aligned}
$$

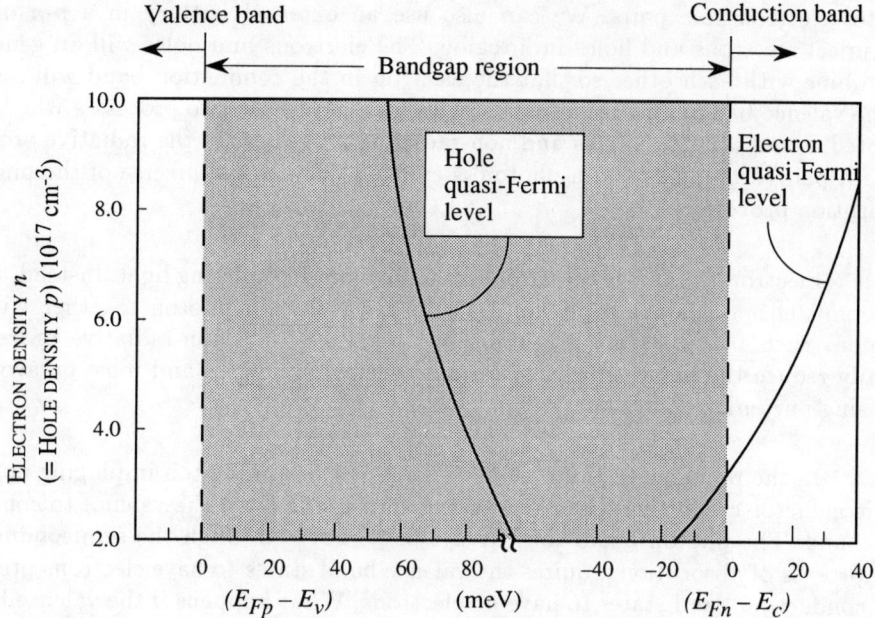

Figure 5.15: Dependence of the quasi-Fermi level positions in GaAs at 300 K on the electron and hole density.

Since in Si, the bandgap is $E_c - E_v = 1.1\text{eV}$, we have

$$E_{Fn} - E_{Fp} = (E_c - E_v) - (0.146 + 0.121)$$
$$= 1.1 - 0.267 = 0.833 \; eV$$

If we had injected only $10^{15}$ cm$^{-3}$ electrons and holes, the difference in the quasi-Fermi levels would be

$$E_{Fn} - E_{Fp} = (E_c - E_v) - (0.266 + 0.24)$$
$$= 1.1 - 0.506 = 0.59 \; eV$$

Thus as the carrier injection is increased, the separation increases.

## 5.7  CHARGE INJECTION AND RADIATIVE RECOMBINATION

$\longrightarrow$ Electrons and holes can be injected into the conduction and valence band in a number of ways. We can shine light on the material and the absorption of photons creates electron-hole pairs. We can also use an external battery in a *p-n* diode and inject electrons and holes in a region. The electrons and holes will, in general, recombine with each other so that the electron in the conduction band will return to the valence band. This recombination process can be via two processes which are denoted as radiative processes and non-radiative processes. In the radiative process the *e-h* pair recombines and a photon is emitted. This is the inverse of the photon absorption process.

Electron-hole pairs can also recombine without emitting light. Instead, they may emit either heat or a phonon or a long-wavelength photon together with a phonon. Such processes are non-radiative processes. The non-radiative processes usually reduce the device efficiency for optoelectronic devices and must be avoided by using pure materials.

In the previous sections we have discussed how a photon impinging upon a semiconductor can be absorbed by moving an electron from the valence to conduction band. The photon beam thus decays as it moves through the semiconductor. The process of absorption requires the valence band states to have electrons present and conduction band states to have no electrons. What happens if the valence band has holes and the conduction band has electrons? Normally, such a situation does not occur, but if electrons are injected into the conduction band and holes into the valence band (as happens for light emitting devices discussed in Chapters 7 and 8), this could occur. *Under such conditions the electron-hole pairs could recombine and emit more photons than are absorbed.* Thus one must talk about the emission coefficient minus the absorption coefficient, or the gain of the material. If the gain is positive, an optical beam will grow as it moves through the material instead of decaying. In the simple parabolic bands we have the gain $g(\hbar\omega)$ given by the generalization of Eqn. 5.40 (gain = emission coefficient − absorption coefficient)

$$g(\hbar\omega) = \frac{\pi e^2 \hbar}{n_r c m_0^2 \hbar\omega} \mid \mathbf{a} \cdot \mathbf{p}_{if} \mid^2 N_{cv} \hbar\omega [f^e(E^e) - (1 - f^h(E^h))] \qquad (5.53)$$

The term in the square brackets arises since the emission of photons is proportional to $f^e \cdot f^h$, while the absorption process is proportional to $(1 - f^e) \cdot (1 - f^h)$. The difference of these terms appears in Eqn. 5.53.

The energies $E^e$ and $E^h$ are (see Fig. 5.16) determined by noting the following (see Eqn. 5.39)

$$\hbar\omega - E_g = \frac{\hbar^2 k^2}{2m_r^*} \qquad (5.54)$$

$$E^e = E_c + \frac{\hbar^2 k^2}{2m_e^*} = E_c + \frac{m_r^*}{m_e^*}(\hbar\omega - E_g) \qquad (5.55)$$

$$E^h = E_v - \frac{h^2 k^2}{2m_h^*} = E_v - \frac{m_r^*}{m_h^*}(\hbar\omega - E_g) \qquad (5.56)$$

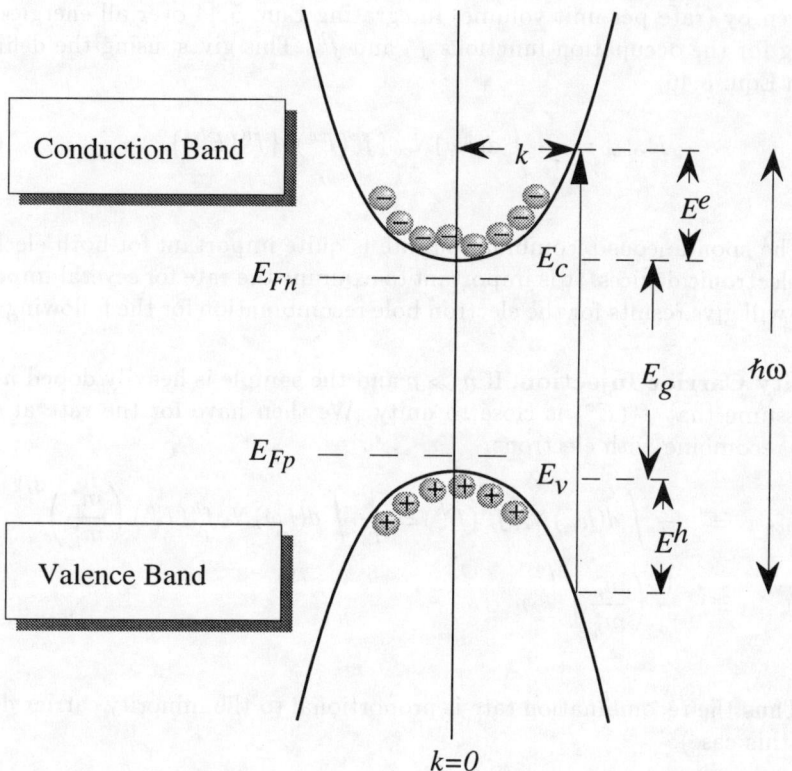

Figure 5.16: The positions of the quasi-Fermi levels, the electron and hole energies at vertical $k$-values. The electron and hole energies are determined by the photon energy and the carrier masses.

If $f^e(E^e) = 0$ and $f^h(E^h) = 0$, i.e., if there are no electrons in the conduction band and no holes in the valence band, we see that the gain is simply $-\alpha(\hbar\omega)$ which we had discussed earlier. A positive value of gain occurs when

$$f^e(E^e) > 1 - f^h(E^h) \qquad (5.57)$$

*a condition that is called inversion.* In this case the light wave passing in the material has the spatial dependence

$$I(z) = I_o \; exp \; (gz) \qquad (5.58)$$

*which grows with distance instead of diminishing* as it usually does if $g(\hbar\omega)$ is negative. The gain in the optical intensity is the basis for the semiconductor laser.

As the electrons and holes are "pumped" into the semiconductor they recombine through the process of spontaneous emission. This process does not require photons to be present for the photon emission process to occur. The spontaneous rate is given by (rate per unit volume) integrating Eqn. 5.44 over all energies after accounting for the occupation functions $f^e$ and $f^h$. This gives, using the definition of $\tau_o$ from Eqn. 5.46,

$$R_{spon} = \frac{1}{\tau_o} \int d(\hbar\omega) N_{cv} \{f^e(E^e)\}\{f^h(E^h)\} \tag{5.59}$$

The spontaneous recombination rate is quite important for both electronic and optoelectronic devices. It is important to examine the rate for several important cases. We will give results for the electron hole recombination for the following cases:

i) **Minority Carrier Injection:** If $n \gg p$ and the sample is heavily doped $n$-type, we can assume that $f^e(E^e)$ is close to unity. We then have for the rate at which holes will recombine with electrons,

$$
\begin{aligned}
R_{spon} &\cong \frac{1}{\tau_o} \int d(\hbar\omega) N_{cv} f^h(E^h) \cong \frac{1}{\tau_o} \int d(\hbar\omega) N_h f^h(E^h) \left(\frac{m_r^*}{m_h^*}\right)^{3/2} \\
&\cong \frac{1}{\tau_o} \left(\frac{m_r^*}{m_h^*}\right)^{3/2} p
\end{aligned}
\tag{5.60}
$$

Thus the recombination rate is proportional to the minority carrier density (holes in this case).

ii) **Strong Injection:** This case is important when a high density of both electrons and holes is injected. We can now assume that both $f^e$ and $f^h$ are sharp step functions and we get approximately

$$R_{spon} = \frac{n}{\tau_o} = \frac{p}{\tau_o} \tag{5.61}$$

iii) **Weak Injection:** In this case we can use the Boltzmann distribution to describe the Fermi functions. We have

$$f^e \cdot f^h \cong exp\left\{-\frac{(E_c - E_{Fn})}{k_B T}\right\} exp\left\{-\frac{(E_{Fp} - E_v)}{k_B T}\right\} \cdot exp\left\{-\frac{(\hbar\omega - E_g)}{k_B T}\right\} \tag{5.62}$$

The spontaneous emission rate now turns out to be

$$R_{spon} = \frac{1}{2\tau_o} \left(\frac{2\pi\hbar^2 m_r^*}{k_B T m_e^* m_h^*}\right)^{3/2} np \tag{5.63}$$

If we write the total charge as equilibrium charge plus excess charge,

$$n = n_o + \Delta n; p = p_o + \Delta n \tag{5.64}$$

we have for the excess carrier recombination (note that at equilibrium the rates of recombination and generation are equal)

$$R_{spon} \cong \frac{1}{2\tau_o} \left( \frac{2\pi\hbar^2 m_r^*}{k_B T m_e^* m_h^*} \right)^{3/2} (\Delta n p_o + \Delta p n_o) \tag{5.65}$$

If $\Delta n = \Delta p$, we can define the rate of a single excess carrier recombination as $\frac{\Delta n}{\tau_r}$, where

$$\frac{1}{\tau_r} = \frac{1}{2\tau_o} \left( \frac{2\pi\hbar^2 m_r^*}{k_B T m_e^* m_h^*} \right) (n_o + p_o) \tag{5.66}$$

At low injection, $\tau_r$ is much larger than $\tau_o$ and is a result of the fact that at low injection, electrons have a low probability to find a hole with which to recombine.

iv) **Inversion Condition:** Another useful approximation occurs when the electron and hole densities are such that $f^e + f^h = 1$. *This is the condition for inversion when the emission and absorption coefficients become equal.* If we assume in this case $f^e \sim f^h = 1/2$, we get the approximate relation

$$R_{spon} \cong \frac{n}{4\tau_o} \cong \frac{p}{4\tau_o} \tag{5.67}$$

The recombination lifetime is approximately $4\tau_o$ in this case. This is a useful result to estimate the threshold current of semiconductor lasers.

The gain and recombination processes discussed here are extremely important in both electronic and optoelectronic devices that will be discussed later. We point out from the above discussion that the recombination time for a single excess carrier can be written, in many situations, in the form

$$\boxed{\tau_r = \frac{\Delta n}{R_{spon}}} \tag{5.68}$$

For minority carrier injection or strong injection $\tau_r \cong \tau_o$. In general, $R_{spon}$ has a strong carrier density dependence as does $\tau_r$. In Fig. 5.17 we show the radiative recombination time for GaAs. In the figure, this time is shown as a function of hole density for the case $n = p$ and also for the case where electrons are injected into a $p$-type material.

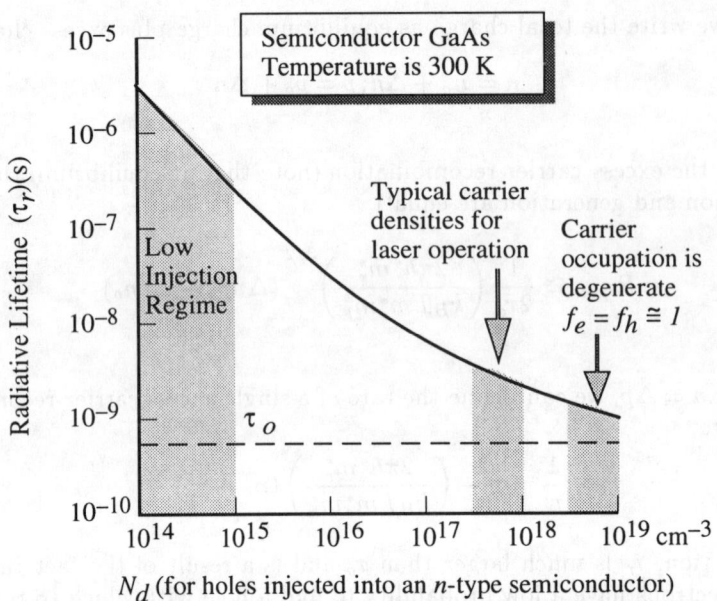

Figure 5.17: The radiative lifetime of electrons in GaAs at 300 K as a function of carrier density. The carrier density could be produced by injection ($n = p$) or by $p$-type doping in which case the lifetime is the minority carrier lifetime.

## 5.7.1  Phosphors and Fluorescence

So far in this section, we have discussed the radiative recombination process for high quality crystalline semiconductors. In the processes discussed here, electrons and holes are injected into the conduction and valence band from which they recombine to emit light. The injection may be due to an electrical current or an optical signal. The general ideas of light emission after excitation can be extended to semiconductors with impurities and to organic materials.

The general problem of light emission as a result of excitation can be represented by the three-level system shown in Fig. 5.18. Level $M$ represents a metastable, or long lived, state, while level $E$ represents an excited state of the system. The level $G$ denotes the ground state of the system. An excitation causes the electronic system to be excited to the level $M$ where there are two distinct possibilities. Electrons excited to the state $M$ can emit photons, as shown by process 3 of Fig. 5.18. In some materials, the transition $M \rightarrow G$ is forbidden due to certain selection rules. In this case, light emission can occur by the electron first being excited to level $E$ due to phonon scattering and then through the $E \rightarrow G$ recombination.

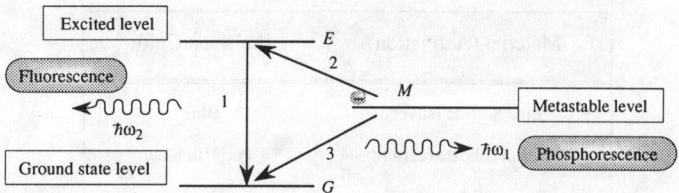

Figure 5.18: A general three-level system to represent light emission through excitation. Light can be emitted by a transition from $M \rightarrow G$ (process 3), and the process is usually known as phosphorescence. It can also be emitted by the process $M \rightarrow E \rightarrow G$, a process called delayed fluorescence.

The processes described by Fig. 5.18 are usually invoked for understanding "phosphors," a term used for light emitting materials used in TV screens, cathode ray tubes, etc. The transition $M \rightarrow G$ mentioned above is called phosphorescence, while the transition $M \rightarrow E \rightarrow G$ is called fluorescence. Since fluorescence requires the initial excitation of electrons to the state $E$ via lattice vibration scattering, it has a strong temperature dependence. As is clear form Fig. 5.18, the light emitted by fluorescence has a different wavelength from the light emitted by phosphorescence.

Phosphors in use in commercial applications range from organic materials used as dyes to inorganic materials used in the cathode ray tubes and TV screens. An interesting example of phosphors is the laundry brighteners that are based on colorless organic phenyl-based dyes. This phosphor is excited by ultraviolet light present in background lighting. Fluorescence occurs emitting blue light which compensates for the loss of blue light thorough absorption in textiles. As a result, the fabric appears "white."

In most inorganic phosphors, the light emission wavelength is determined by impurities that are introduced into the material. The impurity levels produce well defined bandgap states. When electron-hole pairs are created through excitation, the carrier (electron or hole) is trapped at the defect level from where radiative recombination occurs. An important inorganic phosphor is based on the semiconductor ZnS which has a bandgap of ~3.8 eV. When this material is doped with impurities, levels are produced which can be exploited to produce green (copper-activated) or blue (silver-activated) colors. These materials are widely used in TV screen technology along with europium-activated yttrium ortho-vanadate for red color to produce color displays. In Table 5.1 we show the color response of some of the important phosphors used in modern display technology.

At present, in display technology, excitation for the phosphors is provided by electrons emitted from an electron gun. As we will discuss in Chapter 9, this makes the display system quite bulky. As a result, displays based on liquid crystals are dominating applications where size and weight are of importance, as in

| Material (Activation) | Emission Color |
|---|---|
| Zinc Sulfide (silver) | Blue |
| Yttrium Silicate (cerium) | Purplish blue |
| Zinc sulfide (copper) | Green |
| Zinc orthosilicate (manganese) | Yellowish green |
| Gadolium oxysulfide (terbium) | Yellowish green |
| Zinc cadmium sulfide (silver) | Green |
| Yttrium oxysulfide (europium) | Red |

Table 5.1: Characteristics of some phosphor materials used in display screen technology. The impurities used are specified in parenthesis.

the laptop computer industry. However, there is great interest in producing electron emission devices that are compatible with microelectronic technology. As such devices become available, the phosphor-based display will become an important component of "flat panel display."

**EXAMPLE 5.12** Using the Joyce-Dixon approximation, calculate the electron and hole densities needed to cause inversion at the bandedges of GaAs at 300 K. The electron and hole densities are equal.

The effective density of states $N_c$ and $N_v$ for GaAs at 300 K are $4.45 \times 10^{17}$ cm$^{-3}$ and $7.7 \times 10^{18}$ cm$^{-3}$, respectively. The procedure for solving this problem is quite simple. We choose a value of $n(= p)$; use the Joyce-Dixon approximation to calculate $E_{Fn}$ and $E_{Fp}$; calculate $f^e$ and $f^h$ at the bandedges and check if the inversion condition $f^e + f^h = 1$ is satisfied. The carrier density is increased in small steps until the condition is satisfied.

The above exercise gives a value of $n = p = 1.1 \times 10^{18}$ cm$^{-3}$ for the inversion condition.

## 5.8   CHARGE INJECTION: NON-RADIATIVE EFFECTS

↝   The electrons and holes present in a material can recombine not only by emitting light, but also by releasing their energy as heat or lattice vibrations. Such processes are called *non-radiative processes* and can be classified either as defect mediated or as Auger effects, which need not involve defects. We will briefly discuss these two kinds of processes.

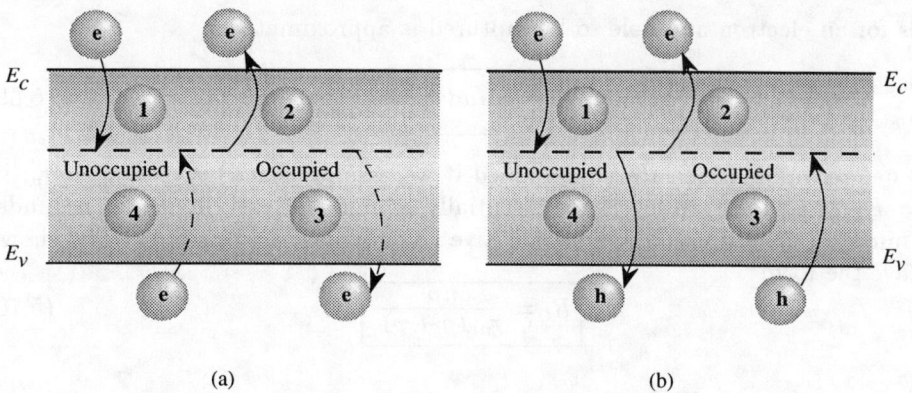

Figure 5.19: Various processes that lead to trapping and recombination via deep levels in the bandgap region (dashed line). The processes 1 and 2 in (a) represent trapping and emission of electrons while 3 and 4 represent the same for holes. The *e-h* recombination is shown in part (b).

## 5.8.1 Defect-Related Non-Radiative Processes

In perfect semiconductors, there are no allowed electronic states in the bandgap region. However, in real semiconductors there are always intentional or unintentional impurities which produce electronic levels which are in the bandgap. These impurity levels can arise from chemical impurities or from native defects such as a vacancy or an anti-site defect (i.e., in compound semiconductors, an atom on the wrong sublattice—Ga on an As site, for example).

The bandgap levels are states in which the electron is "localized" in a finite space near the defect unlike the usual Bloch states which represent the valence and conduction band states, and which are extended in space. As the "free" electrons move in the allowed bands, they can be trapped by the defects (see Fig. 5.19). The defects can also allow the recombination of an electron and hole without emitting a photon as was the case in the previous section. This non-radiative recombination competes with radiative recombination, and can have a positive or negative impact depending upon the device. For example, in a laser the non-radiative recombination is not desirable, but it is purposely increased in $p$-$n$ diodes to increase the speed. We will briefly discuss the non-radiative processes involving a midgap level with density $N_t$.

A defect is characterized by a capture cross-section which describes a certain area associated with the defect. If an electron (hole) comes within this area, the carrier is trapped. If $v_{th}$ is the thermal speed of electrons and holes, the time it takes for an electron and hole to be captured is approximately

$$\tau_n = \frac{1}{N_t v_{th} \sigma_n} \ and \ \tau_p = \frac{1}{N_t v_{th} \sigma_p} \tag{5.69}$$

The net recombination rate is simplified if we assume that i) $\tau_{nr} = \tau_n = \tau_p$; ii) $E_t = E_{Fi}$, i.e., the trap levels are essentially at midgap level; iii) $np \gg n_i^2$ under the injection conditions. The non-radiative recombination rate can then be shown to have the form

$$R_t = \frac{np}{\tau_{nr}(n+p)} \tag{5.70}$$

The time constant $\tau_{nr}$ depends upon the impurity density, the cross-section associated with the defect and the electron thermal velocity as seen in Eqn. 5.69. Typically the cross-sections are in the range $10^{-13}$ to $10^{-15}$ cm$^2$.

The defect-related non-radiative recombination is called *Shockley, Read, Hall* (SRH) recombination. Such recombination of great importance at surfaces of devices, since there is usually a high density of defects at the surfaces.

## 5.8.2   Auger Recombination

In Fig. 5.10, we have shown how impact ionization occurs when a hot electron scatters from a valence band electron producing two electrons in the conduction band and a hole in the valence band. The reverse process can also occur where an electron and a hole recombine, giving up the excess energy to an electron, producing a "hot" electron. The hot electron eventually loses its energy by emitting phonons (i.e., giving up heat.) This process is called the *Auger process* and it is an important non-radiative process, especially in materials with narrow bandgap. The Auger recombination rate is proportional to $np^2$ or $pn^2$, depending upon whether the final hot carrier is an electron or a hole.

## 5.9   THE CONTINUITY EQUATION: DIFFUSION LENGTH

$\longrightarrow$  In our discussion on charge transport, we considered the drift and diffusion processes through the semiconductor, without worrying about the electron-hole recombination. The recombination process removes the electrons and holes and thus alters the charge transport picture. To describe the transport and recombination of injected electrons and holes we develop a continuity equation for the problem. The

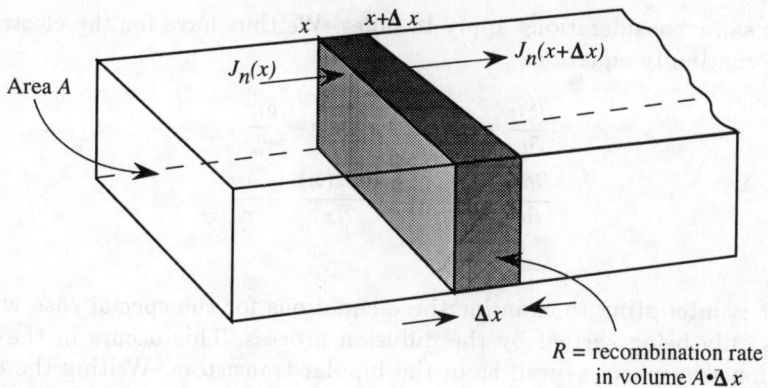

Figure 5.20: A conservation of particles applied to a volume $A \cdot \Delta x$. The difference in the particle currents has to equal the recombination rate.

carrier recombination process is critical for any device that involves both electron and hole flow.

If we consider a volume of space in which charge transport and recombination is taking place, we have the simple equality

Rate of particle flow =
Particle flow rate due to current − Particle loss rate due to recombination.

This equation is simply a statement of conservations of particles. If $\delta n$ is the excess carrier density in the region, the recombination rate in a volume $A \cdot \Delta x$ shown in Fig. 5.20 may be written approximately as

$$R = \frac{\delta n}{\tau_n} \cdot A \cdot \Delta x \qquad (5.71)$$

where $\tau_n$ is the electron recombination time per excess particle and includes radiative and non-radiative components. The particle flow rate into the same volume due to the current $J_n$ is given by

$$\left[ \frac{-J_n(x)}{e} + \frac{J_n(x + \Delta x)}{e} \right] A \cong \frac{1}{e} \frac{\partial J_n(x)}{\partial x} \Delta x \cdot A \qquad (5.72)$$

The total rate of electron build up in the volume $A \cdot \Delta x$ is then

$$A \cdot \Delta x \left[ \frac{\partial n(x,t)}{\partial t} \equiv \frac{\partial \delta n}{\partial t} = \frac{1}{e} \frac{\partial J_n(x)}{\partial x} - \frac{\delta n}{\tau_n} \right] \qquad (5.73)$$

where $\delta n$ is the excess carrier density which is the only part which changes with time. The same considerations apply to holes. We thus have for the electrons and holes, the continuity equations

$$\frac{\partial \delta n}{\partial t} = \frac{1}{e}\frac{\partial J_n(x)}{\partial x} - \frac{\delta n}{\tau_n} \qquad (5.74)$$

$$\frac{\partial \delta p}{\partial t} = -\frac{1}{e}\frac{\partial J_p(x)}{\partial x} - \frac{\delta p}{\tau_p} \qquad (5.75)$$

It is interesting to examine these equations for the special case where the current is only being carried by the diffusion process. This occurs in the cases of p-n junction transports as well as in the bipolar transistors. Writing the diffusion currents as (see Eqn. 5.31)

$$J_n(diff) = eD_n\frac{\partial \delta n}{\partial x} \qquad (5.76)$$

$$J_p(diff) = -eD_p\frac{\partial \delta p}{\partial x} \qquad (5.77)$$

we have

$$\frac{\partial \delta n}{\partial t} = D_n\frac{\partial^2 \delta n}{\partial x^2} - \frac{\delta n}{\tau_n} \qquad (5.78)$$

$$\frac{\partial \delta p}{\partial t} = D_p\frac{\partial^2 \delta p}{\partial x^2} - \frac{\delta p}{\tau_p} \qquad (5.79)$$

In steady state we have (the time derivative is zero)

$$\boxed{\begin{aligned} \frac{d^2 \delta n}{dx^2} &= \frac{\delta n}{D_n\tau_n} = \frac{\delta n}{L_n^2} \\ \frac{d^2 \delta p}{dx^2} &= \frac{\delta p}{D_p\tau_p} = \frac{\delta p}{L_p^2} \end{aligned}} \qquad (5.80)$$

where $L_n(L_p)$ defined as $D_n\tau_n(D_p\tau_p)$ are called the diffusion lengths for reasons that will be clear below.

Consider first the case where due to some external injection mechanism an excess electron density $\delta n(0)$ is maintained at the semiconductor edge $x = 0$ as shown in Fig. 5.21. If $n_o$ is the equilibrium density, we are interested in finding out how the excess density varies with position. The general solution of the second order differential Eqn. 5.80 is

$$\delta n(x) = A_1 e^{x/L_n} + A_2 e^{-x/L_n} \qquad (5.81)$$

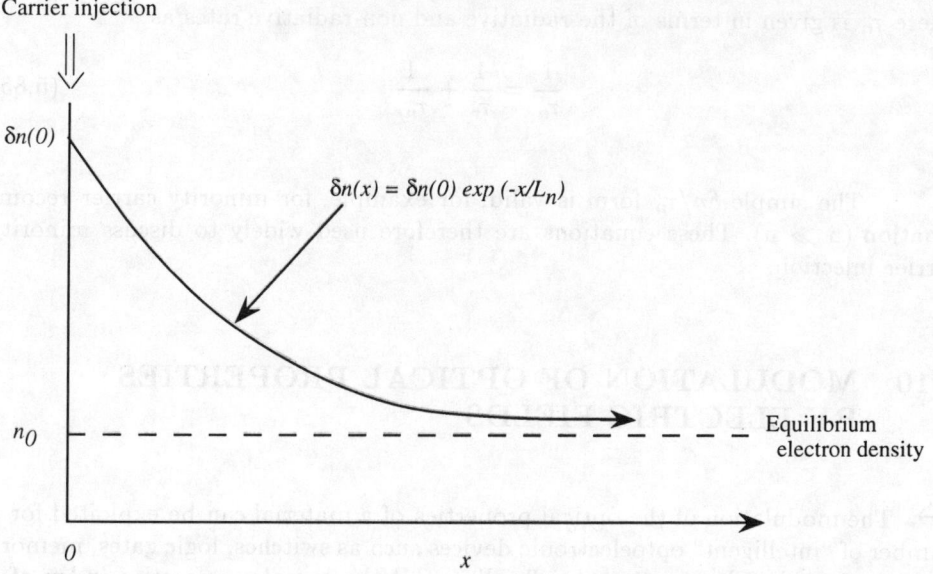

Figure 5.21: Electrons are injected at $x = 0$ into a sample. At $x = 0$, a fixed carrier concentration is maintained. The figure shows how the excess carriers decay into the semiconductor.

Since for large values of $x$, $\delta n(x)$ must go to zero, we must require that $A_1 = 0$. Note that we can only impose this condition if the sample is large. Since $\delta n(0)$ is known and fixed by the injection condition, we have

$$A_2 = \delta n(0) \tag{5.82}$$

The solution is then

$$\delta n(x) = \delta n(0)e^{-x/L_n} \tag{5.83}$$

The diffusion length $L_n$ represents the distance over which the injected carrier density falls to $1/e$ of its original value. It also represents the average distance an electron will diffuse before it recombines with a hole.

This average distance $(= \sqrt{D_n \tau_n})$ depends upon the recombination time and the diffusion constant in the material. In the derivations of this section, we used a simple form of recombination rate

$$R = \frac{\delta n}{\tau_n} \tag{5.84}$$

where $\tau_n$ is given in terms of the radiative and non-radiative rates as

$$\frac{1}{\tau_n} = \frac{1}{\tau_r} + \frac{1}{\tau_{nr}} \qquad (5.85)$$

The simple $\delta n / \tau_n$ form is valid, for example, for minority carrier recombination ($p \gg n$). These equations are therefore used widely to discuss minority carrier injection.

## 5.10  MODULATION OF OPTICAL PROPERTIES BY ELECTRIC FIELDS

$\longrightarrow$  The modulation of the optical properties of a material can be exploited for a number of "intelligent" optoelectronic devices such as switches, logic gates, memory elements, etc. In Chapter 2 we briefly discussed the complex refractive index of a material. The real part of this complex index determines the velocity of light in the medium and the imaginary part determines the attenuation of the light. In principle, any modification in either of these quantities can be used to modulate an optical signal propagating in the medium.

When the refractive index of a material is modified, the effect on a light beam propagating in the material can be classified into two categories, depending upon the photon energy. As shown in Fig. 5.22, if the photon energy is in a region where the absorption coefficient is zero, the effect of the modification of the refractive index is to alter the velocity of propagation of light. On the other hand, if the photon energy is in a region where the absorption coefficient is altered, the intensity of light emerging from the sample will be altered. These two approaches for the modification of the optical properties by an applied electric field are called the electro-optic and the electro-absorption approaches, respectively.

In the electro-optic effect, an applied bias is used to alter the phase velocity of a propagating signal, and this effect can be exploited in an interference scheme to alter the polarization or intensity of the light. This approach is widely used for optical modulators and optical image recording as will be discussed in Chapter 9.

### 5.10.1  The Electro-Optic Effect

$\rightsquigarrow$  The electro-optic effect depends upon the modification of the refractive index of a material by an applied electric field. We discussed in Chapter 2, section 2.2,

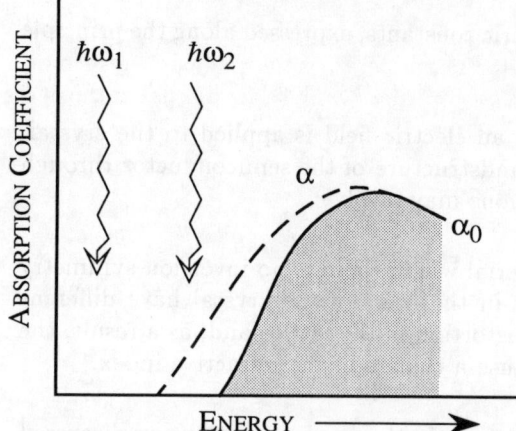

$\alpha_0$ : Absorption coefficient before modulation :

$\alpha$ : Absorption coefficient after modulation

For optical signal at frequency $\omega_1$: propagation velocity $v$ is changed

For optical signal at frequency $\omega_2$: $\Delta\alpha = \alpha - \alpha_0$ is the change in the absorption coefficient

Figure 5.22: A schematic of the effect of a change in optical properties of a material on an optical beam. For energy $\hbar\omega_1$, the main effect of the change in the optical properties is a change in propagation velocity. For $\hbar\omega_2$, the effect is a change in intensity.

how the Maxwell's equations and the material properties $(\epsilon, \mu)$ determine the propagation of light in a material. In general, for an anisotropic medium, the material properties are described by the relation

$$
\begin{aligned}
D_x &= \epsilon_{xx} E_x + \epsilon_{xy} E_y + \epsilon_{xz} E_z \\
D_y &= \epsilon_{yx} E_x + \epsilon_{yy} E_y + \epsilon_{yz} E_z \\
D_z &= \epsilon_{zx} E_x + \epsilon_{zy} E_y + \epsilon_{zz} E_z
\end{aligned}
\tag{5.86}
$$

or in the short form

$$
D_i = \sum_j \epsilon_{ij} E_j
\tag{5.87}
$$

If one examines the energy density of the optical wave and imposes the conditions that the energy must be independent under the transformation $i \to -i$, we get

$$
\epsilon_{ij} = \epsilon_{ji}
\tag{5.88}
$$

Further reduction in the number of the independent $\epsilon_{ij}$ can be achieved, depending upon the symmetry of the crystal structure. A most useful way to describe the propagation of light in the medium is via the index ellipsoid as discussed in Chapter 2, Section 2.5. The equation for the index ellipsoid is

$$
\frac{x^2}{\epsilon_x} + \frac{y^2}{\epsilon_y} + \frac{z^2}{\epsilon_z} = 1
\tag{5.89}
$$

or in terms of the refractive indices,

$$\sum_{i=1}^{3} \frac{x_i^2}{n_i^2} = 1 \qquad (5.90)$$

Here $\epsilon_x, \epsilon_y$ and $\epsilon_z$ are the principal dielectric constants, expressed along the principle axes of the ellipsoid.

Now consider a situation where an electric field is applied to the crystal. The applied electric field modifies the bandstructure of the semiconductor through a number of interactions. These interactions may involve:

i) Strain: In a piezoelectric material where there is no inversion symmetry (e.g., GaAs, CdTe, etc.) the two atoms in the basis of the crystal have different charges. The electric field may cause a distortion in the lattice and, as a result, the bandstructure may change. This may cause a change in the refractive index.

ii) Distortion of the optical absorption: In the previous sections we discussed the optical properties of materials. The presence of an electric field can modify the bandstructure and thus alter the optical spectra of the material.

In general, the change in the refractive index may be written as

$$n_{ij}(E) - n_{ij}(0) = \Delta n_{ij} = r_{ijk}E_k + s_{ijk\ell}E_k E_\ell \qquad (5.91)$$

where $E_i$ is the applied electric field component along the direction $i$ and $r_{ijk}$ and $s_{ijk\ell}$ are the components of the electro-optic tensor. In materials like GaAs and LiNbO$_3$ where the inversion symmetry is missing, $r_{ijk}$ is non-zero, and one has a linear term in the electro-optic effect. The linear effect is called the Pockel effect. In materials like Si where one has inversion symmetry $r_{ijk} = 0$ and the lowest order effect is due to the quadratic effect.

In general, $r_{ijk}$ has 27 elements, but since the tensor is invariant under the exchange of $i$ and $j$, there are only 18 independent terms. It is common to the use of contracted notation $r_{\ell m}$ where $\ell = 1, \ldots 6$ and m = 1, 2, 3. The standard contraction arises from the identification of $i, j = 1,1; 2,2; 3,3; 2,3; 3,1;1,2$ by $\ell = 1$, 2, 3, 4, 5, 6, respectively. The 18 coefficients are further reduced by the symmetry of the crystals. In semiconductors such as GaAs, it turns out that the only non-zero coefficients are

$$r_{41}$$
$$r_{52} = r_{41}$$
$$r_{63} = r_{41} \qquad (5.92)$$

Thus, a single parameter describes the linear electro-optic effect. The value of GaAs is $r_{41} = 1.2 \times 10^{-12}$ m/V. An important class of electro-optic materials are ferroelectric peroskites such as $LiNbO_3$ and $LiTaO_3$ which have trigonal symmetry and materials such as KDP (potassium dihydrogen phosphate) which have tetragonal symmetry). For the trigonal materials the non-zero tensor components are

$$r_{22} = -r_{12} = -r_{61}$$
$$r_{51} = r_{42} = r_{33}$$
$$r_{13} = r_{23} \tag{5.93}$$

For KDP the non-zero elements are

$$r_{41} = r_{52}$$
$$r_{63} \tag{5.94}$$

The second-order electro-optic coefficients $s_{ijk\ell}$ are usually not important for materials unless the optical energy $\hbar\omega$ is very close to the bandgap. In materials like GaAs, the second-order coefficients that are non-zero from symmetry considerations are in the contracted form $s_{pq}, p = 1 \ldots 6, q = 1 \ldots 6$,

$$s_{11} = s_{22} = s_{33}$$
$$s_{12} = s_{13}$$
$$s_{44} = s_{55} = s_{66} \tag{5.95}$$

The electro-optic effect is used to create a modulation in the frequency, intensity or polarization of an optical beam.

In the presence of the electric field, the index ellipsoid is modified, since the refractive indices are altered. A phase change is thus produced in an optical signal traveling through the device. The value of the phase change depends upon the polarization of the light. For example, in GaAs if a field is applied along the $z$-direction, the phase changes are given by

$$\boxed{\begin{aligned}\Delta\phi(01\bar{1}) &= -\frac{\pi L}{\lambda}n_o^3\left[r_{41}E\xi_1 + (s_{12} - s_{11})E^2\xi_2\right]\\[2mm]\Delta\phi(011) &= \frac{\pi L}{\lambda}n_o^3\left[r_{41}E\xi_1 + (s_{12} - s_{11})E^2\xi_2\right]\end{aligned}} \tag{5.96}$$

where $n_o$ is the refractive index in the absence of the field ($= n_x = n_y = n_z$) and $x'$ and $y'$ axes represent light polarized along $< 01\bar{1} > (x')$ and $< 011 > (y')$ directions. The quantity $\xi_1$ and $\xi_2$ represent the overlap of the optical wave with the region where the electric field is present:

$$\xi_1 = \frac{1}{E}\int\int E\mid E_{\text{photon}}\mid^2 dA \tag{5.97}$$

where $E_{photon}$ is the photon field. For bulk devices $\xi_1 \sim 1$

$$\xi_2 = \frac{1}{E^2} \int \int E^2 \mid E_{photon} \mid^2 dA \tag{5.98}$$

The phase charges produced by the electric field can be exploited for a number of important switching or modulation devices. These devices will be discussed in Chapter 9.

**EXAMPLE 5.13** A bulk GaAs device is used as an electro-optic modulator. The device dimension is 1 mm and a phase change of 90° is obtained between light polarized along $< 01\bar{1} >$ and $< 011 >$. The wavelength of the light is 1.5 $\mu$m. Calculate the electric field needed if $\xi_1 = 1$. The value of $r_{41}$ is $1.2 \times 10^{-12}$ m/V.

The phase change produced is

$$\begin{aligned}
\Delta\phi &= \frac{2\pi}{\lambda} n_o^3 r_{41} E L = \frac{\pi}{4} \\
E &= \frac{\lambda}{8 n_o^3 r_{41} L} \\
&= \frac{(1.5 \times 10^{-6} \ m)}{8(3.3)^3 (1.2 \times 10^{-12} \ m/V)(10^{-3} \ m)} \\
&= 4.35 \times 10^6 \ V/m
\end{aligned}$$

If the field is across a 10 $\mu$m thickness, the voltage needed is 4.35 V.

## 5.11  CHAPTER SUMMARY

$\mathcal{R}$  In this chapter we have discussed the basic physical phenomena upon which electronic and optoelectronic devices are based. All devices involve some physical response to external perturbations. These perturbations are usually electric fields or electromagnetic fields. The device performance depends upon how electrons respond to these external stimuli. The summary tables (Tables 5.2-5.4) highlight the concepts discussed in this chapter.

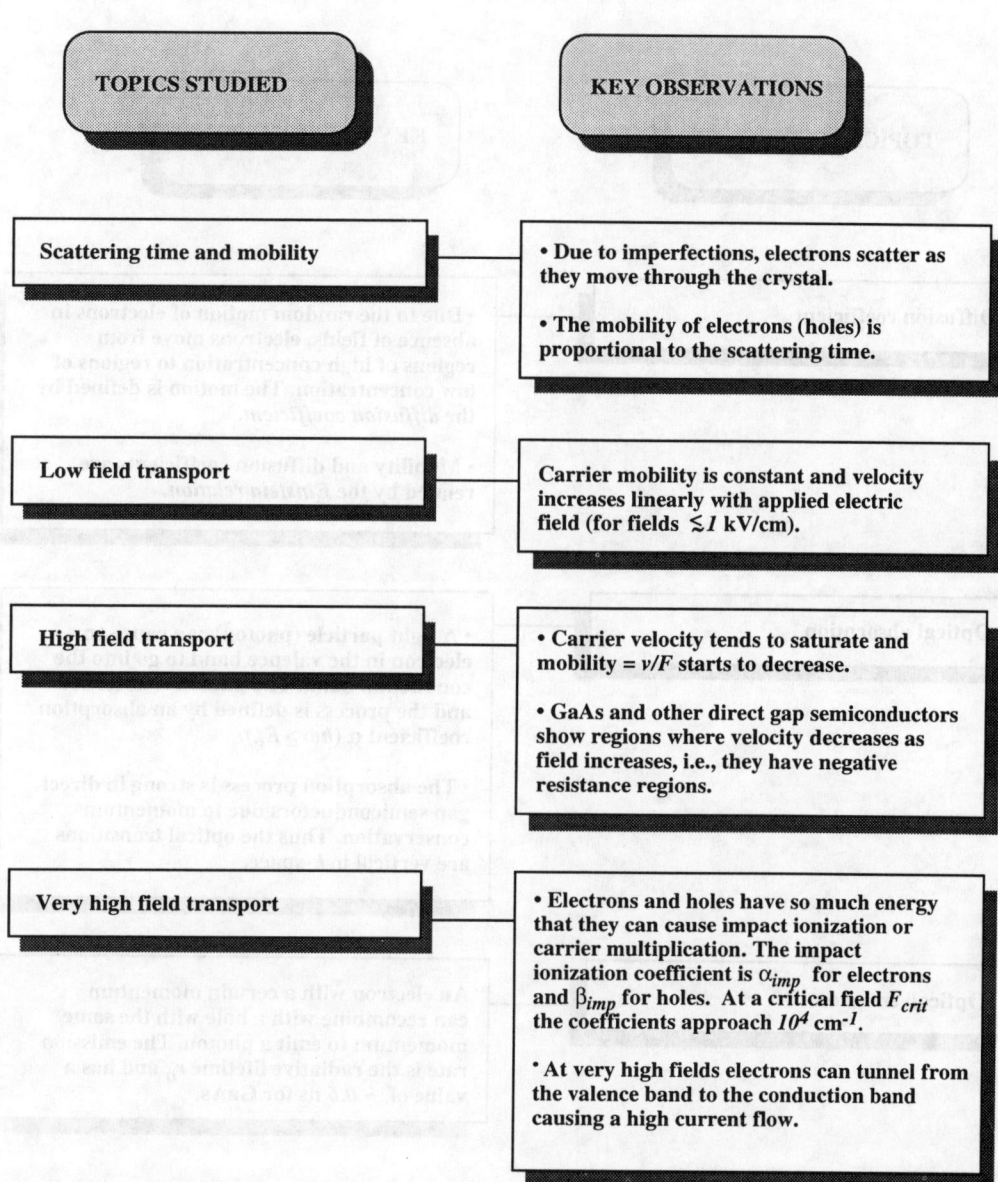

| TOPICS STUDIED | KEY OBSERVATIONS |
|---|---|
| Scattering time and mobility | • Due to imperfections, electrons scatter as they move through the crystal.<br><br>• The mobility of electrons (holes) is proportional to the scattering time. |
| Low field transport | Carrier mobility is constant and velocity increases linearly with applied electric field (for fields $\leq 1$ kV/cm). |
| High field transport | • Carrier velocity tends to saturate and mobility = $v/F$ starts to decrease.<br><br>• GaAs and other direct gap semiconductors show regions where velocity decreases as field increases, i.e., they have negative resistance regions. |
| Very high field transport | • Electrons and holes have so much energy that they can cause impact ionization or carrier multiplication. The impact ionization coefficient is $\alpha_{imp}$ for electrons and $\beta_{imp}$ for holes. At a critical field $F_{crit}$ the coefficients approach $10^4$ cm$^{-1}$.<br><br>• At very high fields electrons can tunnel from the valence band to the conduction band causing a high current flow. |

Table 5.2: Summary table

| TOPICS STUDIED | KEY OBSERVATIONS |
|---|---|
| **Diffusion coefficient** | • Due to the random motion of electrons in absence of fields, electrons move from regions of high concentration to regions of low concentration. The motion is defined by the *diffusion coefficient*.<br><br>• Mobility and diffusion coefficients are related by the *Einstein relation*. |
| **Optical absorption** | • A light particle (photon) can cause an electron in the valence band to go into the conduction band. The photon is absorbed and the process is defined by an absorption coefficient $\alpha$ ($\hbar\omega \geq E_g$).<br><br>• The absorption process is strong in direct gap semiconductors due to momentum conservation. Thus the optical transitions are vertical in $k$-space. |
| **Optical emission** | An electron with a certain momentum can recombine with a hole with the same momentum to emit a photon. The emission rate is the radiative lifetime $r_0$ and has a value of ~ *0.6* ns for GaAs. |

Table 5.3: Summary table

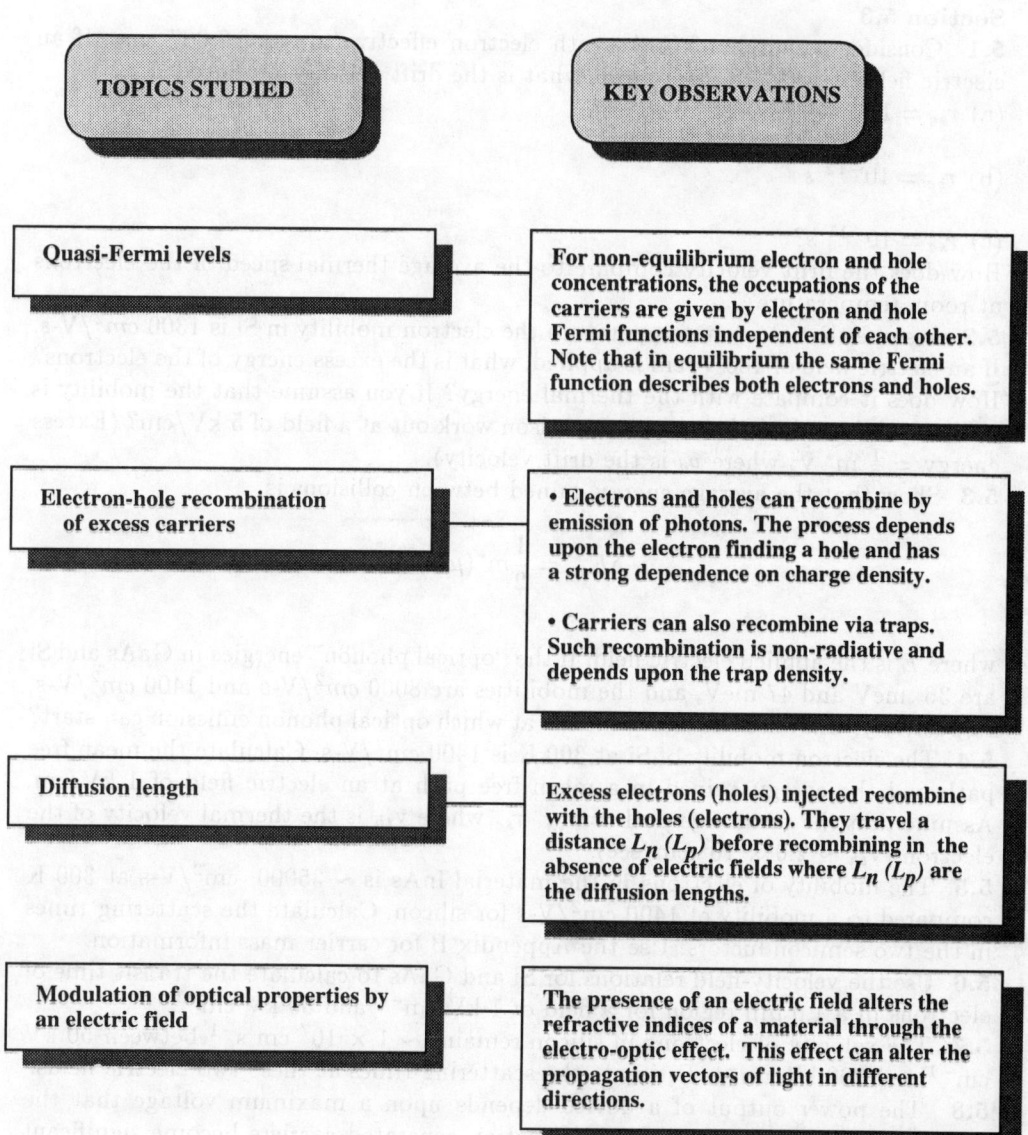

| TOPICS STUDIED | KEY OBSERVATIONS |
| --- | --- |
| **Quasi-Fermi levels** | For non-equilibrium electron and hole concentrations, the occupations of the carriers are given by electron and hole Fermi functions independent of each other. Note that in equilibrium the same Fermi function describes both electrons and holes. |
| **Electron-hole recombination of excess carriers** | • Electrons and holes can recombine by emission of photons. The process depends upon the electron finding a hole and has a strong dependence on charge density. <br><br> • Carriers can also recombine via traps. Such recombination is non-radiative and depends upon the trap density. |
| **Diffusion length** | Excess electrons (holes) injected recombine with the holes (electrons). They travel a distance $L_n$ $(L_p)$ before recombining in the absence of electric fields where $L_n$ $(L_p)$ are the diffusion lengths. |
| **Modulation of optical properties by an electric field** | The presence of an electric field alters the refractive indices of a material through the electro-optic effect. This effect can alter the propagation vectors of light in different directions. |

Table 5.4: Summary table

## 5.12    PROBLEMS

### Section 5.3

**5.1**  Consider a sample of GaAs with electron effective mass of 0.067 $m_0$. If an electric field of 1 kV/cm is applied, what is the drift velocity produced if

(a) $\tau_{sc} = 10^{-13}$ s

(b) $\tau_{sc} = 10^{-12}$ s

(c) $\tau_{sc} = 10^{-11}$ s?
How does the drift velocity compare to the average thermal speed of the electrons at room temperature?

**5.2**  Assume that at room temperature the electron mobility in Si is 1300 $cm^2$/V-s. If an electric field of 100 V/cm is applied, what is the excess energy of the electrons? How does it compare with the thermal energy? If you assume that the mobility is unchanged, how does the same comparison work out at a field of 5 kV/cm? (Excess energy $= \frac{1}{2}$ m* $v_d^2$ where $v_d$ is the drift velocity).

**5.3**  Show that the average energy gained between collisions is

$$\delta E_{av} = \frac{1}{2} m^* (\mu E)^2$$

where $E$ is the applied electric field. If the "optical phonon" energies in GaAs and Si are 36 meV and 47 meV, and the mobilities are 8000 $cm^2$/V-s and 1400 $cm^2$/V-s, respectively, what are the electric fields at which optical phonon emission can start?

**5.4**  The electron mobility of Si at 300 K is 1400 $cm^2$/V-s. Calculate the mean free path and the energy gained in a mean free path at an electric field of 1 kV/cm. Assume that the mean free path $= v_{th} \cdot \tau_{sc}$ where $v_{th}$ is the thermal velocity of the electron ($v_{th} \sim 2.0 \times 10^7$ cm/sec).

**5.5**  The mobility of electrons in the material InAs is $\sim 35000$ $cm^2$/V-s at 300 K compared to a mobility of 1400 cm$^2$/V-s for silicon. Calculate the scattering times in the two semiconductors. Use the Appendix B for carrier mass information.

**5.6**  Use the velocity-field relations for Si and GaAs to calculate the transit time of electrons in a 1.0 $\mu$m region for a field of 1 kV cm$^{-1}$ and 50 kV cm$^{-1}$.

**5.7**  The velocity of electrons in silicon remains $\sim 1 \times 10^7$ cm s$^{-1}$ between 50 kV cm$^{-1}$ and 200 kV cm$^{-1}$. Estimate the scattering times at these two electric fields.

**5.8**   The power output of a device depends upon a maximum voltage that the device can tolerate before impact ionization generated carriers become significant (say 10% excess carriers). Consider a device of length $L$ over which a potential $V$ drops uniformly. What is the maximum voltage that can be tolerated by a Si and a diamond device for $L = 2$ $\mu$m and $L = 0.5$ $\mu$m?

**5.9**  An electron in a silicon device is injected in a region where the field is 500 kV cm$^{-1}$. The length of this region is 1.0 $\mu$m. Calculate the number of impact ionization events that occur for the incident electron.

**5.10** In Problem 5.9, if a hole is injected under the same conditions, how many ionizing events will occur for the incident hole?
The impact ionization coefficient for holes in silicon at a field of 500 kV/cm is

$$\beta_{imp} = 3 \times 10^4 \ cm^{-1}$$

**5.11** In GaAs impact ionization starts to be significant for electrons if the electric field is 350 kV cm$^{-1}$. Calculate the probability for Zener tunneling at this field. Estimate the electric field where the Zener tunneling probability approaches $10^{-6}$.
**5.12** Using the data in this chapter, estimate if impact ionization or tunneling dominates the breakdown in $In_{0.53}Ga_{0.47}As$.

## Section 5.4
**5.13** In a silicon sample at 300 K, the electron concentration drops linearly from $10^{18}$ cm$^{-3}$ to $10^{16}$ cm$^{-3}$ over a length of 2.0 $\mu$m. Calculate the current density due to the electron diffusion current. Use the diffusion constant values given in this chapter.
**5.14** In a GaAs sample, it is known that the electron concentration varies linearly. The diffusion current density at 300 K is found to be 100 A/cm$^2$. Calculate the slope of the electron concentration.
**5.15** In a GaAs sample the electrons are moving under an electric field of 5 kV cm$^{-1}$ and the carrier concentration is uniform at $10^{16}$ cm$^{-3}$. The electron velocity is the saturated velocity of $10^7$ cm s$^{-1}$. Calculate the drift current density. If a diffusion current has to have the same magnitude, calculate the concentration gradient needed. Assume a diffusion coefficient of 100 cm$^2$/s.

## Section 5.5
**5.16** Identify the various semiconductors (including alloys) that can be used for light emission at 1.55 $\mu$m. Remember that light emission occurs at an energy near the bandgap.
**5.17** Using the expressions given in the text, plot the absorption coefficient for GaAs and InP.
**5.18** Calculate the electron-hole recombination time $\tau_o$ for an HgCdTe alloy which has a bandgap of 0.1 eV. The momentum matrix element is the same as GaAs.

## Section 5.6
**5.19** In a GaAs sample at 300 K, equal concentrations of electrons and holes are injected. If the carrier density is n = p = $10^{17}$ cm$^{-3}$, calculate the electron and hole Fermi levels using the Boltzmann and Joyce-Dixon approximations.

## Section 5.7

**5.20** In a $p$-type GaAs doped at $N_a = 10^{18}$ cm$^{-3}$, electrons are injected to produce a minority carrier concentration of $10^{15}$ cm$^{-3}$. What is the rate of photon emission assuming that all $e$-$h$ recombination is due to photon emission ? What is the optical output power? The photon energy is $\hbar\omega = 1.41$ eV.

**5.21** Calculate the electron carrier density needed to push the electron Fermi level to the conduction bandedge in GaAs. Also calculate the hole density needed to push the hole Fermi level to the valence bandedge. Calculate the results for 300 K and 77 K.

## Section 5.8

**5.22** The radiative lifetime of a GaAs sample is 1.0 ns. The sample has a defect at the midgap with a capture cross-section of $10^{-15}$ cm$^2$. At what defect concentration does the non-radiative lifetime become equal to the radiative lifetime at i) 77 K and ii) 300 K?

## Section 5.9

**5.23** Electrons are injected into a $p$-type silicon sample at 300 K. The electron-hole radiative lifetime is 1 $\mu$s. The sample also has midgap traps with a cross-section of $10^{-15}$ cm$^2$ and a density of $10^{16}$ cm$^{-3}$. Calculate the diffusion length for the electrons if the diffusion coefficient is 30 cm$^2$/s.

## Section 5.10

**5.24** Consider a bulk GaAs electro-optic device on which an electric field is applied in the $z$-direction. The field is switched between 0 and $10^5$ V/cm. Calculate the length of the device needed to produce a phase change of $\pi/2$ between the light waves polarized along $< 01\bar{1} >$ and $< 0\bar{1}1 >$. The wavelength of the light is 1.3 $\mu$m.

**5.25** Consider a $KD^*P$ crystal in which the electro-optic coefficient $r_{63}$ is 26.4 $\times$ $10^{-12}$ m/V. Calculate the voltage needed to produce a phase difference of $\pi$ for two polarized waves. Assume the following:

$$\lambda = 1.06 \ \mu m$$
$$n_o = 1.52$$

Use the results given in Eqn. 5.96 with $r_{41}$ replaced by $r_{63}$ and $\xi_1 = 1$. Ignore the quadratic effect.

## 5.13 REFERENCES

- **Transport**

  - R. A. Smith, *Semiconductors*, Cambridge University Press, London (1978).
  - R. B. Adler, A. C. Smith, and R. L. Longini, *Introduction to Semiconductor Physics*, Wiley, New York (1969).
  - J. R. Haynes and W. Shockley, *Phys. Rev.*, **81**, 835 (1951).
  - S. M. Sze, *Physics of Semiconductor Devices*, Wiley, New York (1981).
  - D. L. Rode, *Low Field Electron Transport in Semiconductors and Semimetals*, eds. R. K. Willardson and A. C. Beer, Academic Press, New York (1975), vol. 10.

- **General High Field Transport**

  - M. Lundstrom, *Fundamentals of Carrier Transport*, Modular Series on Solid State Devices, ed. G. W. Neudeck and R. F. Pierret, Addison-Wesley, Reading (1990), vol. X.

- **Optical Processes in Semiconductors**

  - J. I. Pankove, *Optical Processes in Semiconductors*, Prentice-Hall, Englewood Cliffs, NJ (1971).
  - F. Stern, "Elementary Theory of the Optical Properties of Solids," *Solid State Physics*, Academic Press, New York (1963), vol. 15.
  - V. F. Weisskopf, "How Light Interacts with Matter," *Scientific American*, **219**, 60 (1968).

# CHAPTER
# 6

# LIGHT DETECTION
# AND IMAGING

# 6.1 INTRODUCTION

Having discussed the basic physical effects that are present in a variety of materials we are now ready to start our journey into the exciting field of optoelectronic devices. These devices use the interaction between light and electrons in a material to produce a physical effect that can be exploited in information processing. As we have noted before in this text, the information processing by light involves the three manipulations: i) light detection; ii) light generation; and, iii) light modulation. Light detection is used in decoding the information that a light beam contains. This may involve a light signal of an optical communication system which is carrying a voice or data information. It may also involve light coming from a distant star. And it may involve light reflected from the face of an actor in a studio. Optoelectronic devices should be able to convert the optical signal to an electronic signal. The elctronic signal may then be further processed by advanced microelectronic devices. In this chapter we will study the devices that carry out the task of light detection and imaging. In chapters 7 and 8 we will discuss the devices used for light generation and in chapter 9 we will discuss devices that can modulate a light signal for applications in communication and display systems.

# 6.2 OPTICAL ABSORPTION IN A SEMICONDUCTOR

$\longrightarrow$ In order for a semiconductor device to be useful as a detector, some property of the device should be affected by radiation. The most commonly used property is the conversion of light into electron-hole pairs which can be detected in a properly chosen electric circuit.

When light impinges on a semiconductor, it can scatter an electron in the valence band into the conduction band. This process, called the absorption of a photon, was discussed in Chapter 5. In order to take the electron from the fully occupied valence band to the empty conduction band, the photon energy must be at least equal to the bandgap of the semiconductors. We will summarize some of the results discussed in Chapter 5 (Sections 5.5 through 5.8) which the reader should review to understand this and the next chapter.

The photon absorption process is strongest when the photon can directly cause an electron in the valence band to go into the conduction band. Since the photon momentum is extremely small on the scale of the electron momentum, the conservation of momentum requires that the electron-hole transitions are *vertical in k-space*, as shown in Fig. 6.1a (see Section 5.5). Such transitions are only possible near the bandedge for direct bandgap semiconductors. For such semiconductors one

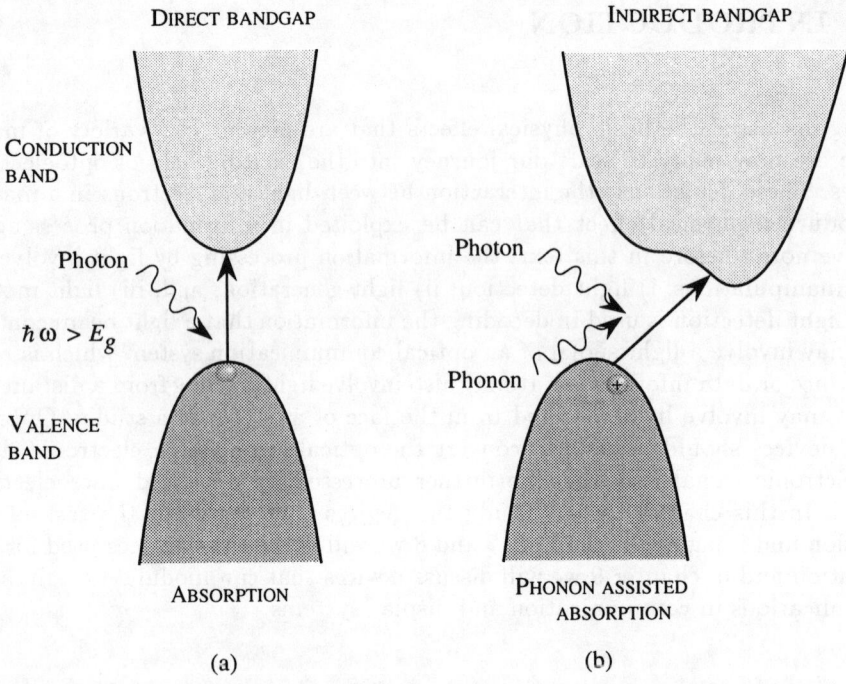

Figure 6.1: Band to band absorption in direct and indirect semiconductors. (a) An electron in the valence band "absorbs" a photon and moves into the conduction band. Momentum conservation ensures that only vertical transitions are allowed. (b) In indirect semiconductors a phonon or lattice vibration must participate to take an electron from the top of the valence band to the bottom of the conduction band. This is a comparatively weak process.

can write the absorption coefficient as

$$\alpha(\hbar\omega) = \frac{\pi e^2 \hbar}{2n_r c m_o^2 \epsilon_o} \frac{|p_{cv}|^2}{\hbar\omega} \frac{\sqrt{2}(m_r^*)^{3/2}(\hbar\omega - E_g)^{1/2}}{\pi^2 \hbar^3} \qquad (6.1)$$

where $m_r^*$ is the reduced $e$-$h$ mass, $n_r$ is the refractive index, $\hbar\omega$ the photon energy, $E_g$ the band gap and $p_{cv}$ is a momentum matrix element which allows the transition to take place. For directgap semiconductors, when the various values for the constants are plugged into Eqn. 6.1, the absorption coefficient turns out to be ($\hbar\omega$, $E_g$ in eV; see Example 5.9)

$$\boxed{\alpha(\hbar\omega) \cong 3 \times 10^6 \left(\frac{m_r^*}{m_o}\right)^{3/2} \frac{(\hbar\omega - E_g)^{1/2}}{\hbar\omega} \, cm^{-1}} \qquad (6.2)$$

When a semiconductor does not have a direct bandgap, vertical $k$ transitions are not possible, and the electrons can absorb a photon only if a phonon (or lattice vibration) participates in the process as shown in Fig. 6.1b. Such processes are not as strong as the ones which do not involve a phonon. The absorption coefficient for indirect gap materials is typically a factor of 100 smaller as compared to the direct gap case for the same value of photon energy above bandgap ($\hbar\omega - E_g$).

As can be seen from Eqn. 6.1, the absorption coefficient is zero above a cutoff wavelength given by $\lambda_c$, where

$$\boxed{\lambda_c = \frac{hc}{E_g} = \frac{1.24}{E_g(eV)} \; (\mu m)} \tag{6.3}$$

where $E_g$ is the semiconductor bandgap. In Fig. 6.2 we show the bandgap and cutoff wavelengths for several semiconductors along with the relative response of the human eye. In Fig. 6.3 we discuss the important semiconductor systems used for detection. Important features of these material systems are also discussed. Materials like GaAs, InP, InGaAs, etc. have strong optical absorption at the bandedges because the optical absorption can occur without a phonon participation. On the other hand, Si and Ge have an indirect bandgap, and the absorption strength is weak near the bandedge. However, this does not mean that these materials cannot be used as detectors (unfortunately they cannot be used as lasers, as we will see in the next chapter). For detection of an optical signal, the light should be absorbed. If $L$ is the length of the sample, the fraction of incident light absorbed in the sample is

$$1 - exp\,(-\alpha L) \tag{6.4}$$

Thus, for strong absorption, we must have

$$L > \frac{1}{\alpha(\hbar\omega)} \tag{6.5}$$

Thus, if Si is to absorb at GaAs laser emission ($\hbar\omega \sim 1.45$ eV), one needs a material thickness of 10-20 $\mu$m. On the other hand, a Ge detector would require an interaction length of only $\sim 1$ $\mu$m, even though Ge is an indirect gap material.

The electron-hole pair generation by "band to band" transition, i.e., an electron being transferred from a valence band to a conduction band, is not the only way of detecting photons. In "extrinsic" detectors, a semiconductor is doped with a particular impurity which creates electron states in the bandgap as shown schematically in Fig. 6.4. Extrinsic detectors form an important class of detectors especially for detection for long wavelength radiation. Intrinsic band to band detectors need a very narrow bandgap for long wavelength radiation detection, and it is difficult to fabricate high-quality devices from such materials. On the other hand, in extrinsic detectors, the radiation energy can be much smaller than the bandgap. In fact, extrinsic detectors can operate at wavelengths up to 120 $\mu$m at low temperatures using certain impurities in Ge or Si. The absorption coefficient for the

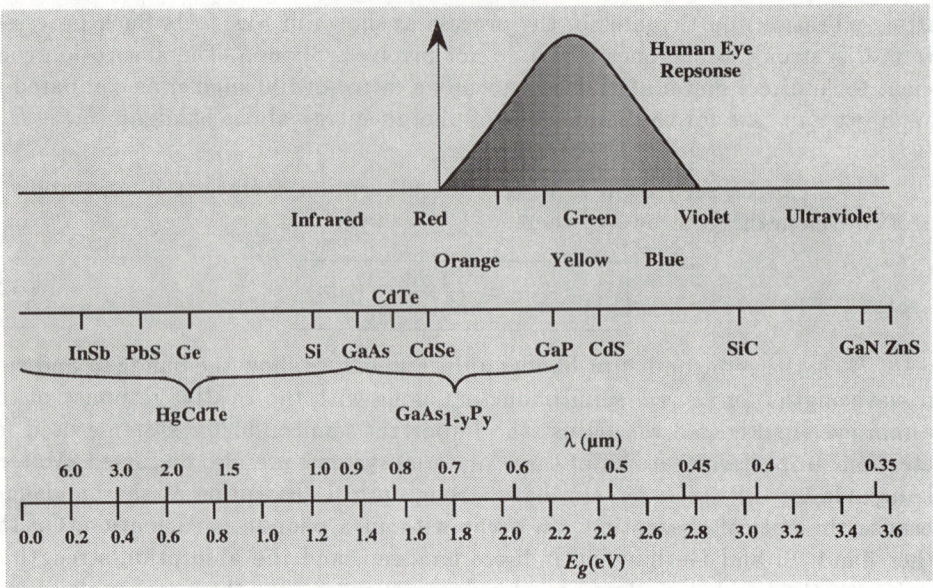

Figure 6.2: The bandgap and cutoff wavelengths for several semiconductors. The semiconductor bandgaps range from 0 (for $Hg_{0.84}Cd_{0.15}Te$) to well above 3 eV, providing versatile detection systems.

extrinsic absorption is however, quite small ($\sim 10$ cm$^{-1}$), so that a thick sample is needed.

Once the absorption coefficient for a semiconductor is known, one needs to know the rate at which electron-hole pairs will be generated. To calculate the rate of e-h pair generation, consider an optical beam with intensity $P_{op}(0)$ impinging upon a semiconductor per unit area. The intensity at a point $x$ is given by (intensity has units of W/cm$^2$)

$$P_{op}(x) = P_{op}(0) \, exp\,(-\alpha x) \tag{6.6}$$

The energy absorbed per second per unit area in a thickness region of thickness $dx$, between points $x$ and $x + dx$ is ($dx$ is very small)

$$
\begin{aligned}
P_{op}(x + dx) - P_{op}(x) &= P_{op}(0) \left[ exp\,(-\alpha(x + dx)) - exp(-\alpha x) \right] \\
&= P_{op}(0) \left[ exp\,(-\alpha x) \right] \alpha dx
\end{aligned}
\tag{6.7}
$$

If this absorbed energy produces e-h pairs of energy $\hbar\omega$, the rate of the carrier generation is $G_L$ (rate per unit volume)

$$G_L = \frac{\alpha P_{op}(x)}{\hbar\omega} = \alpha \Phi_0(x) \tag{6.8}$$

where $\Phi_0$ is the photon flux density impinging at point $x$ (flux has units of cm$^{-2}$ s$^{-1}$).

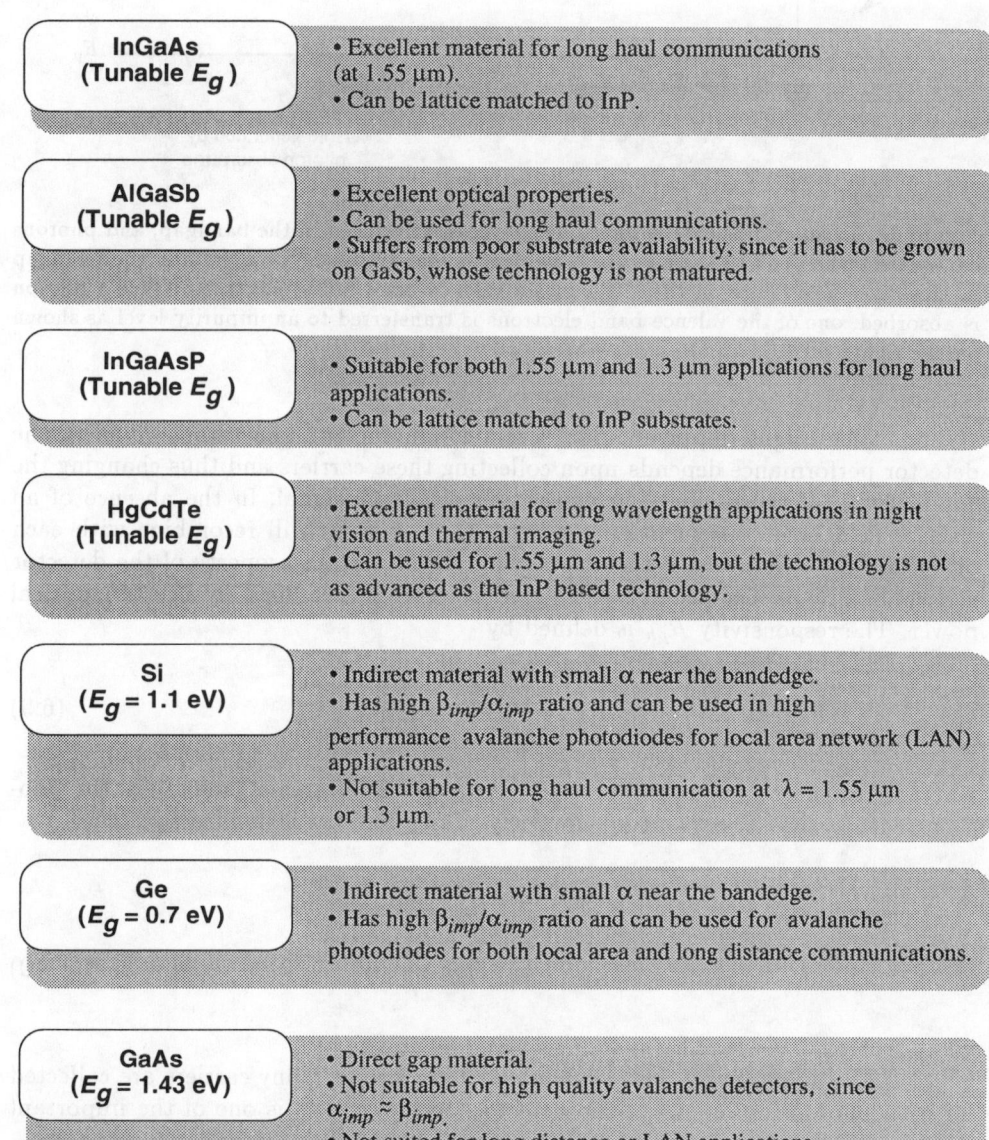

**InGaAs**
**(Tunable $E_g$)**
- Excellent material for long haul communications (at 1.55 µm).
- Can be lattice matched to InP.

**AlGaSb**
**(Tunable $E_g$)**
- Excellent optical properties.
- Can be used for long haul communications.
- Suffers from poor substrate availability, since it has to be grown on GaSb, whose technology is not matured.

**InGaAsP**
**(Tunable $E_g$)**
- Suitable for both 1.55 µm and 1.3 µm applications for long haul applications.
- Can be lattice matched to InP substrates.

**HgCdTe**
**(Tunable $E_g$)**
- Excellent material for long wavelength applications in night vision and thermal imaging.
- Can be used for 1.55 µm and 1.3 µm, but the technology is not as advanced as the InP based technology.

**Si**
**($E_g$ = 1.1 eV)**
- Indirect material with small α near the bandedge.
- Has high $\beta_{imp}/\alpha_{imp}$ ratio and can be used in high performance avalanche photodiodes for local area network (LAN) applications.
- Not suitable for long haul communication at $\lambda$ = 1.55 µm or 1.3 µm.

**Ge**
**($E_g$ = 0.7 eV)**
- Indirect material with small α near the bandedge.
- Has high $\beta_{imp}/\alpha_{imp}$ ratio and can be used for avalanche photodiodes for both local area and long distance communications.

**GaAs**
**($E_g$ = 1.43 eV)**
- Direct gap material.
- Not suitable for high quality avalanche detectors, since $\alpha_{imp} \approx \beta_{imp}$.
- Not suited for long distance or LAN applications.

Figure 6.3: Important semiconductors for detection and their key features.

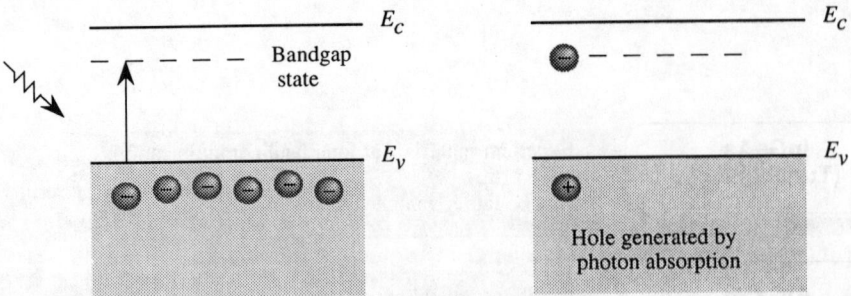

Figure 6.4: In an extrinsic detector, a deep level is introduced in the bandgap, and photons can cause transitions to these levels. The photon energy can be smaller than the bandgap in this case. The figure on the left shows a valence band full of electrons. Once a photon is absorbed, one of the valence band electrons is transferred to an impurity level as shown on the right.

When light impinges upon a semiconductor and generates $e$-$h$ pairs, the detector performance depends upon collecting these carriers and thus changing the conductivity of the material or generating a voltage signal. In the absence of an electric field or a concentration gradient, the $e$-$h$ pairs will recombine with each other and not generate a detectable signal. An important property of the detector is described by its responsivity which gives the current produced by a certain optical power. The responsivity $R_{ph}$ is defined by

$$R_{ph} = \frac{I_L/A}{P_{op}} = \frac{J_L}{P_{op}} \qquad (6.9)$$

where $I_L$ is the photocurrent produced in a device of area $A$ and $J_L$ is the photocurrent density. The quantum efficiency of the detector is defined by

$$\eta_Q = \frac{J_L/e}{P_{op}/\hbar\omega}$$
$$= R_{ph}\frac{\hbar\omega}{e} \qquad (6.10)$$

The quantum efficiency essentially tells us how many carriers are collected for each photon impinging on the detector. Improving $\eta_Q$ is one of the important design considerations for the detector and will be discussed later.

The responsivity of a detector has a strong dependence upon the wavelength of the impinging photons. If the wavelength is above the cutoff wavelength, the photons will not be absorbed and no photocurrent will be generated. When the wavelength is smaller than $\lambda_c$, the photon energy will be larger than the bandgap energy and the difference will be released as heat. Thus, even though the photon energy increases above the bandgap, it still produces the same number of *e-h* pairs. Thus the responsivity starts to decrease.

To collect the electron-hole pairs generated by light one needs an electric field. This can be generated either by simply applying a bias across an undoped semiconductor or by using a *p-n* diode. The former choice leads to the photoconductive detector in which the *e-h* pairs change the conductivity of the semiconductor. The *p-n* (or *p-i-n*) diode is widely used as a detector and exploits the built-in electric fields present at the junction together with an applied reverse bias to collect electrons and holes. The *p-n* diode can be used in a variety of modes, depending upon the applied bias and load configurations.

In addition to the diodes, transistors can also be used to detect optical signals. Phototransistors are widely used in optoelectronic technology. Phototransistors offer high gain due to the transistor gain. In the next few sections we will discuss the various modes of operation.

**EXAMPLE 6.1** The momentum matrix element $p_{cv}$ has a value of 23 eV for GaAs. Calculate the absorption coefficient for GaAs for an incident optical beam with energy of 1.7 eV. Assume that $E_g = 1.43$ eV.

The absorption coefficient for GaAs was calculated in Example 5.9 in Chapter 5. There we found that for GaAs the coefficient is (see also Eqn. 6.2)

$$\alpha(\hbar\omega) = 5.6 \times 10^4 \frac{(\hbar\omega - E_g)^{1/2}}{\hbar\omega} cm^{-1}$$

where $\hbar\omega$ and $E_g$ are expressed in eV. For our case we get

$$\alpha(\hbar\omega = 1.7eV) = 4.21 \times 10^4 \frac{(0.27)}{1.7} = 6.7 \times 10^3 cm^{-1}$$

**EXAMPLE 6.2** A Ge detector is to be used for an optical communication system using a GaAs laser with emission energy of 1.43 eV. Calculate the thickness of the detector needed to be able to absorb 90% of the optical signal entering the detector.

From Fig. 5.14 we see that at $\hbar\omega = 1.43$ eV, $\alpha \cong 2.5 \times 10^4$ cm$^{-1}$. The length of material required to absorb 90% of the light is given by

$$L = -\frac{1}{\alpha}\ell n\,(1 - 0.9) = \frac{2.3}{(2.5 \times 10^4\ cm^{-1})} = 0.92\ \mu m$$

Thus a rather thin region of Ge can absorb a large fraction of light emitted from a GaAs laser. Of course, if the light was emitted by an In$_{0.53}$Ga$_{0.47}$As laser ($\hbar\omega \sim 0.8$ eV), Ge would not be as suitable a material.

**EXAMPLE 6.3** An optical intensity of 10 W/cm$^2$ at a wavelength of 0.75 $\mu$m is incident on a GaAs detector. Calculate the rate at which electron-hole pairs will be produced at this intensity at 300 K. If the $e$-$h$ recombination time is 10$^{-9}$ s, calculate the excess carrier density.

From Fig. 5.14, the absorption coefficient of GaAs at this wavelength is $\sim 7 \times 10^3$ cm$^{-1}$. The $e$-$h$ generation rate is (0.75 $\mu$m wavelength is equivalent to a photon of 1.65 eV)

$$G_L = \frac{\alpha P_{op}}{\hbar \omega} = \frac{(7 \times 10^3 \ cm^{-1})(10 \ W cm^{-2})}{(1.65 \times 1.6 \times 10^{-19} \ J)} = 2.65 \times 10^{23} \ cm^{-3} \ s^{-1}$$

The excess carrier density is

$$\delta n = \delta p = G_L \tau = (2.65 \times 10^{23} \ cm^{-3} \ s^{-1})(10^{-9} \ s)$$
$$= 2.65 \times 10^{14} \ cm^{-3}$$

## 6.3 PHOTOCURRENT IN A P-N DIODE

$\longrightarrow$ When light impinges upon a semiconductor to create electron-hole pairs, some of the carriers are collected at the contact and lead to the photocurrent. In Appendix D we discuss the important properties of a $p$-$n$ diode. The reader is urged to review this appendix in case he or she wants a quick overview of the $p$-$n$ diode. The $p$-$n$ diode is a basic device used in a number of optoelectronic devices (detectors, LEDs, laser diodes, etc.) and the reader should be quite familiar with this device. Let us consider a long $p$-$n$ diode in which excess carriers are generated uniformly at a rate $G_L$. Fig. 6.5 shows a $p$-$n$ diode with a depletion region of width $W$. The electron-hole pairs generated in the depletion region are swept rapidly by the electric field existing in the region. Thus the electrons are swept into the $n$- region while the holes are swept into the $p$-region. The photocurrent arising from the photons absorbed in the depletion region is thus

$$I_{L1} = A \cdot e \int_0^{x'} G_L \cdot dx = A \cdot e G_L W \tag{6.11}$$

where $A$ is the diode area and we have assumed a uniform generation rate in the diode. *Since the electrons and holes contributing to $I_{L1}$ move under high electric fields, the response is very fast, and this component of the current is called the prompt photocurrent.*

In addition to the carriers generated in the depletion region, $e$-$h$ pairs are generated in the neutral $n$-and $p$-regions of the diode. On physical grounds, we

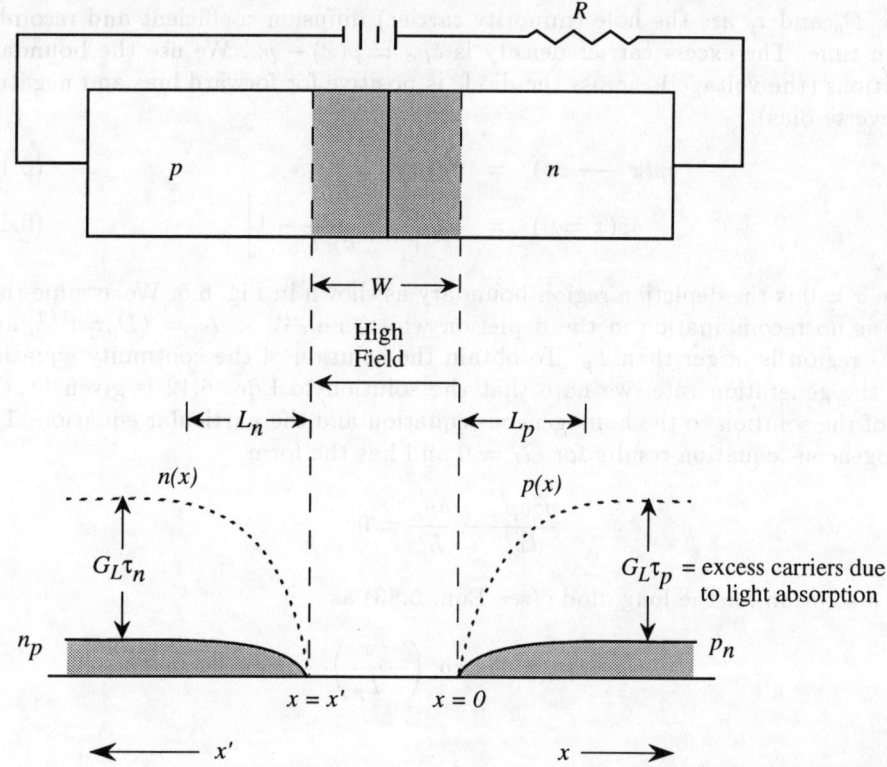

Figure 6.5: A schematic of a p-n diode and the minority carrier concentration in the absence and presence of light. The minority charge goes to zero at the depletion region edge due to the high field which sweeps the charge away. The equilibrium minority charge is $p_n$ and $n_p$ in the n- and p-sides, respectively.

may expect that holes generated within a distance $L_p$ (the diffusion length) of the depletion region edge ($x = 0$ of Fig. 6.5) will be able to enter the depletion region from where the electric field will sweep them into the p-side. Similarly, electrons generated within a distance $L_n$ of the $x' = 0$ side of the depletion region will also be collected and contribute to the current. Thus the photocurrent should come from all carriers generated in a region ($W + L_n + L_p$). A quantitative analysis reaches the same conclusion as shown below.

We will use the diode theory to obtain the photocurrent. We start with the continuity equation assuming that e-h pairs are generated uniformly at a rate $G_L$. The steady state continuity equation for holes in the n-region is (see Eqn. 5.79 and add the generation term $G_L$)

$$D_p \frac{\partial^2 \delta p_n}{\partial x^2} - \frac{\delta p_n}{\tau_p} + G_L = 0 \qquad (6.12)$$

where $D_p$ and $\tau_p$ are the hole (minority carrier) diffusion coefficient and recombination time. The excess carrier density is $\delta p_n = p(x) - p_n$. We use the boundary conditions (the voltage $V$ across the diode is positive for forward bias and negative for reverse bias)

$$\delta p(x \longrightarrow \infty) \;=\; G_L \tau_p \qquad\qquad (6.13)$$

$$\delta p(x = 0) \;=\; p_n \left[ exp \; \frac{eV}{k_B T} - 1 \right] \qquad\qquad (6.14)$$

where $x = 0$ is the depletion region boundary as shown in Fig. 6.5. We assume that there is no recombination in the depletion width, i.e., $W < L_p = (D_p \tau_p)^{1/2}$, and the n- region is larger than $L_p$. To obtain the solution of the continuity equation with the generation rate, we note that the solution to Eqn. 6.12 is given by the sum of the solution to the homogeneous equation and the particular equation. The homogeneous equation results for $G_L = 0$ and has the form

$$\frac{d^2 \delta p_n'}{dx^2} - \frac{\delta p_n'}{L_p^2} = 0$$

with a solution for the long diode (see Eqn. 5.83) as

$$\delta p_n' = A \; exp \left( -\frac{x}{L_p} \right)$$

The particular equation has the form

$$\frac{\delta p_n''}{L_n^2} = \frac{G_L}{D_p}$$

or $\delta p_n'' = G_L \tau_p$.

The resulting solution is thus

$$\delta p_n = A \; exp \left( -\frac{x}{L_p} \right) + G_L \tau_p$$

Using the boundary condition given by Eqn. 6.14 at $x = 0$, we obtain the value for the constant $A$. This gives the final solution for the excess holes in the neutral n-region

$$\delta p(x) = \left[ p_n \left\{ exp \left( \frac{eV}{k_B T} \right) - 1 \right\} - G_L \tau_p \right] exp \left( \frac{-x}{L_p} \right) + G_L \tau_p \qquad (6.15)$$

If the diode is operated in the short circuit mode so that the diode voltage is zero, or in a reverse bias mode, we can assume that $\delta p(x = 0)$ is essentially zero.

This gives us

$$\delta p(x) = G_L \tau_p \left[ 1 - exp\left(\frac{-x}{L_p}\right) \right] \tag{6.16}$$

so that the hole current, due to carriers absorbed in the neutral $n$-region, is

$$I_{pL} = A e D_p \left. \frac{d\delta p}{dx} \right|_{x=0} = e G_L L_p A \tag{6.17}$$

The electron current can be similarly calculated, so that the total current, due to carriers in the neutral region and the depletion region, is

$$\boxed{I_L = I_{nL} + I_{pL} + I_{L1} = e G_L (L_p + L_n + W) A} \tag{6.18}$$

*It is important to note that the photocurrent contribution from the neutral regions has a slower time response since carriers are collected by diffusion under almost no electric fields.* In case the widths of the neutral regions $d_p$ and $d_n$ of the diode are smaller than $L_p$ and $L_n$, and if ohmic boundary conditions are used ($\delta p(d_n) = \delta n(d_p) = 0$), we can assume that half the carriers generated in the neutral regions contribute to the photocurrent. The current is then

$$I_L = e G_L \left( W + \frac{d_n}{2} + \frac{d_p}{2} \right) A \tag{6.19}$$

It must be noted that the $e$-$h$ pair generation is not uniform with depth but decreases with penetration depth. Thus $G_L$ has to be replaced by an average generation rate for an accurate description. *It is also important to note that the photocurrent flows in the direction of the reverse current of the diode.*

The total current in the diode connected to the external load, as shown in Fig. 6.6, is given by the light-generated current and the diode current in the absence of light. In general, if the voltage across the diode is $V$, the total current is (note that the photocurrent flows in the opposite direction to the forward bias diode current)

$$I = I_L + I_0 \left[ 1 - exp\left\{ \frac{e(V + R_s I)}{m k_B T} \right\} \right] \tag{6.20}$$

where $R_s$ is the diode series resistance, $m$ the diode ideality factor, and $V$ is the voltage across the diode. As shown in Fig. 6.6, the photodiode can be used in one of two configurations. In the photovoltaic mode, used for solar cells, there is no external bias applied. The photocurrent passes through an external load to generate power. In the photoconductive mode, used for detectors, the diode is reverse biased and the photocurrent is collected.

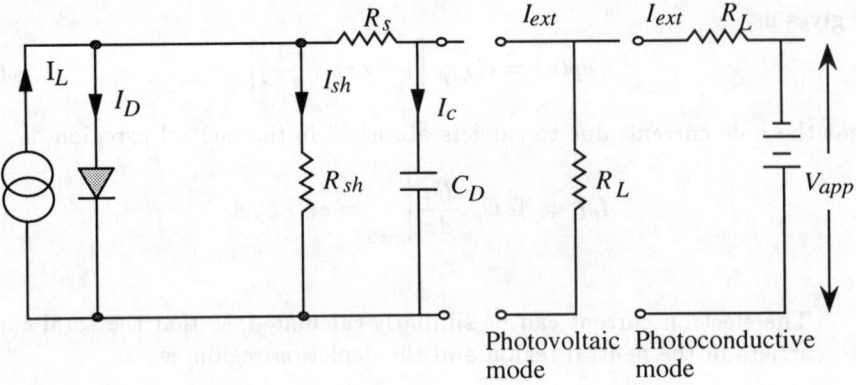

Figure 6.6: The equivalent circuit of a photodiode. The device can be represented by a photocurrent source $I_L$ feeding into a diode. The device's internal characteristics are represented by a shunt resistor $R_{sh}$ and a capacitor $C_D$. $R_s$ is the series resistance of the diode. In the photovoltaic mode (used for solar cells and other devices) the diode is connected to a high resistance $R_L$, while in the photoconductive mode (used for detectors) the diode is connected to a load $R_L$ and a power supply.

## 6.3.1   Application to a Solar Cell

An important use of the *p-n* diode is to convert optical energy to electrical energy as in a solar cell. The solar cell operates without an external power supply and relies on the optical power to generate current and voltage. To calculate the important parameters of a solar cell consider the case where the diode is in the open circuit mode so that the current $I$ is zero. This gives, from Eqn. 6.20

$$I = 0 = I_L - I_0 \left[ exp \left( \frac{eV_{oc}}{mk_BT} \right) - 1 \right] \tag{6.21}$$

where $V_{oc}$ is the voltage across the diode and is known as the open circuit voltage. We get for this voltage

$$\boxed{V_{oc} = \frac{mk_BT}{e} \ln \left( 1 + \frac{I_L}{I_0} \right)} \tag{6.22}$$

At high optical intensities the open circuit voltage can approach the semiconductor bandgap. In the case of Si solar cells for solar illumination (without atmospheric absorption), the value of $V_{oc}$ is roughly 0.7 eV.

A second limiting case in the solar cell is the one where the output is short circuited, i.e., $R = 0$ and $V = 0$. The short circuit current is then

$$I = I_{sc} = I_L \tag{6.23}$$

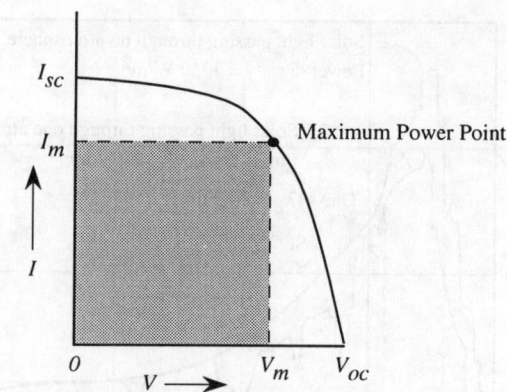

Figure 6.7: The relationship between the current and voltage delivered by a solar cell. The open circuit voltage is $V_{oc}$ and the short circuit current is $I_{sc}$. The maximum power is delivered at the point shown.

A plot of the diode current in the solar cell as a function of the diode voltage then provides the curve shown in Fig. 6.7. In general, the electrical power delivered to the load is given by

$$P = I \times V = I_L V - I_0 \left[ exp \left( \frac{eV}{k_B T} \right) - 1 \right] V \qquad (6.24)$$

The maximum power is delivered at a voltage and current value of $V_m$ and $I_m$ as shown in Fig. 6.7.

The conversion efficiency of a solar cell is defined as the rate of the output electrical power to the input optical power. When the solar cell is operating under maximum power conditions, the conversion efficiency is

$$\eta_{conv} = \frac{P_m}{P_{in}} \times 100 (\text{percent}) = \frac{I_m V_m}{P_{in}} \times 100 (\text{percent}) \qquad (6.25)$$

Another useful parameter in defining solar cell parameters is the fill factor $F_f$, defined as

$$F_f = \frac{I_m V_m}{I_{sc} V_{oc}} \qquad (6.26)$$

In most solar cells the fill factor is $\sim 0.7$.

In the solar cell conversion efficiency, it is important to note that photons which have an energy $\hbar\omega$ smaller than the semiconductor bandgap will not produce any electron-hole pairs. *Also, photons with energy greater than the bandgap will produce electrons and holes with the same energy difference $(E_g)$ regardless of how*

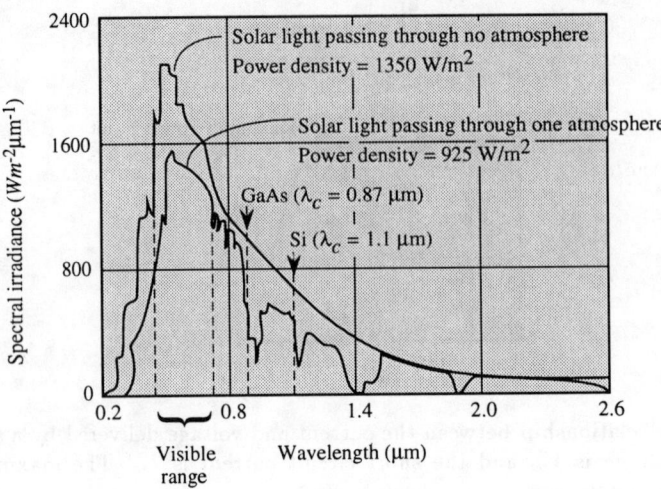

Figure 6.8: The spectral irradiance of the solar energy. The spectra are shown for no absorption in the atmosphere and for the sea level spectra. Also shown are the cutoff wavelengths for GaAs and Si.

*large* $\hbar\omega - E_g$ *is.* The excess energy $\hbar\omega - E_g$ is simply dissipated as heat. Thus the solar cell efficiency depends quite critically on how the semiconductor bandgap matches with the solar energy spectra. In Fig. 6.8 we show the solar energy spectra. Also shown are the cutoff wavelengths for silicon and GaAs. GaAs solar cells are better matched to the solar spectra and provide greater efficiencies. However, the technology is more expensive when compared to Si technology. Thus GaAs solar cells are used for space applications while silicon (or amorphous silicon) solar cells are used for applications where cost is a key factor.

**EXAMPLE 6.4** Consider a long Si $p$-$n$ junction that is reverse biased with a reverse bias voltage of 2 V. The diode has the following parameters (all at 300 K):

| | | |
|---|---|---|
| Diode area, | $A$ | $= 10^4 \ \mu m^2$ |
| $p$-side doping, | $N_a$ | $= 2 \times 10^{16} \ cm^{-3}$ |
| $n$-side doping, | $N_d$ | $= 10^{16} \ cm^{-3}$ |
| Electron diffusion coefficient | $D_n$ | $= 20 \ cm^2/s$ |
| Hole diffusion coefficient, | $D_p$ | $= 12 \ cm^2/s$ |
| Electron minority carrier lifetime, | $\tau_n$ | $= 10^{-8} \ s$ |
| Hole minority carrier lifetime, | $\tau_p$ | $= 10^{-8} \ s$ |
| Electron-hole pair generation rate by light, | $G_L$ | $= 10^{22} \ cm^{-3}s^{-1}$ |

Calculate the photocurrent.

The electron diffusion length is

$$L_n = \sqrt{D_n \tau_n} = \left[ (20)(10^{-8}) \right]^{1/2} = 4.5 \ \mu m$$

The hole diffusion length is

$$L_p = \sqrt{D_p \tau_p} = \left[ (12)(10^{-8}) \right]^{1/2} = 3.46 \ \mu m$$

To calculate the depletion width, we need to find the built-in voltage (see Appendix D)

$$V_{bi} = \frac{k_B T}{e} \ell n \left( \frac{N_a N_d}{n_i^2} \right) = 0.026 \ \ell n \ \left( \frac{(2 \times 10^{16})(10^{16})}{(1.5 \times 10^{10})^2} \right) = 0.715 \ V$$

The depletion width is now

$$
\begin{aligned}
W &= \left\{ \frac{2\epsilon_s}{e} \left( \frac{N_a + N_d}{N_a N_d} \right) (V_{bi} + V_R) \right\}^{1/2} \\
&= \left\{ \frac{2(11.9)(8.85 \times 10^{-14})}{(1.6 \times 10^{-19})} \left( \frac{(2 \times 10^{16} + 10^{16})}{(2 \times 10^{16})(10^{16})} \right) (2.715) \right\}^{1/2} \\
&= 0.73 \ \mu m
\end{aligned}
$$

We see in this case that $L_n$ and $L_p$ are larger than $W$. The prompt photocurrent is thus a small part of the total photocurrent. The photocurrent is now

$$
\begin{aligned}
I_L &= eAG_L(W + L_n + L_p) \\
&= (1.6 \times 10^{-19} C)(10^4 \times 10^{-8} \ cm^2)(10^{22} \ cm^{-3} \ s^{-1})(0.73 \times 10^{-4} \ cm + 4.5 \times 10^{-4} \ cm + 3.46 \times 10^- \\
&= 0.137 \ mA
\end{aligned}
$$

The photocurrent is much larger than the reverse saturation current $I_0$ and its direction is the same as the reverse current.

**EXAMPLE 6.5** Consider an Si solar cell at 300 K with the following parameters:

| | | |
|---|---|---|
| Area, | $A$ = | $1.0 \ cm^2$ |
| Acceptor doping, | $N_a$ = | $5 \times 10^{17} \ cm^{-3}$ |
| Donor doping, | $N_d$ = | $10^{16} \ cm^{-3}$ |
| Electron diffusion coefficient, | $D_n$ = | $20 \ cm^2/s$ |
| Hole diffusion coefficient, | $D_p$ = | $10 \ cm^2/s$ |
| Electron recombination time, | $\tau_n$ = | $3 \times 10^{-7} \ s$ |
| Hole recombination time, | $\tau_p$ = | $10^{-7} \ s$ |
| Photocurrent, | $I_L$ = | $25 \ mA$ |

Calculate the open circuit voltage of the solar cell.

To find the open circuit voltage, we need to calculate the saturation current $I_0$, which is given by (see Appendix D)

$$I_0 = A \left[ \frac{eD_n n_p}{L_n} + \frac{eD_p P_n}{L_p} \right] = A e n_i^2 \left[ \frac{D_n}{L_n N_a} + \frac{D_p}{L_p N_d} \right]$$

Also,

$$L_n = \sqrt{D_n \tau_n} = \left[(20)(3 \times 10^{-7})\right]^{1/2} = 24.5 \ \mu m$$

$$L_p = \sqrt{D_p \tau_p} = \left[(10)(10^{-7})\right]^{1/2} = 10.0 \ \mu m$$

Thus,

$$I_0 = (1)(1.6 \times 10^{-19})(1.5 \times 10^{10})^2 \left[\frac{20}{(24.5 \times 10^{-4})(5 \times 10^{17})} + \frac{10}{(10 \times 10^{-4})(10^{16})}\right]$$

$$= 3.66 \times 10^{-11} A$$

The open circuit voltage is now

$$V_{oc} = \frac{k_B T}{e} \ell n \left(1 + \frac{I_L}{I_0}\right) = (0.026)\ell n \left(1 + \frac{25 \times 10^{-3}}{3.66 \times 10^{-11}}\right) = 0.53 \ V$$

**EXAMPLE 6.6** A single solar cell of area 1 cm$^2$ has a photocurrent of $I_L = 25$ mA and a diode saturation current of $3.66 \times 10^{-11}$ A at 300 K. a) Calculate the open circuit voltage and short circuit current of the solar cell; b) calculate the power extracted from each cell if the fill factor is 0.8; c) if a solar power system requires a power of 10 W at a voltage level of 10 V, calculate the number of solar cells needed in series and the number of rows in parallel for such a solar cell array.

(This diode has the same features as the diode considered in Example 6.5.)

The open circuit voltage was calculated in Example 6.5 and is 0.53 V.

The short circuit current is simply $I_L = 25$ mA.

The power per solar cell is

$$P = 0.8 I_{sc} V_{oc} = 0.8(25 \times 10^{-3})(0.53) = 1.06 \ mW$$

The number of solar cells needed in series to produce an output voltage of 10 V is (each cell produces approximately $V_M \sim (F_f)^{1/2} \sim 0.9 V_{oc}$)

$$N(\text{series}) = \frac{10}{0.9 \times 0.53} \sim 24 \text{ cells}$$

The number of rows needed to produce a power of 10 W is now ($I_m \sim 0.9 I_{sc}$)

$$N(\text{parallel}) = \frac{10 \ W}{10 \ V(25 \times 10^{-3} \times 0.9 \ A)} = 45 \text{ rows}$$

Thus the system needs a total of 1080 solar cells to meet the specifications.

## 6.4   THE PHOTOCONDUCTIVE DETECTOR

$\longrightarrow$ The photoconductive detector is the simplest of the detectors and consists of a simple region of semiconductor across which a bias is applied as shown in Fig. 6.9a. When light with a proper wavelength impinges upon the semiconductor, $e$-$h$ pairs are created which are then collected by the electric field. The change in current is detected by a circuit of the form shown in Fig. 6.9b. *An important benefit of the photoconductive detector is the gain in the device, i.e., one can collect more than one electron (or hole) for each photon impinging.* Let us examine the operation of the photoconductive detector and the gain mechanism.

When light impinges on the $i$-region, $e$-$h$ pairs are generated which change the material conductivity. The electric field in the device causes the electrons and holes to move in opposite directions, leading to current. *The carriers remain in the system until they either recombine or are collected at the contacts.* Consider the case where we have an $n$-$i$-$n$ structure with an $e$-$h$ recombination rate $R_{eh}$ which is equal to the photogeneration rate

$$R_{eh} = \delta n / \tau_p = G_L \qquad (6.27)$$

where $\tau_p$ is the effective recombination time for the excess carriers. Let us assume that we have a lightly doped $n$-type device where the electrons dominate the conductivity. In the absence of the light signal, the conductivity is ($n_0$ and $p_0$ are the electron and hole densities in dark)

$$\sigma_o = e\left(\mu_n n_0 + \mu_p p_0\right) \qquad (6.28)$$

If the optical signal generates an excess carrier density of $\delta n = \delta p$, the conductivity becomes

$$\sigma = e\left[\mu_n(n_0 + \delta n) + \mu_p(p_0 + \delta p)\right] \qquad (6.29)$$

The excess carrier density is given by (see Eqn. 6.27)

$$\delta p = \delta n = G_L \tau_p \qquad (6.30)$$

where $G_L$ is the $e$-$h$ pair generation rate. The change in the conductivity of the material due to the optical signal is called the photoconductivity and is given by Eqn. 6.29 as

$$\Delta\sigma = e\delta p(\mu_n + \mu_p) \qquad (6.31)$$

In the presence of an electric field $E$, the current density is given by

$$J = (J_d + J_L) = (\sigma_0 + \Delta\sigma)E \qquad (6.32)$$

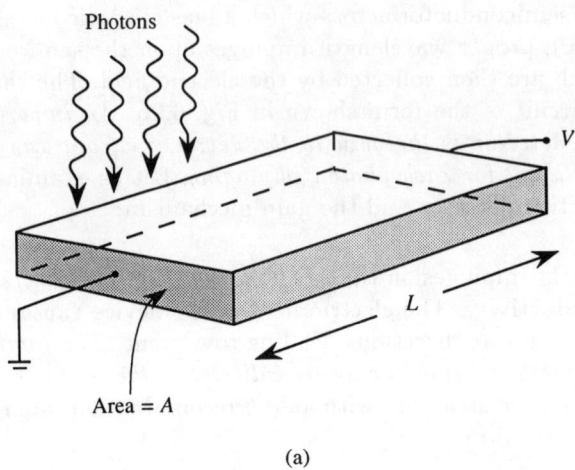

(a)

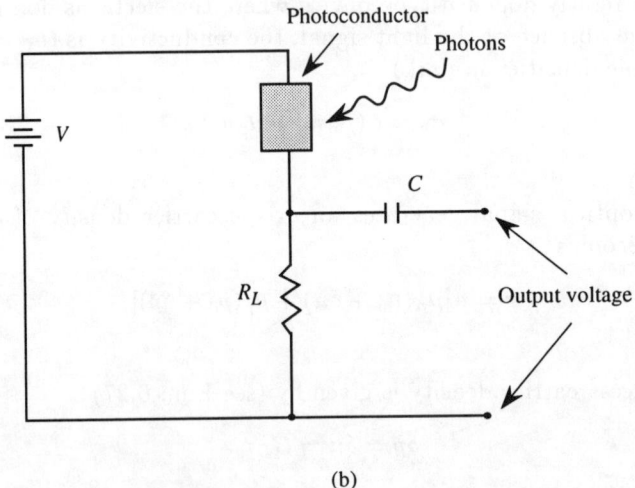

(b)

Figure 6.9: (a) Geometry of a photoconductor of length $L$ and area $A$. (b) A typical bias circuit for a photodetector. Light causes a change in the resistance of the photoconductor. A blocking capacitor may be used if only the ac signal is to be detected.

where $J_d$ is the dark current density of the detector. The photocurrent is thus

$$
\begin{aligned}
I_L = J_L \cdot A &= e\delta p(\mu_n + \mu_p)AE \\
&= eG_L\tau_p(\mu_n + \mu_p)AE
\end{aligned}
\tag{6.33}
$$

One must keep in mind that $\mu_n E$ and $\mu_p E$ represent the electron and hole velocities and may not increase linearly with the electric field. In high electric fields, $\mu_n E$ and $\mu_p E$ simply are the saturation velocities independent of the field. Let us define the transit time of the electrons in the device by

$$
t_{tr} = \frac{L}{\mu_n E}
\tag{6.34}
$$

The photocurrent now becomes, after expressing $\mu_n E$ in terms of $t_{tr}$ and $L$ using Eqn. 6.34,

$$
I_L = eG_L\left(\frac{\tau_p}{t_{tr}}\right)\left(1 + \frac{\mu_p}{\mu_n}\right)AL
\tag{6.35}
$$

This is the photocurrent generated in the circuit. We may define the primary photocurrent as

$$
I_{Lp} = eG_L AL
\tag{6.36}
$$

*This would be the photocurrent if each e-h pair simply contributed one charge at the contact, i.e., if there was no gain in the device.* The gain of the photoconductive detector is now

$$
\boxed{G_{ph} = \frac{I_L}{I_{Lp}} = \frac{\tau_p}{t_{tr}}\left(1 + \frac{\mu_p}{\mu_n}\right)}
\tag{6.37}
$$

The gain in the device arises because the electron goes around the circuit several times before it can recombine with a photogenerated hole. Each time the electron goes through the circuit it contributes to the current as shown schematically in Fig. 6.10.

If $\tau_p$ is large and $t_{tr}$ is small, a very large gain can be produced. In Si based devices where $\tau_p$ can be very large, gains of 1000 or more can be obtained. However, the improved gain comes at the expense of speed, since the speed is controlled by $\tau_p$ and not $t_{tr}$. Thus the gain bandwidth product is essentially constant.

While the photoconductive $n$-$i$-$n$ detector can produce a large gain, it suffers from the presence of a large dark current noise in the detector. As can be seen from

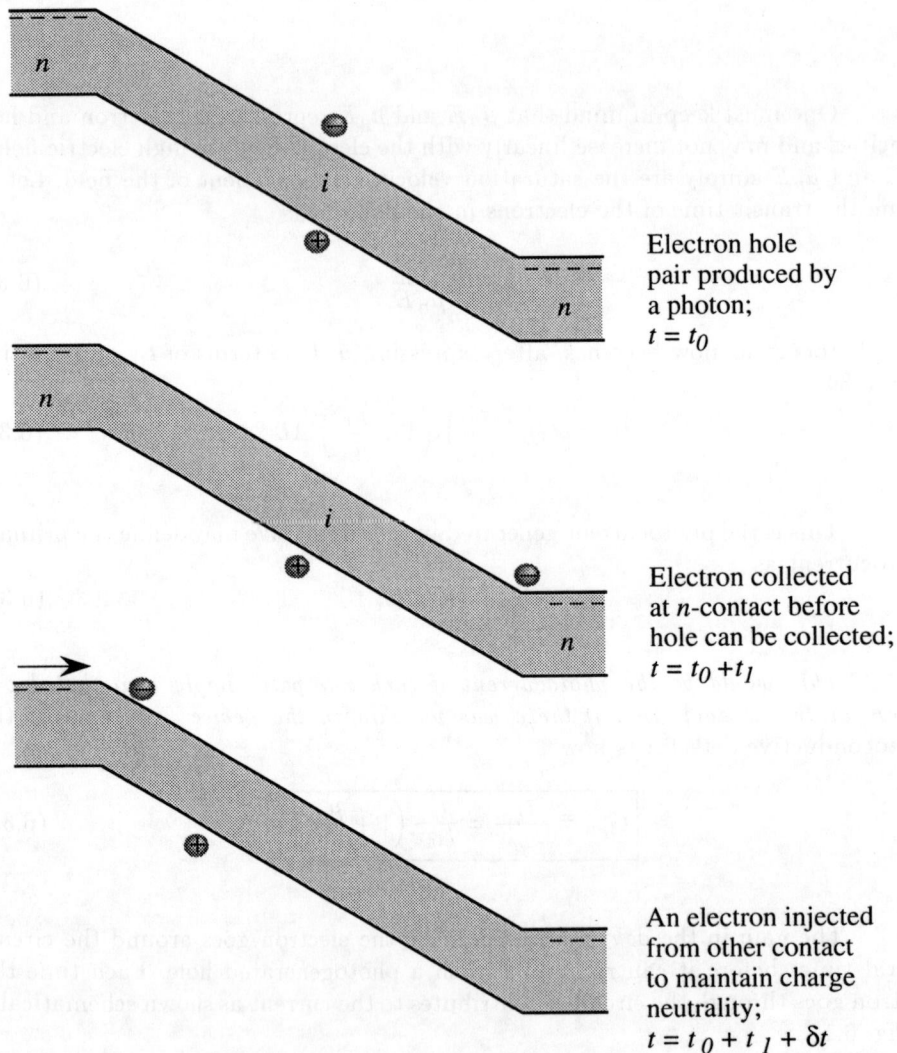

Electron hole
pair produced by
a photon;
$t = t_0$

Electron collected
at $n$-contact before
hole can be collected;
$t = t_0 + t_1$

An electron injected
from other contact
to maintain charge
neutrality;
$t = t_0 + t_1 + \delta t$

Figure 6.10: A schematic picture of how gain is produced in an $n$-$i$-$n$ detector due to several "round trips" an electron can make before the hole recombines either at the contact or with an electron in the semiconductor $i$-region.

Eqn. 6.29, even in dark the device can have a high conductivity and thus have a large dark current. In contrast, a reverse biased *p-n* or *p-i-n* diode has a very low dark current, thus allowing a large signal to noise ratio. Next we examine the photodiode detectors.

**EXAMPLE 6.7** Consider a GaAs photoconductor (*n*-type) with a length of 25 $\mu$m and an area of $10^{-6}$ cm$^2$. The minority carrier lifetime is $10^{-7}$ s. A voltage of 5 V is applied across the detector. Calculate the gain of the device using a constant mobility model with $\mu_n = 8000$ cm$^2$/V-s and $\mu_p = 1000$ cm$^2$/V-s. Also calculate the gain if an appropriate velocity-field relation is used.

The electron transit time is (for a constant mobility model)

$$t_{tr} = \frac{L}{\mu_n F} = \frac{L^2}{\mu_n V} = \frac{(25 \times 10^{-4} \ cm)^2}{(8000 \ cm^2 V^{-1} \ s^{-1})(5 \ V)} = 1.56 \times 10^{-10} \ s$$

The gain is

$$G_{ph} = \frac{\tau_p}{t_{tr}} \left(1 + \frac{\mu_p}{\mu_n}\right) = \frac{10^{-7} s}{1.56 \times 10^{-10} s} \left(1 + \frac{1000}{8000}\right) = 721.15$$

If we assume a proper velocity-field relation, $v(e) \cong 1.5 \times 10^7$ cm/s; $v(h) \cong 2 \times 10^6$ cm/s at the applied field of 2 kV/cm, the transit time is

$$t_{tr} = \frac{L}{v(e)} = \frac{25 \times 10^{-4}}{1.5 \times 10^7} = 1.67 \times 10^{-10} \ s$$

The gain is

$$G_{ph} = \frac{10^{-7}}{1.67 \times 10^{-10}} \left(1 + \frac{2 \times 10^6}{1.5 \times 10^7}\right) = 678.6$$

The error produced is minimal at the low fields of the detector. If the applied field was higher, the error in using the constant mobility model would become larger.

## 6.5   THE P-I-N PHOTODETECTOR

$\longrightarrow$ An important mode of operation of the *p-n* (or *p-i-n*) diode under illumination is when the diode is under reverse bias conditions. The reverse bias is, however, not so strong that there are breakdown effects as in the avalanche photodiode to be discussed in the next section.

A schematic of the band profile of a $p$-$i$-$n$ detector is shown in Fig. 6.11. Since the device is in reverse bias, the diode current in dark is $I_0$ and is independent of the applied bias. The photocurrent $I_L$ is essentially due to the carriers generated in the depletion region ($i$- region) that are collected. The diode is reverse biased so that the entire $i$-region is depleted and has a strong electric field. The device response is fast since the photocurrent is primarily due to the prompt photocurrent discussed in Section 6.3.1. The maximum current that can be collected is (we assume that the intrinsic region is larger than the diffusion lengths of the electrons or holes)

$$I_L = eA \int_0^W G_L(x)dx \tag{6.38}$$

where $W$ is the depletion width. In this expression we will account for the fact that as photons penetrate a material, their intensity decreases through absorption. The generation rate at a point $x$ is given from Eqn. 6.8 by

$$G_L(x) = \alpha \Phi_0(0) \, exp(-\alpha x) \tag{6.39}$$

where $\Phi_0(0)$ is the photon flux (number per cm$^2$ per second) at $x = 0$. The photocurrent is then from Eqns. 6.38 and 6.39

$$I_L = eA\Phi_0(0) \left[1 - exp\left(-\alpha W\right)\right] \tag{6.40}$$

If $R$ is the reflectivity of the surface (i.e., the fraction of photons that actually go into the device is $1 - R$), the photocurrent is

$$I_L = eA\Phi_0(0)(I - R) \left[1 - exp(-\alpha W)\right]$$

One measure of the detector efficiency is the ratio of the photocurrent density to the incident flux,

$$\eta_{det} = \frac{I_L}{eA\Phi_0(0)} = (I - R) \left[1 - exp(-\alpha W)\right] \tag{6.41}$$

For high efficiency one must have a small $R$ (by placing anti-reflective coatings) and a long $W$. However, if $W$ is too long the electron transit time that controls the device speed becomes too large, reducing the device speed. High speed devices have $W$ of about a micron or less and can operate at speeds in excess of 10 GHz.

## 6.5.1 Material Choice and Frequency Response of a P-I-N Detector

$\longrightarrow$ The foremost issue in the detector design is to work with a material which has a good absorption coefficient for the frequencies to be detected. For communication

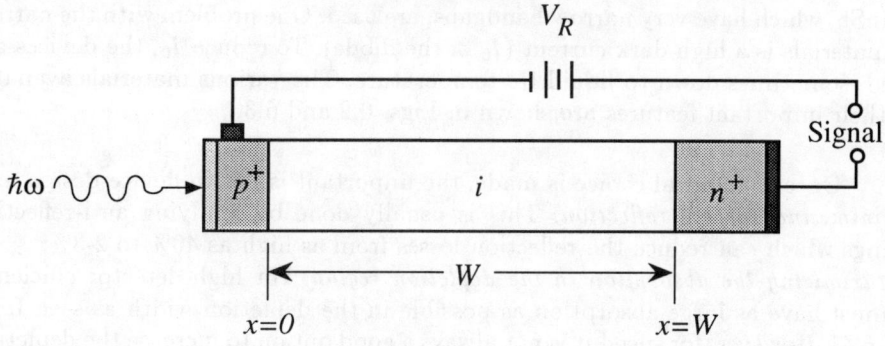

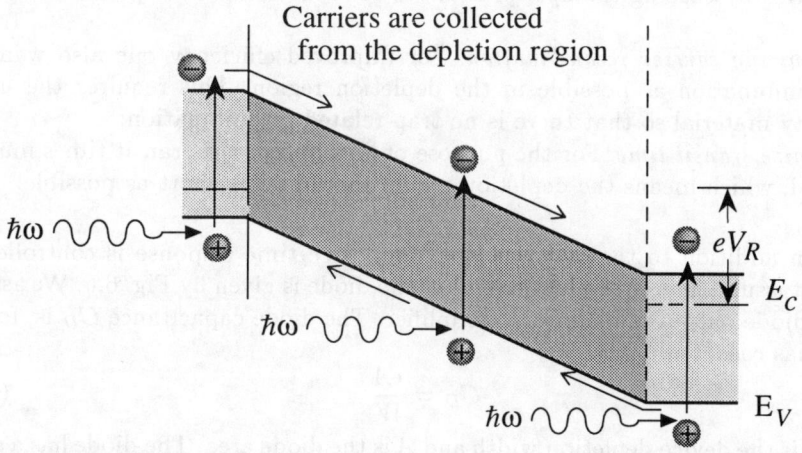

Figure 6.11: A cross-section and energy band profile of a *p-i-n* detector structure. Carriers generated in the depletion region are collected and contribute to the current. If the intrinsic region is thick, the photocurrent is dominated by carriers collected from the depletion region since the carriers generated in the neutral regions contribute a smaller fraction of photocurrent. Since the photocurrent is dominated by the prompt photocurrent, the device response is fast.

applications where GaAs/AlGaAs ($\hbar\omega \sim 1.45$ eV) sources are used (usually for local area networks), Si detectors are adequate unless high speeds are required. The Si detectors must have an absorption length of $> 10\mu$m. For longer wavelength applications, Ge detectors are used. An important wavelength is the 1.55 $\mu$m used for long haul communications since the fiber propagation loss is a minimum. For high speed applications one requires direct gap detectors so that the active absorption length can be brought down to a micron or less. Thus InGaAs detectors are now widely used for long haul communication applications.

In the case of night vision applications, materials like $Hg_xCd_{1-x}Te$, InAs, and InSb, which have very narrow bandgaps, are used. One problem with the narrow gap materials is a high dark current ($I_0$ of the diode). To reduce $I_0$, the devices are cooled, sometimes down to liquid He temperature. The various materials available and their important features are shown in Figs. 6.2 and 6.3.

Once a material choice is made, the important issues in device design are:
i) *Minimizing surface reflection:* This is usually done by applying anti-reflective coatings which can reduce the reflection losses from as high as 40% to 2-3%;
ii) *Maximizing the absorption in the depletion region:* For high detector efficiency one must have as large absorption as possible in the depletion width as seen from Eqn. 6.41. However, for speed it is not always a good option to increase the depletion width. Often metal mirrors can be used to increase the optical interaction length of the device by causing the optical beam to take more than one path through the device;
iii) *Minimizing carrier recombination:* For improved efficiency one also wants as little recombination as possible in the depletion region. This requires the use of high purity material so that there is no trap-related recombination;
iv) *Minimize transit time:* For the purpose of high speed, the transit times must be minimized, which means the depletion region should be as short as possible.

In addition to the above issues, the device time response is controlled by the circuit issues. The equivalent circuit of the diode is given by Fig. 6.6. We assume that the diode output is fed into an amplifier. The diode capacitance $C_D$ is, for the reverse bias case,

$$C_D = \frac{\epsilon A}{W} \qquad (6.42)$$

where $W$ is the device depletion width and $A$ is the diode area. The diode has a series resistance $R_s$ and conductance $G_D$. For high frequency response the capacitance and resistance are each to be minimized, which usually means reducing the area $A$ since if $W$ is increased too much the device is limited by transit time effects.

If the device capacitance and resistance are optimized, the transit time limits the device response. The transit time is controlled by the width of the depletion region and the saturation velocity, and is given by

$$t_{tr} = \frac{W}{v_s} \qquad (6.43)$$

Thus high frequency performance requires one to work with narrow depletion widths.

**EXAMPLE 6.8** Consider a silicon $p$-$i$-$n$ photodiode with an intrinsic region of width 10 $\mu$m. Light from a GaAs laser at energy $\hbar\omega = 1.43$ eV impinges upon the diode. The optical power is 1 W/cm$^2$. Calculate the photocurrent density in the detector.

The photon flux incident on the detector is

$$\Phi_0 = \frac{P_{op}}{\hbar\omega} = \frac{1 \ W/cm^{-2}}{1.43(1.6 \times 10^{-19} \ J)} = 4.37 \times 10^{18} \ cm^{-2} \ s^{-1}$$

The absorption coefficient for Si at GaAs wavelength (i.e., photons with energy 1.43 eV) is $\sim$700 cm$^{-1}$. The photocurrent density is (assuming no reflection losses)

$$
\begin{aligned}
J_L &= e\Phi_o \{1 - exp - (\alpha W)\} \\
&= (1.6 \times 10^{-19})(4.37 \times 10^{18}) \{1 - exp \ (-700 \times 10^{-3})\} \\
&= 0.352 \ A/cm^2
\end{aligned}
$$

One can see from this example that Si detectors are capable of producing acceptable response to GaAs photons. Since Si technology is so advanced, one uses Si detectors for GaAs lasers.

## 6.6   THE AVALANCHE PHOTODETECTOR

$\longrightarrow$ In addition to the *p-i-n* detector discussed in the previous section, an important class of detectors uses the impact ionization or avalanche process to obtain very high gain devices. While in the *p-i-n* detector the gain of the detector can, at most, be unity; in the avalanche photodetector (APD), very large gains can be achieved.

In Chapter 5, Section 5.3.2, we discussed the basis for the avalanche process in which a high energy electron (hole) creates an electron-hole pair. Usually this process, occurring at high electric fields, limits the high power operation of electronic devices, but in APDs it is exploited to multiply carriers generated by a photon.

It must be noted that the avalanche process requires the initial electron to have an energy somewhat greater than the bandgap energy since both energy and momentum are to be conserved as discussed in Chapter 5. The impact ionization coefficients for the electrons and holes are denoted by $\alpha_{imp}$ and $\beta_{imp}$. In Fig. 5.11 we show the value of $\alpha_{imp}$ and $\beta_{imp}$ for some important semiconductors.

Because of carrier multiplication, the APD has a very high gain and is thus used widely for optical communication systems. However, since the multiplication process is random, the device is quite noisy. The noise level depends upon the carrier multiplication factor and the $\alpha_{imp}/\beta_{imp}$ ratio. A number of material systems including Ge, Si, and many III-V compound semiconductors have been used in designing photodetectors. In the next subsection we will discuss some of the design issues for APDs.

## 6.6.1    APD Design Issues

As in the $p$-$i$-$n$ detector, the first design issue for an APD is to have a depletion region which is thick enough to allow absorption of the optical signal. This region has to be $\sim 1/\alpha(\hbar\omega)$ and can range from a micron for direct gap semiconductors to several tens of microns for indirect gap materials. The absorbing region and the avalanche region are kept distinct in general (especially if the absorbing region is larger than a micron) because of the difficulty of maintaining a constant high electric field over a long region. Typically electric fields of $\gtrsim 10^5$ V/cm are needed for the avalanche process. If the field is nonuniform, local charge oscillations can develop and the device output becomes difficult to control and predict.

An important structure for APDs is the "reach through" structure shown in Fig. 6.12. The structure has an $n^+$-$p$-$\pi$-$p^+$ configuration. Photon absorption occurs in the undoped $\pi$ region with thickness $W_{abs}$, while the avalanche region has a thin extent in the $n^+$-$p$ junction with width $W_{av}$. One can design the structure so that the field in the absorbing region is high enough that all the carriers move with saturation velocity ($v_s(e)$ or $v_s(h)$). Also, either electrons or holes can be chosen to be injected into the avalanche region. The avalanche process should be initiated by the carrier with the high impact ionization coefficient (electrons in our discussion) to optimize the device response.

In the APD, the current gain is very much dependent upon not only the bias applied, but also thermal fluctuations. Thus heat sinking is crucial in such devices. Also, guard rings are introduced to minimize the electric fields around the $p$-$n$ junction edges of the device. The guard ring involves an $n$-dopant which produces a $\pi$ region in the $p$ part of the device and thus there is a lower field at the $n^+\pi$ region of the guard ring and breakdown is avoided at the edges.

Use of heterojunction APDs has been increasing, and III-V compound semiconductors have produced some of the highest performance devices. In case of direct bandgap semiconductors, in principle, one does not need the long absorbing region, and devices can be built with thin absorbing and avalanching regions (which are the same physical region). However, for narrow gap materials, the strong field needed for impact ionization can cause a large leakage current due to band-to-band tunneling. To avoid this, one uses the separate absorption and multiplication (SAM) APD. In such devices, the multiplication occurs in a larger gap material, while the absorption occurs in the narrow gap material. In Fig. 6.13 we show such a structure for a long wavelength APD based on $In_{0.53}Ga_{0.47}As/InP$ technology. Note that a graded region is used so that holes generated in absorption are not trapped at the valence band discontinuity of InGaAs and InP.

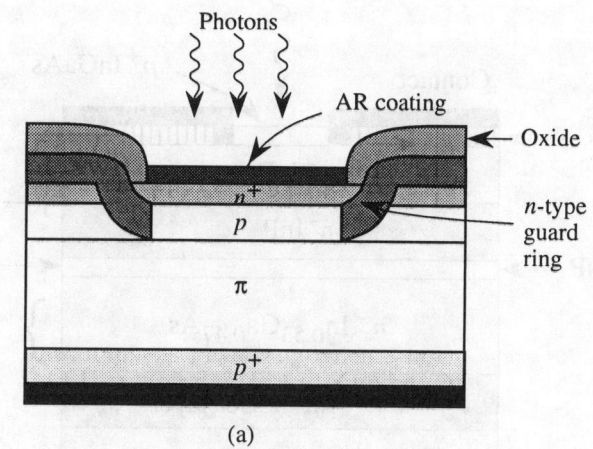

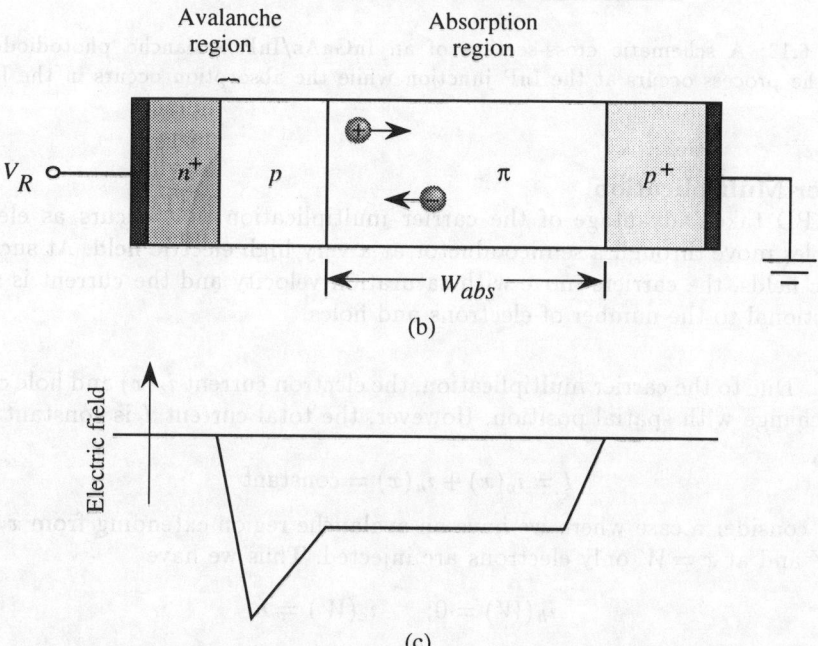

Figure 6.12: (a) A schematic of the reach through APD. (b) The cross-section of the APD showing the regions for absorption and avalanching. In the structure shown, the electrons are responsible for starting the multiplication process. (c) The electric field profile in the APD structure. The strong field at the $n^+p$ junction causes the avalanche process.

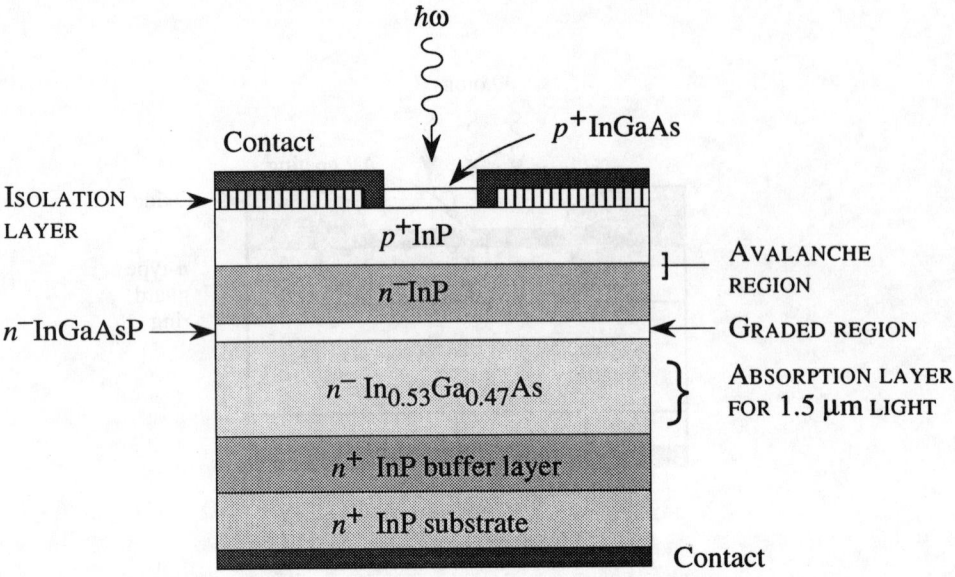

Figure 6.13: A schematic cross-section of an InGaAs/InP avalanche photodiode. The avalanche process occurs at the InP junction while the absorption occurs in the InGaAs region.

### Carrier Multiplication

The APD takes advantage of the carrier multiplication that occurs as electrons and holes move through a semiconductor at a very high electric field. At such high electric fields, the carriers move with saturation velocity and the current is simply proportional to the number of electrons and holes.

Due to the carrier multiplication, the electron current $i_e(x)$ and hole current $i_h(x)$ change with spatial position. However, the total current $I$ is constant in the device,

$$I = i_e(x) + i_h(x) = \text{constant} \tag{6.44}$$

Let us consider a case where we have an avalanche region extending from $x = 0$ to $x = W$ and at $x = W$ only electrons are injected. Thus we have

$$i_h(W) = 0; \qquad i_e(W) = I \tag{6.45}$$

The multiplication factor for the device can be defined by the relation

$$M_e = \frac{I}{i_e(0)} \tag{6.46}$$

i.e., the ratio of the electron current injected at $x = 0$ and the total current in the device.

If the avalanche process is occuring under a uniform electric field, the values of $\alpha_{imp}$ and $\beta_{imp}$ have no spatial dependence. A straightforward, but rather tedious analysis needs to be done to find the multiplication factor. We will simply give the result of such an analysis (see, for example, Gowar (1989) for more details) as

$$
\begin{aligned}
M_e &= \frac{1}{1 - \alpha_{imp} \int_0^W exp\left\{-(\alpha_{imp} - \beta_{imp})x\right\}dx} \\
&= \frac{1}{1 - \frac{\alpha_{imp}}{\alpha_{imp} - \beta_{imp}}\left[1 - exp\left\{-(\alpha_{imp} - \beta_{imp})W\right\}\right]}
\end{aligned}
\tag{6.47}
$$

If $\alpha_{imp}$ and $\beta_{imp}$ are the same, we simply get

$$
M_e \longrightarrow \frac{1}{1 - \alpha_{imp}W}
\tag{6.48}
$$

In an experimental setup, two factors related to circuit parameters limit the multiplication level reached. One is the series resistance $R_s$ between the junction and the diode terminals. The second factor comes from the fact that once multiplication starts, the device temperature increases and this reduces $\alpha_{imp}$ and $\beta_{imp}$ and thus limits $M_e(M_h)$. In a diode with a breakdown voltage $V_B$, the experimentally observed multiplication factor can be fitted to the following relation:

$$
M = \frac{1}{1 - \left(\frac{V - IR}{V_B}\right)^{n'}}
\tag{6.49}
$$

where $n'$ is a parameter depending upon the device design, $R$ is an effective resistance, and $V$ is the applied bias. A key attraction of the APDs is the high gain that can be achieved in the device. Thus the device is suitable for detection of very low photon intensities. However, a price has to be paid in terms of the device bandwidth and noise. The gain-bandwidth product remains essentially constant for a device. This outcome is similar to that observed in the case of photoconductors as well. Thus, if one wants a very high detectivity for very weak optical signals, one has to sacrifice the device speed.

While APDs do produce very efficient detectors, particularly for low level optical signals, they suffer from a number of problems. They require a very high voltage power supply which is difficult to provide especially for long haul undersea communication operations. They also need temperature stabilization circuitry. Also, their reliability is not as high as that of a p-i-n detector. Thus their use is presently limited to applications where high gain is of paramount importance.

**EXAMPLE 6.9** Consider a typical avalanche photodiode with the following parameters:

| | | | |
|---|---|---|---|
| Incident optical power, | $P_{op} \cdot A$ | = | 50 $mW$ |
| Efficiency, | $\eta_{det}$ | = | 90% |
| Optical frequency, | $\nu$ | = | $4.5 \times 10^{14}\ Hz$ |
| Breakdown voltage, | $V_B$ | = | 35 $V$ |
| Diode voltage, | $V$ | = | 34 $V$ |
| Dark current, | $I_0$ | = | 10 $nA$ |
| Parameter $n'$ for the multiplication | | = | 2 |

Assume that the series resistance is negligible. Calculate the a) multiplication factor; b) photon current; c) photocurrent.

a) The multiplication factor from Eqn. 6.49 is

$$M = \left[1 - \left(\frac{34}{35}\right)^2\right]^{-1} = 16.67$$

b) The photon current is

$$I_{ph} = A\Phi_0 = \frac{P_{op}A}{h\nu} = \frac{(50 \times 10^{-3}\ W)}{(6.625 \times 10^{-34}\ J-s)(4.5 \times 10^{14}\ Hz)}$$

$$= 1.68 \times 10^{17}\ s^{-1}$$

c) The unmultiplied photocurrent is from Eqn. 6.41

$$I_L = e\eta_{det}I_{ph} = (1.6 \times 10^{-19}\ C)(1.51 \times 10^{17}\ s^{-1})$$

$$= 24.16\ mA$$

The multiplied photocurrent is

$$M \cdot I_L = (24.16\ mA)(16.67) = 0.4\ A$$

## 6.7  THE PHOTOTRANSISTOR

$\longrightarrow$ While the APD discussed above provides very high gain detection, it is an inherently high noise device due to the random nature of the carrier multiplication process. Another device which can produce gain and function as a detector is the bipolar transistor. The phototransistor, the name for the bipolar device used for optical detection, provides high gain due to the transistor action. The device is also a low noise device when compared to an APD.

The phototransistor is shown schematically in Fig. 6.14. The device is usually operated with the base open-circuited. Focusing on the *npn* phototransistor of Fig. 6.14, the electrons and holes produced in the reverse bias base-collector junction are swept in the field to produce the photocurrent $I_L$. The holes are injected into the base and provide the base current, causing electrons to be injected from the emitter. The currents have the following relation,

$$I_E = \alpha I_E + I_L \tag{6.50}$$

where $\alpha$ is the common base current gain. Since the base is open- circuited, $I_C = I_E$ and we have

$$I_C = \alpha I_C + I_L \tag{6.51}$$

$$\frac{I_C}{I_L} = \frac{1}{1 - \alpha} \tag{6.52}$$

Since $\alpha$ is close to unity, there is a large current gain because of the transistor action. The overall gain of the device has to be obtained by multiplying the above result by a factor $\eta$ which represents the fraction of light absorbed in the base-collector junction of the transistor. Losses due to reflection and due to transmission through the device cause $\eta$ to be less than unity.

The phototransistor does not have an extremely good high frequency response due to the very large capacitance associated with the base-collector junction. However, it finds important uses due to its low noise and high gain.

## 6.8 METAL-SEMICONDUCTOR DETECTORS

$\longrightarrow$ An extremely important class of photodetectors involves the use of a Schottky barrier produced between a metal and a lightly doped semiconductor. A key advantage of the Schottky barrier device is that being a majority carrier device (i.e., based upon only electrons or upon holes), it does not suffer from speed delays arising from minority carrier lifetime issues.

Schottky-barrier-based devices involve two kinds of configuration. In Fig. 6.15a we show a device which is a simple mesa structure with an $n^+$ layer on a semi-insulating substrate. The active absorbing layer is lightly doped ($N_d \sim 10^{15}$ cm$^{-3}$) and a thin semitransparent metal layer is deposited on it. The metal film is thick

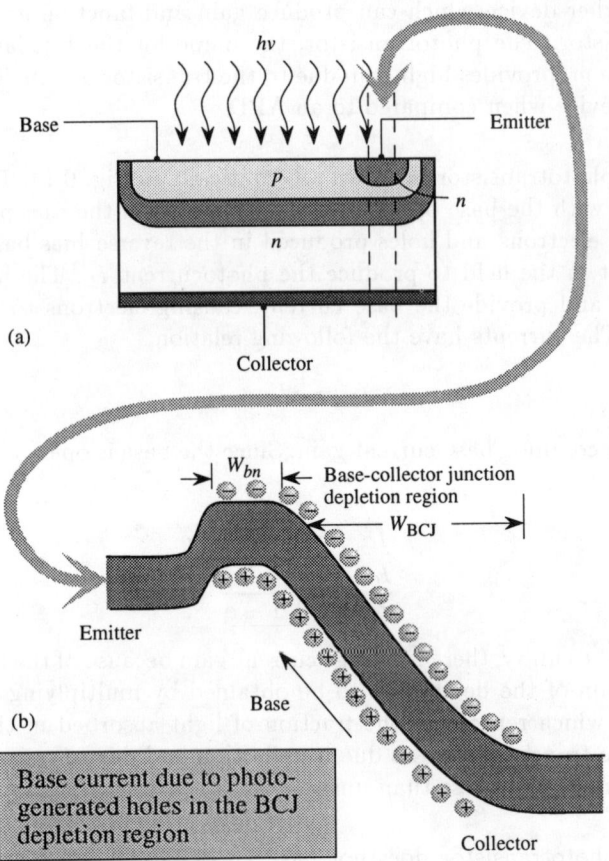

Figure 6.14: (a) A schematic of the phototransistor. (b) Band diagram of the phototransistor in the open base mode. Holes generated in the reverse biased base-collector junction region provide a base current signal which causes the electrons to be injected from the emitter. Due to the transistor action, a small photocurrent induced base current produces a large collector current.

enough to allow the Schottky barrier formation ($\sim$300-400 Å) but thin enough to allow light to pass through. For high performance the metal film is coated with dielectric anti-reflection coatings and the device area is kept as small as $10^{-5}$ cm$^2$ ($\sim$50 $\mu$m diameter mesa diodes).

The band profile of the Schottky barrier diode is shown in Fig. 6.15b. Also shown are the Schottky barrier height $e\phi_{bn}$ and the potential drop across the barrier. When light impinges upon the diode, the diode can respond in two important regimes:

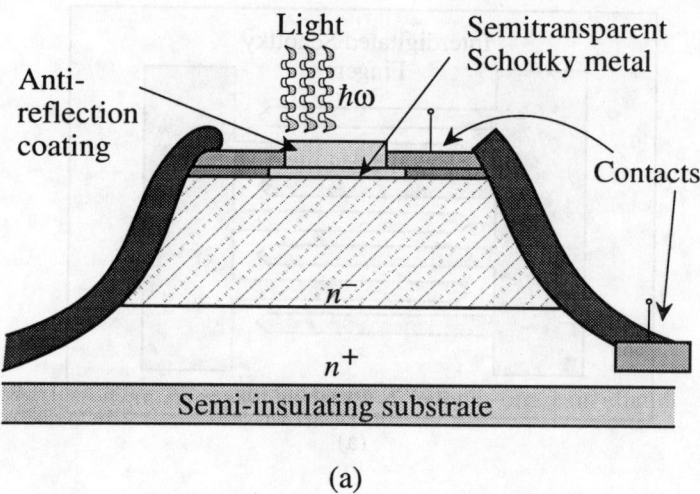

(a)

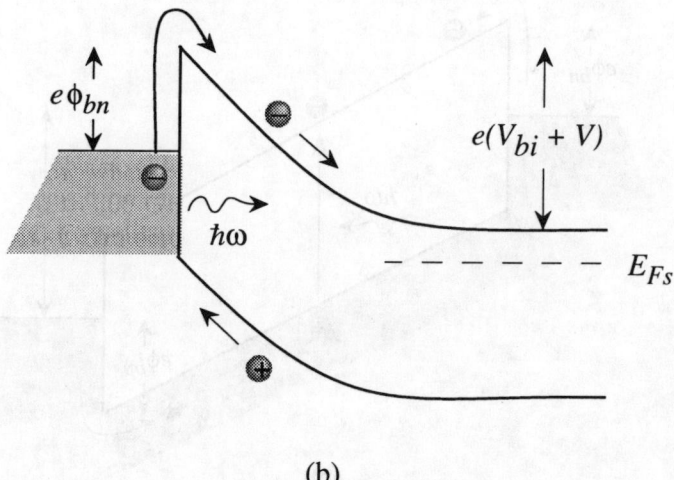

(b)

Figure 6.15: (a) A schematic of the Schottky barrier detector. (b) The band profile of the detector. $V_{bi}$ is the built-in voltage and $V$ is the applied bias.

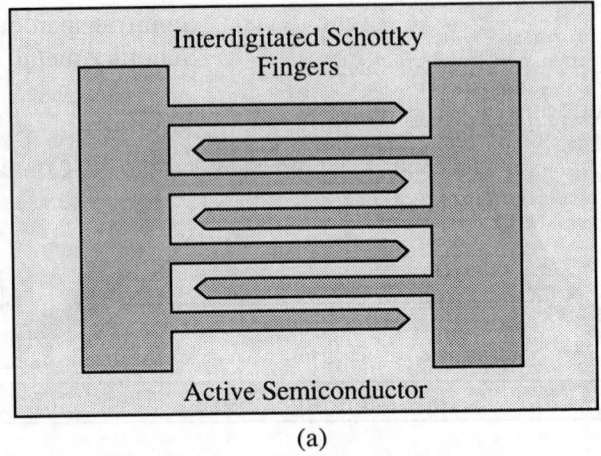

(a)

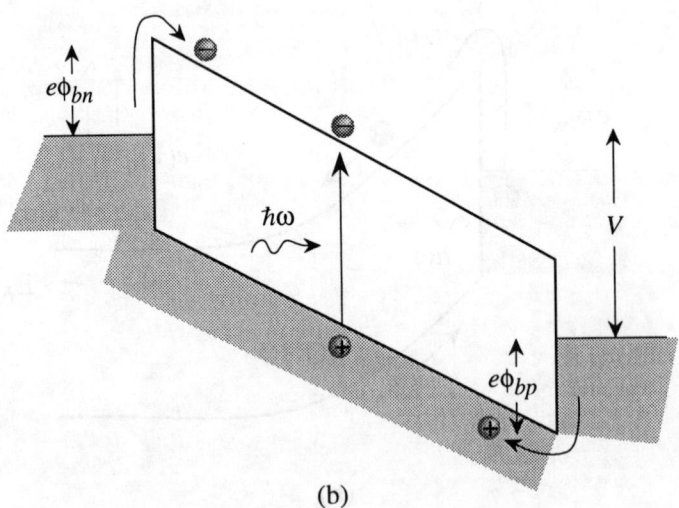

(b)

Figure 6.16: (a) A schematic of the MSM detector using interdigitated Schottky fingers. (b) Band profile of the MSM photodiode under bias.

i) $\hbar\omega > e\phi_{bn}$: In this case, electrons can be excited in the metal barrier to overcome the Schottky barrier height. As a result, a photocurrent will flow in the device. This current will add to the dark current in the reverse bias diode.

ii) $\hbar\omega > E_g$: In this case, e-h pairs will be created in the semiconductor. As in the case of the photodiode, the carriers generated in the depletion region will be swept out to produce photocurrent.

In high-speed devices the depletion region is less than a micron so that device speeds can be extremely high. With proper design Schottky barrier diodes can operate up to 150 GHz.

A second class of metal semiconductor detectors is the metal-semiconductor-metal (MSM) detector in which two Schottky barriers are placed in a planar geometry close to each other. In actual design the approach used is the interdigitated scheme shown in Fig. 6.16a. The spacing between the fingers is ~1-5 $\mu$m so that when a bias is applied between the contacts, the region between the fingers can be completely depleted.

As seen in Fig. 6.16b, when a bias is applied across the fingers, one junction becomes reverse biased, while the other one becomes forward biased. However, since the semiconductor is depleted, the current in the forward biased junction is not the usual high electron forward bias current. Instead, the dark current in the forward biased junction is due to the hole current injected from the metal over the barrier $e\phi_{bp}$ as shown in Fig. 6.16b. As a result, under a strong applied bias, the dark current of the device is equal to the reverse saturation currents from electrons and holes. The dark current density of a Schottky barrier is given by

$$J = A_n^* T^2 e^{-e\phi_{bn}/k_B T} + A_p^* T^2 e^{-\phi_{bp}/k_B T}$$

where $A_n^*$ and $A_p^*$ are the electron and hole effective Richardson constants. The dark current density is usually higher than that achievable in $p$-$i$-$n$ diodes. However, sufficiently low dark current can be achieved for most applications.

The MSM detectors are found to have internal gain, often at even low applied biases where impact ionization cannot occur. This suggests the possibility of photoconductive gain enhanced by traps which may capture and re-emit either electrons or holes. MSM diodes have been fabricated in both GaAs and InGaAs systems. Thus these devices can be applicable in both local area networks and long haul communication systems. It is also important to point out that MSM detectors are very attractive for optoelectronic integrated circuit (OEIC) applications.

## 6.9  NOISE AND DETECTION LIMITS

~→  So far, in this chapter, we have discussed a number of important optical detectors. To assess the quality of the photodetectors, important figure of merit parameters have been developed. An important parameter refers to the weakest source of radiation that can be detected.

To understand the detection limits we will discuss the noise of the detectors. The optical signal must generate a current signal larger than the device noise. The electrons which make up the device current are discrete particles carrying discrete charge. When current flows in a device, the electrons are distributed in energy and momentum according to some distribution function. *As a result, all electrons do not travel with identical velocity and energy.* Thus, if one sits at an electrode and counts the number of electrons coming in a certain time interval $\Delta t$, the number will vary as shown schematically in Fig. 6.17. *The shorter the time interval of interrogation, the larger the variation.*

We assume that the variation in the number of electrons coming in during a time interval $\Delta t$ is given by a Poisson distribution. According to this statistical distribution, if $a\Delta t$ is the *average number of particles passing in a time interval $\Delta t$,* the probability of getting $N$ particles in the interval $\Delta t$ is (for $N$ much larger than one)

$$P(N, \Delta t) = \frac{1}{\sqrt{2\pi(a\Delta t)}} exp \left( -\frac{(N - a\Delta t)^2}{2a\Delta t} \right) = \frac{1}{\sqrt{2\pi\overline{N}}} exp \left( -\frac{\Delta N^2}{2\overline{N}} \right) \quad (6.53)$$

where $\overline{N}$ is the average value $(= a\Delta t)$ and $\Delta N$ is the fluctuation from the average value. This probability function is shown in Fig. 6.17b. This function maximizes when

$$N = \overline{N} = a\Delta t \quad (6.54)$$

as expected. The rms deviation of the Poisson distribution, i.e., the noise, is found, from statistics, to be

$$\sqrt{(\Delta N)^2} = \sqrt{(N - \overline{N})^2} = \sqrt{\overline{N}} \quad (6.55)$$

*It is important to point out that this noise, called the shot noise, would occur in the photon stream that is impinging upon a detector, or the current flowing in the detector resulting from the e-h pair generation since both events involve discrete particles.*

The average signal in the device is given by $\overline{N}$ $(= a\Delta t)$. An important device parameter is the ratio between the signal generated and the random noise.

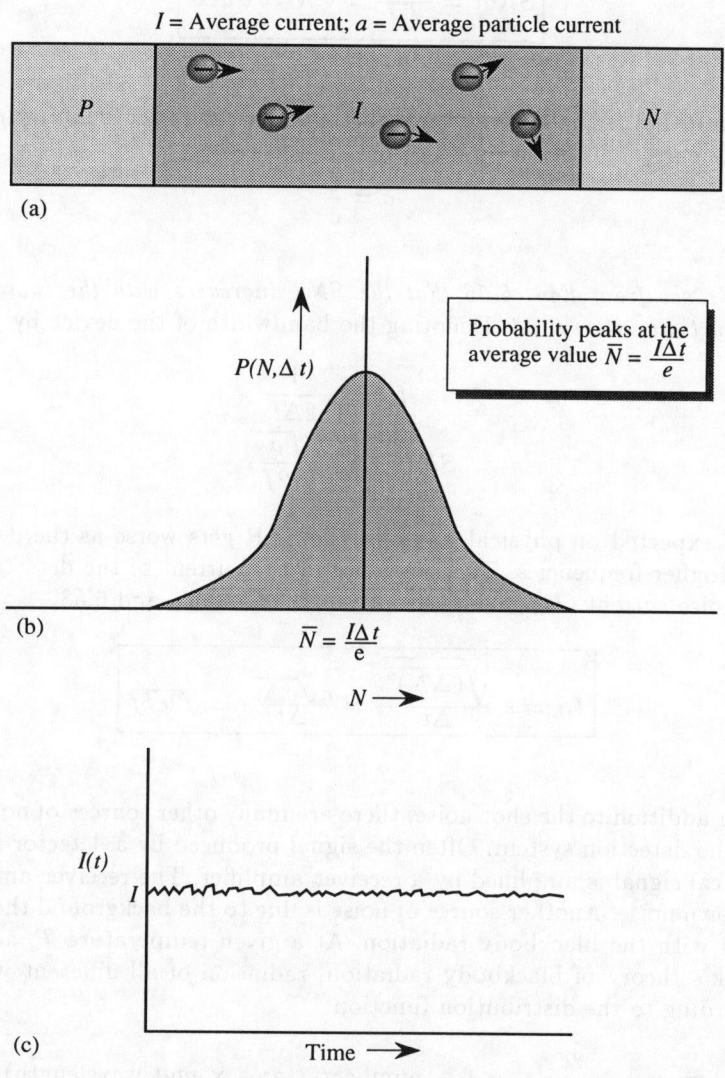

Figure 6.17: (a) The electrons in a semiconductor are moving randomly with a certain distribution function; (b) the probability of finding $N$ electrons crossing an area $A$ in time interval $\Delta t$. The average particle current is $a$ so that the mean value of the particle number is $\overline{N} = a\Delta t$; (c) schematic of the current flow in a device. The statistical variations result in noise in the current.

The signal to noise ratio (SNR) for the shot noise limited detector becomes

$$SNR = \frac{\overline{N}}{\sqrt{\overline{N}}} = \sqrt{\overline{N}} = \sqrt{a\Delta t} \qquad (6.56)$$

Note that if the current through the device is $I$, the quantity $a$, which is the particle current, is given by

$$a = \frac{I}{e} \qquad (6.57)$$

*We see from Eqn. 6.56 that the SNR increases with the increase in the observation time interval* $\Delta t$. Denoting the bandwidth of the device by $f$, we have

$$f \cong \frac{1}{2\Delta t} \qquad (6.58)$$

$$SNR = \sqrt{\frac{a}{2f}} \qquad (6.59)$$

As can be expected on physical grounds, the SNR gets worse as the device is operated at higher frequencies. The rms noise in the current of the detector is called the shot noise current ($I_{sh}$) and is, from Eqns. 6.55. 6.56, and 6.58,

$$I_{sh} = e\frac{\sqrt{(\Delta N)^2}}{\Delta t} = \frac{e\sqrt{a\Delta t}}{\Delta t} = \sqrt{2eIf} \qquad (6.60)$$

In addition to the shot noise, there are many other sources of noise encountered by the detection system. Often the signal produced by a detector in response to an optical signal is amplified by a receiver amplifier. The receiver amplifier produces its own noise. Another source of noise is due to the background thermal noise associated with the blackbody radiation. At a given temperature $T$, according to the Planck's theory of blackbody radiation, radiation of all different wavelengths exist according to the distribution function

$$N(\lambda) = \frac{2c}{\lambda^4 \left[exp \frac{\hbar\omega}{k_BT} - 1\right]} \quad \text{number/s/(area} \times \text{unit wavelength)} \qquad (6.61)$$

*The photons with energy* $\hbar\omega > E_g$ *will create noise by creating e-h pairs. However, if* $k_BT$ *is small compared to the bandgap, the noise level due to the thermal background is negligible. Thus at most optical frequencies, the thermal noise is not significant.* However, the thermal noise plays an important role in the receiver amplifier resistance noise.

The effect of the noise is reflected in the bit error rate (BER) in a digital communication system. Consider a system operating at a rate of $B$ bits/s. Let us assume that there are $\overline{N}$ photons impinging when the data bit is 1 and zero photons when the data bit is 0. *What is the probability that no photons come during a period when the data bit is 1?* This is given by the Poisson function of Eqn. 6.53. *The ratio of the probabilities $P(N = 0)$ to $P(N = \overline{N})$ during the bit 1 transmission gives the bit error rate.* This rate is from Eqn. 6.53,

$$BER = \frac{P(N = 0)}{P(N = \overline{N})} = exp\left(-\overline{N}\right) \tag{6.62}$$

If $E(1)$ is the average energy carried by the optical pulse in a data bit 1, the probability becomes $(\overline{N} = E(1)/\hbar\omega)$

$$\boxed{BER = exp\left(-E(1)/\hbar\omega\right)} \tag{6.63}$$

In the shot noise limit, this gives the bit error rate of one part in 1/BER. A BER of $\sim 10^{-9}$ is usually required in an optical communication system. This requires a certain minimum energy in the optical pulse, as examined below in Example 6.12.

An important figure of merit for a detector is the minimum detectable signal that would produce the same rms output as that generated by the noise. This is given by the noise equivalent power (NEP). Let us consider the case of the detector that is limited by the shot noise. The optical power density $P_{op}$ needed to produce a photocurrent $I_L$ in a detector with efficiency $\eta_Q$ is (see Eqn. 6.10)

$$P_{op} \cdot A = \frac{I_L \, \hbar\omega}{\eta_Q e} \tag{6.64}$$

*For the noise equivalent power we equate the noise current $I_{sh}$ to $I_L$.* This gives, from Eqn. 6.60,

$$I_L = I_{sh} = [2e(I_L + I_0)f]^{1/2} \tag{6.65}$$

where $I_0$ is the dark current of the detector. If $I_0 \ll I_L$, we get

$$I_L = 2ef \tag{6.66}$$

The optical power required at bandwidth $f$ is now from Eqn. 6.64

$$\boxed{P_{op} \cdot A = NEP = \frac{2\hbar\omega f}{\eta_Q}} \tag{6.67}$$

The NEP of the ideal quantum detector is given by the above equation with the quantum efficiency $\eta_Q = 1$.

If $I_0 \gg I_L$ we have, from Eqn. 6.65,

$$I_L \sim (2eI_0 f)^{1/2} \tag{6.68}$$

and

$$P_{op} \cdot A = NEP = \frac{(2eI_0 f)^{1/2} \hbar \omega}{\eta_Q e} \tag{6.69}$$

The detectivity of the detector is defined by

$$D = \frac{1}{NEP} \tag{6.70}$$

The detectivity or NEP depends upon the area of the detector as well as the bandwidth of the detector. A quantity called the specific detectivity, $D^*$, is defined which accounts for the variable bandwidth and detector area:

$$D^* = \frac{(Af)^{1/2}}{NEP} \tag{6.71}$$

In the choice of a detector, *once the bandwidth requirements are met, one chooses a detector with the highest $D^*$ values.* In Example 6.13 below, we discuss typical values of the parameters discussed above.

In the discussion above we have not included the effects of the random carrier multiplication processes that occur in an avalanche device. In an APD, the absorbed photocarriers are injected into an avalanche region and not only is the injected photocurrent multiplied, the noise is also amplified. The net effect of the carrier multiplication is that the total noise in the device is given by a noise current given by (compare this with the result in Eqn. 6.60)

$$(I_{sh})^2 = 2eMF_n I f \tag{6.72}$$

where $M$ is the carrier multiplication and $F_n$ is a noise factor. If the multiplication is initiated by electrons, we denote $F_n$ by $F_e$, and the noise factor turns out to have a form (see, for example, J. Gowar, 1989)

$$F_e = M_e \left[ 1 - \frac{(1 - \tilde{k})(M_e - 1)^2}{M_e^2} \right] \tag{6.73}$$

while if it is initiated by holes it becomes

$$F_h = M_h \left[ 1 + \frac{(1 - \tilde{k})(M_h - 1)^2}{\tilde{k} \, M_h^2} \right] \tag{6.74}$$

Here $\tilde{k}$ is the ratio $\beta_{imp}/\alpha_{imp}$ of the hole and electron impact ionization coefficient $M_e(M_h)$ is the electron (hole) multiplication factor. If we consider the electron initiated multiplication, for large $M_e$ we get

$$F_e \longrightarrow \tilde{k} M_e \tag{6.75}$$

Thus a material with a small $\tilde{k}$-value (i.e., $\alpha_{imp} \gg \beta_{imp}$) has a lower noise factor. If holes were being injected for multiplication, the noise factor would be low if $\beta_{imp} >> \alpha_{imp}$. It is thus optimum to use the injection of the carrier for which the ionization is highest.

**EXAMPLE 6.10** A flux of $8 \times 10^6$ particles/s impinges on a detector. Calculate the maximum bandwidth at which the SNR for the device is unity for a shot noise limited case.

From Eqn. 6.59,

$$SNR \;=\; 1 = \sqrt{\frac{8 \times 10^6 \ s^{-1}}{2fs^{-1}}}$$

$$\text{or} \qquad f = 4 \times 10^6 \ Hz.$$

**EXAMPLE 6.11** A shot noise limited detector is to operate at 1 GHz. Calculate the current level needed to ensure an SNR of 100 (or 40 dB).

The particle current is given by Eqn. 6.59 as

$$\begin{aligned} a \;&=\; (SNR)^2(2f) \\ &=\; (100)^2(2 \times 10^9 \ s^{-1}) = 2 \times 10^{13} \text{particles/s} \end{aligned}$$

The electric current is

$$I = ea = (1.6 \times 10^{-19} \ C)(2 \times 10^{13} \ s^{-1}) = 3.2 \times 10^{-6} \ A$$

**EXAMPLE 6.12** In an optical data communication system, a bit error rate of $10^{-9}$ is desired. The transmission is to occur at 1 Gbit/s. Calculate the minimum optical energy needed per pulse to code a 1 bit if the detector is shot-noise limited. The transmission is to occur with a GaAs laser. Calculate also the average power needed.

A bit error rate of $10^{-9}$ requires that the number of photons needed to code the 1 bit are (see Eqn. 6.63)

$$\overline{N} = -\ell n \ (10^{-9}) = 21$$

The energy needed is

$$E(1) = \overline{N}\hbar\omega = 21(1.43)eV = 4.8 \times 10^{-18} \ J$$

The average power needed for a 1.0 Gbit/s transmission (assuming equal number of 0 and 1 bits) is

$$Power = \frac{(4.8 \times 10^{-18} \ J)}{2}(10^9 \ s^{-1}) = 2.9 \times 10^{-9} \ watts$$

**EXAMPLE 6.13** A detector has the following parameters:

| | | |
|---|---|---|
| Detector area, | $A$ | $= 1 \ cm^2$ |
| Detection wavelength, | $\lambda$ | $= 1.0 \ \mu m$ |
| Quantum efficiency, | $\eta_Q$ | $= 20\%$ |
| Bandwidth, | $f$ | $= 1 \ Hz$ |
| Noise current, | $I_{sh}$ | $= 10 \ pA$ |

Calculate the a) responsivity, b) noise equivalent power, c) detectivity, and d) specific detectivity of the device.

a) The responsivity is given by

$$R = \frac{\eta_Q e}{\hbar \omega} = \frac{(0.2)(1.6 \times 10^{-19} \ C)}{1.242 \times 1.6 \times 10^{-19} \ J} = 0.16 \ A/W$$

b) The noise equivalent power is from Eqns. 6.66 and 6.67 (equating $I_L - I_{sh}$) as

$$NEP = \frac{I_{sh}\hbar\omega}{\eta_Q e} = \frac{(10 \times 10^{-12} \ A)(1.242 \times 1.6 \times 10^{-19} \ J)}{(0.2)(1.6 \times 10^{-19} \ C)}$$

$$= 62 \ pW$$

c) The detectivity is

$$D = \frac{1}{NEP} = 1.61 \times 10^{10} \ W^{-1}$$

d) The specific detectivity is

$$D^* = D(Af)^{1/2} = 1.61 \times 10^{10}(1)^{1/2}$$

$$= 1.61 \times 10^{10} \ cm - Hz^{1/2}/W$$

## 6.10   THE RECEIVER AMPLIFIER

$\mathcal{R}$  The photodetectors discussed above are rarely used independently in an optical information processing system. Usually the optical signal coming in and the

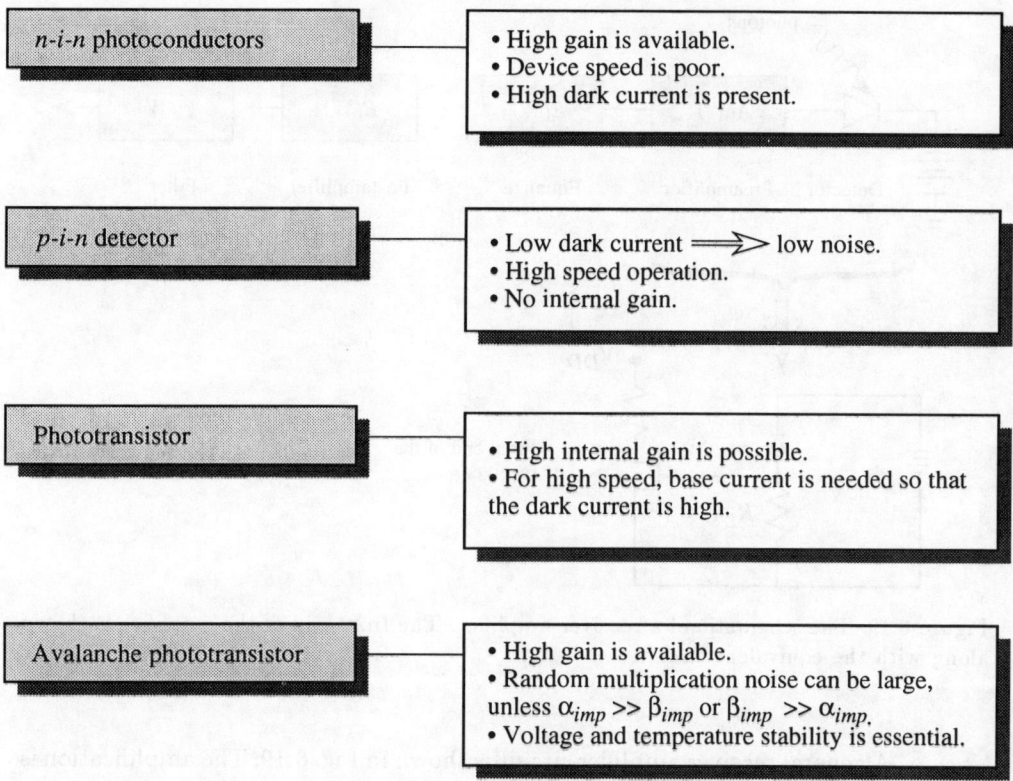

Figure 6.18: A comparison of the advantages and disadvantages of various detectors. The *p-i-n* and APD devices are chosen for high-speed applications. An amplifier is used in the receiver circuit.

resultant photocurrent is quite weak and must be amplified before it can be used for further processing. In detectors such as *p-i-n* diodes, there is no gain in the device which makes it essential to have an amplification circuit along with the detector. For high sensitivity, internal gain can be provided by a phototransistor, an *n-i-n* photoconductive detector or an avalanche photodiode. In Fig. 6.18 we show a comparison of the advantages and disadvantages of the various detectors.

The *p-i-n* detector and the APD have emerged as the most important detectors for high speed high sensitivity applications. The *p-i-n* is based on direct gap materials to provide a short absorption region and high speed. The APD's can be based on Si, Ge and compound semiconductors such as InGaAs. In an APD one has to maintain very stable voltage and temperature values which makes the system costly and somewhat unreliable especially if it is placed in a region that is difficult to access. For example, *p-i-n* diodes and not APDs are used in undersea regenerators for long distance optical communications.

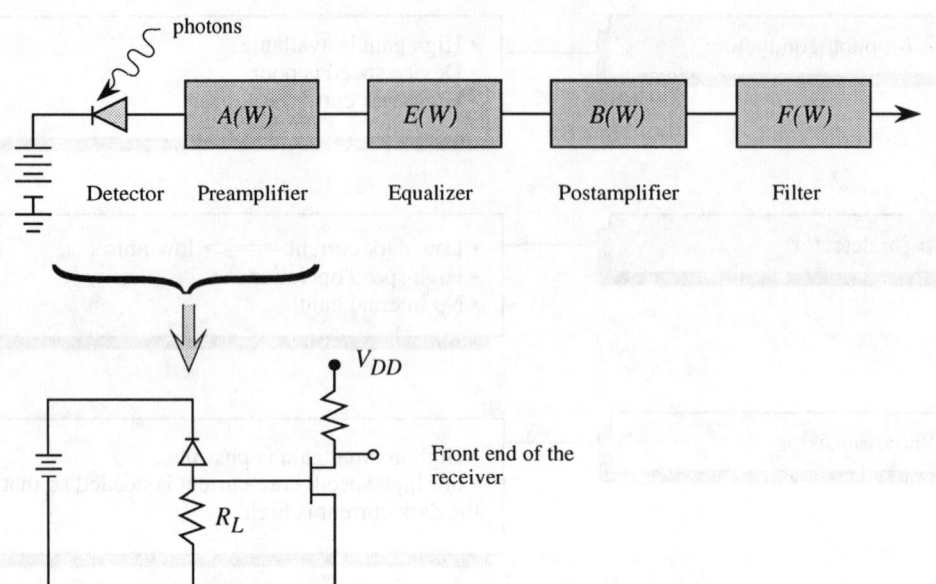

Figure 6.19: The schematic of a receiver amplifier. The front end of the amplifier is shown along with the equivalent circuit.

A general receiver amplifier circuit is shown in Fig. 6.19. The amplification is provided by a transistor which is either a field effect transistor or a bipolar transistor. The choice of the transistor is a very important issue in the receiver design. One obviously wants a device with high gain and low noise. A field effect transistor (FET) is usually used in the receiver amplifier because of the very high frequencies up to which the FET can operate and the low noise level that can be achieved. Advances in FET technology have come from the use of the compound semiconductors and from the use of heterostructure based devices such as the MODFET. An HBT can also be used in the photoreceiver, and recent advances in HBT technology are being exploited for integrated HBT-detector receivers.

We will not discuss the details of the noise of the receiver amplifier, but we will summarize some of the important findings. In Fig. 6.19 we show the block diagram of the photoreceiver. The equalizer shown in the figure is used to reshape the input pulse which is usually distorted during the detection process. The post amplifier further amplifies the pulse and the filter sets the bandwidth of the receiver.

In addition to the detector noise, a number of additional noise sources limit the receiver performance. These include the noise in the load resistor $R_L$, the noise in the transistor channel, and flicker noise due to defects in the semiconductor used. *At low frequencies ($\sim$ 500 MHz), the dominant noise source is usually the noise in the load resistor. At higher frequencies, the transistor channel noise dominates the* *detector noise.*

## 6.11  THE CHARGE COUPLED DEVICE

$\longrightarrow$  In our discussions so far on various detectors, we have considered a single detector element. In many applications, an entire optical image is to be detected and stored. In such imaging applications, arrays of detectors are used and the electrical information generated by the image is stored in an electronic memory for further manipulation. A most important technology that has developed in imaging applications is the charge coupled device (CCD) technology. This technology has revolutionized the video camera industry, essentially replacing the camera tubes by semiconductor chips capable of very high resolution images. These days, CCD chips with up to 2 million pixels are available. The CCD technology is also responsible for the home video industry which has allowed the average consumers with no special expertise to make video movies of their bouncing babies.

The basic device upon which the CCD technology is based is the metal-insulator-semiconductor (MIS) or metal-oxide-semiconductor (MOS) capacitor. In Si technology, the MOS structure is used. Such a structure is shown in Fig. 6.20a. In Fig. 6.20b, we show the MOS capacitor under the situation where a positive bias is applied (for the *n*-Si semiconductor; a negative bias is needed for *p*-Si) to the gate to create a depletion region in the semiconductor. The depletion creates a "well" under the gate so that if an electron is placed in this well, it is trapped in the well.

Now, if an optical signal shines on the MOS capacitor, it creates *e-h* pairs as in any other semiconductor and the electrons flow into the well created under the gate. The holes that are created are pushed away from the oxide-semiconductor interface due to the presence of the electric field. As a result, as shown in Fig. 6.20c, a pocket of electrons is accumulated under the gate. The charge that is developed is proportional to the intensity of the light that has impinged on the device and is thus a measure of the "gray level" of the image.

In a CCD, an array of MOS capacitors are connected, as shown in Fig. 6.21. The approach that is used in CCD imaging is to expose the array to the image for a certain time, and then transfer the charges produced at each MOS capacitor to an electronic storage or memory. To do this, the gate potentials have to be changed by a properly clocked sequence, as shown in Fig. 6.22.

Referring to Fig. 6.21, we have a situation where the middle device has a collection of excess electrons under the gate. If the voltage $V_3$ on the next gate is increased to a value greater than $V_2$ (say $V_1 = 5$ V; $V_2 = 10$ V, and $V_3 = 15$ V), the

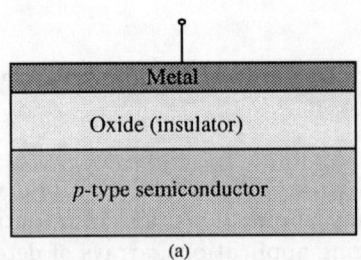

(a)

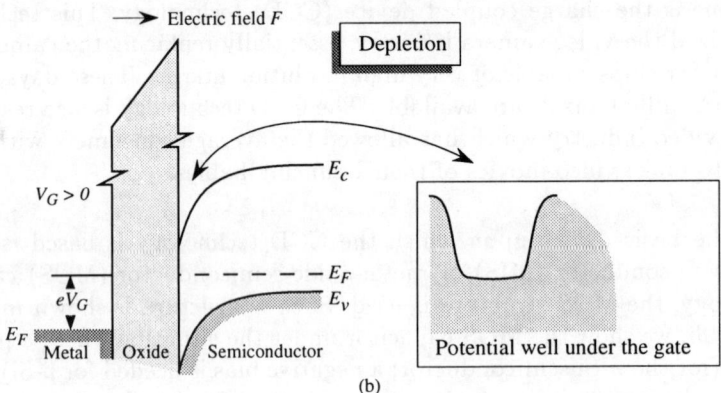

(b)

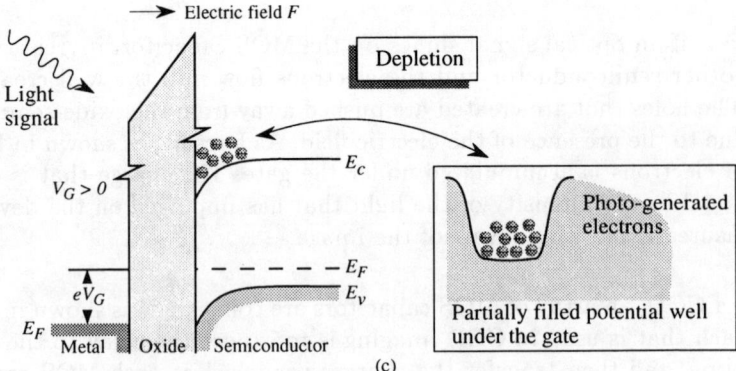

(c)

Figure 6.20: (a) A schematic of an MOS capacitor. (b) Band profile for the MOS capacitor under a positive bias with no light signal. (c) The effect of an optical signal is to create a pocket of electrons under the gate.

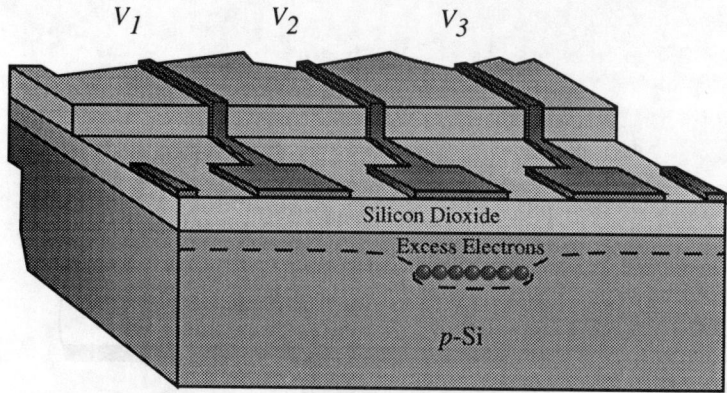

Figure 6.21: An arrangement of metal-insulator-semiconductor (MIS) capacitors to produce a CCD array for information transfer. By choosing a proper bias level, the charge in the channel can be transferred from one capacitor to another in a sequential manner. In the figure shown, an MOS device is used. However, other semiconductor technologies can also be exploited.

barrier to the charge on the right hand side of the electrons will disappear and the electrons will be "dumped" into the well under the third MOS capacitor. Thus, a transfer of charge has occurred. This sequence of charge packets creates an electrical signal that can then be stored in a semiconductor (or other) memory, to be recalled later, when needed.

**EXAMPLE 6.14** A three-phase CCD has a length of 12 $\mu m$. An electric field of 1 kV/cm is produced to cause the transfer of the electron charge packet. If the mobility of electrons is 1000 cm$^2$/V-s, calculate the signal transit time.

The transit time across one device is

$$t_{tr} = \frac{L}{v} = \frac{L}{\mu E}$$

$$= \frac{12 \times 10^{-4} \ cm}{(1000 \ cm^2/V - s)(1000 \ V/cm)}$$

$$= 1.2 \ ns$$

## 6.12  ADVANCED DETECTORS

$\mathcal{R}$  Important issues in detectors are: i) *tunability*; ii) *speed*; and iii) *integration*. The issue of tunability is an important one and involves mainly the use of different

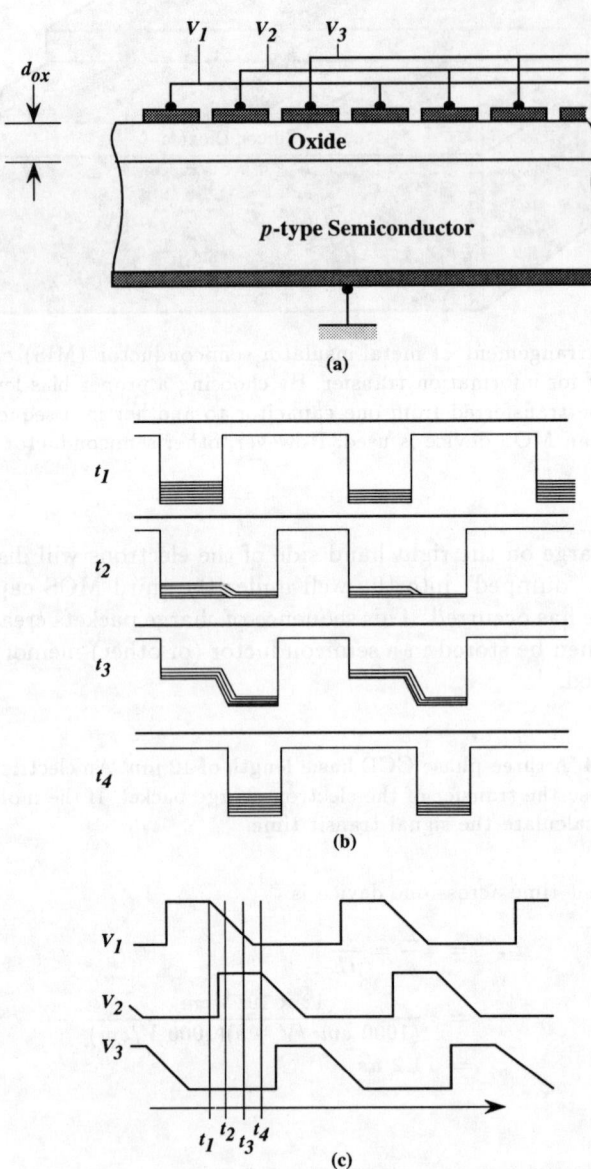

Figure 6.22: (a) A schematic of a CCD array in which each capacitor is connected to its third neighbor. (b) The time evolution of the capacitor charge due to the voltage sequences shown in part (c).

material systems with bandgaps that fit a given range of photon energies. A key challenge in this context is the detection of long wavelength photons ($\lambda \sim 10$ - $14\mu$m). This range is important for night vision applications, thermal imaging (for medical applications), and vision through fog. Two approaches are being pursued for long wavelength detection. These involve the use of narrow gap materials such as HgCdTe alloy or InAsSb alloy or multiquantum well structures. In narrow gap alloys the difficulty arises due to the inherent "softness" of the material which makes it easier to produce defects and makes processing difficult. Nevertheless, focal plane arrays are now made from HgCdTe alloys and their heterostructures.

The quantum well structure discussed in Chapter 4, Section 4.7 is also exploited for light detection. The intersubband separation in the quantum well formed in the conduction band or the valence band can be adjusted by choosing a proper geometry of the structure. If the lower level is filled with electrons, photons can be detected by exciting electrons to upper subbands from which they can be collected. Such detectors are acquiring increasing importance.

The speed of detectors is essentially controlled by the $RC$ time constant and the transit time for the carriers. The device speed design issues are thus similar to those for electronic devices. The devices are made as small as possible, and for high speed materials such as InGaAs, it is possible to have bandwidths approaching 150 GHz with current technology.

An important class of detectors for high speed and integration is the Schottky metal-semiconductor detector. This detector is easy to fabricate and has an extremely high speed. Schottky detectors have been shown to have 3-dB bandwidths approaching 160 GHz.

Integration of the detector with electronic devices such as bipolar transistors or FETs is an important area of research. The issue of integration is, however, quite complicated due to the very different structure of the detectors and the transistors. This incompatibility forces one to use epitaxial growth techniques together with "etch back and regrowth" techniques, discussed in Chapter 11. Regrowth techniques, in principle, allow one to grow a *p-i-n* structure (for example) next to a MODFET structure. However, during the etching process, the surface of the crystal develops defects. When regrowth is carried out, these defects can cause serious device degradation. Thus, device integration and fabrication of OEICs depend critically on developing better etching and regrowth techniques. In the early nineties, the performance of detectors integrated with transistors (amplifiers) still lags behind hybrid circuits. Hopefully, with advances in processing techniques, this situation will change.

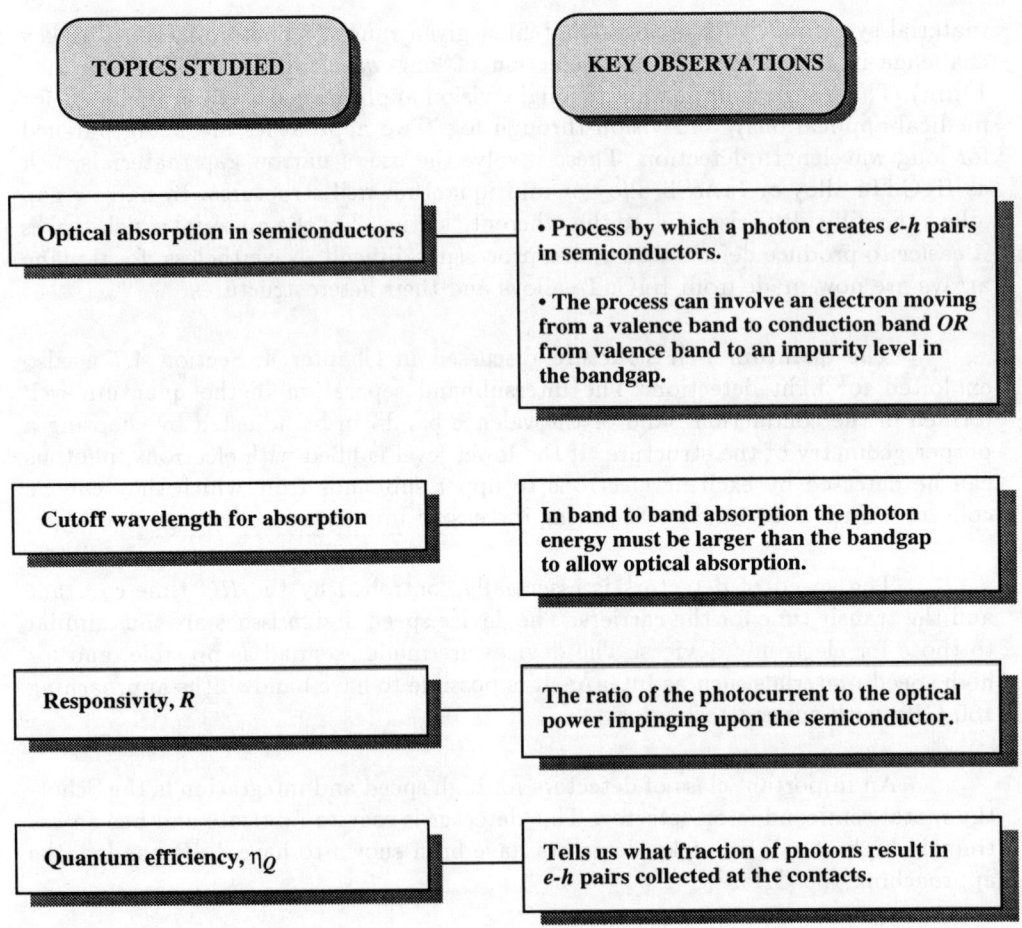

Table 6.1: Summary table

## 6.13   CHAPTER SUMMARY

$\mathcal{R}$  In this chapter we have examined optoelectronic devices which can convert an optical signal to an electrical signal. Most of these devices depend upon band to band transitions in which electrons are transferred from the valence band to the conduction band by the photons. The key findings of this chapter are summarized by the chapter summary tables (Tables 6.1-6.4).

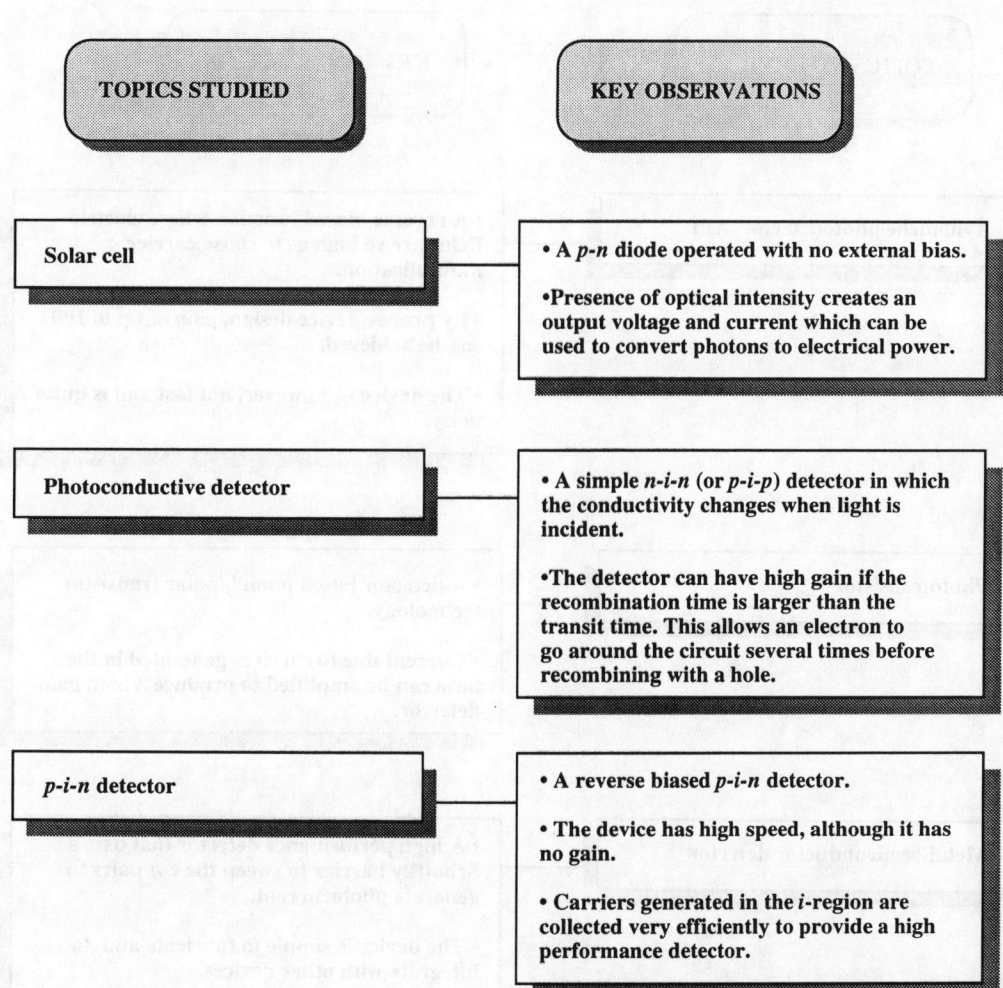

| TOPICS STUDIED | KEY OBSERVATIONS |
| --- | --- |
| **Solar cell** | • A *p-n* diode operated with no external bias.<br><br>•Presence of optical intensity creates an output voltage and current which can be used to convert photons to electrical power. |
| **Photoconductive detector** | • A simple *n-i-n* (or *p-i-p*) detector in which the conductivity changes when light is incident.<br><br>•The detector can have high gain if the recombination time is larger than the transit time. This allows an electron to go around the circuit several times before recombining with a hole. |
| ***p-i-n* detector** | • A reverse biased *p-i-n* detector.<br><br>• The device has high speed, although it has no gain.<br><br>• Carriers generated in the *i*-region are collected very efficiently to provide a high performance detector. |

Table 6.2: Summary table

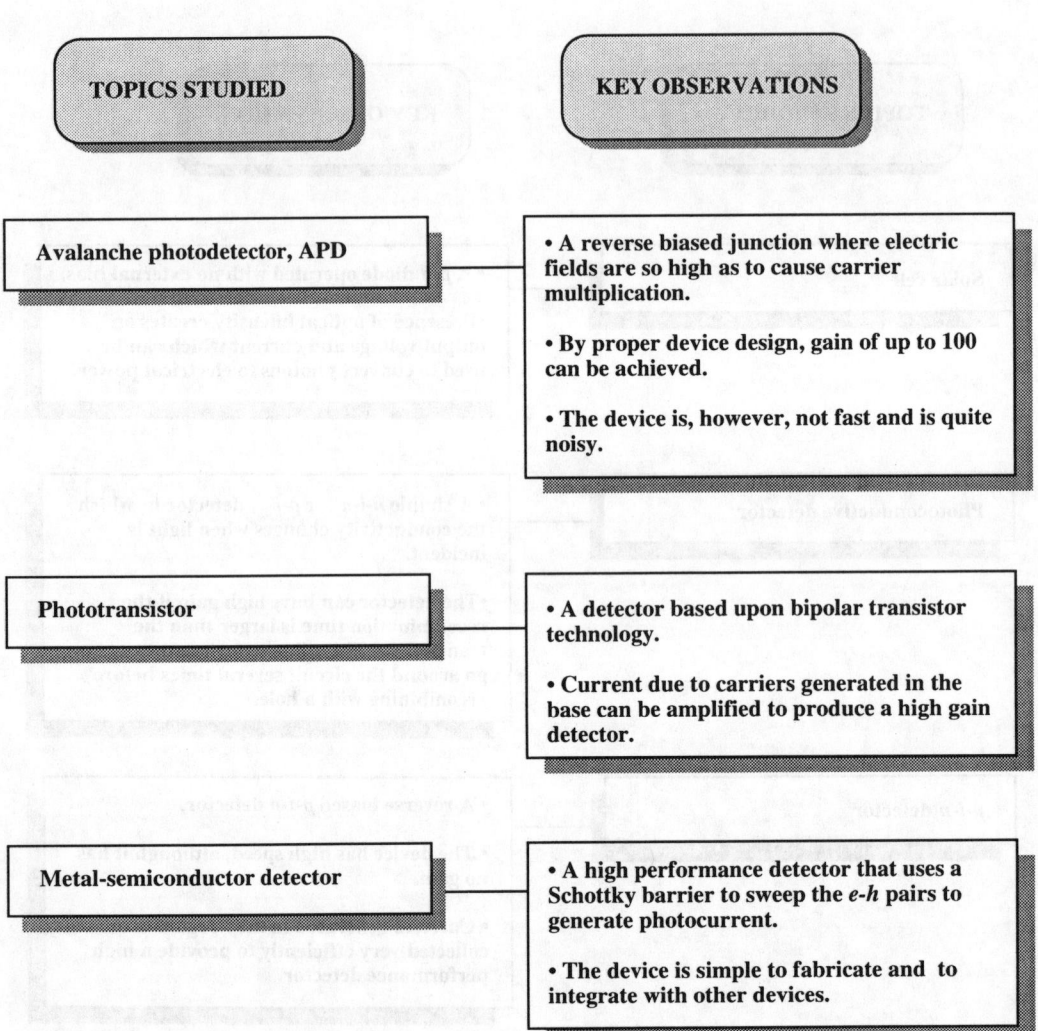

Table 6.3: Summary table

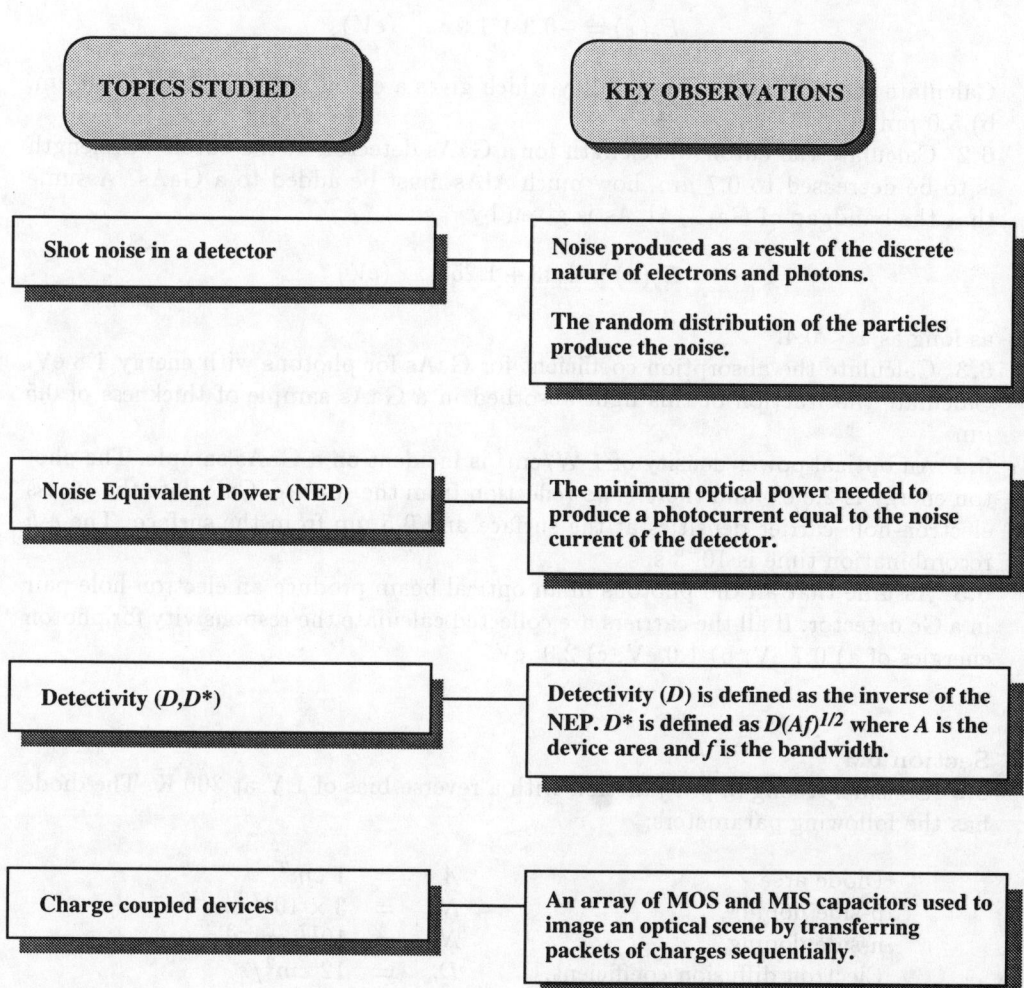

Table 6.4: Summary table

## 6.14   PROBLEMS

**Section 6.2**

**6.1** The bandgap of the $Hg_{1-x}Cd_xTe$ alloy is given by the expression

$$E_g(x) = -0.3 + 1.9x \quad (eV)$$

Calculate the composition of an alloy which gives a cutoff wavelength of a) 10 $\mu$m;
b) 5.0 $\mu$m.

**6.2** Calculate the cutoff wavelength for a GaAs detector. If the cutoff wavelength is to be decreased to 0.7 $\mu$m, how much AlAs must be added to a GaAs? Assume that the bandgap of $Ga_{1-x}Al_xAs$ is given by

$$E_g(x) = 1.43 + 1.25x \quad (eV)$$

as long as $x \leq 0.4$.

**6.3** Calculate the absorption coefficient for GaAs for photons with energy 1.8 eV. Calculate the fraction of this light absorbed in a GaAs sample of thickness of 0.5 $\mu$m.

**6.4** An optical power density of 1 W/cm$^2$ is incident on a GaAs sample. The photon energy is 2.0 eV and there is no reflection from the surface. Calculate the excess electron-hole carrier densities at the surface and 0.5 $\mu$m from the surface. The e-h recombination time is $10^{-8}$ s.

**6.5** Assume that all the photons in an optical beam produce an electron-hole pair in a Ge detector. If all the carriers are collected calculate the responsivity for photon energies of a) 0.7 eV; b) 1.0 eV; c) 2.0 eV.

**Section 6.3**

**6.6** Consider a long Si $p$-$n$ junction with a reverse bias of 1 V at 300 K. The diode has the following parameters:

| | | |
|---|---|---|
| Diode area, | $A$ | $= 1\ cm^2$ |
| p-side doping, | $N_a$ | $= 3 \times 10^{17}\ cm^{-3}$ |
| n-side doping, | $N_d$ | $= 10^{17}\ cm^{-3}$ |
| Electron diffusion coefficient, | $D_n$ | $= 12\ cm^2/s$ |
| Hole diffusion coefficient, | $D_p$ | $= 8\ cm^2/s$ |
| Electron minority carrier lifetime, | $\tau_n$ | $= 10^{-7}\ s$ |
| Hole minority carrier lifetime, | $\tau_p$ | $= 10^{-7}\ s$ |
| Optical absorption coefficient, | $\alpha$ | $= 10^3\ cm^{-1}$ |
| Optical power density, | $P_{op}$ | $= 10\ W/cm^2$ |
| Photon energy, | $\hbar\omega$ | $= 1.7\ eV$ |

Calculate the photocurrent in the diode.

**6.7** Consider a long Si $p$-$n$ junction solar cell with an area of 4 cm$^2$ at 300 K. The solar cell has the following parameters:

$$
\begin{aligned}
\text{n-type doping,} && N_d &= 10^{18} \ cm^{-3} \\
\text{p-type doping,} && N_a &= 3 \times 10^{17} \ cm^{-3} \\
\text{Electron diffusion coefficient,} && D_n &= 15 \ cm^2/s \\
\text{Hole diffusion coefficient,} && D_p &= 7.5 \ cm^2/s \\
\text{Electron minority carrier lifetime,} && \tau_n &= 10^{-7} \ s \\
\text{Hole minority carrier lifetime,} && \tau_p &= 10^{-7} \ s \\
\text{Photocurrent,} && I_L &= 1.0 \ A \\
\text{Diode ideality factor,} && m &= 1.25
\end{aligned}
$$

Calculate the open circuit voltage of the diode. If the fill factor is 0.75, calculate the maximum power output.

**6.8** Consider the solar cell of problem 6.7. A solar system is to be developed from such cells to deliver a power of 15 W at a voltage level of 5 V. Calculate the total number of solar cells needed.

## Section 6.4

**6.9** Consider a GaAs photoconductor in which carriers are generated at a rate of $G_L = 10^{20} \ cm^{-3} \ s^{-1}$. The device area is (10 $\mu$m $\times$ 10 $\mu$m) and the length is 10 $\mu$m. The material parameters are:

$$
\begin{aligned}
\text{Background doping,} && N_d &= 10^{16} \ cm^{-3}; N_a = 0 \\
\text{Applied voltage,} && V_{app} &= 1.0 \ V \\
\text{Electron velocity at the applied field,} && v_e &= 6 \times 10^6 \ cm/s \\
\text{Hole velocity at the applied field,} && v_h &= 10^6 \ cm/s \\
\text{Electron lifetime,} && \tau_n &= 10^{-7} \ s \\
\text{Hole lifetime,} && \tau_p &= 10^{-8} \ s
\end{aligned}
$$

Calculate the excess carrier concentration, the steady state photocurrent, and the photoconductive gain.

**6.10** Consider a silicon photoconductor at 300 K with the following parameters:

$$
\begin{aligned}
\text{Background doping,} && N_d &= 10^{15} \ cm^{-3} \\
\text{Electron mobility,} && \mu_n &= 1200 \ cm^2/V-s \\
\text{Hole mobility,} && \mu_p &= 400 \ cm^2/V-s \\
\text{Electron lifetime,} && \tau_n &= 10^{-6} \ s \\
\text{Hole lifetime,} && \tau_p &= 5 \times 10^{-7} \ s \\
\text{Detector area,} && A &= 10^{-4} \ cm^2 \\
\text{Detector length,} && L &= 100 \ \mu m
\end{aligned}
$$

A bias of 5 V is applied to the detector. Calculate the dark current. If light falls on the detector to produce a generation rate of $10^{21} \ cm^{-3} \ s^{-1}$, calculate the excess concentration, the photoconductivity, and device gain.

**6.11** Consider the detector of Problem 6.10. Calculate the gain-bandwidth product of the photoconductive detector.

## Section 6.5

**6.12** Consider a long silicon $p$-$n$ photodiode at 300 K on which light from a GaAs laser ($\hbar\omega = 1.43$ eV) is impinging. The optical power density is $10^{-2}$ W/cm$^2$. The diode has the following parameters:

| | | |
|---|---|---|
| Device area, | $A$ | $= 10^{-6}\ cm^2$ |
| n-type doping, | $N_d$ | $= 5 \times 10^{16}\ cm^{-3}$ |
| p-type doping, | $N_a$ | $= 10^{17}\ cm^{-3}$ |
| Electron diffusion coefficient, | $D_n$ | $= 20\ cm^2/s$ |
| Hole diffusion coefficient, | $D_p$ | $= 12\ cm^2/s$ |
| Electron minority carrier lifetime, | $\tau_n$ | $= 1.5 \times 10^{-7}\ s$ |
| Hole minority carrier lifetime, | $\tau_p$ | $= 10^{-7}\ s$ |
| Absorption coefficient, | $\alpha$ | $= 700\ cm^{-1}$ |

A reverse bias voltage of 5 V is applied to the diode. Assume that the carrier generation rate is uniform. Calculate the prompt photocurrent and the total photocurrent in the detector.

**6.13** Consider a GaAs $p$-$i$-$n$ detector with an intrinsic layer width of 1.0$\mu$m. Optical power density (photon energy 1.6 eV) of 0.1 Watt/cm$^2$ impinges upon the detector. The absorption coefficient for the active region is $10^4$cm$^{-1}$. Calculate the prompt photocurrent of the device. The device area is $10^{-4}$ cm$^2$.

**6.14** Consider a silicon $p$-$i$-$n$ photodetector in which the $i$ layer is 10.0 $\mu$m thick. Calculate the maximum quantum efficiency of this detector if only light absorbed in the undoped region contributes to the photocurrent. The absorption coefficient is $10^3$ cm$^{-1}$. Also calculate the minimum thickness of the $i$-region needed to ensure a quantum efficiency of 0.8. There are no reflection losses.

## Section 6.6

**6.15** Using the data for $\alpha_{imp}$ and $\beta_{imp}$ in the text (Chapter 5, Fig. 5.11), calculate the ratio $\tilde{k} = \beta_{imp}/\alpha_{imp}$ for Si, Ge, and GaAs as a function of electric field.

**6.16** An avalanche photodetector has an avalanche region of 5 $\mu$m and the electric field is such that $\alpha_{imp} = \beta_{imp} = 10^4$cm$^{-1}$. Calculate the multiplication factor of the device.

## Section 6.9

**6.17** A detector receives digital data from an optical source. On an average, the bit 1 has 50 photons in it and the bit 0 has 25 photons. Assuming that the photons obey a Poisson distribution, estimate the bit error rate arising from the random noise.

**6.18** A $p$-$i$-$n$ detector which is limited by shot noise is to be operated at 10 GHz. Calculate the photo current needed to ensure a signal to noise ratio of 60 dB.

**6.19** The $D^*$ value for an HgCdTe detector for $10\mu$m radiation is found to be $5 \times 10^{10}$ cm $\text{Hz}^{1/2}W^{-1}$ at 77 K. The detector is to be used in an imaging array which is to operate at a bandwidth of 100 Hz. The area of the detector is $10^{-4}$ cm$^2$. Calculate the minimum power that can be detected.

**6.20** A detector material is to be selected for a night vision imaging application. The detector area is to be $10^{-4}$ cm$^2$, and it should be able to detect radiation with power levels one millionth of the daytime power levels. The imaging system is to operate at a bandwidth of 100 Hz. The daytime power levels are 0.1 $W/$cm$^2$. Calculate the $D^*$ of the detector needed.

## 6.15 REFERENCES

● **General**

- J. Gowar, *Optical Communication Systems*, Prentice-Hall, Englewood Cliffs, NJ (1989).

- J. I. Pankove, *Optical Processes in Semiconductors*, Dover Publications, New York (1977).

- J. Singh, *Physics of Semiconductors and Their Heterostructures*, McGraw-Hill, New York (1993).

- J. Wilson and J. F. B. Hawkes, *Optoelectronics: An Introduction*, Prentice-Hall, Englewood Cliffs, NJ (1983).

# CHAPTER
# 7

# THE LIGHT
# EMITTING DIODE

## 7.1 INTRODUCTION

An important component of optical information processing is the generation of an optical signal. Optical signals are used for communication where the superior overall performance of optical fibers as a carrier of information is rapidly replacing metallic cables. Optical signals are needed for display devices to project information. Optical beams are required for memory systems based on optical reading.

The light emitting diode (LED) is one of the simplest optoelectronic devices which has found important applications as a display device as well as an optical signal generator for optical communication. Compared to the laser diode (to be discussed in the next chapter), the fabrication of the LED is very simple since it does not require any special optical cavity for its operation. However, one pays the price in terms of low optical output, broad and incoherent spectra, and slow device response. In this chapter we will explore the physics behind the operation of the LED. The reader is advised to review our discussions in Chapter 5 on optical processes in semiconductors (Sections 5.5 through 5.7).

## 7.2 MATERIAL SYSTEMS FOR THE LED

$\mathcal{R}$ The simplicity of the light emitting diode (LED) makes it a very attractive device for display and communication applications. The LED can operate up to modulation speeds of $\sim 1$ GHz. The spectral width of the optical output of an LED is of the order of $k_B T$ which translates into a wavelength spread of 200-300 Å at room temperature. Although this is a large value, the LED produces a single color to the human eye. Thus the LEDs can be used very effectively in color displays. An important recent application of LEDs is in providing tail lights in automobiles, an application that could make LEDs very important commercial devices.

The basic LED is a *p-n* junction which is forward biased to inject electrons and holes into the *p-* and *n*-sides respectively. The injected minority charge from the *n-* and *p*-sides recombines with the majority charge in the depletion region or the neutral region. *In direct band semiconductors, this recombination leads to light emission since radiative recombination dominates in high quality materials. In indirect gap materials, the light emission efficiency is quite poor and most of the recombination paths are non-radiative which generate heat rather than light.*

Light emitting devices are one class of devices that have given impetus to the compound semiconductor industry. Since Si is an indirect gap material, and the radiative recombination is very poor, this material, which dominates all other areas of electronics, finds itself handicapped when it comes to light emission.

In Appendix B, Fig. B.1, we show the bandgaps of a variety of compound semiconductors along with their lattice constants. The direct gap regions are denoted by the solid line while the indirect gap materials are denoted by the dashed line. As can be seen, a wide range of combinations is available to the device designer.

We will briefly outline some of the important considerations in choosing a semiconductor for LEDs or laser diodes. The reader should revisit the discussion on materials used for detectors (Section 6.2).

**Emission Energy**: The light emitted from the device is very close to the semiconductor bandgap since the injected electrons and holes are described by quasi-Fermi distribution functions. The desire to have a particular emission energy may arise from a number of motivations. In Fig. 6.2 we show the response of the human eye to radiation of different wavelengths. Also shown are the bandgaps of some semiconductors. If a color display is to be produced that is to be seen by people, one has to choose an appropriate semiconductor. Very often one has to choose an alloy since there is a greater flexibility in the bandgap range available. In Fig. 7.1 we show the loss characteristics of an optical fiber discussed in Chapter 3. As can be seen, the loss is lowest at 1.55 $\mu$m and 1.3 $\mu$m. If optical communication sources are desired, one must choose materials which can emit at these wavelengths. This is especially true if the communication is long haul, i.e., over hundreds or even thousands of kilometers. Materials like GaAs which emit at 0.8 $\mu$m can still be used for local area networks (LANs) which involve communicating within a building or local areas.

**Substrate Availability**: Almost all optoelectronic light sources depend upon epitaxial crystal growth techniques where a thin active layer (a few microns) is grown on a substrate (which is $\sim 200$ $\mu$m). The availability of a high quality substrate is extremely important in epitaxial technology. If a substrate that lattice matches to the active device layer is not available, the device layer may have dislocations and other defects in it. These can seriously hurt the device performance. The important substrates that are available for light emitting technology are GaAs and InP. A few semiconductors and their alloys can match with these substrates. The lattice constant of an alloy is the weighted mean of the lattice constants of the individual components, i.e., the lattice constant of the alloy $A_x B_{1-x}$ is

$$a_{all} = x a_A + (1 - x) a_B \qquad (7.1)$$

where $a_A$ and $a_B$ are the lattice constants of $A$ and $B$. Semiconductors that cannot lattice match with GaAs or InP have an uphill battle for technological success.

Important semiconductor materials exploited in optoelectronics are the alloy $Ga_x Al_{1-x} As$ which is lattice matched very well to GaAs substrates; $In_{0.53}Ga_{0.47}As$ and $In_{0.52}Al_{0.48}As$ which are lattice matched to InP; InGaAsP which is a quaternary material whose composition can be tailored to match with InP and can emit at 1.55 $\mu$m; and GaAsP which has a wide range of bandgaps available. Recently

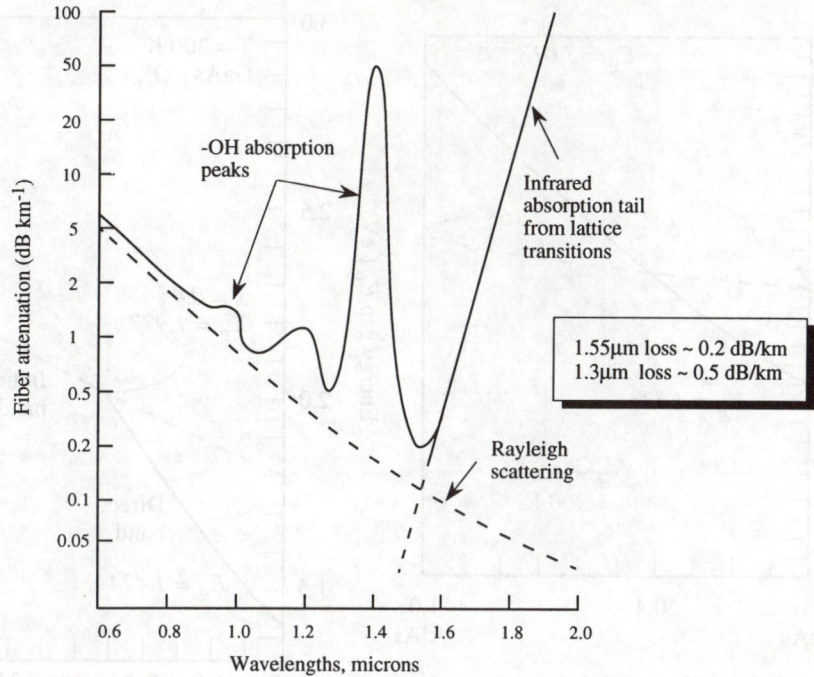

Figure 7.1: Optical attenuation vs. wavelength for an optical fiber. Primary loss mechanisms are identified as absorption and scattering.

there has been a considerable interest in large bandgap materials such as ZnSe, ZnS, SiC, AlInGaP, and GaN to produce devices that emit blue or green light. The motivation is for superior display technology and for high density optical memory applications (a shorter wavelength allows reading of smaller features). Reliable SiC and GaN LEDs are now available in the commercial market, although only a few suppliers can meet the technology challenge.

It is important to keep in mind that alloys like GaAlAs and GaAsP become indirect at certain compositions as shown in Fig. 7.2. For efficient light emission, one needs to work in the direct gap region. However, with a suitable impurity, one can obtain light emission in an indirect bandgap material. In Fig. 7.3 we show some important material systems used in light emitting devices, along with some of their special properties.

As we saw in Chapter 5, the momentum conservation causes strong radiative transitions to occur only in direct gap semiconductors. *Some indirect gap materials can, however, have a reasonable radiative efficiency if they are doped with certain impurities.* The impurities create levels in the bandgap and photon absorption is allowed by moving electrons into these levels. The absorption and emission rates

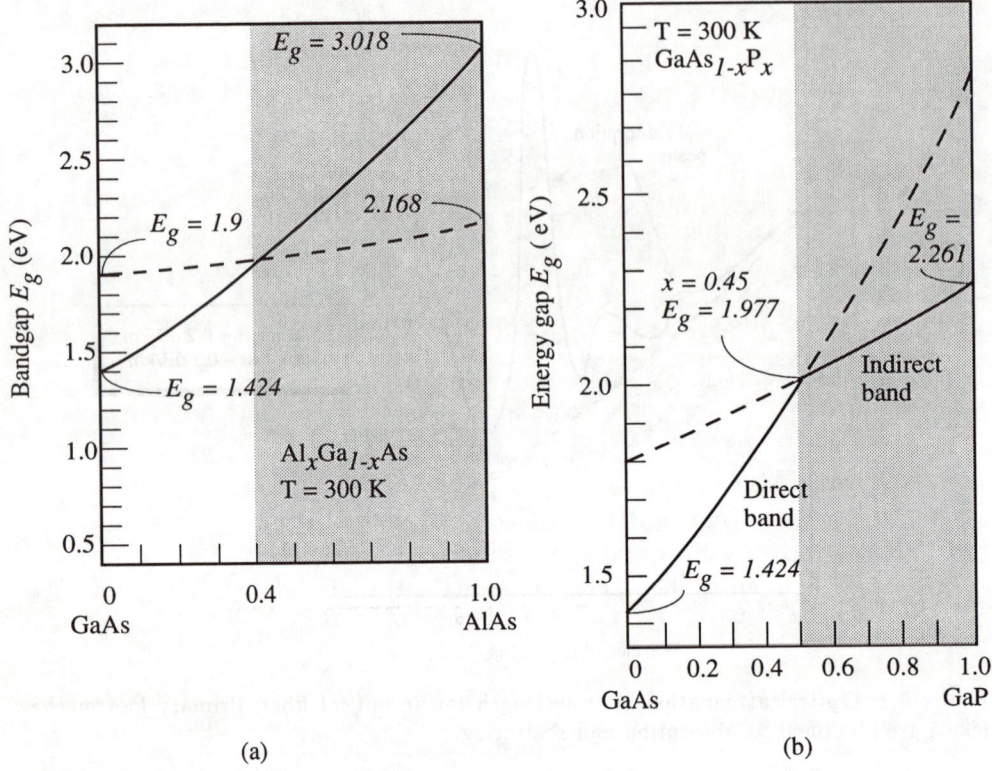

Figure 7.2: Bandgap of (a) $Al_xGa_{1-x}As$ and (b) $GaAs_{1-x}P_x$ as a function of alloy composition. Note that the bandgap changes from direct to indirect as shown. (After H. C. Casey and M. B. Panish, *Heterostructure Lasers*, Academic Press, New York (1978).)

are, however, smaller than those for direct gap semiconductors. The GaAsP alloy system is one semiconductor system in which impurity levels have been widely used to produce LEDs. However, since the light emission efficiency is poor, it has not been possible to use these impurity levels to produce laser diodes. The need for higher radiative efficiencies for laser diodes will become clear when we discuss their physics in the next chapter.

## 7.3   OPERATION OF THE LED

$\longrightarrow$ The LED is a forward biased $p$-$n$ diode in which electrons and holes are injected into a region where they recombine. In general, the electron-hole recombination process can occur by radiative and nonradiative channels. Under the condition of minority carrier recombination or high injection recombination, as shown in Chapter

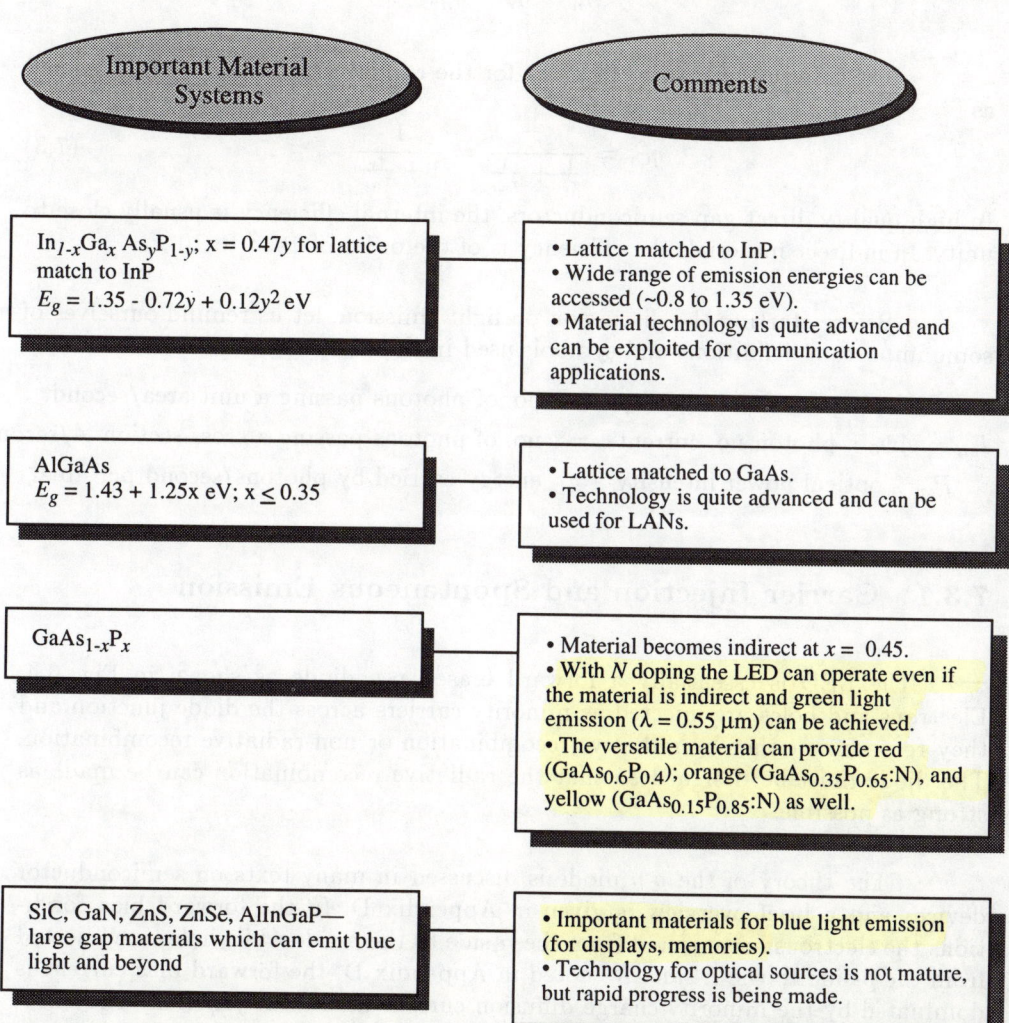

Figure 7.3: Important material systems for light emission. LEDs can be made from indirect materials with an appropriate impurity, but the emission efficiency is low.

5 (Sections 5.7 and 5.8), one can define a lifetime for carrier recombination. If $\tau_r$ and $\tau_{nr}$ are the radiative and non-radiative lifetimes, the total recombination time is (for, say, an electron)

$$\frac{1}{\tau_n} = \frac{1}{\tau_r} + \frac{1}{\tau_{nr}} \tag{7.2}$$

The internal quantum efficiency for the radiative processes is then defined as

$$\eta_{Qr} = \frac{\frac{1}{\tau_r}}{\frac{1}{\tau_r} + \frac{1}{\tau_{nr}}} = \frac{1}{1 + \frac{\tau_r}{\tau_{nr}}} \tag{7.3}$$

In high quality direct gap semiconductors, the internal efficiency is usually close to unity. In indirect materials the efficiency is of the order of $10^{-2}$ to $10^{-3}$.

Before starting the discussion on light emission, let us remind ourselves of some important definitions and symbols used in this chapter:

$\Phi_0$ : photon current density $=$ no. of photons passing a unit area/second.

$I_{ph} = A\Phi_0$ : photon no. current $=$ no. of photons passing a cross-section $A$/secon

$P_{op}$ : optical power intensity $=$ energy carried by photons/second per area.

## 7.3.1  Carrier Injection and Spontaneous Emission

$\longrightarrow$  The LED is essentially a forward biased $p$-$n$ diode as shown in Fig. 6.4. Electrons and holes are injected as minority carriers across the diode junction and they recombine either by radiative recombination or non-radiative recombination. The diode must be designed so that the radiative recombination can be made as strong as possible.

The theory of the $p$-$n$ diode is discussed in many texts on semiconductor devices and a brief overview is given in Appendix D. In the forward bias conditions the electrons are injected from the $n$-side to the $p$-side while holes are injected from the $p$-side to the $n$-side. As noted in Appendix D, the forward bias current is dominated by the minority charge diffusion current across the junction. The diffusion current, in general, consists of three components: i) minority carrier electron diffusion current, ii) minority carrier hole diffusion current; and iii) trap assisted recombination current in the depletion region of width $W$. These current densities have the following forms respectively (see Eqn. D.13):

$$J_n = \frac{eD_n n_p}{L_n}\left[exp\left(\frac{eV}{k_B T}\right) - 1\right] \tag{7.4}$$

$$J_p = \frac{eD_p p_n}{L_p}\left[exp\left(\frac{eV}{k_B T}\right) - 1\right] \tag{7.5}$$

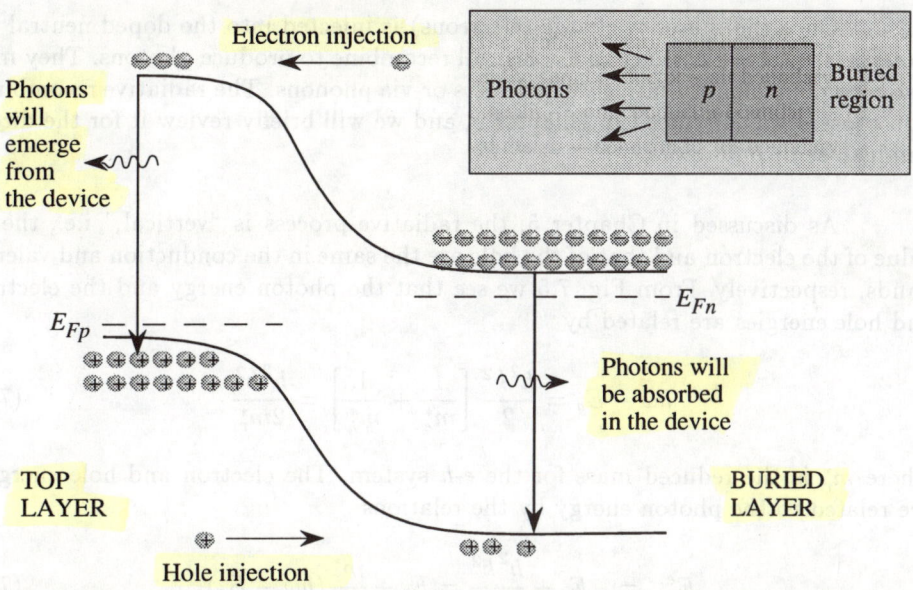

Figure 7.4: In a forward biased *p-n* junction, electrons and holes are injected as shown. In the figure, the holes injected into the buried n region will generate photons which will not emerge from the surface of the LED. The electrons injected will generate photons which are near the surface and have a high probability to emerge.

$$J_{GR} = \frac{e n_i W}{2\tau} \left[ exp\left( \frac{eV}{2 k_B T} \right) - 1 \right] \tag{7.6}$$

Here $J_n$ and $J_p$ are the electron and hole current densities, $D_n$ and $D_p$ are the electron and hole diffusion constants in the $p$ and $n$ regions, $n_p$ and $p_n$ are the electron and hole densities in the $p$ and $n$ sides, $V$ is the applied bias, $W$ is the depletion width, and $\tau$ is the recombination time in the depletion region and depends upon the trap density. *The LED is designed so that the photons are emitted close to the top layer and not in the buried layer as shown in Fig. 7.4. The reason for this choice is that photons emitted deep in the device have a high probability of being reabsorbed. Thus one prefers to have only one kind of carrier injection for the diode current.* Usually the top layer of the LED is $p$-type, and for photons to be emitted in this layer one must require the diode current to be dominated by the electron current (i.e., $J_n \gg J_p$). The ratio of the electron current density to the total diode current density is called the injection efficiency $\gamma_{inj}$. Thus we have

$$\gamma_{inj} = \frac{J_n}{J_n + J_p + J_{GR}} \tag{7.7}$$

If the diode is $pn^+$, $n_p \gg p_n$, and, as can be seen from Eqns. 7.4 and 7.5, $J_n$ becomes much larger than $J_p$. If, in addition, the material is high quality so that the recombination current is small, the injection efficiency approaches unity.

   Once the minority charge (electrons) is injected into the doped neutral region (p-type), the electrons and holes will recombine to produce photons. They may
also recombine non-radiatively via defects or via phonons. The radiative recombination process was discussed in Chapter 5, and we will briefly review it for the direct
bandgap semiconductors.

   As discussed in Chapter 5, the radiative process is "vertical," i.e., the k-
value of the electron and that of the hole are the same in the conduction and valence
bands, respectively. From Fig. 7.5 we see that the photon energy and the electron
and hole energies are related by

$$\hbar\omega - E_g = \frac{\hbar^2 k^2}{2}\left[\frac{1}{m_e^*} + \frac{1}{m_h^*}\right] = \frac{\hbar^2 k^2}{2m_r^*} \tag{7.8}$$

where $m_r^*$ is the reduced mass for the e-h system. The electron and hole energies
are related to the photon energy by the relations

$$E^e = E_c + \frac{\hbar^2 k^2}{2m_e^*} = E_c + \frac{m_r^*}{m_e^*}(\hbar\omega - E_g) \tag{7.9}$$

$$E^h = E_v - \frac{\hbar^2 k^2}{2m_h^*} = E_v - \frac{m_r^*}{m_h^*}(\hbar\omega - E_g) \tag{7.10}$$

   *If an electron is available in a state k, and a hole is also available in the
state k (i.e., the Fermi functions for the electrons and holes satisfy $f^e(k) = f^h(k)$
= 1), the radiative recombination rate is given by* (see Eqns. 5.44 and 5.46)

$$W_{em} = \frac{1}{\tau_o} = \frac{e^2 n_r \hbar\omega}{3\pi\epsilon_o m_o^2 c^3 \hbar^2}|p_{cv}|^2 \tag{7.11}$$

where $n_r$ is the refractive index of the semiconductor, $m_0$ the free electron mass,
and $p_{cv}$ the momentum matrix elements between the conduction and valence bands.
It turns out that $p_{cv}$ does not vary too much between semiconductors and has a
value given by the equation

$$\frac{2p_{cv}^2}{m_0} \cong 22 \; eV \tag{7.12}$$

Thus the emission rate turns out to be (see Examples 5.10 or 7.1)

$$\boxed{W_{em} \sim 1.14 \times 10^9 \hbar\omega(eV) \qquad s^{-1}} \tag{7.13}$$

and the recombination time becomes ($\hbar\omega$ is expressed in electron volts)

$$\boxed{\tau_o = \frac{0.88}{\hbar\omega(eV)} \qquad ns} \tag{7.14}$$

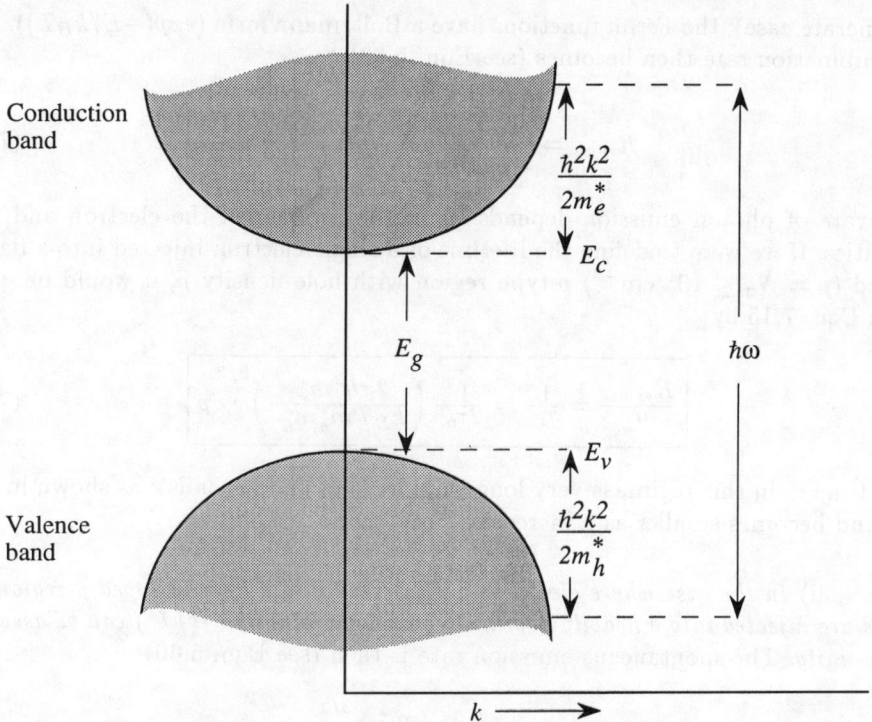

Figure 7.5: A schematic of the $E$-$k$ diagram for the conduction and valence bands. Optical transitions are vertical; i.e., $k$-vector of the electron in the valence band and in the conduction band is the same.

*The recombination time discussed above is the shortest possible spontaneous emission time since we have assumed that the electron has a unit probability of finding a hole with the same k-value.*

When carriers are injected into the semiconductors (see Section 5.7) the occupation probabilities for the electron and hole states are given by the appropriate quasi-Fermi levels. *The e-h recombination process is determined by the spontaneous emission which implies that the photon density of the emission is quite low so that stimulated emission is not significant.* The emitted photons leave the device volume so that the photon density never becomes high in the $e$-$h$ recombination region. In a laser diode the situation is different, as we shall see later. The photon emission rate is given by integrating the emission rate $W_{em}$ over all the electron-hole pairs after introducing the appropriate Fermi functions. There are several important limits of the spontaneous rate which were discussed in Chapter 5. The reader is advised to review the discussions in Section 5.7. We will summarize them here:

i) *In the case where the electron and hole densities n and p are small* (non-degenerate case), the Fermi functions have a Boltzmann form ($exp(-E/k_BT)$). The recombination rate then becomes (see Eqn. 5.63)

$$R_{spon} = \frac{1}{2\tau_o} \left( \frac{2\pi\hbar^2 m_r^*}{k_BT m_e^* m_h^*} \right)^{3/2} n\,p \tag{7.15}$$

The rate of photon emission depends upon the product of the electron and hole densities. If we were to define the lifetime of a single electron injected into a lightly doped ($p = N_a \le 10^{17}\text{cm}^{-3}$) p-type region with hole density $p$, it would be given from Eqn. 7.15 by

$$\boxed{\frac{R_{spon}}{n} = \frac{1}{\tau_r} = \frac{1}{2\tau_o} \left( \frac{2\pi\hbar^2 m_r^*}{k_BT m_e^* m_h^*} \right)^{3/2} p} \tag{7.16}$$

The time $\tau_r$ in this regime is very long (hundreds of nanoseconds), as shown in Fig. 7.6 and becomes smaller as $p$ increases.

   ii) *In the case where electrons are injected into a heavily doped p-region (or holes are injected into a heavily doped n-region), the function $f^h(f^e)$ can be assumed to be unity.* The spontaneous emission rate is then (see Eqn. 5.60)

$$R_{spon} \sim \frac{1}{\tau_o} \left( \frac{m_r^*}{m_h^*} \right)^{3/2} n \tag{7.17}$$

for electron concentration $n$ injected into a heavily doped p-type region and

$$R_{spon} \sim \frac{1}{\tau_o} \left( \frac{m_r^*}{m_e^*} \right)^{3/2} p \tag{7.18}$$

for hole injection into a heavily doped n-type region.

   The minority carrier lifetimes (i.e., $n/R_{spon}$) play a very important role not only in LEDs, but also in diodes and bipolar devices. In this regime the lifetime of a single electron (hole) is independent of the holes (electrons) present since there is always a unity probability that the electron (hole) will find a hole (electron). The lifetime is now essentially $\tau_o$, as shown in Fig. 7.6.

   iii) *Another important regime is that of high injection where $n = p$ is so high that one can assume $f^e = f^h = 1$ in the integral for the spontaneous emission rate.* The spontaneous emission rate is (see Eqn. 5.61)

$$\boxed{R_{spon} \sim \frac{n}{\tau_o} \sim \frac{p}{\tau_o}} \tag{7.19}$$

and the radiative lifetime ($n/R_{spon} = p/R_{spon}$) is $\tau_o$.

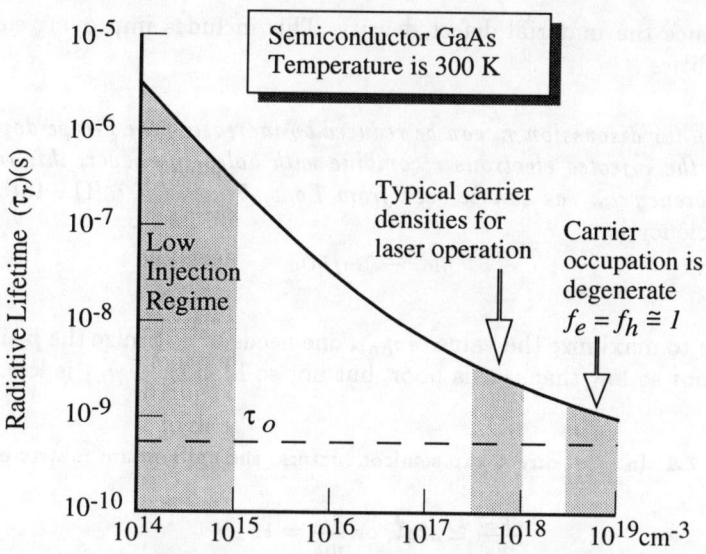

$N_d$ (for holes injected into an $n$-type semiconductor)

$n = p$ (for excess electron hole pairs injected into a region)

Figure 7.6: Radiative lifetimes of electrons or holes in a direct gap semiconductor as a function of doping or excess charge. The figure gives the lifetimes of a minority charge (a hole) injected into an $n$-type material. The figure also gives the lifetime behavior of electron-hole recombination when excess electrons and holes are injected into a material as a function of excess carrier concentration.

iv) *A regime that is quite important for laser operation is one where suffi-cient electrons and holes are injected into the semiconductor to cause "inversion."* This occurs if $f^e + f^h \geq 1$. If we make the approximation $f^e \sim f^h = 1/2$, for all the electrons and holes at inversion, we get the relation (see Eqn. 5.67)

$$R_{spon} \sim \frac{n}{4\tau_o} \qquad (7.20)$$

or the radiative lifetime at inversion is

$$\boxed{\tau_r \sim \frac{\tau_o}{4}} \qquad (7.21)$$

This value is a reasonable choice to calculate for the spontaneous emission rate in lasers near threshold.

The radiative recombination depends upon the radiative lifetime $\tau_r$ and the non-radiative lifetime $\tau_{nr}$. To improve the efficiency of photon emission one needs a value of $\tau_r$ as small as possible and $\tau_{nr}$ as large as possible. To increase $\tau_{nr}$

one must reduce the material defect density. This includes improving surface and interface qualities.

*From our discussion $\tau_r$ can be reduced by increasing the p-type doping in the region where the injected electrons recombine with holes. However, this reduces the injection efficiency $\gamma_{inj}$ as can be seen from Eqns. 7.4 and 7.7. The total internal quantum efficiency is*

$$\eta_{int} = \gamma_{inj}\eta_{Qr} \tag{7.22}$$

Thus to maximize the value of $\eta_{int}$, one needs to optimize the p-side doping so that it is not so low that $\eta_{Qr}$ is poor, but not so high that $\gamma_{inj}$ is low.

**EXAMPLE 7.1** In most direct gap semiconductors, the momentum matrix element $p_{cv}$ is given by

$$\frac{2p_{cv}^2}{m_0} \cong 22eV \text{ or } \frac{p_{cv}^2}{m_0} = 11 \ eV$$

Calculate the electron-hole recombination time $\tau_o$ for GaAs. Remember that $\tau_o$ is defined as the recombination time for an electron in state $k$ to recombine with a hole in state $k$, i.e., $f^e = 1$, $f^h = 1$.

The spontaneous emission rate is given by

$$W_{em} = \frac{1}{\tau_o} = \frac{4}{3}\frac{e^2}{4\pi\epsilon_o}\frac{n_r\hbar\omega|p_{cv}|^2}{m_o^2c^3\hbar^2}$$

Using $\hbar\omega = 1.5$ eV, $n_r = 3.5$, and $|p_{cv}|^2$ as given, we get

$$\frac{1}{\tau_o} = \frac{4}{3}\frac{\left(1.6\times10^{-19}C\right)^2\times(3.5)\times\left(1.5\times1.6\times10^{-19} \ J\right)}{\left(4\times3.1416\times8.85\times10^{-12} \ F/m\right)\times\left(9.1\times10^{-31} \ kg\right)}$$

$$\times \frac{\left(11.0\times1.6\times10^{-19} \ J\right)}{\left(3\times10^8 \ m/s\right)^3\times\left(1.05\times10^{-34} \ Js\right)^2}$$

$$= 1.67\times10^9 \ s^{-1}$$

or

$$\tau_o = 0.6 \ ns$$

**EXAMPLE 7.2** Calculate the e-h recombination time when an excess electron and hole density of $10^{15}$ cm$^{-3}$ is injected into a GaAs sample at room temperature.

Since $10^{15}$ cm$^{-3}$ or $10^{21}$ $m^{-3}$ is a very low level of injection, the recombination time is given by Eqn. 7.16 as

$$\frac{1}{\tau_r} = \frac{1}{2\tau_o}\left(\frac{2\pi\hbar^2 m_r^*}{k_B T m_e^* m_h^*}\right)^{3/2} p$$

$$= \frac{1}{2\tau_o} \left( \frac{2\pi\hbar^2}{k_B T m_e^* + m_h^*} \right)^{3/2} p$$

Using $\tau_o = 0.6$ ns and $k_B T = 0.026$ eV, we get for $m_e^* = 0.067\ m_0, m_h^* = 0.45\ m_0,$

$$\frac{1}{\tau_r} = \frac{10^{21} m^{-3}}{2 \times (0.6 \times 10^{-9}\ s)} \left[ \frac{2 \times 3.1416 \times (1.05 \times 10^{-34}\ Js)^2}{(0.026 \times 1.6 \times 10^{-19}\ J) \times (0.517 \times 9.1 \times 10^{-31}\ kg)} \right]^{3/2}$$

$$\tau_r = 5.7 \times 10^{-6}\ s \cong 9.5 \times 10^3 \tau_o$$

We see from this example, that at low injection levels, the carrier lifetime can be very long. Physically, this occurs because at such a low injection level, the electron has a very small probability of finding a hole to recombine with.

**EXAMPLE 7.3** In two $n^+ p$ GaAs LEDs, $n^+ >> p$ so that the electron injection efficiency is 100% for both diodes. If the non-radiative recombination time is $10^{-7}$s, calculate the 300 K internal radiative efficiency for the diodes when the doping in the $p$-region for the two diodes is $10^{16}$ cm$^{-3}$ and $5 \times 10^{17}$ cm$^{-3}$.

When the $p$-type doping is $10^{16}$ cm$^{-3}$, the hole density is low and the $e$-$h$ recombination time for the injected electrons is given by Eqn. 7.16 as

$$\frac{1}{\tau_r} = \frac{1}{2\tau_o} \left( \frac{2\pi\hbar^2 m_r^*}{k_B T m_e^* m_h^*} \right)^{3/2} p$$

From the previous example, we can see that for $p$ equal to $10^{16}$ cm$^{-3}$, we have (in the previous example the value of $p$ was ten times smaller)

$$\tau_r = 5.7 \times 10^{-7}\ s$$

In the case where the $p$ doping is high, the recombination time is given by the high density limit (see Eqn. 7.17) as

$$\frac{1}{\tau_r} = \frac{R_{spon}}{n} = \frac{1}{\tau_o} \left( \frac{m_r^*}{m_h^*} \right)^{3/2}$$

$$\tau_r = \frac{\tau_o}{0.05} \sim 20\tau_o \sim 12\ ns$$

For the low doping case, the internal quantum efficiency for the diode is

$$\eta_{Qr} = \frac{1}{1 + \frac{\tau_r}{t_{nr}}} = \frac{1}{1 + (5.7)} = 0.15$$

For the heavier doped $p$-region diode, we have

$$\eta_{Qr} = \frac{1}{1 + \frac{10^{-7}}{20 \times 10^{-9}}} = 0.83$$

Thus, there is an increase in the internal efficiency as the $p$ doping is increased. Of course, as discussed in the text, this increase cannot continue with $p$ doping since eventually the injection efficiency will decrease.

**EXAMPLE 7.4** Consider a GaAs $p$-$n$ diode with the following parameters at 300 K:

| | | |
|---|---|---|
| Electron diffusion coefficient, | $D_n$ | $= \quad 30 \; cm^2/V - s$ |
| Hole diffusion coefficient, | $D_p$ | $= \quad 15 \; cm^2/V - s$ |
| p-side doping, | $N_a$ | $= \quad 5 \times 10^{16} \; cm^{-3}$ |
| n-side doping, | $N_d$ | $= \quad 5 \times 10^{17} \; cm^{-3}$ |
| Electron minority carrier lifetime, | $\tau_n$ | $= \quad 10^{-8} \; s$ |
| Hole minority carrier lifetime, | $\tau_p$ | $= \quad 10^{-7} \; s$ |

Calculate the injection efficiency of the LED assuming no recombination due to traps.

The intrinsic carrier concentration in GaAs at 300 K is $2 \times 10^6$ cm$^{-3}$. This gives

$$n_p = \frac{n_i^2}{N_a} = \frac{(2 \times 10^6)^2}{5 \times 10^{16}} = 8 \times 10^{-5} \; cm^{-3}$$

$$p_n = \frac{n_i^2}{N_d} = \frac{(2 \times 10^6)^2}{5 \times 10^{17}} = 8 \times 10^{-6} cm^{-3}$$

The diffusion lengths are

$$L_n = \sqrt{D_n \tau_n} = \left[(30)(10^{-8})\right]^{1/2} = 5.47 \; \mu m$$

$$L_p = \sqrt{D_p \tau_p} = \left[(15)(10^{-7})\right]^{1/2} = 12.25 \; \mu m$$

The injection efficiency is now (assuming no recombination via traps)

$$\gamma_{inj} = \frac{\frac{eD_n n_{po}}{L_n}}{\frac{eD_n n_{po}}{L_n} + \frac{eD_p p_{no}}{L_p}} = 0.98 \tag{7.23}$$

**EXAMPLE 7.5** Consider the $p$-$n^+$ diode of the previous example. The diode is forward biased with a forward bias potential of 1 V. If the radiative recombination efficiency $\eta_{Qr} = 0.5$, calculate the photon flux and optical power generated by the LED. The diode area is 1mm $^2$.

The electron current injected into the $p$-region will be responsible for the photon generation. This current is

$$
\begin{aligned}
I_n &= \frac{AeD_n n_{po}}{L_n} \left[ exp \left( \frac{eV}{k_B T} \right) - 1 \right] \\
&= \frac{(10^{-2}\ cm^2)(1.6 \times 10^{-19}\ C)(30\ cm^2/s)(8 \times 10^{-5}\ cm^{-3})}{5.47 \times 10^{-4}\ cm} \left[ exp \left( \frac{1}{0.026} \right) - 1 \right] \\
&= 0.35\ mA
\end{aligned}
$$

The photon particle current generated per second is

$$
\begin{aligned}
I_{ph} = \frac{I_n}{e} \cdot \eta_{Qr} &= \frac{(0.35 \times 10^{-3}\ A)(0.5)}{1.6 \times 10^{-19}\ C} \\
&= 1.09 \times 10^{15}\ s^{-1}
\end{aligned}
$$

Each photon has an energy of 1.41 eV (= bandgap of GaAs). The optical power is thus

$$
\begin{aligned}
Power &= (1.09 \times 10^{15}\ s^{-1})(1.41)(1.6 \times 10^{-19}\ J) \\
&= 0.25\ mW
\end{aligned}
$$

It must be kept in mind that not all of this light will be useful due to reabsorption or reflection from the GaAs-air interface. The next section discusses the overall efficiency of the diode.

## 7.4 EXTERNAL QUANTUM EFFICIENCY

$\longrightarrow$ In the previous section we have seen how photons are created in an LED structure. For these photons to emerge from the device, great care must be taken in the design of the LED. There are three main loss mechanisms for the emitted photons: i) the emitted photons can be reabsorbed in the semiconductor by creating an electron-hole pair; ii) a certain fraction of photons will be reflected back at the semiconductor-air interface; and iii) some photons impinge upon the surface with angles greater than the critical angle thus suffering total internal reflection.

To minimize the absorption of the photons, it is essential that the photons be emitted near the surface so that a good fraction of the photons do not have to travel long distances to the surface. This criterion was considered in our discussion of the injection efficiency $\gamma_{inj}$ in the previous subsection. Note that for direct gap materials, a photon can only travel a micron or so before getting absorbed. *It must be noted, however, that the active emission volume cannot be placed too close to the surface; otherwise non-radiative recombination processes mediated by surface defects will reduce the device efficiency.*

Photons that are able to make it to the semiconductor-air surface have to suffer reflection from the surface, as shown in Fig. 7.7a. Those that are reflected are lost. If $n_{r2}$ is the refractive index of the semiconductor and $n_{r1}$ the index of air, (as discussed in Chapter 2, Section 2.4) the reflection coefficient is (for vertical incident light),

$$R = \left(\frac{n_{r2} - n_{r1}}{n_{r2} + n_{r1}}\right)^2 \qquad (7.24)$$

This loss is called the Fresnel loss. For a GaAs LED, if we choose $n_{r2} = 3.66$, $n_{r1} = 1.0$, we get a loss of 0.33, i.e., 33% of the photons cannot get through. To avoid this excessive loss, the device is usually encapsulated in a dielectric dome, as shown in Fig. 7.7b. The dielectric has a refractive index of $\sim 1.6$ and this allows a greater fraction of photons to emerge.

Finally, one has the loss of photons due to total internal reflection. If light impinges at a surface from a region of high refractive index ($n_{r2} > n_{r1}$), it is totally reflected back if the angle of incidence is greater than a critical angle $\theta_C$ where

$$\theta_C = sin^{-1}\left(\frac{n_{r1}}{n_{r2}}\right) \qquad (7.25)$$

For the GaAs-air surface, the critical angle is 15.9°. Once again, use of the dome encapsulation suppresses this loss.

In addition to the three loss mechanisms discussed above, for many applications, the photons have to be coupled into special devices. For example, for optical communications, the light must be coupled into an optical fiber. If light is to be coupled into an optical fiber it must enter it on an angle so that it suffers total internal reflection, as shown in Fig. 7.8. This problem has been discussed in Chapter 3. If $n_{r1}$ and $n_{r2}$ are the refractive indices of the core and the cladding region, the maximum angle of acceptance, $\theta_A$, is given by (the result is derived in Chapter 3, Section 3.4)

$$\theta_A = sin^{-1}\left(n_{r1}^2 - n_{r2}^2\right)^{1/2} = sin^{-1}(A_n) \qquad (7.26)$$

where $A_n$ is called the numerical aperture of the fiber.

The photons emerging from the LED have an angular distribution between $\theta = 0$ and $\theta = \pi/2$. Let us assume that the distribution has a form

$$I_{ph}(\theta) = I_0 cos\theta \qquad (7.27)$$

so that there is higher probability of photons emerging normally where the photons travel the least distance through the semiconductor and thus suffer the fewest losses.

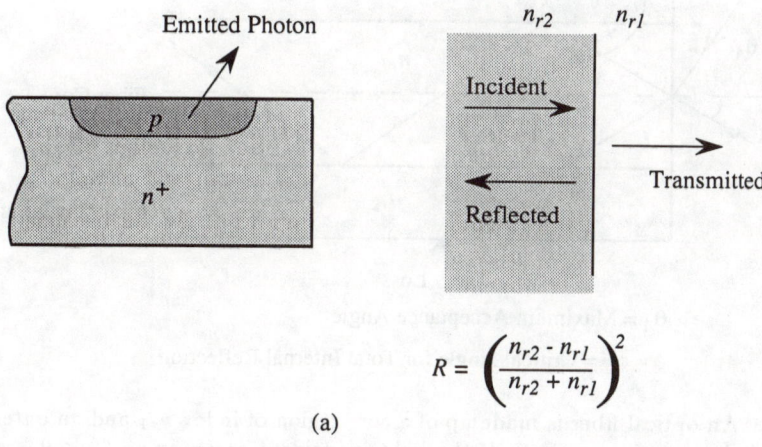

$$R = \left( \frac{n_{r2} - n_{r1}}{n_{r2} + n_{r1}} \right)^2$$

(a)

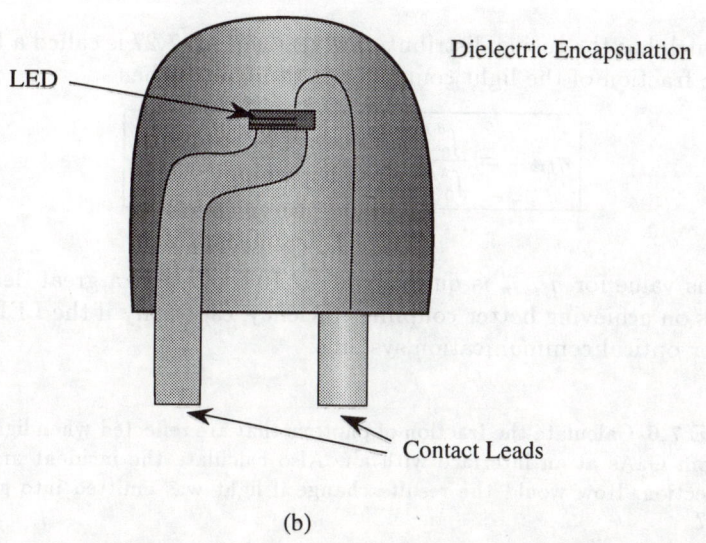

(b)

Figure 7.7: (a) The LED structure along with a schematic of the reflection and transmission of light at the semiconductor surface. (b) The dielectric encapsulation used to improve the transmission of photons generated. The presence of the dielectric reduces the reflection losses from the GaAs-air surface.

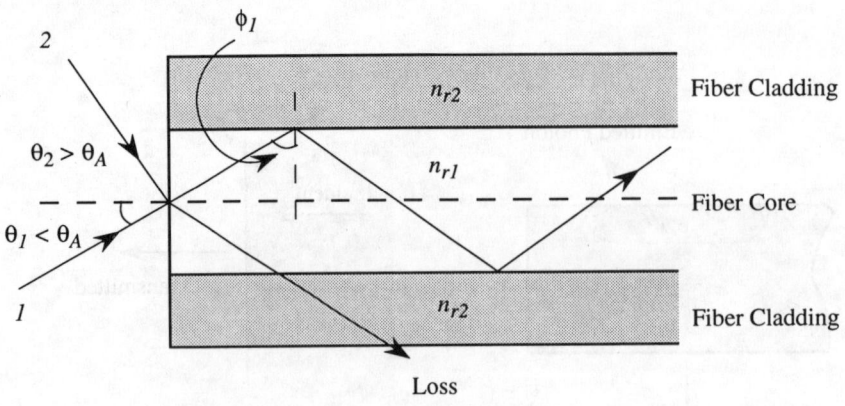

$\theta_A$ = Maximum Acceptance Angle

$\phi = \phi_C$ = Critical Angle for Total Internal Reflection

Figure 7.8: An optical fiber is made up of a core region of index $n_{r1}$ and an outer cladding region of index $n_{r2}$ ($n_{r1} > n_{r2}$). If the light is incident at an angle $\theta_1 < \theta_A$ so that the light suffers total internal reflection in the fiber, the light is coupled successfully into the fiber. Optical waves coming at an angle greater than $\theta_A$ are not able to propagate in the fiber.

A source which has the cosine distribution given by Eqn. 7.27 is called a Lambertian source. The fraction of the light coupled into the fiber is then

$$\eta_{fiber} = \frac{\int_0^{\theta_A} I_{ph}(\theta) sin\theta d\theta}{\int_0^{\pi/2} I_{ph}(\theta) sin\theta d\theta} = sin^2\theta_A \qquad (7.28)$$

This value for $\eta_{fiber}$ is quite small ($\sim 10\%$) so that a great deal of design has to focus on achieving better coupling efficiency, especially if the LEDs are to be used in fiber optical communication systems.

**EXAMPLE 7.6** Calculate the fraction of photons that are reflected when light is incident normally from GaAs at an interface with air. Also calculate the incident angle for total internal reflection. How would the results change if light was emitted into glass with an index of 1.5?

The refractive index of GaAs is 3.66 while that of air is 1.0. The reflection coefficient for normal incidence is

$$R = \frac{(n_{r2} - n_{r1})^2}{(n_{r2} + n_{r1})^2} = \left(\frac{2.66}{4.66}\right)^2 = 0.33$$

Thus 33% of the photons are reflected back and do not escape to the outside of the device. The angle of total internal reflection is

$$\theta_c = sin^{-1}\left(\frac{n_{r1}}{n_{r2}}\right) = sin^{-1}\left(\frac{1.0}{3.66}\right) = 15.9°$$

If $n_{r1}$ was to change to 1.5 we get

$$R = 0.18; \theta_c = 24.2°$$

Thus a much higher fraction of photons would escape to the outside.

**EXAMPLE 7.7** Consider an LED in which the photon output obeys a cosine law for its intensity. The light is to be coupled to an optical fiber which has a refractive index of 1.5 for the core and 1.40 for the outer cladding layer. Calculate the maximum angle of acceptance and the coupling efficiency for the fiber.

The maximum angle of acceptance is

$$\theta_A = sin^{-1}\left(n_{r1}^2 - n_{r2}^2\right)^{1/2} = 14.1°$$

The coupling efficiency is

$$\eta_{fiber} = sin^2\theta_A = 0.06$$

Thus only 6% of the optical output is able to couple into the optical fiber.

## 7.5 ADVANCED LED STRUCTURES

$\longrightarrow$   As we have discussed in the previous sections, important issues for LED technology are internal and external quantum efficiency, spectral purity of the light output, and temporal response. Although it is difficult to improve spectral purity and temporal response without using laser diodes, a number of advances have been made in the LED technology.

### 7.5.1   Heterojunction LED

If the LED is made from a single semiconductor, there are a number of problems that reduce the device efficiency. An important problem is that in a homojunction LED (i.e., a device based on a single semiconductor), the photon emission volume must be close to the surface so that the emitted photons are not reabsorbed. Since near the surface the semiconductor quality is usually not very good due to the presence of defect states, this causes a great deal of surface state mediated non-radiative recombination. In addition to this problem, the electrons injected from the $n^+$ side into the $p$-region can diffuse over long distances before recombining with holes. Thus the effective volume from which photons are emerging is quite large.

The heterojunction LED resolves these problems by injecting charge from a larger bandgap material in a narrow gap active region. Fig. 7.9 gives a schematic of such an LED. Electrons and holes are injected from the wide gap $n$- and $p$-regions into the narrow gap active region. The electrons cannot enter the wide gap $p$-region below the active region and thus do not suffer from poor surface conditions. *The photons emitted are also not absorbed in the top or bottom region since the photon energy is smaller than the bandgap of the n- or p-region.*

The heterojunction LEDs are made by epitaxial processes, and the active region is kept to ~0.1-0.2 $\mu$m. The materials commonly used are GaAs/AlGaAs grown on GaAs and InGaAsP/InP and InGaAs/InGaAsP grown on InP substrates.

## 7.5.2    Edge-Emitting LED

An important issue in optical communication is the efficiency with which the light emitted by an LED couples into an optical fiber. We had discussed this coupling efficiency and seen that one needs a highly collimated beam for efficient coupling. The heterostructure technology is exploited to fabricate the edge-emitting LED shown schematically in Fig. 7.10. As we shall see later, the device looks almost like a laser diode. The difference is that in the laser diode exceptional care is taken to produce a high quality optical cavity to ensure optical feedback.

*An important ingredient of the edge-emitting LED is the wide gap cladding layers which confine not only the electrons and holes to the active layer, but also cause the emitted photons to travel along the LED axis and emerge from the edge of the device.* This waveguide optical cavity has been discussed in Chapter 3 and will be further discussed in Chapter 8.

Due to the superior collimation of the edge-emitting LED (~30° width perpendicular to the layer and ~ 120° parallel to the layer) the coupling efficiency to a fiber is greatly improved.

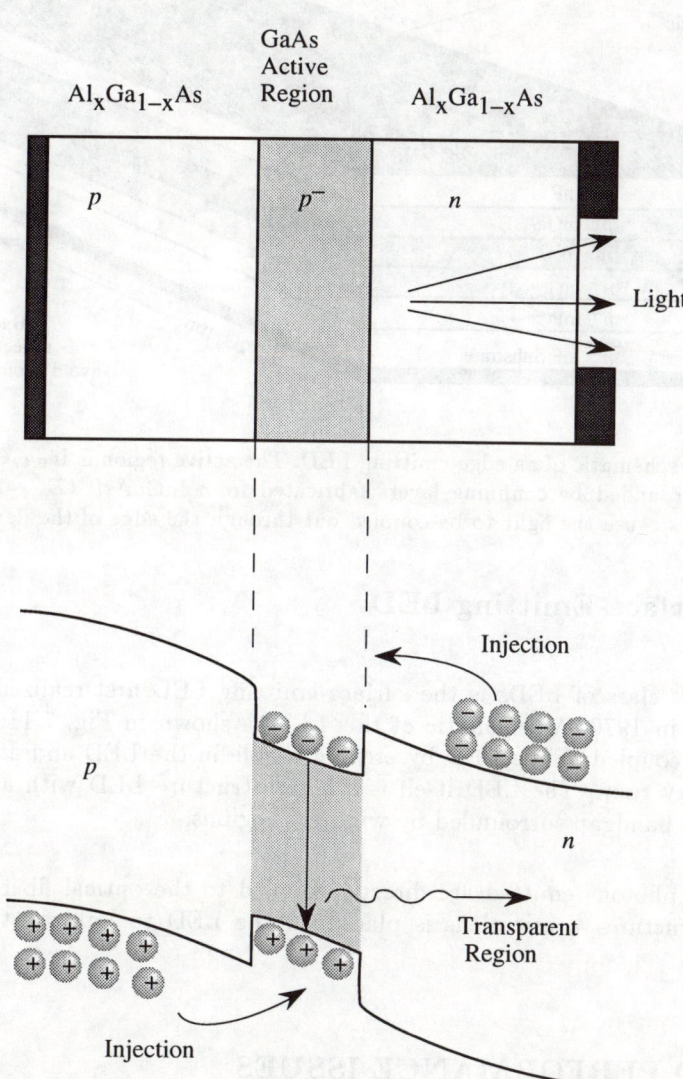

Figure 7.9: A heterojunction LED uses a narrow gap semiconductor for the active region. The photons that are emitted are not reabsorbed in the top or bottom layers, which are transparent to the emitted radiation.

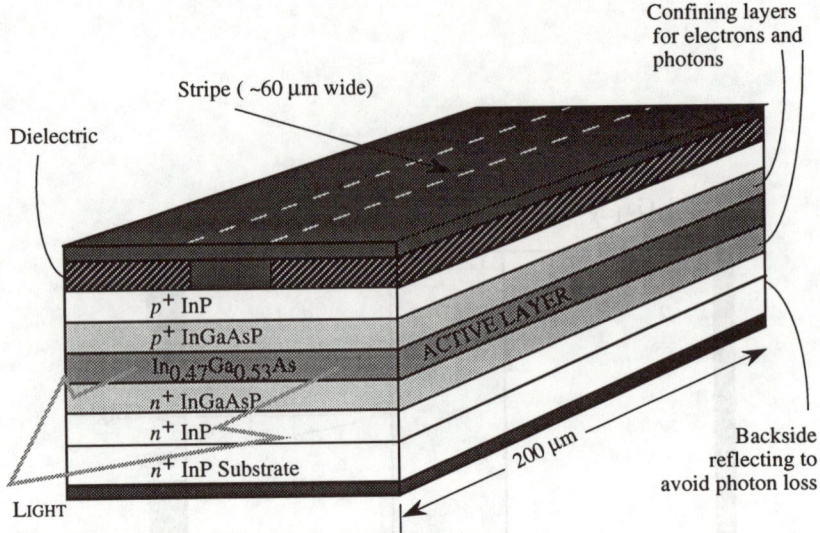

Figure 7.10: A schematic of an edge-emitting LED. The active region is $In_{0.47}Ga_{0.53}As$ ($E_g$ = 0.8 eV) surrounded by confining layers fabricated from InGaAsP ($E_g \sim 1.0$ eV). The confining layers cause the light to be coupled out through the edge of the device.

### 7.5.3 Surface-Emitting LED

An important class of LEDs is the surface emitting LED first realized by Burrus and Dawson in 1970. A schematic of this LED is shown in Fig. 7.11. An optical fiber is butt coupled to the LED by etching a hole in the LED and attaching the fiber by epoxy resin. The LED itself is a heterostructure LED with a thin active region of low bandgap surrounded by wide gap regions.

The photons emitted are directly coupled to the optical fiber. In various advanced structures a microlens is placed on the LED to improve the coupling efficiency.

## 7.6   LED PERFORMANCE ISSUES

$\longrightarrow$ The LED depends upon the spontaneous emission process to provide light from the injected electrons and holes. As a result there are simplifications in the fabrication and design of the LED, but one has to pay a price in performance when an LED is compared to a laser diode. This comparison will be clear after our discussions on the laser diode. For the LED the important performance issues are represented by the light-current characteristics, the spectral purity of the light output, the time

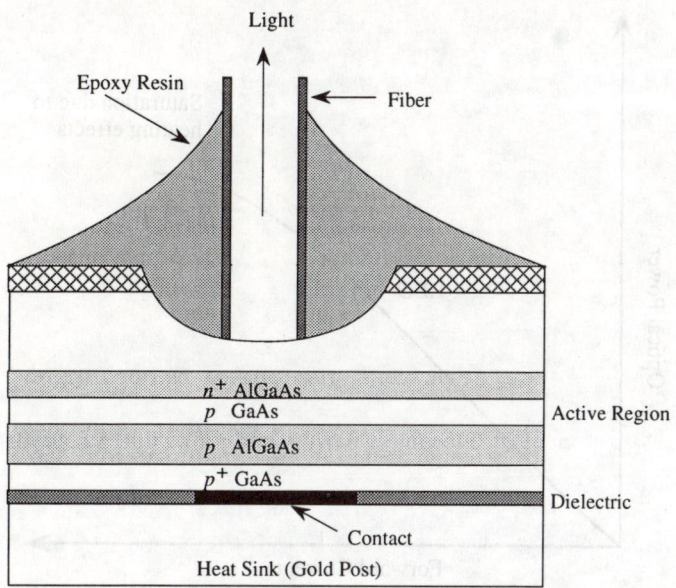

Figure 7.11: A schematic of a surface-emitting LED. A heterostructive LED has an optical fiber butt coupled to it with an epoxy resin. The light generated in the active $p$-GaAs is fed into the fiber directly. In advanced designs, a microlens is used to improve the coupling.

response of the LED to external electrical signals and the temperature dependence of the output. The spectral purity (i.e., spread in wavelengths of the output light) is quite critical for high performance optical communication systems, as will be discussed in Chapter 10.

## 7.6.1 Light-Current Characteristics

$\longrightarrow$ When a current $I$ is passing through the forward biased diode, a certain fraction of the current is converted to light. If $\eta_{Tot}$ represents the total efficiency of this conversion, the photon particle current that emerges from the diode is

$$
\begin{aligned}
I_{ph} &= \text{Number of photons per second} \\
&= \eta_{Tot} \cdot \frac{I}{e}
\end{aligned}
\tag{7.29}
$$

In general, $\eta_{Tot}$ depends upon the injected current since the carrier radiative lifetime $\tau_r$ depends upon carrier injection level. However, in an LED this dependence is quite weak so that the $I_{ph} - I$ characteristics are essentially linear as shown in Fig. 7.12. At very high injection the light output starts to saturate as the device heats and the radiative recombination efficiency decreases.

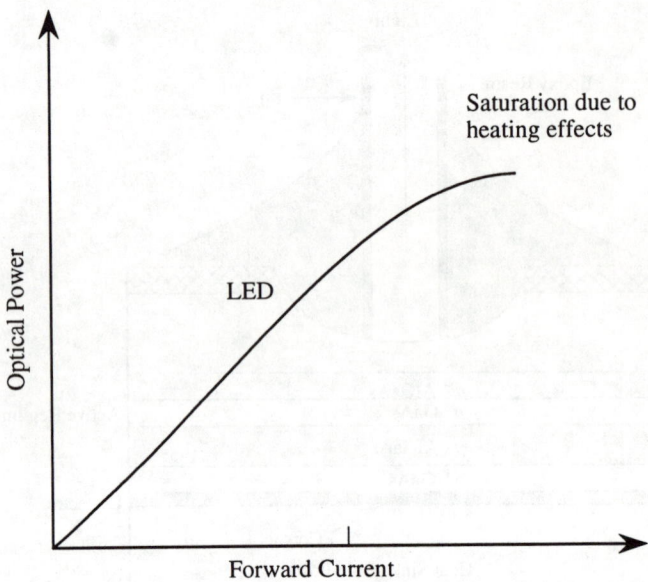

Figure 7.12: The output power of an LED is essentially linear with the injected currents.

In surface-emitting LEDs, it is often found that the light output decreases sublinearly with current, an effect that cannot be simply explained by heating effects. It appears that at high drive current the photon density in the LED becomes large enough that stimulated emissions start to occur. This emission is in the plane of the LED so that the photons, emitted perpendicularly to the surface, decrease. Such LEDs are called superluminescent LEDs and behave similarly to the laser diode discussed in the next chapter.

## 7.6.2  Spectral Purity of LEDs

$\longrightarrow$ The spectral purity or the linewidth of the emitted radiation is an important characteristic of optical devices. The importance of the spectral purity of the emitted light depends upon applications. If the LED is to be used in a display device, the spectral purity is not an issue. However, in optical communication applications the spectral purity is a critical issue. Light pulses of different wavelengths travel through an optical fiber at different speeds, as discussed in Chapter 10. Thus a signal gets distorted if the optical beam has a large wavelength spread. As can be seen from the discussion in Chapter 5, Section 5.7, the emission spectrum is essentially determined by the product $(\hbar\omega - E_g)^{1/2} \, f^e(E^e) f^h(E^h)$. This is the convoluted product of the electron and hole occupation probabilities. At low injection, this width is of the

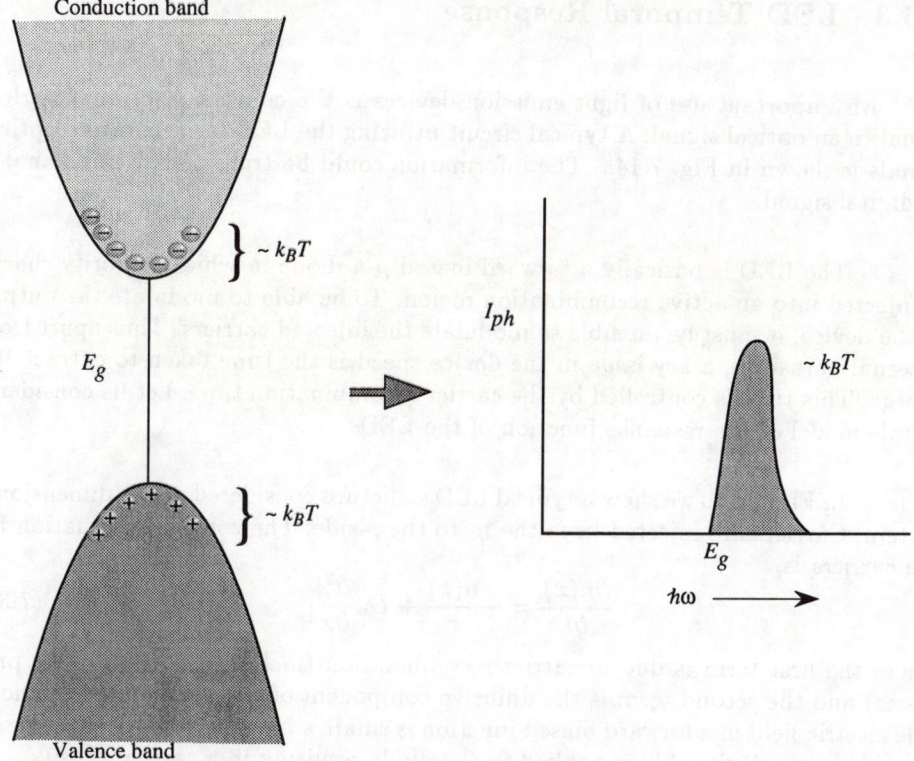

Figure 7.13: In an LED, the electrons and holes are distributed over an energy width of ~ $k_BT$. Since all the $e$-$h$ pairs contribute to the optical output, the LED output is quite broad with a width roughly equal to $k_BT$. The shape of the output depends upon the carrier occupation function and the density of states function.

order of $k_BT$. At high injection the width is ($n$ is the total charge density)

$$\Delta E \sim \frac{n}{N_c} k_BT \qquad (7.30)$$

where $N_c$ is the effective bandedge density of states. A typical emission spectrum for an LED is shown in Fig. 7.13. The linewidth (full width at half maxima) is seen to be of the order of 20 nm (200 Å) at room temperature. This is obviously a broad spectrum. However, for many applications this width is adequate. *In fact, LEDs are widely used in optical communication as long as the signal does not have to be sent over long distances.* However, for long distance communication, the LED output is not adequate and a laser diode must be used.

## 7.6.3   LED Temporal Response

〜→   An important use of light emission devices is the conversion of an electrical signal to an optical signal. A typical circuit utilizing the LED for generating optical signals is shown in Fig. 7.14a. The information could be transmitted as an analog or digital signal.

The LED is basically a forward biased p-n diode in which minority charge is injected into an active recombination region. To be able to modulate the output of the device, it must be possible to modulate the injected carriers. Thus apart from external parasitics, a key issue in the device speed is the time taken to extract the charge. This time is controlled by the carrier recombination time. Let us consider a simple model of the response function of the LED.

In Fig. 7.14b we show a typical LED structure considered as a 1-dimensional system. Carriers are injected from the n- to the p-side. The continuity equation for the carriers is

$$\frac{\partial n(x)}{\partial t} = -\frac{n(x)}{\tau} + D_n \frac{\partial^2 n}{\partial x^2} \tag{7.31}$$

where the first term is due to carrier recombination (including non-radiative processes) and the second term is the diffusive component of the particle current flow. The electric field in a forward biased junction is small, so we ignore the drift current. Consider a small signal bias applied to the diode resulting in a carrier density

$$n(x,t) = n_o(x) + \tilde{n}(x) exp(i\omega t) \tag{7.32}$$

Substituting this in Eqn. 7.31, we get, comparing the dc and ac components

$$D_n \frac{\partial^2 n_o}{\partial x^2} - \frac{n_o}{\tau} = 0 \tag{7.33}$$

$$D_n \frac{\partial^2 \tilde{n}}{\partial x^2} - \frac{\tilde{n}(1 + i\omega\tau)}{\tau} = 0 \tag{7.34}$$

Defining the lengths

$$L_n = (D_n \tau)^{1/2} \tag{7.35}$$

$$L_n(\omega) = \left[\frac{D_n \tau}{1 + i\omega\tau}\right]^{1/2} \tag{7.36}$$

we get

$$\frac{\partial^2 \tilde{n}(x)}{\partial x^2} = \frac{\tilde{n}(x)}{L_n^2(\omega)} \tag{7.37}$$

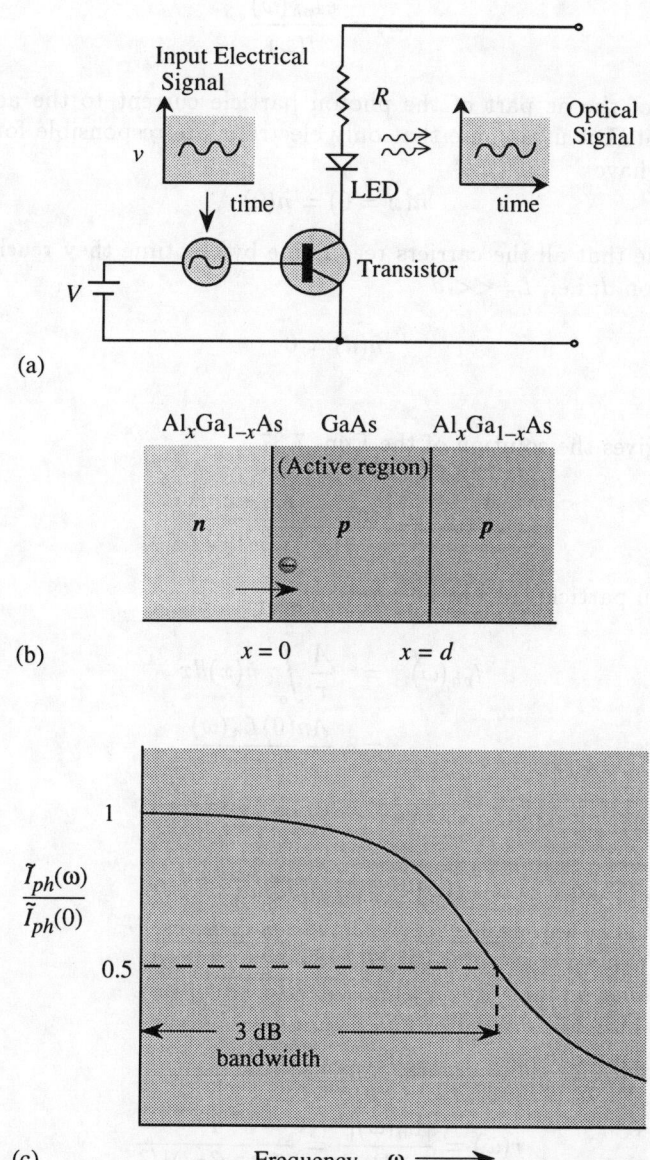

Figure 7.14: (a) A schematic of a circuit for modulation of the output of an LED. (b) Geometry of the LED used to study the intrinsic temporal response to an ac signal. (c) The response of the LED falls with frequency as shown. The 3dB bandwidth is indicated.

We can define the temporal response of the LED by the response

$$r(\omega) = \frac{e\tilde{I}_{ph}(\omega)}{\tilde{I}(\omega)} \tag{7.38}$$

i.e., the ratio of the ac part of the photon particle current to the ac part of the electron current. Let us assume that only electrons are responsible for the current flow. Then we have

$$\tilde{n}(x = 0) = \tilde{n}(0) \tag{7.39}$$

We also assume that all the carriers recombine by the time they reach the edge of the active region $d$; i.e., $L_n \ll d$

$$\tilde{n}(d) = 0 \tag{7.40}$$

This gives the solution of the Eqn. 7.37

$$\tilde{n}(x) = \tilde{n}(0) exp\left(\frac{-x}{L_n(\omega)}\right) \tag{7.41}$$

and the photon particle current becomes

$$\tilde{I}_{ph}(\omega) \quad = \quad \frac{A}{\tau} \int_o^d \tilde{n}(x) dx \tag{7.42}$$

$$= \quad \frac{A\tilde{n}(0) L_n(\omega)}{\tau} \tag{7.43}$$

Also,

$$\tilde{I}(\omega) \quad = \quad eAD_n \frac{\partial \tilde{n}(x)}{\partial x} \tag{7.44}$$

$$= \quad -eA \frac{D_n \tilde{n}(0)}{L_n(\omega)} \tag{7.45}$$

This gives for the temporal response function

$$r(\omega) = \frac{|L_n(\omega)|^2}{\tau D_n} = \frac{1}{(1 + \omega^2 \tau^2)^{1/2}} \tag{7.46}$$

This expression shows the importance of the recombination time $\tau$ in the bandwidth limits that can be reached in LEDs. The modulated bandwidth $f_c$ is defined at the frequency where the power is one-half that of the zero frequency value

$$f_c = \frac{\omega_c}{2\pi} = \frac{1}{2\pi\tau} \tag{7.47}$$

and

$$\frac{1}{\tau} = \frac{1}{\tau_r} + \frac{1}{\tau_{nr}} \tag{7.48}$$

where $\tau_r$ and $\tau_{nr}$ are the radiative and non-radiative recombination times. For high quality devices, $\tau \sim \tau_r$. The LED response as a function of frequency is shown in Fig. 7.14c.

In Fig. 7.6 we had shown the dependence of the radiative lifetime on the carrier density or the doping of the active region. As the LED drive current increases, the recombination time decreases and the modulation bandwidth thus increases. This dependence is, indeed, seen experimentally as shown in Fig. 7.15. Note that the carrier density in the active region is proportional to $J/d$, where $J$ is the current density and $d$ is the active region thickness.

The modulation bandwidth can also be increased by increasing the doping of the active region. The higher $p$-doping allows the injected electrons to recombine with holes in a shorter time.

The ultimate bandwidth is controlled by the time $\tau_o$ for an $e$-$h$ recombination (with $f_e = f_h = 1$) which, as discussed earlier, has a value of $\sim 0.5$ ns. Thus, the LED cutoff frequency can approach a gigahertz.

The limits arising from the radiative recombination is a key difference between LEDs and laser diodes. The laser diodes operate not under conditions of spontaneous emission, but stimulated emission. As we shall see later, the stimulated emission depends upon the photon density present and can result in recombination times approaching 10 ps.

**EXAMPLE 7.8** An AlGaAs($n$) - GaAs($i$) - AlGaAs($p$) LED is operated at 300 K at current density levels of 100 A/cm$^2$, 500 A/cm$^2$ and 1000 A/cm$^2$. The width of the active region is 1.0 $\mu$m. Calculate the 3 dB cutoff frequency for the two operating points. Assume that the temporal response is limited by the $e$-$h$ recombination time and the efficiency is unity. Use the information given in Fig. 7.6.

The current density in the LED is (for unity efficiency)

$$J = \frac{end}{\tau_r}$$

or

$$\frac{n}{\tau_r} = \frac{J}{ed} = \frac{p}{\tau_r}$$

For a current density of 100 A/cm$^2$ we have

$$\frac{n}{\tau_r} = \frac{(100 \ A/cm^2)}{(1.6 \times 10^{-19} \ C)(1.0 \times 10^{-4} cm)} = 6.25 \times 10^{24} \ cm^{-3} \ s^{-1}$$

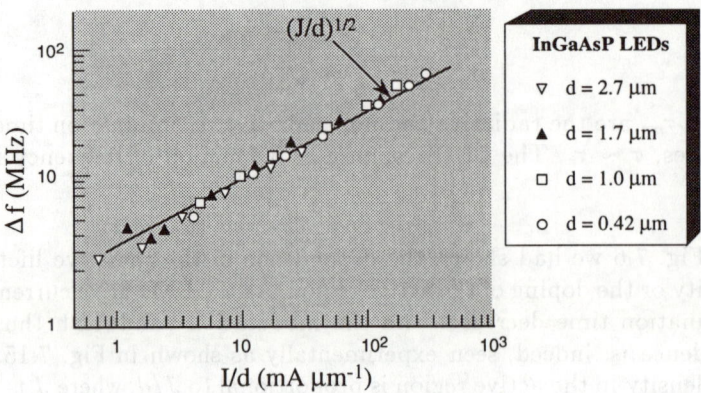

Figure 7.15: Modulation bandwidth as a function of normalized drive current density for InGaAsP LEDs with different active region thicknesses. (After O. Wada, Y. Yamakoshi, M. Abe, Y. Yishitoni, and T. Sakwai, *IEEE J. Quantum Electronics*, QE-17, 174 (1981).)

From Fig. 7.6 we see that this value occurs for the combination

$$n \cong 6 \times 10^{16}\ cm^{-3}; \tau_r \sim 10^{-8}\ s$$

Thus the cutoff frequency is

$$f_c = \frac{1}{2\pi\tau_r} = 15.9\ MHz$$

When the drive current is 500 A/cm$^{-2}$, we get

$$\frac{n}{\tau_r} = 3.38 \times 10^{25}\ cm^{-3}\ s^{-1}$$

Once again from Fig. 7.6, we see that this occurs when

$$n \cong 3 \times 10^{17}\ cm^{-3}; \tau_r \cong 8 \times 10^{-9}\ s$$

The cutoff frequency is now

$$f_c = 20\ MHz$$

When the current density is 1000 A/$cm^2$, we get

$$f_c \cong 35\ MHz$$

## 7.6.4  Temperature Dependence of LED Emission

$\longrightarrow$ In the previous subsection on light output as a function of current, we noted that the output power saturates at high current due to heating effects. The temperature of the high quality diodes with negligible defect densities influences the *e-h* recombination efficiency through two phenomenon: i) Leakage of injected carriers into the contact regions at high temperatures; ii) Auger processes (see Section 5.8) that contribute to non-radiative recombination.

In the LED, which is a forward biased *p-n* diode, carriers are injected from the doped sides into an active region where they recombine. Ideally, the carriers should thermalize in the active region as shown schematically in Fig. 7.16a and recombine to emit photons. However, as temperature increases, the distribution of the injected charge is wider as shown in Fig. 7.16b, so that an increasing fraction of the charge can leak across the active region. The leakage current will not contribute to the photon emission process and it thus reduces the optical power.

As an LED is pumped harder by increasing the drive current, the device heats, and as a result, the leakage current increases. The leakage current depends upon the device design. For example, if the active region is wide, the leakage may be small. The leakage current can be quite serious in most devices and contribute as much as 20-30% of the total current. One can suppress the LED heating at high drive current by using a pulsed source of current.

The Auger process is another important mechanism for non-radiative recombination. As discussed in Chapter 5, Section 5.8, the Auger process involves three carriers in the initial state which could be 2 electrons and a hole or one electron and two holes. The end product after the Auger recombination is one "hot" electron or hole with no emission of a photon. The recombination rate is strongly dependent on carrier density ($\propto n^3$), the bandgap of the material operating temperature and details of the bandstructure. The net effect is that for narrow gap based LEDs ($E_g < 1.0\ eV$), the Auger process can be an important source of non-radiative recombination. The process also has a strong temperature dependence.

The net effect of the carrier leakage and Auger processes produces an optical output from a LED which has a form

$$I_{ph} = I_{ph}(0)exp\left(-\frac{T}{T_1}\right) \tag{7.49}$$

where $T_1$ is a temperature that depends upon the material bandgap and the LED design parameters. The temperature $T_1$ should be as large as possible to ensure temperature independence of LEDs. For 1.3 $\mu$m LEDs based on InGaAsP, the value of $T_1$ is in the range of 180-200 K while for GaAs based LEDs it is 300-350 K.

In addition to the optical power dependence upon temperature, it is important to note that the bandgap of all semiconductors decreases with temperature. As a result, the peak of the emission spectrum of the LED will shift to higher wavelengths as temperature increases. This shift is an approximately 3.5 $\text{Å}K^{-1}$ for GaAs LEDs and 6 $\text{Å}K^{-1}$ for InGaAsP LEDs.

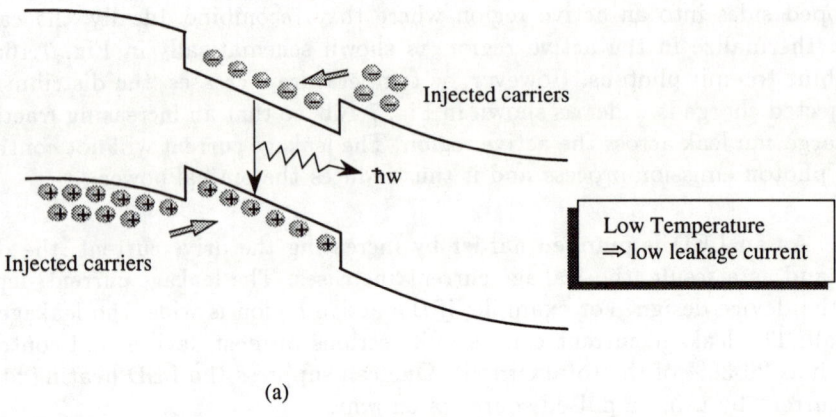

(a)

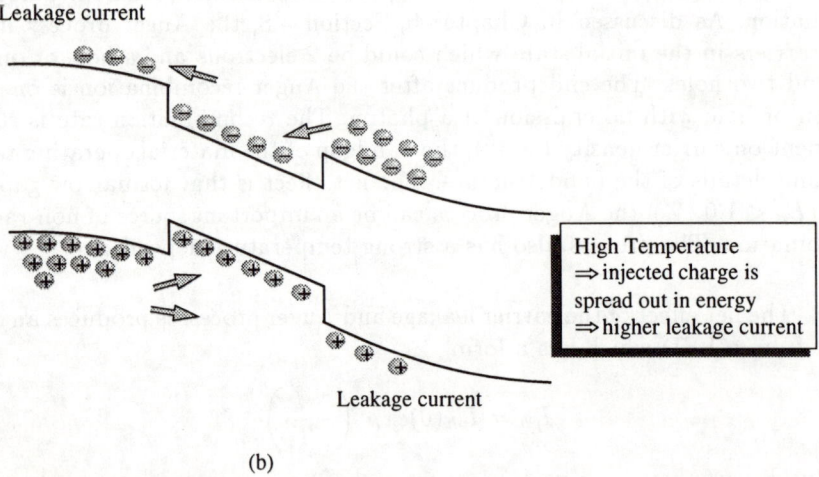

(b)

Figure 7.16: (a) A schematic of a charge injection in a LED at low temperature. All the injected carriers recombine in the active region. (b) At higher temperatures, due to the energy spread ($\sim k_B T$) of the injected carriers, a greater fraction of charge can leak, reducing the radiative efficiency.

## 7.7  APPLICATION OF LEDs

$\mathcal{R}$  Light-emitting diodes find a variety of important uses in optical information processing systems. These applications include displays where one or more LEDs are used to generate an optical signal that is visible to the eye. LEDs are also used in optical communication systems for local area networks. Another important use of LEDs is in the optical isolators where two electrical circuits are "connected" by the light from an LED. We will briefly discuss the important issues in these three areas.

**LEDs as Display Elements**
An important application of LEDs is in the area of display technology. The LED display is an "active" display, unlike the liquid crystal display where the light is not generated by the liquid crystal cell. As a result, the LED displays can be extremely bright, generating enough power for applications such as tail lights of automobiles, daytime programmable displays for advertisements, etc.

Since the displays are to be seen by the human eye, the light should be generated to cover the entire visible spectrum if full color displays are to be produced. Red light ($\lambda \sim 6600$ Å) LEDs are made from $GaAs_{0.6}P_{0.4}$ and AlGaAs alloy systems. Green light can be emitted by GaP ($\lambda \sim 5200$ Å). It is extremely difficult to get blue light emission from LEDs, since the bandgap of the semiconductor needed is $\sim 2.1$-$2.4$ eV and the technology for such materials is not mature. LEDs from GaN and ZnSe are available for blue light generation, but very few suppliers can provide these diodes.

The LED may be used as a single lamp (often to indicate the status of an electronic system) or in arrays. The arrays are electronically wired to generate alphanumeric characters. A typical display for characters involves a seven-segment array which is made up of seven different LED segments, as shown in Fig. 7.17. This figure shows how a 14 pin connection (with 4 omitted pins) can be used to drive the display.

LED-based displays are especially useful in applications where the display brightness is of importance. The power output of the LEDs is high enough to allow the displays to be visible in sunlight. Also, since the LED is a simple $p$-$n$ diode, it is straightforward to integrate it with other semiconductor devices. While in principle, it is possible to make LED-based flat panel displays for laptop computers and other such applications, cost and power consumption considerations give liquid crystal displays, discussed in Chapter 9, the advantage. However, the progress in LED technology is likely to allow them to replace many displays based on light bulbs. Examples may include automobile brake lights and traffic signals.

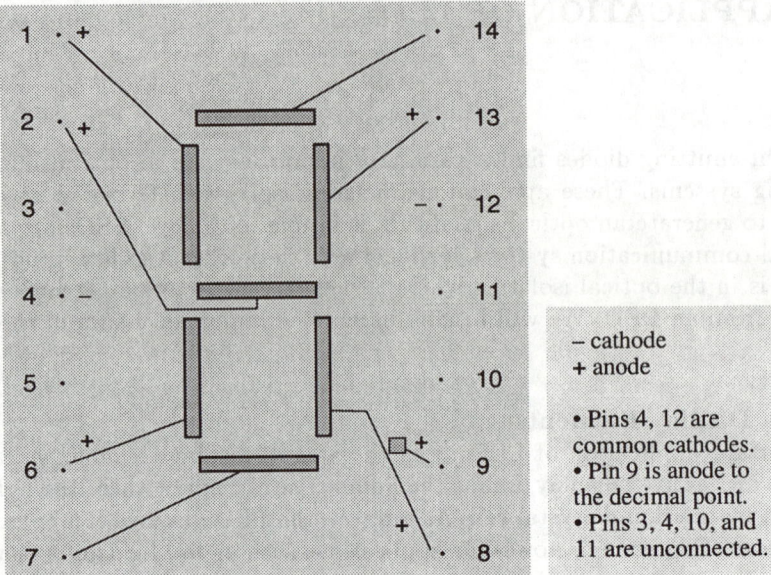

**Figure 7.17:** A schematic of a seven segment LED display with a decimal point. A 14 pin package is shown, along with typical pin connections.

## LEDs in Communication

To the human eye, the output of an LED appears to be of a single wavelength. However, as discussed in the previous section, the output has a fairly broad spectral content. Typically the width of the output is ~200 Å in wavelength. In Chapter 10, we discuss the problems arising in optical fiber communications from material dispersion. Material dispersion represents the fact that light beams of different wavelengths will travel at different speeds in the fiber. As a result, if an optical pulse of a certain temporal width is injected into a fiber, and if the pulse has many wavelengths in it, it will start to broaden temporally. This limits the bandwidth-distance product as discussed in Chapter 10. Thus, LEDs cannot be used for long distance communications and are used primarily for local area networks.

Another limitation of LEDs in communication arises from the speed at which transmission can occur. At present, optical communication is done by intensity variation of the light source. As discussed in the previous section, LEDs cannot by modulated much beyond 1 GHz. While for most applications, 1 GHz represents a perfectly acceptable frequency, the technology is rapidly moving towards higher frequencies and LEDs cannot compete. As a result, laser diodes are used in most demanding optical communication applications.

## LEDs in Opto-Isolators

An important application of LEDs is in opto-isolators. The opto-isolator is a device

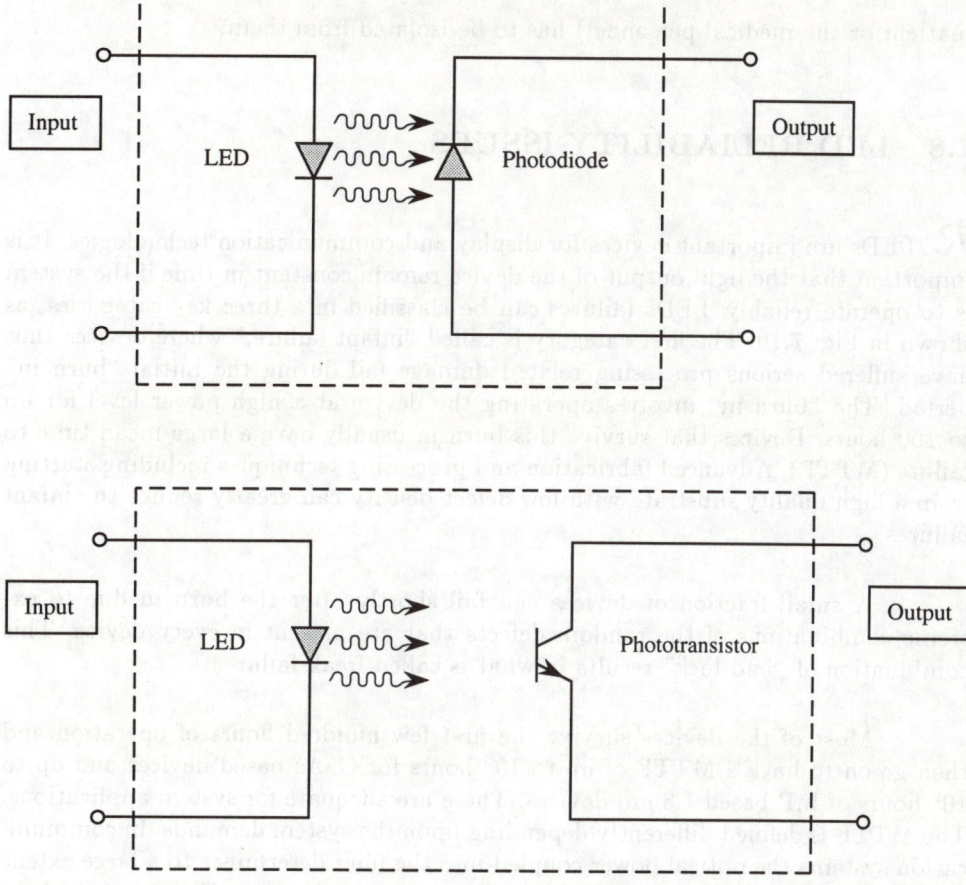

Figure 7.18: Schematic of two opto-isolators using a photodiode and a phototransistor in the output circuit.

consisting of a light emitter and a sensor. Signals can be transmitted while maintaining an extremely high degree of isolation between input and output. In the past, isolation of this degree was provided by relays, isolation transformers and blocking capacitors.

The basic opto-isolator circuit is shown in Fig. 7.18. The LED is part of the input circuit and depending upon the state of the input, could be emitting light. The light is detected by a detector which can be a photo diode, a phototransistor, or some other semiconductor detector. The connection between the input and output is through free space, thus providing a tremendous isolation. Systems which obviously need this degree of isolation are, for example, electrocardiograph amplifiers, which need a very high degree of isolation to ensure patient safety during medical treatment. In such medical devices, very often, high voltages are used and the user (patient or the medical personnel) has to be isolated from them.

## 7.8   LED RELIABILITY ISSUES

$\mathcal{R}$ LEDs are important devices for display and communication technologies. It is important that the light output of the device remain constant in time if the system is to operate reliably. LEDs failures can be classified into three key categories, as shown in Fig. 7.19. The first category is called "infant failure," where devices that have suffered serious processing related damage fail during the initial "burn in" period. The "burn in" involves operating the device at a high power level for up to 100 hours. Devices that survive this burn in usually have a large mean time to failure (MTTF). Advanced fabrication and processing techniques including starting from a high quality substrate with low defect density can greatly reduce the infant failures.

A small fraction of devices can fail shortly after the burn in due to extreme combinations of the random defects that are present in every device. This combination of "bad luck" results in what is called freak failure.

Most of the devices survive the first few hundred hours of operation and then go on to have a MTTF of up to $10^6$ hours for GaAs based devices and up to $10^9$ hours of InP based 1.3 $\mu$m devices. These are adequate for system applications. The MTTF is defined differently depending upon the system demands. In communication systems the optical power coupled into the fiber determines to a large extent (along with the optical wavelength) the repeater spacing (see Chapter 10). As a result MTTF is defined in terms of the time taken for the power output to decrease to a certain fraction of the original value. In a system with enough tolerance MTTF may be defined as the time taken for the power to drop by 50% (-3 dB). However, in systems designed with high repeater spacings the MTTF may be defined as the time taken for a 20% (-1 dB) power drop.

Failure in LEDs is gradual in contrast to laser diodes where the failure is catastrophic. The failure mechanism in LEDs involves an increase in the nonradiative recombination due to various kinds of defects. In the early stages of LED and LD development, an important defect mechanism involved migration of dislocations into the active device region. These defects which contribute strong nonradiative centers are called dark line defects (they appear as dark lines in electroluminescence). These defects create a catastrophic failure in LDs, but in LEDs the effect is a gradual decrease in light output. The dark line defect formation process involves a climb of substrate defects into the active region. This is found to occur in GaAs based devices but not in the InP based 1.3 $\mu$m devices. As a result, InP based devices have a longer MTTF.

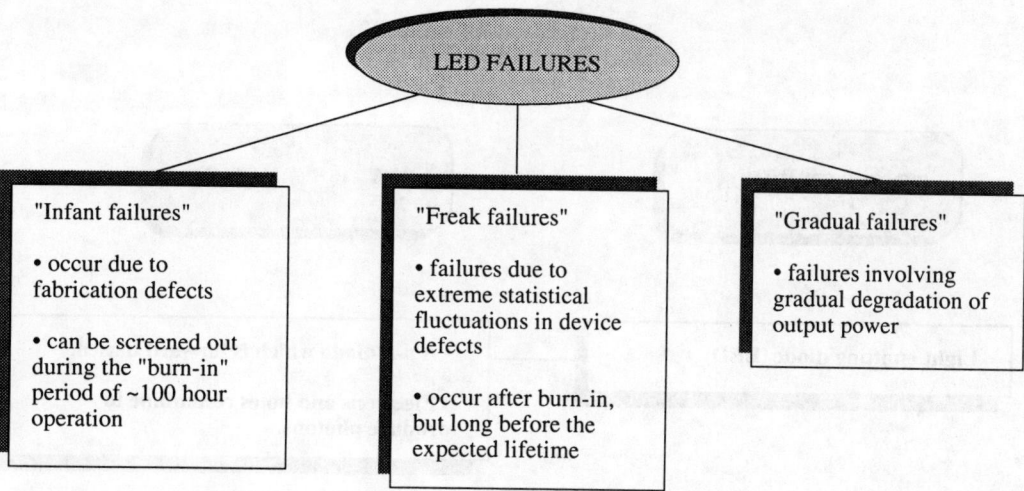

Figure 7.19: LED failures can be classified into three categories as shown in this chart.

Testing and controlling reliability of LEDs (and lasers) remains an important challenge. Researchers are constantly looking for testing techniques that allow elimination of devices that have a poor MTTF without long time testing. Also, the influence of substrate and processing (including metals used in contact formation ) is still being actively studied.

## 7.9 CHAPTER SUMMARY

The LED is an important optical source which finds uses in many applications, including those in display systems and optical communication. The main advantages of the LED are the simplicity of the fabrication process and the easy incorporation of the device (which is a simple *p-n* diode) in most circuitry. The key drawbacks of the LED are the broad spectrum of the emitted light and the difficulty in pushing the modulation bandwidths above a few gigahertz.

The summary tables (Tables 7.1-7.3) highlight the important areas discussed in this chapter.

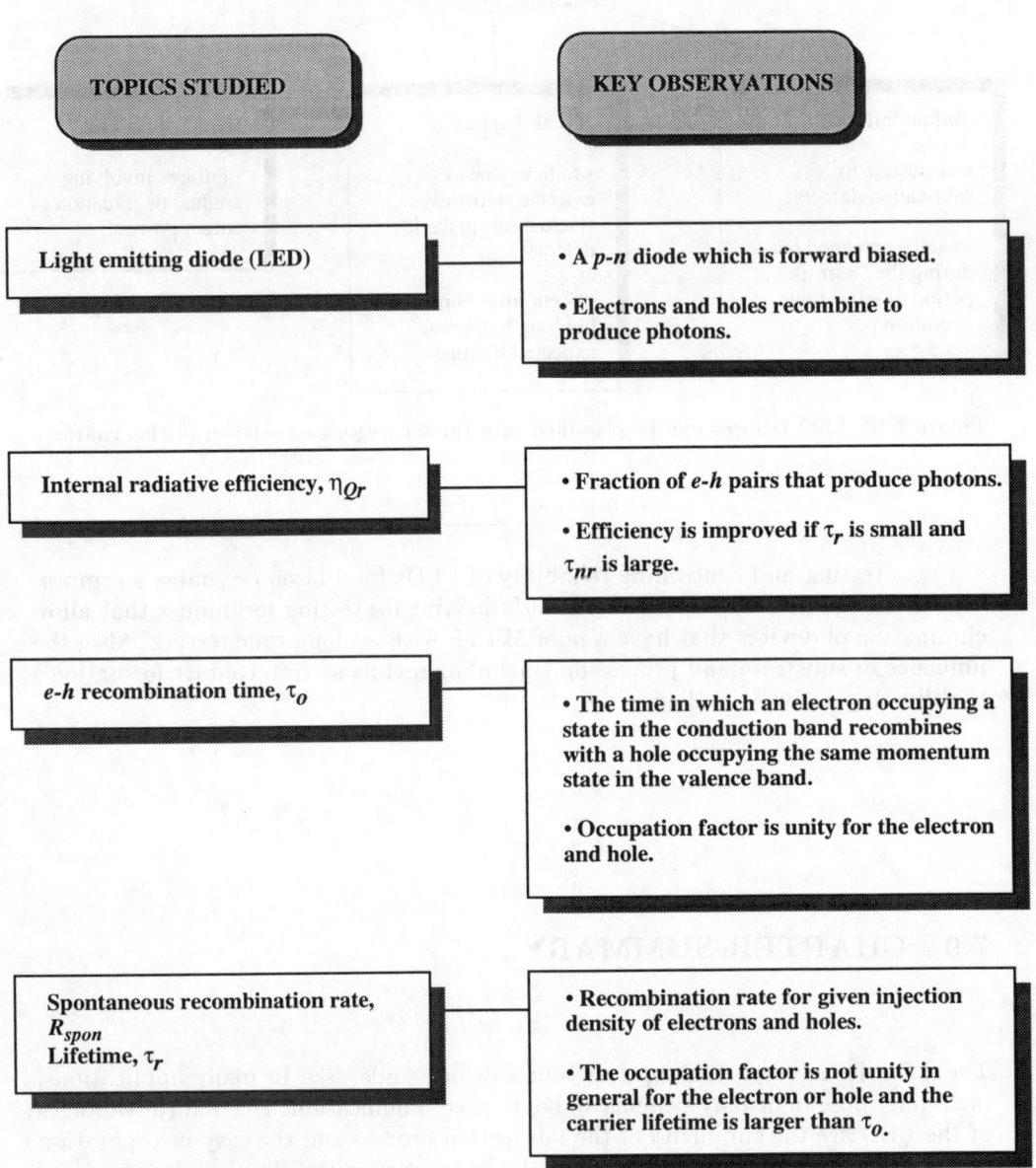

**TOPICS STUDIED**

**KEY OBSERVATIONS**

**Light emitting diode (LED)**

- A *p-n* diode which is forward biased.

- Electrons and holes recombine to produce photons.

**Internal radiative efficiency, $\eta_{Qr}$**

- Fraction of *e-h* pairs that produce photons.

- Efficiency is improved if $\tau_r$ is small and $\tau_{nr}$ is large.

***e-h* recombination time, $\tau_o$**

- The time in which an electron occupying a state in the conduction band recombines with a hole occupying the same momentum state in the valence band.

- Occupation factor is unity for the electron and hole.

**Spontaneous recombination rate, $R_{spon}$
Lifetime, $\tau_r$**

- Recombination rate for given injection density of electrons and holes.

- The occupation factor is not unity in general for the electron or hole and the carrier lifetime is larger than $\tau_o$.

Table 7.1: Summary table.

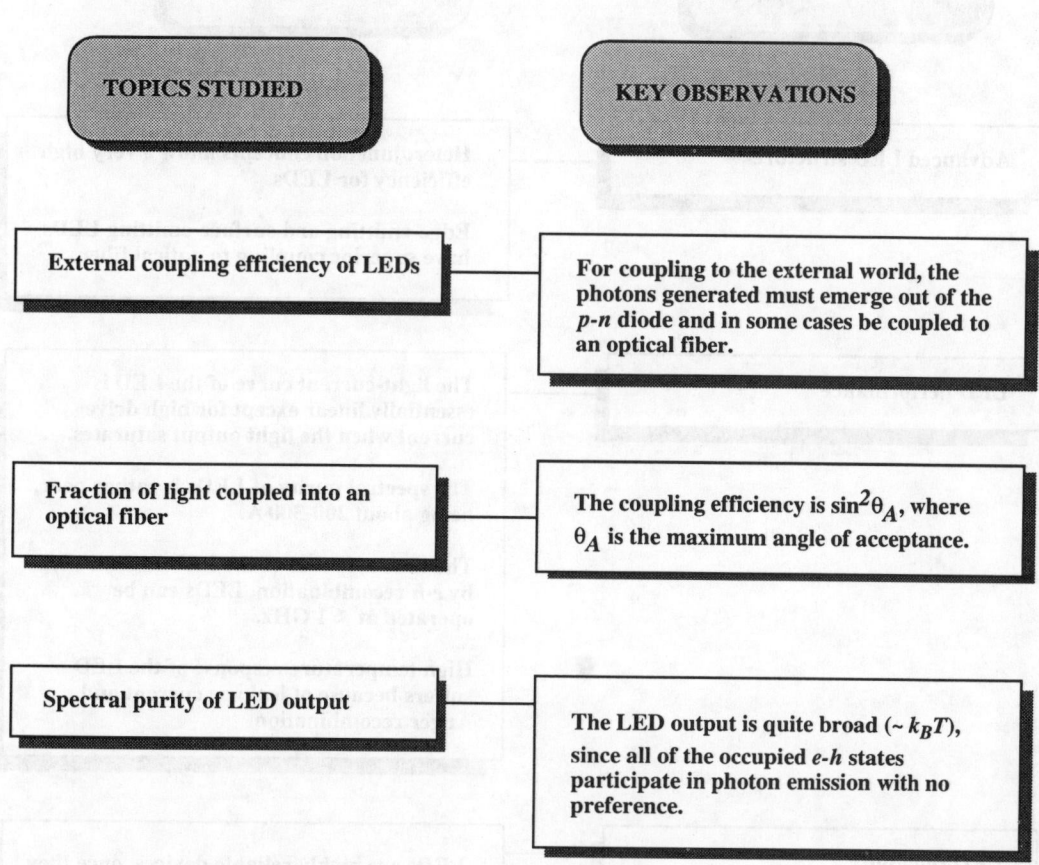

**TOPICS STUDIED**

**KEY OBSERVATIONS**

**External coupling efficiency of LEDs**

For coupling to the external world, the photons generated must emerge out of the *p-n* diode and in some cases be coupled to an optical fiber.

**Fraction of light coupled into an optical fiber**

The coupling efficiency is $\sin^2\theta_A$, where $\theta_A$ is the maximum angle of acceptance.

**Spectral purity of LED output**

The LED output is quite broad ($\sim k_B T$), since all of the occupied *e-h* states participate in photon emission with no preference.

Table 7.2: Summary table.

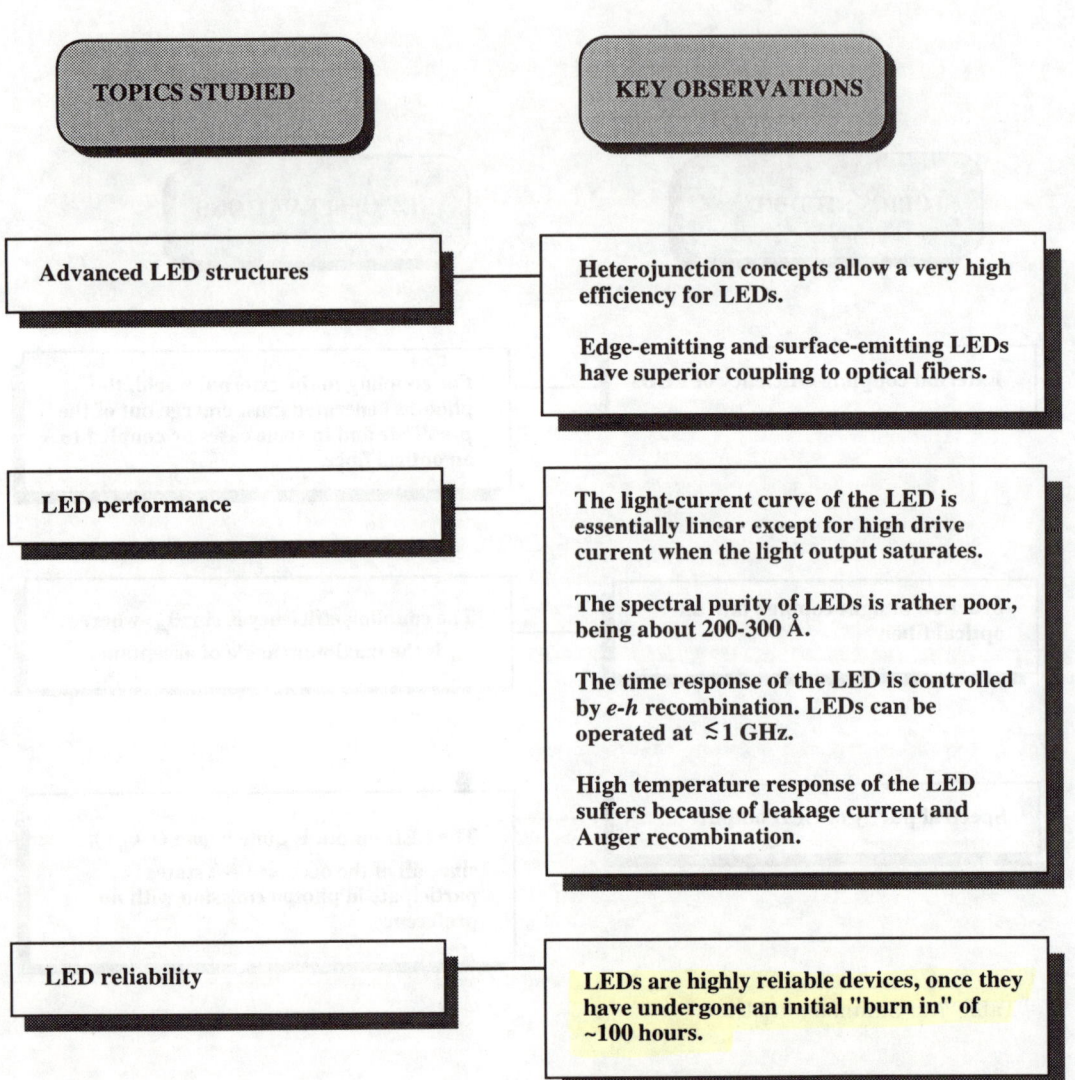

Table 7.3: Summary table.

## 7.10 PROBLEMS

### Section 7.2

**7.1** Identify a composition of $In_x Ga_{1-x} As_y P_{1-y}$ which has a bandgap corresponding to the photon wavelength of 1.55 $\mu$m and which has a lattice constant equal to that of InP. Assume that the lattice constants and bandgaps scale linearly with the composition of the alloy. Use the bandgap and lattice constant values given in the text.

**7.2** Identify the semiconductors (direct and indirect) that can be used for blue light emission.

**7.3** Consider the alloys $Hg_x Cd_{1-x} Te$, $In_x Ga_{1-x} As$ and $GaAs_x Sb_{1-x}$. All of these alloys can be exploited for LEDs emitting near the minima of the optical fiber attenuation. Calculate the change in the emission energy for a 1% change in the alloy composition for each of these systems.

**7.4** An important emerging semiconductor for LEDs is the material SiC. Calculate the emission wavelength for LEDs made from SiC. Discuss the improvement possible in the memory density if SiC were to be used, instead of GaAs, in reading optical discs.

### Section 7.3

**7.5** Calculate the $e$-$h$ radiative recombination time $\tau_o$ (i.e., for $f^e = 1 = f^h$) for carriers in $Hg_{1-x} Cd_x Te$ ($E_g(\text{eV}) = -0.3 + 1.9x$) for $x$ between 0.5 and 1.0. The momentum matrix element is given by

$$\frac{2p_{cv}^2}{m_0} = 22 \ eV$$

Assume that the refractive index is 3.7 and is independent of composition.

**7.6** Consider an $In_{0.53}Ga_{0.47}As$ sample at 300 K in which excess electrons and hole, are injected. The excess density is $10^{15}$ cm$^{-3}$. Calculate the rate at which photons are generated in the system. The bandgap is 0.8 eV and carrier masses are $m_e^* = 0.042m_0$; $m_h^* = 0.4m_0$. Also calculate the photon emission rate if the same density is injected at 77 K. (Use the low injection approximation.) Assume that the refractive index and the momentum matrix element is the same as in GaAs (given in the text).

**7.7** Determine the error in the position of the Fermi levels $E_{Fn}$ and $E_{Fp}$ calculated from the Boltzmann and Joyce-Dixon approximations for GaAs when the free carrier density $n = p$ ranges from $10^{15}$ cm$^{-3}$ to $10^{18}$ cm$^{-3}$. Calculate the results at 77 K and 300 K.

**7.8** Consider a GaAs $p$-$n^+$ junction LED with the following parameters at 300 K:

| | | |
|---|---|---|
| Electron diffusion coefficient, | $D_n$ = | $25\ cm^2/s$ |
| Hole diffusion coefficient, | $D_p$ = | $12\ cm^2/s$ |
| n-side doping, | $N_d$ = | $5 \times 10^{17}\ cm^{-3}$ |
| p-side doping, | $N_a$ = | $10^{16}\ cm^{-3}$ |
| Electron minority carrier lifetime, | $\tau_n$ = | $10\ ns$ |
| Hole minority carrier lifetime, | $\tau_p$ = | $10\ ns$ |

Calculate the injection efficiency of the LED assuming no trap related recombination.

**7.9** The diode in Problem 7.8 is to be used to generate an optical power of 1 mW. The diode area is 1 mm$^2$ and the external radiative efficiency is 20%. Calculate the forward bias voltage required.

**7.10** Consider the GaAs LED of Problem 7.8. The LED has to be used in a communication system. The binary data bits 0 and 1 are to be coded so that the optical pulse output is 1 nW and 50 $\mu$W. If the external efficiency factor is 10%, calculate the forward bias voltages required to send the 0's and 1's. The LED area is 1 mm$^2$.

## Section 7.4

**7.11** Light from a GaAs LED is to be coupled to an external transmission system. A dome of a dielectric is placed over the LED. Calculate the refractive index of the dome if the reflection coefficient for the normally incident photons from the GaAs into the dome is to be 10%.

**7.12** The light from a GaAs LED is coupled into an optical fiber which has refractive indices of 1.51 and 1.47 for the core and the cladding layers. Calculate the maximum angle of acceptance for the fiber. The LED has a Lambertian (cosine) output. Calculate the coupling efficiency for the diode.

## Section 7.5

**7.13** Discuss why in a homojunction LED it is important to ensure that the electron injection current is dominant, but in a heterojunction LED this is not important.

**7.14** Discuss the reasons why the edge emitting LED has a better coupling efficiency to an optical fiber than a surface emitting LED.

## Section 7.6

**7.15** Consider an AlGaAs/GaAs heterojunction LED. The injection densities for electrons and holes are equal and are both $10^{17}\ cm^{-3}$ in the active GaAs region. Calculate the position of the emission peak energy if $E_g$(GaAs) is 1.43 eV. Calculate the shift in the peak position if the injection density is increased to $10^{18}\ cm^{-3}$. The temperature is 300 K.

**7.16** In the previous problem, calculate the spectral width at the two injection densities at 300 K and 77 K.

**7.17** Discuss whether a GaAs LED or an Si LED will have a broader spectral width if the light output power is the same.

**7.18** A heterojunction LED based on GaAs is biased at 100 A/cm$^2$ current density at 300 K. The active layer of the LED is 0.5 $\mu$m. Calculate the cutoff frequency of the diode. If the temperature of operation changes to 400 K, calculate the change in the cutoff frequency. Assume that the cutoff frequency is limited by the radiative lifetime. Use the information provided by Fig. 7.6.

**7.19** Consider two heterojunction LEDs based on GaAs biased at 100 A/cm$^2$. In one case, the active region is undoped while, in the other, the region is $p$-type doped at 10$^{18}$ cm$^{-3}$. The active layer thickness is 1.0 $\mu$m. Compare the radiative lifetime limited modulation bandwidth of the two LEDs at 300 K.

**7.20** A GaAs LED is to be used in a local area network. The fiber system used puts a restriction on the light emitter that the peak wavelength should not shift by more than $\pm$50 Å. Calculate the level of temperature control needed for this device.

**7.21** A GaAs LED is to be designed with an output power of 5.0 mW. The maximum device area that can be allowed is 100 $\mu$m$^2$. Estimate the thickness of the active region needed. The efficiency of the device is 20%. The maximum injection density is 10$^{18}$ cm$^{-3}$.

**7.22** A 0.1 mm$^2$ SiC LED has a total overall optical efficiency of 2%. If the maximum current density is limited to 50 kA/cm$^2$, calculate the output power of this device. How does this value compare to the optical power of a typical flashlight you may have at home?

## 7.11 REFERENCES

- **General**

    - H. Kressel and J. K. Butler, *Semiconductor Lasers and Heterojunction LEDs*, Academic Press, New York (1977).

    - R. Baets, "Heterostructures in III-V Optoelectronic Devices," *Solid State Electronics*, 30, 1175 (1987).

    - Articles in *Semiconductors and Semimetals*, ed. W. T. Tsang, Volume 22, part C, Academic Press, New York (1985).

    - W. T.Tsang, "High Speed Photonic Devices," *High Speed Semiconductor Devices*, ed. S. M. Sze, Wiley-Interscience, New York (1990).

# CHAPTER
# 8

# THE LASER DIODE

## 8.1  INTRODUCTION

The LED discussed in the previous chapter is an important optical source which finds uses in many applications, including those in display systems and optical communication. The LED is not, however, a device of choice for many high performance applications. The main advantages of the LED are the simplicity of the fabrication process and the easy incorporation of the device (which is a simple $p$-$n$ diode) in most circuitry. The key drawbacks of the LED are the broad spectrum of the emitted light and the difficulty in pushing the modulation bandwidths above a gigahertz. We have already discussed the sources of these limitations. In Fig. 8.1 we show the important limitations of the LED. The laser diode is able to overcome these limitations by exploiting special properties of optical cavities and stimulated emission as discussed in Chapter 5. As a result the semiconductor laser diode provides an extremely sharp emission line with linewidth up to two orders of magnitude narrower than that of an LED. The modulation bandwidth of the laser diode approaches 50 GHz and can be even higher in principle. Also, because of its superior spatial coherence, the laser beam does not spread as much as beams from other sources and can thus be focused to give a very high intensity.

## 8.2  SPONTANEOUS AND STIMULATED EMISSION

$\longrightarrow$  The semiconductor laser diode operates as a forward bias $p$-$n$ junction just as the LED studied in the previous sections. *However, while the structure appears similar to the LED as far as the electron and holes are concerned, it is quite different from the point of view of the photons.*

As in the case of the LED, electrons and holes are injected into an active region by forward biasing the laser diode. At low injection, these electrons and holes recombine radiatively via the spontaneous emission process to emit photons. However, the laser structure is so designed that at higher injections the emission process occurs by stimulated emission. In Chapter 5, Section 5.2, we have discussed the difference between spontaneous and stimulated emission. The stimulated emission process provides spectral purity to the photon output, provides coherent photons, and offers high speed performance. *Thus the key difference between the LED and the laser diode arises from the difference between spontaneous and stimulated emission.*

Let us develop an understanding of this difference using Fig. 8.2. Consider an electron with wave vector $k$ and a hole with a wave vector $k$ in the conduction and valence bands, respectively, of a semiconductor. In the case shown in Fig. 8.2a, initially there are no photons in the semiconductor. The electron and hole recombine to emit a photon as shown, and this process is the spontaneous emission. The

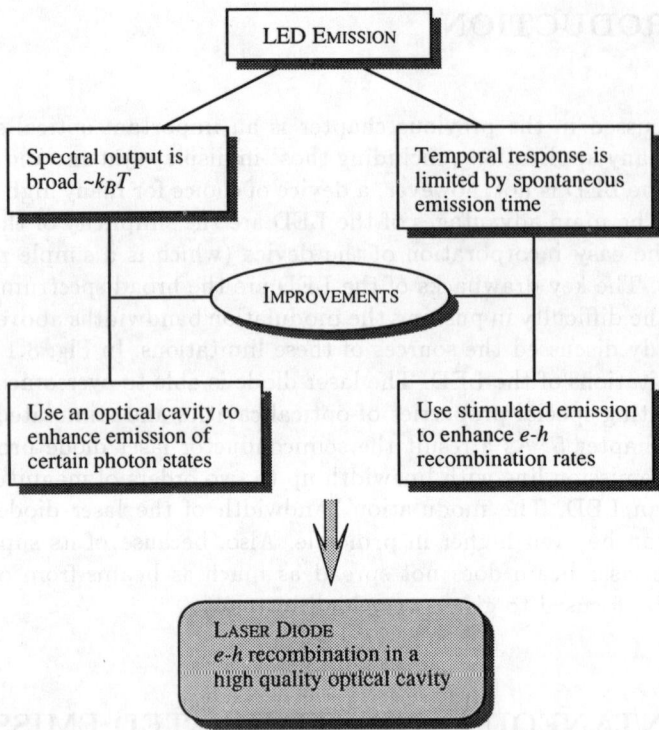

Figure 8.1: A schematic description of how the LED performance can be improved by exploiting an optical cavity.

spontaneous emission rate was discussed in the context of the LED.

In the case of Fig. 8.2b, we show the electron-hole pair along with photons of energy $\hbar\omega$ equal to the electron-hole energy difference. In this case, in addition to the spontaneous emission rate, one has an additional emission rate called the stimulated emission process. The stimulated emission process is proportional to the photon density (of photons with the correct photon energy to cause the $e\text{-}h$ transition). *The photons that are emitted are in phase (i.e., same energy and wave vector) with the incident photons.* As discussed in Chapter 5, the rate for stimulated emission is (see Eqns. 5.44 and 5.45)

$$\boxed{W_{em}^{st}(\hbar\omega) = W_{em}(\hbar\omega) \cdot n_{ph}(\hbar\omega)} \qquad (8.1)$$

where $n_{ph}(\hbar\omega)$ is the photon occupation number and $W_{em}$ is the spontaneous emission rate discussed earlier. In the LED, when photons are emitted by spontaneous emission, they are lost either by reabsorption or simply leave the structure. Thus $n_{ph}(\hbar\omega)$ remains extremely small and stimulated emission cannot get started.

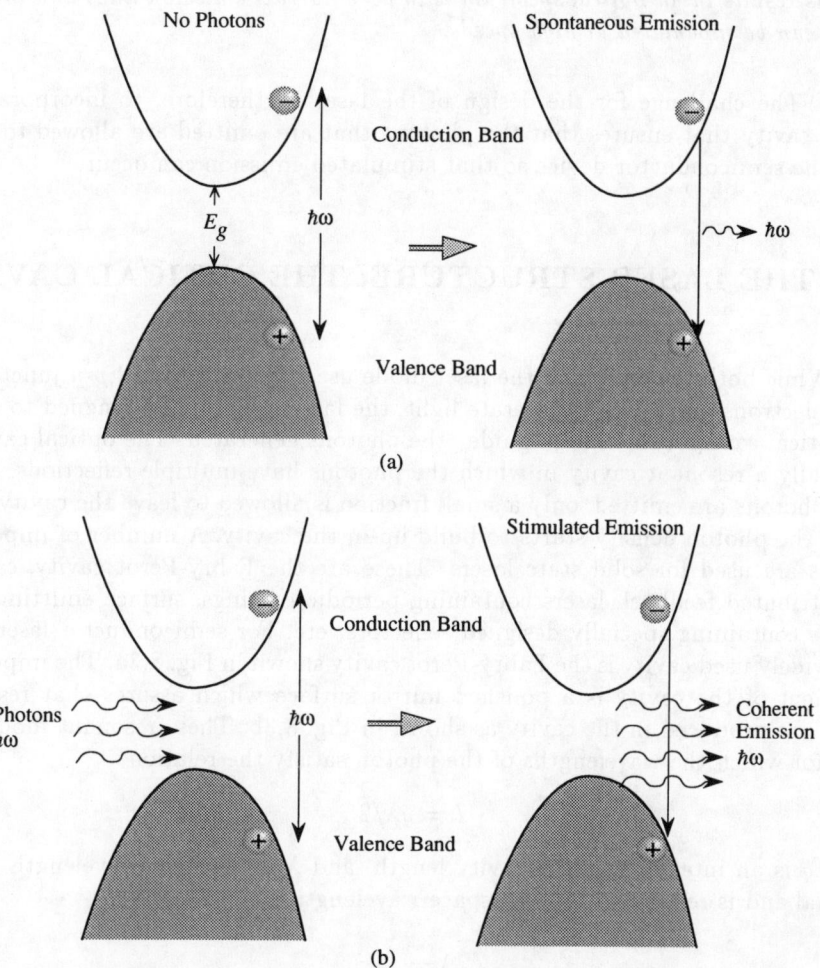

Figure 8.2: (a) In spontaneous emission, the $e$-$h$ pair recombines in the absence of any photons present to emit a photon. (b) In simulated emission, an $e$-$h$ pair recombines in the presence of photons of the correct energy $\hbar\omega$ to emit coherent photons. In coherent emission the phase of the photons emitted is the same as the phase of the photons causing the emission.

*Consider now the possibility that when photons are emitted via spontaneous emission, an optical cavity is designed so that photons with a well defined energy are selectively confined in the semiconductor structure. This would increase the photon occupation $n_{ph}(\hbar\omega)$ and result in increased stimulated emission. As shown in Fig. 8.1, this results in an optical spectrum with very narrow emission lines and the light output can be modulated at high speeds.*

The challenge for the design of the laser is, therefore, to incorporate an optical cavity that ensures that the photons that are emitted are allowed to build up in the semiconductor device so that stimulated emission can occur.

## 8.3   THE LASER STRUCTURE: THE OPTICAL CAVITY

$\longrightarrow$ While both the LED and the laser diode use a forward biased *p-n* junction to inject electrons and holes to generate light, the laser structure is designed to create an "optical cavity" which can "guide" the photons generated. The optical cavity is essentially a resonant cavity in which the photons have multiple reflections. Thus, when photons are emitted, only a small fraction is allowed to leave the cavity. As a result, the photon density starts to build up in the cavity. A number of important cavities are used for solid state lasers. These are the Fabry-Perot cavity, cavities for distributed feedback lasers containing periodic gratings, surface emitting laser cavities containing specially designed reflectors, etc. For semiconductor lasers, the most widely used cavity is the Fabry-Perot cavity shown in Fig. 8.3a. The important ingredient of the cavity is a polished mirror surface which assures that resonant modes are produced in the cavity as shown in Fig. 8.3b. These resonant modes are those for which the wavelengths of the photon satisfy the relation

$$L = q\lambda/2 \qquad (8.2)$$

where $q$ is an integer, $L$ is the cavity length, and $\lambda$ is the light wavelength in the material and is related to the free space wavelength by

$$\lambda = \frac{\lambda_o}{n_r} \qquad (8.3)$$

where $n_r$ is the refractive index of the cavity. The spacing between the stationary modes is given by

$$\Delta k = \frac{\pi}{L} \qquad (8.4)$$

As can be seen from Fig. 8.3a, the Fabry-Perot cavity has mirrored surfaces on two sides. The other sides are roughened so that photons emitted through these sides are not reflected back and are not allowed to build up. Thus only the resonant modes are allowed to build up and participate in the stimulated emission process.

In Chapter 3, Section 3.6, we have discussed the modes of a planar waveguide. It will be useful for the reader to review that discussion. While the optical cavity can confine the photons with certain characteristics, it must be noted that the active region of the laser in which electron-hole pairs are recombining may only occupy a small fraction of the optical cavity. *It is important that a large fraction of the optical waveform overlap with the active region since only this fraction will be responsible for stimulated emission. As a result, it is important to design the laser structure so that the optical wave has a high probability of being in the region where e-h pairs are recombining.*

If a planar waveguide of the form shown in Fig. 8.3c is used to confine the optical wave in the $z$-direction, the optical equation has the form (discussed in Chapter 3)

$$\frac{d^2 E_k(z)}{d^2 z} + \left( \frac{\epsilon(z)\omega^2}{c^2} - k^2 \right) E_k(z) = 0 \qquad (8.5)$$

where $E$ is the electric field representing the optical wave. The dielectric constant $\epsilon(z)$ is chosen to have a $z$-direction variation so that the optical wave is confined in the $z$-direction as shown in Fig. 8.3c. This requires the cladding layers to be made from a large bandgap material. This leads to a structure similar to the one discussed for the heterostructure LED in Section 7.4.1.

Many advances in laser physics are being driven by superior optical cavities. In the above discussion the optical confinement is improved by the heterostructure cladding layers. This is straightforward to do in an epitaxial process. It is somewhat difficult to produce dielectric constant variation in the plane of the laser, i.e., in the $y$-direction. Thus, usually, the laser is fabricated as a stripe of width $\sim 10~\mu m - 50~\mu m$. The stripe is produced by etching. It is also possible to produce "buried" lasers where the $y$-direction optical confinement is produced by doping or defect introduction since these processes can also change the dielectric constant.

An important parameter of the laser cavity is the optical confinement factor $\Gamma$, which gives the fraction of the optical wave in the active region (see Section 6.1),

$$\Gamma = \frac{\int_{\text{active region}} |E(z)|^2 dz}{\int |E(z)|^2 dz} \qquad (8.6)$$

This confinement factor is almost unity for "bulk" double heterostructure lasers where the active region is $\stackrel{>}{\sim} 1.0~\mu m$, while it is as small as 1% for advanced quantum well lasers. However, in spite of the small value of $\Gamma$, quantum well lasers have superior performance because of their superior electronic properties owing to their 2-dimensional density of states.

**EXAMPLE 8.1** Consider a GaAs laser with a cavity length of 200 $\mu m$. Calculate the frequency separation of the resonant modes.

The frequency separation is given by

$$\Delta\nu = \frac{c}{2n_r L} = \frac{3 \times 10^{10}}{2(3.66)(200 \times 10^{-4})} = 2 \times 10^{11}\ Hz$$

The energy separation of the modes is

$$h\Delta\nu = \frac{(6.64 \times 10^{-34})(2 \times 10^{11})}{1.6 \times 10^{-19}} = 0.83\ meV$$

The linewidth of each of these lines under lasing conditions is smaller than this separation, so that one can see single modes.

## 8.3.1   Optical Absorption, Loss and Gain

$\longrightarrow$  The photon particle current (proportional to the optical intensity) associated with an electromagnetic wave traveling through a semiconductor is described by

$$I_{ph} = I_{ph}^0\ exp\ (-\alpha x) \qquad (8.7)$$

where $\alpha$ (the absorption coefficient) is usually positive, and $I_{ph}^0$ is the incident photon current at $x = 0$. The optical intensity which is the photon current multiplied by the photon energy $\hbar\omega$ falls as the wave travels if $\alpha$ is positive. However, if electrons are pumped in the conduction band and holes in the valence band, the electron-hole recombination process (photon emission) can be stronger than the reverse process of electron-hole generation (photon absorption). In general, as discussed in Section 5.7, the gain coefficient is defined by gain = emission coefficient−absorption coefficient. If $f^e(E^e)$ and $f^h(E^h)$ denote the electron and hole occupation, the emission coefficient depends upon the product of $f^e(E^e)$ and $f^h(E^h)$ while the absorption coefficient depends upon the product of $(1-f^e(E^e))$ and $(1-f^h(E^h))$. Here the energies $E^e$ and $E^h$ are related to the photon energy by the condition of vertical $k$-transitions. In Section 7.3.1 we had found that (see Eqns. 7.9 and 7.10)

$$E^e = E_c + \frac{m_r^*}{m_e^*}(\hbar\omega - E_g)$$

$$E^h = E_v - \frac{m_r^*}{m_h^*}(\hbar\omega - E_g) \qquad (8.8)$$

The occupation probabilities $f^e$ and $f^h$ are determined by the quasi-Fermi levels for electrons and holes as discussed in Section 5.7 (it would be very useful for the reader to review Section 5.7).

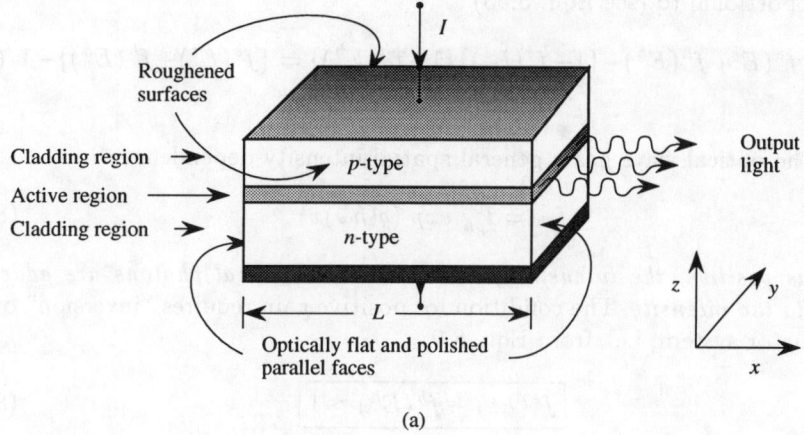

(a)

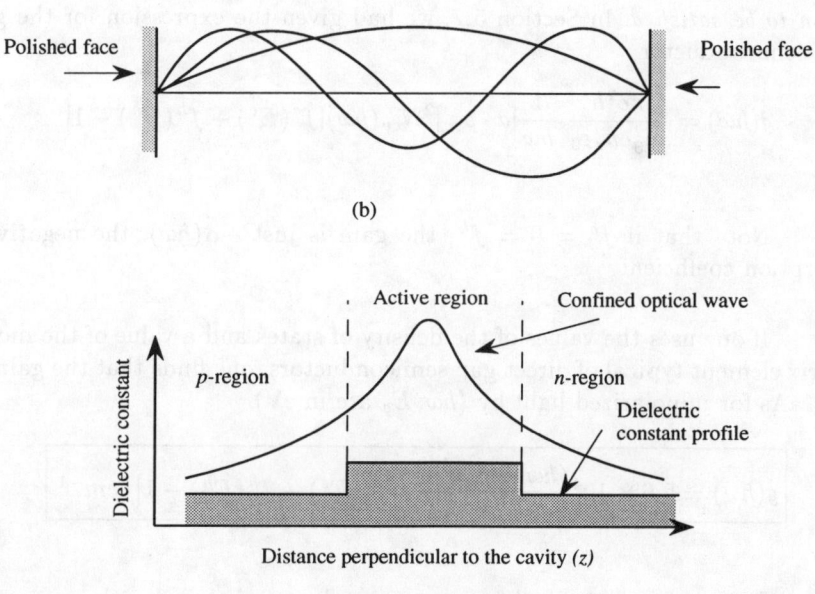

(b)

Figure 8.3: (a) A typical laser structure showing the cavity and the mirrors used to confine photons. The active region can be quite simple as in the case of double heterostructure lasers or quite complicated as in the case of quantum well lasers. (b) The stationary states of the cavity. The mirrors are responsible for these resonant states. (c) The variation in dielectric constant is responsible for the optical confinement. The structure for the optical cavity shown in this figure is called the Fabry-Perot cavity.

The gain which is the difference of the emission and absorption coefficient is now proportional to (see Eqn. 5.53)

$$g(\hbar\omega) \sim f^e(E^e) \cdot f^h(E^h) - \{1 - f^e(E^e)\}\{1 - f^h(E^h)\} = \{f^e(E^e) + f^h(E^h)\} - 1 \quad (8.9)$$

The optical wave has a general spatial intensity dependence

$$I_{ph} = I_{ph}^0 \; exp \; (g(\hbar\omega)x) \qquad\qquad (8.10)$$

and *if g is positive, the intensity grows because additional photons are added by emission to the intensity.* The condition for positive gain requires "inversion" of the semiconductor system, i.e., from Eqn. 8.9,

$$\boxed{f^e(E^e) + f^h(E^h) > 1} \qquad\qquad (8.11)$$

*The quasi-Fermi levels must penetrate their respective bands for this condition to be satisfied.* In Section 5.7, we had given the expression for the gain in a bulk semiconductor

$$g(\hbar\omega) = \frac{\pi e^2 \hbar}{m_0^2 c n_r \epsilon_0} \frac{1}{\hbar\omega} |a \cdot p_{cv}|^2 N_{cv}(\hbar\omega)[f^e(E^e) + f^h(E^h) - 1] \qquad (8.12)$$

Note that if $f^e = 0 = f^h$, the gain is just $-\alpha(\hbar\omega)$, the negative of the absorption coefficient.

If one uses the values of the density of states and a value of the momentum matrix element typical of direct gap semiconductors, one finds that the gain is given for GaAs for unpolarized light by ($\hbar\omega, E_g$ are in eV)

$$\boxed{g(\hbar\omega) \cong 5.6 \times 10^4 \frac{(\hbar\omega - E_g)^{1/2}}{\hbar\omega} \left[f^e(E^e) + f^h(E^h) - 1\right] \; cm^{-1}} \qquad (8.13)$$

The prefactor for a different semiconductor with reduced mass $m_r^*(A)$ can be obtained by multiplying the ratio $(m_r^*(A)/m_r^*(GaAs))^{3/2}$. To evaluate the actual gain in a material as a function of carrier injection $n(= p)$, one has to find the electron and hole quasi-Fermi levels and the occupation probabilities $f^e(E^e)$ and $f^h(E^h)$, where $E^e$ and $E^h$ are related to $\hbar\omega$ by Eqn. 8.8. The procedure is described in more detail through Examples 8.3 and 8.4.

It must be noted that the laser operates under conditions where $f^e$ and $f^h$ are quite large. In this high injection limit, the occupation probabilities are not given accurately by the Boltzmann statistics. A useful approach is to use the Joyce-Dixon approximation for the position of the Fermi levels. For a given injection density $n(= p)$, the position of the quasi-Fermi levels is given by (see Section 5.6)

$$E_{Fn} = E_c + k_B T \left[ \ell n \frac{n}{N_c} + \frac{1}{\sqrt{8}} \frac{n}{N_c} \right] \tag{8.14}$$

$$E_{Fp} = E_v - k_B T \left[ \ell n \frac{p}{N_v} + \frac{1}{\sqrt{8}} \frac{p}{N_v} \right] \tag{8.15}$$

where $N_c$ and $N_v$ are the effective density of states at the conduction and valence bands.

With these expressions the gain can be calculated as a function of photon energy for various levels of injection densities $n(= p)$. At low injections, $f^e$ and $f^h$ are quite small and the gain is negative. However, as injection is increased, for electrons and holes near the bandedges, $f^e$ and $f^h$ increase and gain can be positive. However, even at high injections, for $\hbar\omega >> E_g$, the gain is negative. The general form of the gain-energy curves for different injection levels is shown in Fig. 8.4. Note that at energies well above the bandgap energy, the gain is negative, even for very high injection densities, since the occupation of these states is always small. However, just above the bandgap energy, the gain becomes positive, as the injection density is increased. The shape of the gain curves is determined by a convolution of the density of states and the Fermi function.

The gain discussed above is called the material gain and comes only from the active region where the recombination is occurring. Often this active region is of very small dimensions. In this case, one needs to define the cavity gain which is given by

$$\boxed{\text{Cavity gain} = g(\hbar\omega)\Gamma} \tag{8.16}$$

where $\Gamma$ is the fraction of the optical intensity overlapping with the gain medium, given by Eqn. 8.6. The value of $\Gamma$ is almost unity for double heterostructure lasers and $\sim 0.01$ for quantum well lasers. In quantum well lasers, the overall cavity gain can still be very high since the gain in the quantum well is very large for a fixed injection current density when compared to bulk semiconductors.

In order for the laser oscillations to start, it is essential that when photons are emitted in the laser cavity, the gain associated with the cavity is able to surmount the loss suffered by the photons. The photon loss consists of two parts: i) loss because of absorption of the photons in the cladding regions and contacts of the laser; ii) loss due to the photons emerging from the cavity.

The cavity loss $\alpha_{loss}$ is primarily due to free carrier absorption of the light. Compared to the band to band optical absorption process, this is a very weak process, and in high quality materials this loss can be as low as 10 cm$^{-1}$. It must

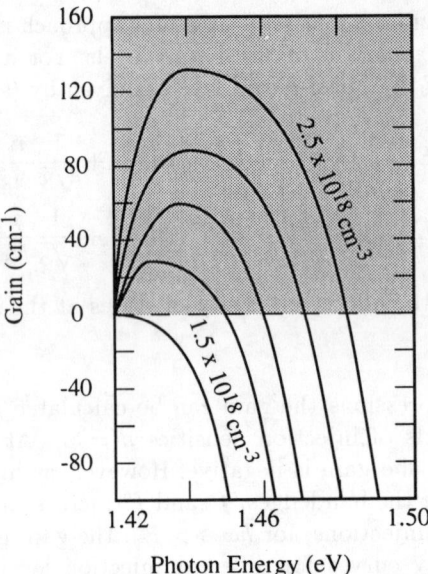

Figure 8.4: Gain vs. photon energy curves for a variety of carrier injections for GaAs at 300 K. The electron and hole injections are the same. The injected carrier densities are increased in steps of $0.25 \times 10^{18}$ cm$^{-3}$ from the lowest value shown.

be noted that the loss is dependent upon doping and defects in the material and, therefore, the material quality should be very good, especially in regions where the optical wave is confined.

To study the photon losses by reflection and transmission from the cavity, let us consider a Fabry-Perot cavity whose reflection coefficient and transmission coefficient is shown in Fig. 8.5a. Let us consider a wave with field $E_0$ incident on one edge of the cavity as shown in Fig. 8.5b and let us follow this wave as it moves through the cavity.

It is straightforward to see that the transmitted and the reflected fields are given by (refer to Fig. 8.5b)

$$E_{\text{trans}} = E_4 + E_{10} + \ldots = \frac{t_1 t_2 A}{1 - A^2 r_1^2} E_0$$

$$E_{\text{ref}} = E_2 + E_7 + \ldots = \left( r_2 + \frac{r_1 t_1 t_2 A^2}{1 - A^2 r_1^2} \right) E_0 \qquad (8.17)$$

where $A$ is the gain of the wave when it moves a distance $L$ and is given by

$$A = exp \left[ \left( \frac{g_{tot}}{2} + ik \right) L \right] \qquad (8.18)$$

$r_1$ : Amplitude reflected at the semiconductor – air boundary

$t_1$ : Amplitude transmitted at the semiconductor – air boundary

$r_2$ : Amplitude reflected at the air – semiconductor boundary

$t_2$ : Amplitude transmitted at the air – semiconductor boundary

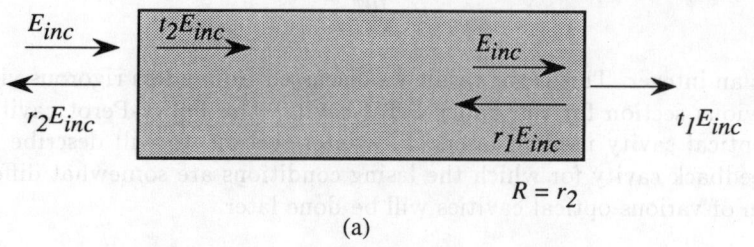

(a)

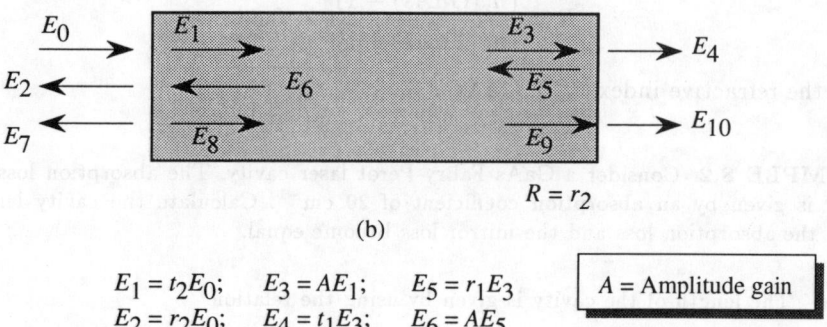

(b)

| | | |
|---|---|---|
| $E_1 = t_2 E_0$; | $E_3 = A E_1$; | $E_5 = r_1 E_3$ |
| $E_2 = r_2 E_0$; | $E_4 = t_1 E_3$; | $E_6 = A E_5$ |

$A$ = Amplitude gain

Figure 8.5: (a) A schematic of the Fabry-Perot cavity showing the reflectance and transmittance of waves. (b) The path of a light wave as it moves through the cavity.

where $g_{tot}$ consists of gain in the cavity and any loss term $\alpha_{\text{loss}}(g_{tot} = \Gamma g - \alpha_{\text{loss}})$.

Laser action occurs when non-zero $E_{\text{trans}}$ and $E_{\text{ref}}$ exist when $E_0$ is zero, i.e., photon generation in the cavity is sufficient to create photons outside the cavity. This requires a certain value of $g_{tot} = \Gamma g_{th}$. For lasings to start, we must have (from Eqn. 8.17)

$$A^2 r_1^2 = 1 \tag{8.19}$$

The real part of this condition gives (using Eqn. 8.18)

$$g_{tot}(th) = \Gamma g_{th} - \alpha_{\text{loss}} = \frac{1}{L}\ell n r_1^{-2} \tag{8.20}$$

or $(R = r_1^2)$

$$\Gamma g_{th} = \alpha_{\text{loss}} - \frac{1}{L} \ell n R \qquad (8.21)$$

The phase part of the lasing condition (Eqn. 8.19) requires that

$$k = \frac{m\pi}{L} \qquad (8.22)$$

where $m$ is an integer. This is the result we discussed from a less rigorous viewpoint in the previous section for the Fabry-Perot cavity. The Fabry-Perot cavity is not the only optical cavity used in lasers. In a later section we will describe the distributed feedback cavity for which the lasing conditions are somewhat different. A comparison of various optical cavities will be done later.

For the GaAs-air interface, the value of the reflection coefficient is

$$R = \frac{\left(n_r(GaAs) - 1\right)^2}{\left(n_r(GaAs) + 1\right)^2} \sim 0.33 \qquad (8.23)$$

since the refractive index $n_r$ of GaAs is 3.66.

**EXAMPLE 8.2** Consider a GaAs Fabry-Perot laser cavity. The absorption loss in the cavity is given by an absorption coefficient of 20 cm$^{-1}$. Calculate the cavity length at which the absorption loss and the mirror loss become equal.

The length of the cavity is given by using the relation

$$\alpha_R = \alpha_{loss} = -\frac{1}{L} \ell n R$$

$$\text{or } L = \frac{-1}{20} \ell n(0.33) = 554 \ \mu m$$

## 8.4 THE LASER BELOW AND ABOVE THRESHOLD

$\longrightarrow$ In Fig. 8.6 we show the light output as a function of injected current density in a laser diode. If we compare this with the output from an LED shown in Fig. 7.12 we notice an important difference. The light output from a laser diode displays a rather abrupt change in behavior below the "threshold" condition and above this condition. The threshold condition is usually defined as the condition where the cavity gain overcomes the cavity loss for any photon energy, i.e., when

$$\boxed{\Gamma g(\hbar\omega) = \alpha_{\text{loss}} - \frac{\ell n R}{L}} \qquad (8.24)$$

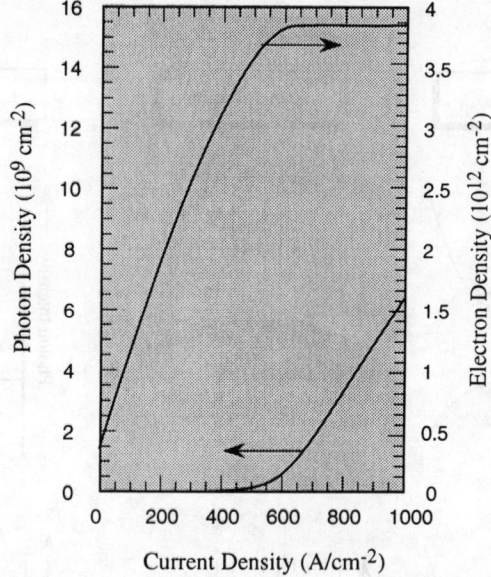

Figure 8.6: Typical light output as a function of current injection in a semiconductor laser. Above threshold, the presence of a high photon density causes stimulated emission to dominate. Electron and photon densities are given as areal densities.

In high quality lasers $\alpha_{\text{loss}} \sim 10$ cm$^{-1}$ and the reflection loss may contribute a similar amount. Another useful definition in the laser is the condition of transparency when the light suffers no absorption or gain, i.e.,

$$\boxed{\Gamma g(\hbar\omega) = 0} \tag{8.25}$$

When the p-n diode making up the semiconductor laser is forward biased, electrons and holes are injected into the active region of the laser. These electrons and holes recombine to emit photons. It is important to identify two distinct regions of operation of the laser. Referring to Fig. 8.7, when the forward bias current is small, the number of electrons and holes injected is small. As a result, the gain in the device is too small to overcome the cavity loss. The photons that are emitted are either absorbed in the cavity or lost to the outside. Thus, in this regime there is no buildup of photons in the cavity. However, as the forward bias increases, more carriers are injected into the device until eventually the threshold condition is satisfied for some photon energy. As a result, the photon number starts to build up in the cavity. As the device is further biased beyond threshold, stimulated emission starts to occur and dominates the spontaneous emission. The light output in the photon mode for which the threshold condition is satisfied becomes very strong.

Below the threshold the device essentially operates as an LED except that

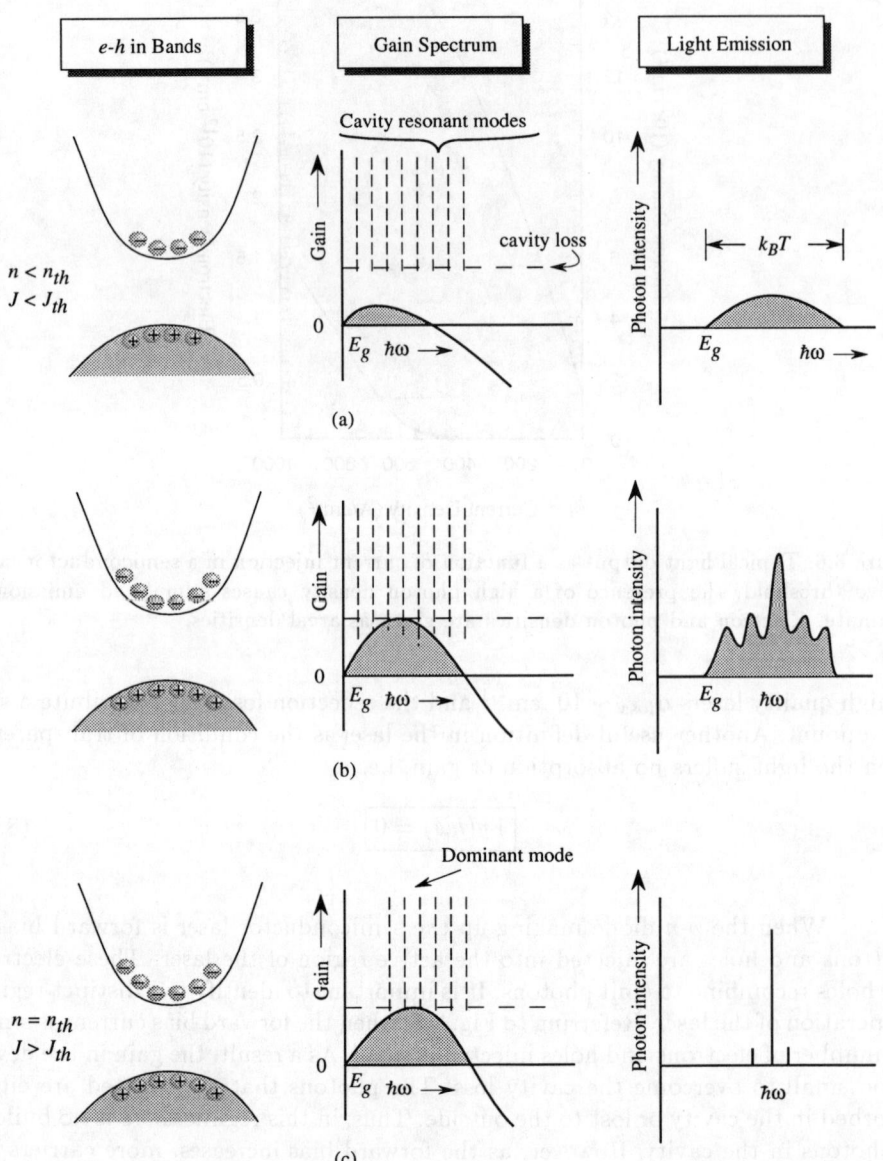

(a)

(b)

(c)

Figure 8.7: (a) The laser below threshold. The gain is less than the cavity loss and the light emission is broad as in an LED. (b) The laser at threshold. A few modes start to dominate the emission spectrum. (c) The laser above threshold. The gain spectrum does not change, but due to the stimulated emission, a dominant mode takes over the light emission.

there is a higher cavity loss in the laser diode since photons cannot escape from the device due to the superior laser mirrors. Let $\beta_{\text{loss}}$ be the fraction of photons that cannot escape from the device. The photon particle current output is given by (the reader must not confuse this with the photocurrent $I_L$ of a detector)

$$
\begin{aligned}
I_{ph} &= (1 - \text{loss})\,(\text{total } e{-}h \text{ recombination per second}) \\
&= (1 - \text{loss})\,(\text{electron particle current})
\end{aligned}
\tag{8.26}
$$

i.e.,

$$
I_{ph} = (1 - \beta_{\text{loss}})\,(R_{\text{spon}} A d_{las}) = (1 - \beta_{\text{loss}})\frac{I}{e}
\tag{8.27}
$$

where $A$ is the laser cavity area, $d_{las}$ is the thickness of the active layer where the recombination is occurring, and $I$ is the injected current. The light output $I_{ph}$ is quite low, due to the high value of the photon loss term $\beta_{\text{loss}}$. This situation is shown schematically in Fig. 8.7a.

Once the carrier density of the electrons and holes is high enough that the threshold condition given by Eqn. 8.24 is met, the photons generated in the laser cavity grow in intensity after emission. Of course, out of all the optical modes that are allowed in the cavity, one or two will have the highest gain since the gain curves have a peak at some energy as seen from Fig. 8.7b. Since the gain is positive, the photon density in the laser cavity starts to increase rapidly. As a result, the stimulated emission process starts to grow. As noted in Section 8.2, the stimulated emission rate is simply related to the spontaneous emission rate by

$$
W^{st}_{em}(\hbar\omega) = W_{em}(\hbar\omega) \cdot n_{ph}(\hbar\omega)
\tag{8.28}
$$

where $n_{ph}(\hbar\omega)$ is the photon density in the mode.

In order to study the laser characteristics around and above threshold, let us establish a simple relation between the injected current density, radiative lifetime, dimensions of the active region where recombination occurs, and the carrier density ($n = p$) in the active region. The rate of arrival of electrons (holes) into the active region is

$$
\frac{JA}{e}
$$

The rate at which the injected $e$-$h$ pairs recombine is

$$
\frac{nAd_{las}}{\tau_r(J)}
$$

where $\tau_r(J)$ is the current density dependent radiative lifetime. Assuming a radiative efficiency of unity, we equate the two results given above to get

$$
\boxed{n = \frac{J\tau_r(J)}{ed_{las}}}
\tag{8.29}
$$

At threshold we have

$$n_{th} = \frac{J_{th}\tau_r(J_{th})}{ed_{las}} \qquad (8.30)$$

As discussed in Section 7.2.2, at threshold $\tau_r(J_{th}) \sim 4\tau_o$ ($\sim 2$ ns for a GaAs laser).

As the current density exceeds $J_{th}$, the photon density in the dominant mode builds up as discussed above and the value of $\tau_r$ starts to become smaller. *As a result, even though the injected charge density increases, the carrier density in the active region saturates close to the threshold density $n_{th}$.* This is shown in Fig. 8.6.

The light output is given by ($n = n_{th}$; use Eqn. 8.29 for the current density)

$$I_{ph} = \frac{I}{e} = \frac{n_{th}Ad_{las}}{\tau_r} \qquad (8.31)$$

*Upon examining this equation with the results for light output from an LED, it may be seen that the photon current is similar for the same injected current for an LED and LD (biased above threshold). However, in the case of the laser diode, the entire photon output emerges only in one or two photon modes rather than in a broad spectrum of width $k_BT$.* This spectral purity that arises because of the importance of stimulated emission distinguishes the LD from an LED. Also, the light output is highly collimated and coherent for similar reasons.

An important relationship is given by Eqn. 8.30 between $n_{th}, J_{th}$, and the active region thickness $d_{las}$. This relationship will be further discussed to optimize the laser for low threshold current applications. From Eqn. 8.30 it is clear that for low threshold current density, the active layer thickness should be decreased. However, from Eqn. 8.31 it is seen that for higher optical output the active layer thickness should be large. Depending upon the application, one designs a semiconductor laser according to these results.

**EXAMPLE 8.3** According to the Joyce-Dixon approximation, the relation between the Fermi level and carrier concentration is given by Eqns. 8.14 and 8.15. Calculate the carrier density needed for the transparency condition in GaAs at 300 K and 77 K. The transparency condition is defined at the situation where the maximum gain is zero (i.e., the optical beam propagates without loss or gain). Calculate the transparency condition at $\hbar\omega = E_g$.

At room temperature the valence and conduction band effective density of states is

$$N_v = 7 \times 10^{18} \ cm^{-3}$$

$$N_c = 4.7 \times 10^{17} \; cm^{-3}$$

The values at 77 K are

$$N_v = 0.91 \times 10^{18} \; cm^{-3}$$
$$N_c = 0.61 \times 10^{17} \; cm^{-3}$$

In the semiconductor laser, an equal number of electrons and holes are injected into the active region. We will look for the transparency conditions for photons with energy equal to the bandgap. The approach is very simple: i) choose a value of $n$ or $p$; ii) calculate $E_{Fn}$ and $E_{Fp}$ from the Joyce-Dixon approximation; iii) calculate $f^e + f^h - 1$ and check if it is positive at the bandedge. The same approach can be used to find the gain as a function of $\hbar\omega$.

For 300 K we find that the material is transparent when $n \sim 1.1 \times 10^{18}$ cm$^{-3}$ at 300 K and $n \sim 2.5 \times 10^{17}$ cm$^{-3}$ at 77 K. Thus a significant decrease in the injected charge occurs as temperature is decreased.

**EXAMPLE 8.4** Consider a GaAs double heterostructure laser at 300 K. The optical confinement factor is unity. Calculate the threshold carrier density assuming that it is 20% larger than the density for transparency. If the active layer thickness is 2.0 $\mu$m, calculate the threshold current density.

From Example 8.3 we see that at transparency

$$n = 1.1 \times 10^{18} \; cm^{-3}$$

The threshold density is then

$$n_{th} = 1.32 \times 10^{18} \; cm^{-3}$$

The radiative recombination time is approximately four times $\tau_o$, i.e., $\sim$2.4 ns. The current density then becomes

$$J_{th} = \frac{e \cdot n_{th} \cdot d_{las}}{\tau_r} = \frac{(1.6 \times 10^{-19} \; C)(1.32 \times 10^{18} \; cm^{-3})(2 \times 10^{-4} \; cm)}{2.4 \times 10^{-9} \; s}$$
$$= 1.76 \times 10^4 \; A/cm^2$$

**EXAMPLE 8.5** In Chapter 4 (also in Appendix C) we have discussed the density of states for electrons. Calculate the density of states for photons using similar arguments.

Just like the electron states, the photon states also have a phase dependence $exp \, (ik \cdot r)$. However, the dependence of the energy of the $k$-vector is different. The photon "dispersion relation" is ($k = 2\pi/\lambda$)

$$\hbar\omega = \frac{h\upsilon}{\lambda} = \hbar\upsilon k$$

where $v$ is the light velocity.

As in the case of the electrons, the $k$-space occupied by each photon state is $V/(2\pi)^3$, where $V$ is the volume of the space in which photons are confined. The number of states having energy between $\hbar\omega$ and $\hbar\omega + d(\hbar\omega)$ is then ($\rho(\hbar\omega)$ is the density of photon states)

$$\rho(\hbar\omega)d(\hbar\omega)V = \frac{4\pi k^2 dk}{(2\pi)^3}V$$

$$= \frac{\omega^2 d(\hbar\omega)V}{2\pi^2 v^3 \hbar}$$

The density of states is then

$$\rho(\hbar\omega) = \frac{\omega^2}{2\pi^2 v^3 \hbar}$$

In general, there are two different polarization modes for the light (both the modes are transverse, i.e., if the light is travelling in the $z$ direction, the polarization can be in $x$ or $y$ direction). Thus the total photon density of states is twice the result obtained.

**EXAMPLE 8.6** Consider a GaAs optical cavity which has a length of 200 $\mu$m and the reflectivity of the mirrors is 0.33. The absorption loss in the cavity is 10 cm$^{-1}$. Calculate the time spent by a photon in the cavity before it is absorbed or emitted. The time is called the photon lifetime $\tau_{ph}$.

The loss coefficient for the photon is

$$\alpha_{tot} = \alpha_{loss} + \alpha_R = \alpha_{loss} - \frac{1}{L}\ell nR = 10 - \frac{\ell n(0.33)}{2 \times 10^{-2}}$$

$$= 65.43 \ cm^{-1}$$

This represents the inverse distance travelled by the photon before it is either absorbed or emitted from the cavity. The lifetime is therefore ($v$ = velocity of light)

$$\tau_{ph} = \frac{1}{v\alpha_{tot}} = \frac{3.6}{3 \times 10^{10} \times 65.43} = 0.51 \ ps$$

## 8.5    THE TIME RESPONSE OF A LASER

$\longrightarrow$  The optical output of a laser diode must be modulated to be useful for information transmission. The most important and simplest approach to modulation is direct modulation in which the current through the laser is modulated. Depending upon the application, one can divide the modulation techniques into the three categories outlined below and shown in Fig. 8.8.

**Large Signal Modulation**

In this approach the laser is turned ON and OFF, i.e., the current goes from well above the threshold value to a value below threshold. This kind of modulation may be used for "optical interconnects," or for certain logic applications. The response of the laser to such a modulation is rather slow ($\sim$ 10 ns) as we shall see later. Large signal modulation is not used for optical communication due to the slow response and due to the spectral width of the output. In fact, the large signal response of a laser is not much better than that of a LED.

**Small Signal Modulation**

In the small signal modulation approach the laser is biased at a point well above threshold and a small ac signal is applied to it. This approach produces the highest frequency response for the laser. It also provides insight into the fundamental limitations of the laser. Modulation bandwidth approaching 50 GHz can be achieved in a small signal modulation approach.

**Pulse Code Modulation**

Pulse code modulation is the most widely used scheme for modern optical communication. The approach falls between the large signal and small signal modulation approaches. The laser is biased well above threshold and a current (voltage) pulse is applied so that the current goes to a high value and a low value as shown in Fig. 8.8. The important point to note is that even in the low state, the diode current remains above the threshold current. In the pulse code modulation scheme the laser can operate at up to ten GHz.

The overall modulation response of semiconductor lasers is controlled by both extrinsic and intrinsic factors. The extrinsic limits arise from a variety of sources. An important restriction is due to the laser heating produced when the laser is biased at a high current. The high current biasing is necessary to operate the device at high speeds as will be discussed later. The heat produced causes a deterioration in the device parameters such as gain spectra, threshold current, etc.

Another important issue in lasers biased at high power is the catastrophic degeneration produced by facet or mirror damage. This produces a catastrophic failure of the laser due to the damage produced in the mirror of the cavity. Thus the laser has an upper limit on injection, at which it can be operated safely.

A final and important extrinsic limit to the laser speed is due to the extrinsic parasitics of the laser diode. The laser must be designed so that the resistance, capacitance and the inductance of the device do not limit the device response.

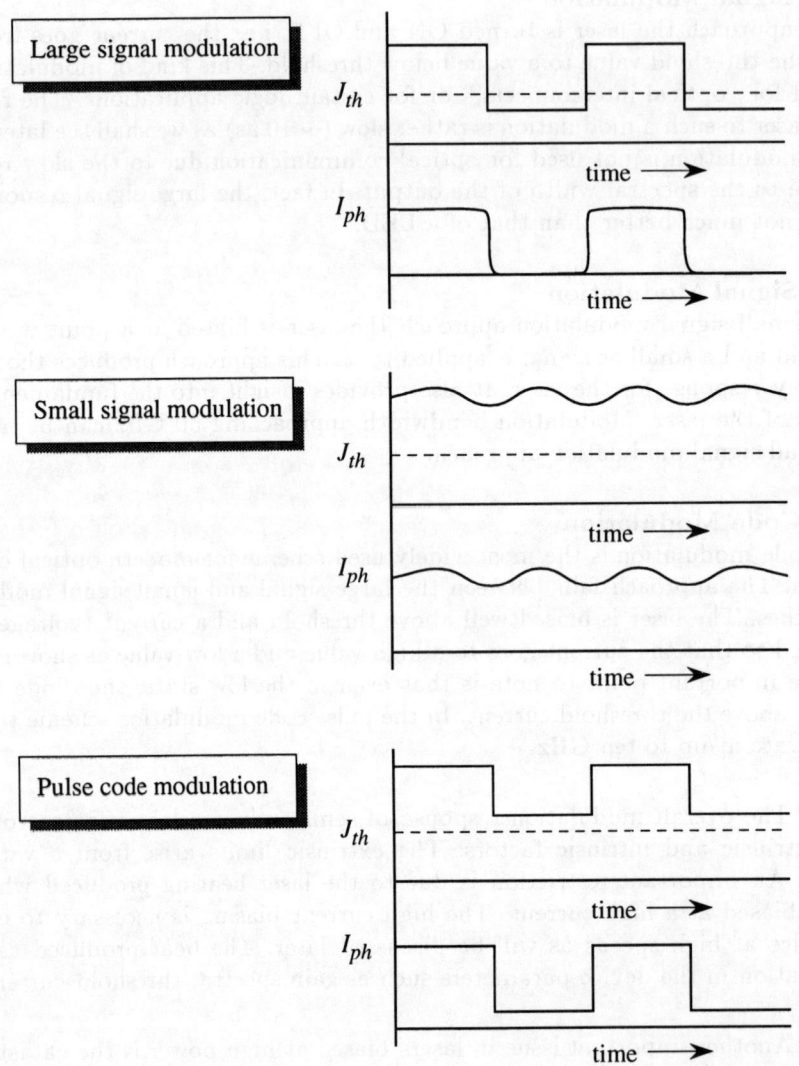

Figure 8.8: Three different modulation approaches used for direct modulation of lasers.

The intrinsic modulation limits of the laser come from several different sources. These include limits arising from the cavity design, carrier drift and diffusion as well as effects which limit the gain and speed of small signal modulation. We start with the large signal response of the laser.

## 8.5.1   Large Signal Switching of a Laser

$\longrightarrow$   As discussed above, the large signal switching of a laser involves changing the diode current from a value below threshold to a value above threshold. The initial value of the current may be zero. Before carrying out a mathematical analysis of the problem, let us see what happens physically when a laser diode is subjected to a step pulse of current.

Before the current step, the carrier density in the active region of the laser is essentially zero. As the current pulse is turned on, the carrier density increases. As a result, the gain in the device also starts to increase. However, until the gain reaches the cavity loss value, there are very few photons emitted out of the cavity. Thus, for a time $t_d$, no photons emerge from the device. Once the carrier concentration reaches $n_{th}$, stimulated emission starts. However, the carrier concentration overshoots the value $n_{th}$ thus increasing the photon output beyond the steady state value. The high photon output, in turn, reduces the carrier density through higher stimulated $e$-$h$ recombination. Thus oscillations are produced in the carrier density and the photon output. The entire process is shown schematically in Fig. 8.9a.

Let us calculate the delay time $t_d$ by examining a simple rate equation for the carrier density. This equation can be written for the two dimensional carrier density $(n_{2D} = n \, d_{las})$ as

$$\frac{dn_{2D}}{dt} = \frac{J}{e} - \frac{n_{2D}}{\tau} - R_{stim} \qquad (8.32)$$

where $\tau$ is the total $e$-$h$ recombination time. The first term on the right hand side represents the rate of flow (two dimensional or areal) of particles into the active region; the second term represents the carrier loss rate due to spontaneous recombination; and the third term represents loss rate due to stimulated emission. If the current density is changed from 0 to a value $J$, during the time $n_{2D} < n_{2D}(th)$, no photons are present in the cavity and $R_{stim} \sim 0$. Integrating this equation from $t = 0$ to $t = t_f$, and $n_{2D} = n_{2D}(i)$ to $n_{2D}(f)$, we get

$$t_f = \tau \, ln \left( \frac{J - en_{2D}(i)/\tau}{J - en_{2D}(f)/\tau} \right) \qquad (8.33)$$

The photon density starts to change when $n_{2D}(f) = n_{2D}(th)$. Thus, the delay time is (for $n_{2D}(i) = 0$)

$$t_d = \tau \ln \left( \frac{J}{J - e n_{2D}(th)/\tau} \right)$$

$$= \tau \ln \left( \frac{J}{J - J_{th}} \right) \tag{8.34}$$

The time $\tau$ is due to radiative and non-radiative processes below threshold. If the non-radiative processes are negligible, $\tau = \tau_r$, and thus there is a delay of several nanoseconds between the time the current is switched and the photons emerge from the cavity. This and the relaxation oscillations produced after photons start to emerge is a serious handicap of lasers for many applications.

## 8.5.2  Small Signal Response of a Laser

$\rightsquigarrow$ The second important time response in a laser is the small signal modulation response to the current density modulation,

$$J(t) = J_o + J_s \sin \omega t \tag{8.35}$$

where $J_o > J_{th}$ and $J_s$ is small.

In order to understand the laser operation above threshold we will consider the interaction of the photons and the electrons via the rate equations. We will write the equations relevant to two dimensional (areal) quantities. Thus, in the following $S_m$ and $n_{2D}$ are the photon population per unit area in the mode $m$ and the areal carrier density, respectively. The rate equation for the photons is ($E_M$ is the energy of the mode $m$)

Rate of change of photon density  =  Stimulated emission − cavity loss rate

         +   spontaneous emission

This gives

$$\frac{dS_m}{dt} = [\Gamma g(n_{2D}, E_M) - \alpha_c] \frac{c}{n_r} S_m + \beta R_{sp}(n_{2d}) \tag{8.36}$$

where the cavity loss $\alpha_c$ is

$$\alpha_c = \alpha_{loss} + \frac{1}{L} \ln R \tag{8.37}$$

The parameter $\beta$ is the spontaneous emission factor which represents the fraction of total spontaneous emission photons that are emitted into a particular mode. Since the total spontaneous emission is over an energy range of $\sim k_B T$ while the mode

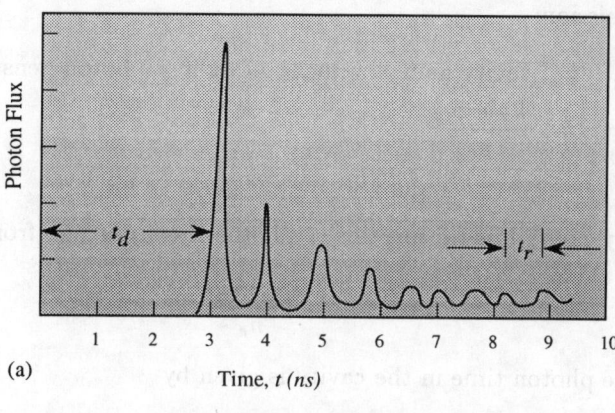

(a)

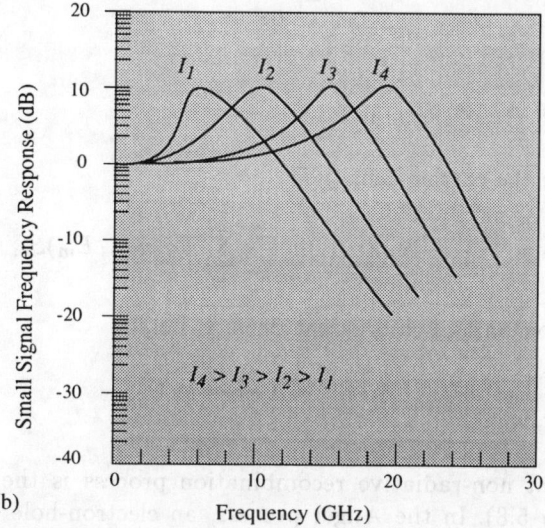

(b)

Figure 8.9: (a) The temporal response of light output from a laser for large signal switching from below threshold to above threshold. The response is characterized by a delay time $t_d$ and relaxation oscillations. The large delay time causes serious limitations in many laser applications. (b) A typical frequency response for a semiconductor laser being operated at a current level well above the threshold current. Unlike the LED, the laser response is not limited by the spontaneous recombination time of electrons and holes. The small signal response improves as the laser is pumped at higher powers.

linewidth is only a few $\mu eV$, the factor $\beta$ is typically $10^{-4}$ to $10^{-5}$ for Fabry-Perot cavities. The various factors in the rate equation come as follows:

- Stimulated emission:

$$
\begin{aligned}
R_{stim} &= \text{cavity gain} \times \text{velocity of light} \times \text{photon density} \\
&= \frac{\Gamma g \; c \; S_m}{n_r}
\end{aligned}
$$

- Photon loss by cavity loss (absorption + photon loss by escape from the cavity):

$$
\text{Loss rate} = \frac{\alpha_c c S_m}{n_r}
$$

We note that the photon time in the cavity is given by

$$
\frac{1}{\tau_{ph}} = \frac{\alpha_c c}{n_r}
$$

- Spontaneous emission rate $\beta R_{spon}$

The rate equation for the carrier density is

$$
\frac{dn_{2D}}{dt} = \frac{J_{rad}}{e} - R_{sp}(n_{2D}) - \frac{c}{n_r} \sum_m \Gamma g(n_{2D}, E_m) S_m \tag{8.38}
$$

Here $J_{rad}$ the radiative part of the current density, i.e.,

$$
J_{rad} = J - J_{nr} \tag{8.39}
$$

An important non-radiative recombination process is the *Auger process* (discussed in Section 5.8). In the Auger process, an electron-hole pair recombine with each other and give their energy to a thick electron or hole. As a result, no photons are emitted. The Auger processes are particularly important in narrow gap lasers and their importance increases with temperature.

To understand the response of the laser to a small signal with frequency $\omega$, let us return to the photon and carrier density rate equations for the mode $m$

$$
\begin{aligned}
\frac{dS_m}{dt} &= [\Gamma g(n_{2D}, E_m) - \alpha_c] \frac{c}{n_r} S_m + \beta R_{sp}(n_{2D}) \\
\frac{dn_{2D}}{dt} &= \frac{J}{e} - \frac{J_a}{e} - R_{sp}(n_{2D}) - \frac{c}{n_r} \sum_m \Gamma g(n_{2D}, E_m) S_m
\end{aligned}
$$

Here we have explicitly included the Auger current $J_a$ (we assume that $J_{nr} = J_a$). We will initially assume that $J_a$ is zero and later comment on the influence of the Auger processes. In the small signal theory we apply a current signal

$$J = \bar{J} + \tilde{J} exp(i\omega t) = \bar{J} + \Delta J \tag{8.40}$$

which causes a variation in the carrier density and the photon density

$$n_{2D} = \bar{n}_{2D} + \tilde{n}_{2D} exp(i\omega t) = \bar{n}_{2D} + \Delta n_{2D} \tag{8.41}$$

$$S_m = \bar{S}_m + \tilde{S}_m exp(i\omega t) = \bar{S}_m + \Delta \tilde{S}_m \tag{8.42}$$

We also assume that the gain and spontaneous rate can also be linearized:

$$g(\bar{n}_{2D} + \Delta n_{2D}, E_m) \simeq g(\bar{n}_{2D}, E_m) + \frac{\partial g(\bar{n}_{2D}, E_m)}{\partial n_{2D}} \Delta n_{2D} \tag{8.43}$$

$$R_{sp}(\bar{n}_{2D} + \Delta n_{2D}) \simeq R_{sp}(\bar{n}_{2D}) + \frac{\partial R_{sp}(\bar{n}_{2D})}{\partial n_{2D}} \Delta n_{2D} \tag{8.44}$$

In the small signal modulation, the laser is usually biased well above threshold for maximum available bandwidth (we will see this below), and at such high biasing, the main mode of the laser dominates. Thus it is sufficient to carry out a single mode study for the small signal response. Substituting the small signal variations into the rate equations and retaining only the first order terms in the small signals, we get the equations

$$\tilde{S}\left[i\omega - \Gamma g \frac{c}{n_r} + \alpha_c \frac{c}{n_r}\right] = \tilde{n}_{2D}\left[\Gamma \frac{\partial g}{\partial n_{2D}} \frac{c}{n_r} \bar{S} + \beta \frac{\partial R_{sp}(n_{2D})}{\partial n_{2D}}\right] \tag{8.45}$$

$$\tilde{n}_{2D}\left[i\omega + \frac{\partial R_{sp}(n_{2D})}{\partial n_{2D}} + \frac{c}{n_r}\Gamma \frac{\partial g}{\partial n_{2D}} \bar{S}\right] = \frac{\tilde{J}}{e} - \frac{c}{n_r}\Gamma \bar{g}\tilde{S} \tag{8.46}$$

Eliminating $\tilde{n}_{2D}$ we get a relation between the photon signal $\tilde{S}$ and the current signal $\tilde{J}$

$$\tilde{S}\left[\Gamma \frac{cg(\bar{n}_{2D})}{n_r} + \frac{i\omega + \gamma}{\zeta}\left\{i\omega - \Gamma g(\bar{n}_{2D}, E_p) - \alpha_c\right\}\frac{c}{n_r}\right] = \frac{\tilde{J}}{e} \tag{8.47}$$

where

$$\gamma = \frac{\partial R_{sp}(\bar{n}_{2D})}{\partial n_{2D}} + \frac{\Gamma c}{n_r}\frac{\partial g(\bar{n}_{2D}, E_p)}{\partial n_{2D}}\bar{S} \tag{8.48}$$

$$\zeta = \beta \frac{\partial R_{sp}(\bar{n}_{2D})}{\partial n_{2D}} + \frac{\Gamma c}{n_r}\frac{\partial g(\bar{n}, E_p)}{\partial n_{2D}}\bar{S} \tag{8.49}$$

we may also denote the photon lifetime $\tau_{ph}$ by

$$\tau_{ph} = \frac{1}{\Gamma \frac{c}{n_r} g(\bar{n}_{2D})} = \frac{n_r}{\alpha_c c} \tag{8.50}$$

The laser response is given by the transfer function relating $\tilde{S}$ to $\tilde{J}$. We get

$$\left| \frac{\tilde{S}}{\tilde{J}} \right| = R(\omega) = \frac{\omega_r^2}{(\omega_r^2 - \omega^2) + i\omega\gamma} \tag{8.51}$$

where

$$\omega_r^2 = \frac{\beta}{\tau_{ph}} \frac{\partial R_{sp}(\bar{n}_{2D})}{\partial n_{2D}} + \frac{c\Gamma\bar{S}}{n_r \tau_{ph}} \frac{\partial g(\bar{n}_{2D})}{\partial n_{2D}} \tag{8.52}$$

At high biasing where stimulated emission dominates, the resonance frequency is

$$\boxed{ f_r = \frac{\omega_r}{2\pi} = \frac{1}{2\pi} \sqrt{\frac{c\Gamma\bar{S}}{n_r \tau_{ph}} \frac{\partial g(n_{2D})}{\partial n_{2D}}} } \tag{8.53}$$

The term $\gamma$ defined above is the damping rate and has the value in terms of the resonance frequency and photon lifetime

$$\gamma = \frac{\partial R_{sp}(\bar{n}_{2D})}{\partial n_{2D}} + \omega_r^2 \tau_{ph} \tag{8.54}$$

The general response of the laser is shown in Fig. 8.9b. The response function has a peak at $\omega_r$ (or $f_r$) as shown and the magnitude of the resonance peak is

$$\frac{R(\omega_r)}{R(0)} = \frac{\omega_r}{\gamma} \tag{8.55}$$

Having derived the response function, let us briefly examine the factors that affect the frequency response of lasers.

- **The Injection Current and Threshold Current**

It is clear from the transfer function calculation that a laser's high frequency response will improve as the photon density $\bar{S}$ becomes higher. Equivalently, if the laser is biased at high injection current value, the device will operate better. *A low threshold current laser is important since the photon density will be higher at the same injection current in a lower threshold laser.*

It must be appreciated that one cannot simply drive the laser at a very high current value due to extrinsic effects. At high injection, heating and high photon density induced effects will degrade the laser performance.

- **Auger Effects**

The Auger effects have two important effects on the laser performance. Since a fraction of the current is not available for creating photons, one has to drive the laser at a higher current to reach a certain value of $\bar{S}$ (compared to the case of $J_a = 0$). In addition, the damping factor becomes

$$\gamma(Auger) = \gamma(J_a = 0) + \frac{1}{e}\frac{\partial J(\bar{n}_{2D})}{\partial n_{2D}} \tag{8.56}$$

As a result, the damping factor increases and the device response suffers.

- **Photon Lifetime**
The expression for the frequency response suggests that the photon lifetime should be made as small as possible. However, one has to be very careful since the photon lifetime is strongly coupled to the gain and the threshold current. The photon lifetime can be decreased by using a shorter length cavity, but this increases the cavity loss and, as a result, a higher $n_{2D}(th)$ is needed to reach the same gain. Thus, there is an optimum cavity length for a given laser. Typical optimum cavity lengths for lasers are around 100 $\mu$m.

- **Differential Gain**
The appearance of the differential gain $\frac{\partial g(\bar{n}_{2D})}{\partial n_{2D}}$ shows that the device response can be improved by choosing the active region of the laser to have a high differential gain.

     An obvious question is: how high is the cutoff frequency in real laser diodes? Unfortunately, the cutoff frequency has not reached very high values, in spite of a tremendous amount of effort on pushing this limit. Lasers have been demonstrated to perform up to $\sim 40$ GHz. State of the art quantum well lasers have been shown to operate up to $\sim 35$ GHz. While these numbers are impressive, they are rather embarrassingly low when compared to state of the art microwave devices. Cutoff frequencies of up to 300 GHz have been demonstrated in modulation doped FETs based on InGaAs technology.

## 8.6 SEMICONDUCTOR LASER DESIGN: ELECTRONIC STRUCTURE DESIGN

$\rightsquigarrow$ Since one of the most important applications of semiconductor lasers is in the area of optical communications, a key driving force for superior laser design is low threshold current and high modulation bandwidth. Other motivations include lasers with emission frequencies that are important for particular applications. These include long wavelength lasers for communication, short wavelength lasers for optical memory applications, blue and green light lasers and LEDs for displays, etc. Let us consider some of the design considerations that are important for optimum laser performance.

## 8.6.1    Low Threshold Current

The first successful lasers involved use of very thick active layers of thickness $d_{las} \gtrsim$ $1\mu$m. The optical confinement factors in these layers were very high ($\Gamma \sim 1.0$). *It is important to note that the 3-D density of electrons (holes) required to produce the transparency condition is a value n (transparency) which is independent of the active layer thickness if $\Gamma \sim 1.0$.* The value of $n$ (transparency) is very close to $n_{th}$ in high quality structures. *The current density needed at threshold is, however, related to the 2-D carrier density (see Eqn. 8.30) as*

$$\boxed{J_{th} = \frac{en_{th}d_{las}}{\tau_r(J_{th})} = \frac{en_{2D}(\text{threshold})}{\tau_r(J_{th})}} \tag{8.57}$$

Thus, in these devices the threshold current depends directly on the active layer thickness. This is, indeed, seen in actual devices, as seen in Fig. 8.10.

Once $d_{las}$ becomes much smaller than the emission wavelength, the value of the optical confinement starts to decrease. Also for very small values of $d_{las}$ ($\sim 100$ Å) quantum effects become important. It can be shown that the lowest threshold current can be achieved in $\sim$50-100  Å quantum well devices. Typical quantum well laser structures are shown in Fig. 8.11. In GaAs/AlGaAs structures, threshold currents at room temperature as low as 100 A/cm$^2$ are achieved. This means that a laser cavity of 10 $\mu$m $\times$ 200 $\mu$m can be turned on by a current of 2 mA. Such current levels can be generated by electronic devices which can be modulated at high speeds to provide a high speed modulation in the optical output.

An important new development in quantum well lasers is the use of strained quantum wells. The strain is produced by growing a very thin layer of a material on a substrate with a different lattice constant. Strained layers with lattice constant differences of up to 3% can be grown by epitaxial crystal growth methods, such as MBE or MOCVD (these techniques are discussed in Chapter 11). The strain in the quantum well causes a strong reduction in the valence band density of states (or hole mass). Hole density of states mass can be reduced by up to a factor of 3 by a strain of $\sim 2\%$. *As a result of the lower hole mass, the inversion condition ($f^e + f^h - 1 > 0$) can be reached for a lower hole injection since the hole Fermi level can penetrate deeper into the valence band for a given hole injection.* Threshold current reduction of up to a factor of 3 can be achieved by the use of strained quantum wells.

The laser performance can also be improved by doping the active region $p$-type. This is useful, since it allows the inversion condition to be reached with less injection density as can be seen from Example 8.7, below.

**EXAMPLE 8.7** Consider two double heterostructure GaAs/AlGaAs lasers at 300 K. One

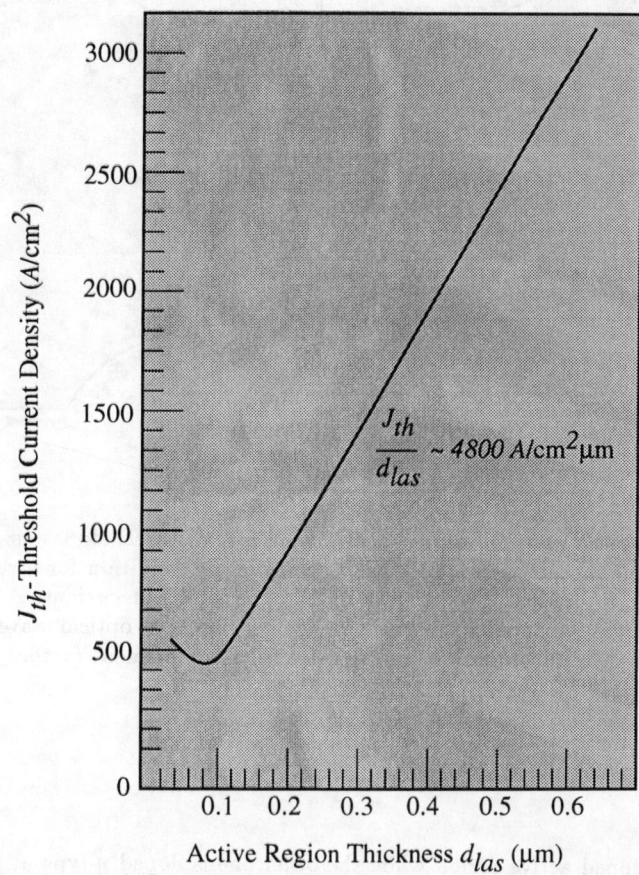

Figure 8.10: Dependence of threshold current in double heterostructure lasers on width of the active region. The threshold current density decreases with the active layer thickness because the 2-dimensional sheet charge density of the injected charge needed for the threshold condition decreases inversely with the active layer thickness. At very small active layer thicknesses $d \lesssim 50$ Å, the threshold current density increases because the optical wave confinement factor goes toward zero so that the cavity gain is almost zero.

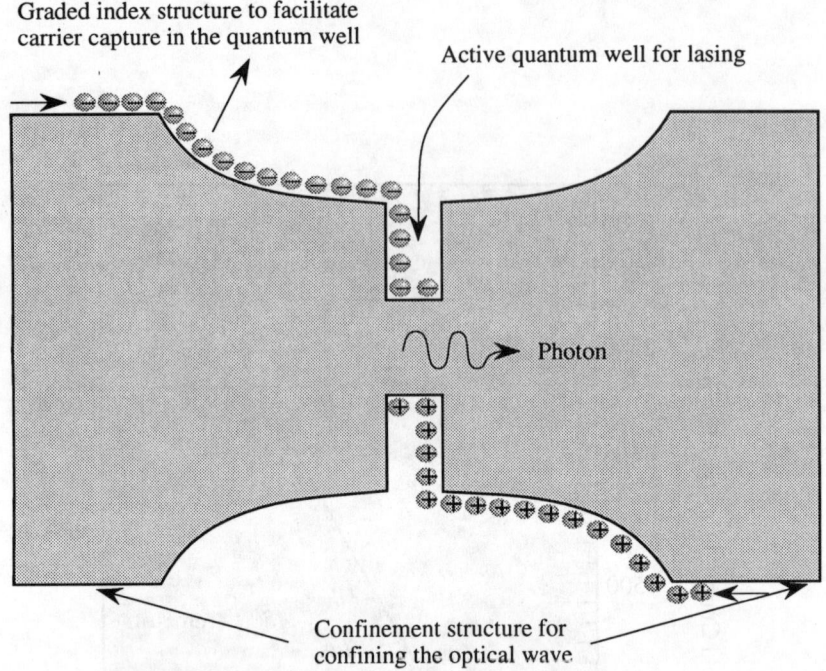

Figure 8.11: A typical quantum well laser structure for low threshold lasers. The density of states in the 2-D quantum well allows one to achieve the condition for inversion of bands at a lower injection density. This results in a lower threshold current. A cladding layer with a high bandgap surrounds the quantum well so that the optical wave is confined as much as possible near the quantum well to get a high confinement factor.

laser has an undoped active region while the other one is doped $p$-type at $8 \times 10^{17}$ cm$^{-3}$. Calculate the threshold current densities for the two lasers if the cavity loss is 50 cm$^{-1}$ and the radiative lifetime at lasing is 2.4 ns for both lasers. The active region width is 0.1 $\mu$m.

This example is chosen to demonstrate the advantages of $p$-type doping in threshold current reduction. Since holes are already present in the active region, one does not have to inject as much charge to create gain. However, it must be noted that too much $p$-doping can cause an increase in cavity loss and even non-radiative traps (some dopants can be incorporated on unintended sites in the crystal).

To solve this problem, a computer program should be written. This program should calculate the quasi-Fermi levels for the electrons and holes and then evaluate the gain.

We have

$$E_{Fn} = E_c + k_B T \left[ \ell n \frac{n}{N_c} + \frac{1}{\sqrt{8}} \frac{n}{N_c} \right]$$

$$E_{Fp} = E_v - k_B T \left[ \ell n \frac{p_{tot}}{N_v} + \frac{1}{\sqrt{8}} \frac{p_{tot}}{N_V} \right]$$

where $n$ is the electron (and hole) density injected and

$$p_{tot} = p + p_A$$

where $p_A$ is the acceptor density. For the undoped laser, one finds that at approximately $1.1 \times 10^{18}$ cm$^{-3}$ the laser reaches the threshold condition. For the doped laser we get a value of $n = p = 8.5 \times 10^{17}$ cm$^{-3}$. The threshold current densities in the two cases are

$$J(\text{undoped}) = \frac{(1.1 \times 10^{18} \ cm^{-3})(0.1 \times 10^{-4} \ cm)(1.6 \times 10^{-19} \ C)}{(2.4 \times 10^{-9} \ s)}$$

$$= 733 \ A/cm^2$$

$$J(\text{doped}) = \frac{(8.5 \times 10^{17} \ cm^{-3})(0.1 \times 10^{-4} \ cm)(1.6 \times 10^{-19} \ C)}{(2.4 \times 10^{-9} \ s)}$$

$$= 566 \ A/cm^2$$

## 8.7  ADVANCED STRUCTURES: TAILORING THE CAVITY

### 8.7.1  Issues in a Fabry-Perot Cavity

$\rightsquigarrow$ A commonly used laser cavity structure is the Fabry-Perot cavity having a waveguiding structure of length $L$ and width $d_T$ as shown in Fig. 8.3. In this cavity, there are a number of optical modes which can form stationary states as discussed before. As a laser is pumped, the gain becomes positive and eventually the cavity loss is overcome. At this point, a number of the optical longitudinal modes can start to have stimulated emission. As a cavity is pumped harder, a few modes, whose frequency lies close to the peak energy in the gain curve, start to get much stronger as discussed before and shown in Fig. 8.7. Nevertheless, several modes are usually emitted during the laser operation. Thus the modal purity of the Fabry-Perot laser is not very good. The spacing between these Fabry-Perot modes is given by (see Section 8.3)

$$\Delta k = \pi/L; \quad \Delta \omega = \frac{v\pi}{L} \tag{8.58}$$

where $v$ is the velocity of light. The photon modes that the $e$-$h$ recombination can emit in a Fabry-Perot cavity are essentially the same as in a bulk semiconductor. This is because of the large size of the cavity compared to the wavelength of light.

The Fabry-Perot cavity is one of the simplest optical cavities used for semiconductor lasers. It is simply produced by cleaving a semiconductor wafer along the cleavage planes (110) or ($\bar{1}$10) for (001) direction grown wafer as shown in Fig. 8.12. The cleavage produces high quality mirrors with reflectivity of $\sim$ 0.3 to 0.4 depending upon the semiconductor. Typical lengths of the cavity range from 150 $\mu$m to 1 mm depending upon the application of the laser diode.

The Fabry-Perot cavity also has a certain lateral dimension, $d_T$, which determines the transverse modes of the light that is emitted. As a result, the output of the Fabry-Perot mode not only has a number of longitudinal modes present, but in general may also have several transverse modes present as shown in Fig. 8.13.

As the laser injection current is increased, the relative strength of the various modes changes. The longitudinal mode which has the highest photon density is the one which has a photon wavelength closest to the peak in the gain spectrum. Since the peak in the gain spectrum shifts with carrier injection, the peak mode shifts.

Additionally, if the transverse confinements of the optical wave is weak (i.e., $d_T$ is large), there may be several transverse modes whose frequency is closely spaced. The laser output may involve different such modes as the injection current is increased. This results in a "kink" in the output power as shown in Fig. 8.14. To avoid the "kink" which produces noise in the optical transmitter, it is important to ensure a strong transverse confinement in the structure. This confinement can be achieved by two approaches: i) gain-guided cavities, and ii) index-guided cavities. We will now discuss these two approaches.

### Gain-Guided Cavities

The most common technique to confine the optical wave and to guide it along the cavity length is to fabricate a stripe geometry laser as shown in Fig. 8.15a. The semiconductor structure is covered with a thin layer of oxide ($SiO_2$) into which a thin strip of width $d_T = 2 - 10$ $\mu$m is etched. The contact is now made through this stripe as shown. This allows the injected current to be confined into a very narrow region. A similar effect is created in the ridge laser shown in Fig. 8.15b.

The current is injected through a narrow opening of width, $d_T$, as shown in Fig. 8.15c, but the current spreads out under the stripe as shown in Fig. 8.15d. This spreading is due to the diffusive nature of the current flow and is controllable by the device design. The carrier concentration under the stripe has a nonuniform behavior as shown in Fig. 8.15e. Since the refractive index of a material is dependent upon the carrier density, the index profile also becomes nonuniform as shown in Fig. 8.15f. Finally, the gain of the device has a nonuniform behavior shown schematically in Fig. 8.15g.

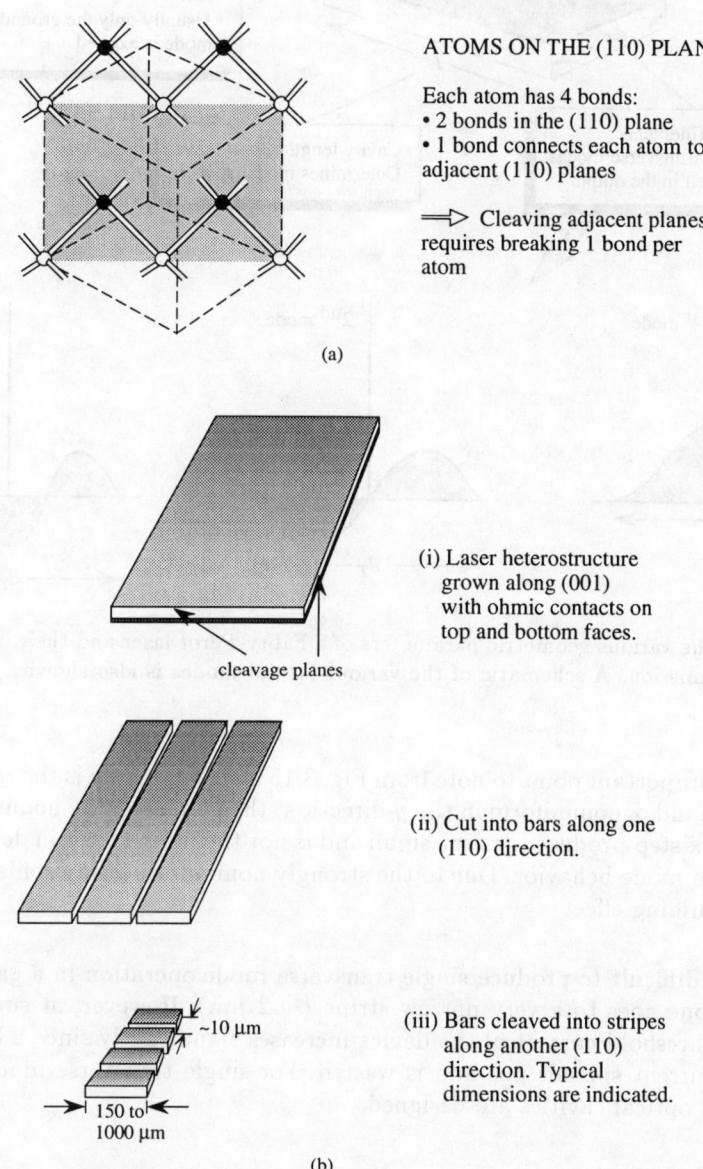

**ATOMS ON THE (110) PLANE**

Each atom has 4 bonds:
• 2 bonds in the (110) plane
• 1 bond connects each atom to adjacent (110) planes

⟹ Cleaving adjacent planes requires breaking 1 bond per atom

(a)

(i) Laser heterostructure grown along (001) with ohmic contacts on top and bottom faces.

cleavage planes

(ii) Cut into bars along one (110) direction.

~10 μm

(iii) Bars cleaved into stripes along another (110) direction. Typical dimensions are indicated.

150 to 1000 μm

(b)

Figure 8.12: (a) The cleaving plane of zinc-blende structures has adjacent planes connected by a single bond. (b) The approach used to produce a Fabry-Perot optical cavity involves cleaving a wafer containing the laser diode structure.

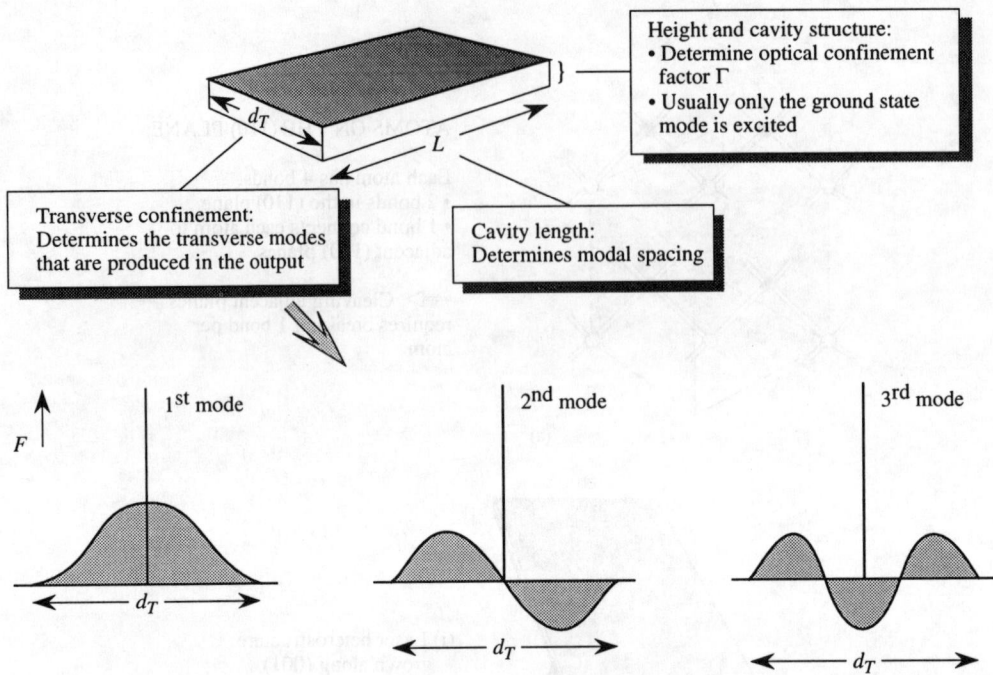

Figure 8.13: The various geometric parameters of a Fabry-Perot laser and their importance for the laser emission. A schematic of the various lateral modes is also shown.

The important point to note from Fig. 8.15 is that not only is the real part of the refractive index nonuniform in the $y$-direction, the gain is highly nonuniform. In fact, the index step produced is very small and is not the main factor in determining the transverse mode behavior. Due to the strongly nonuniform gain profile, the gain produces a guiding effect.

It is difficult to produce single transverse mode operation in a gain-guided laser unless one goes to a very narrow stripe ($\sim 2~\mu$m). However, at such narrow stripes, the threshold current of the device increases significantly since a large fraction of the current spreads out and is wasted. For single transverse mode output, index-guided optical cavities are designed.

**Index-Guided Cavities**

Index-guided cavities rely on a step in the index profile in the lateral direction. An example is shown in Fig. 8.16. Such a device is also called a buried heterostructure laser. The device fabrication is much more complex. To produce a lateral index step, one requires to first grow an epitaxial layer with a structure similar to the normal laser. The structure is then etched down leaving a few micron regions. Regrowth is then carried out to surround the active region by a large bandgap material.

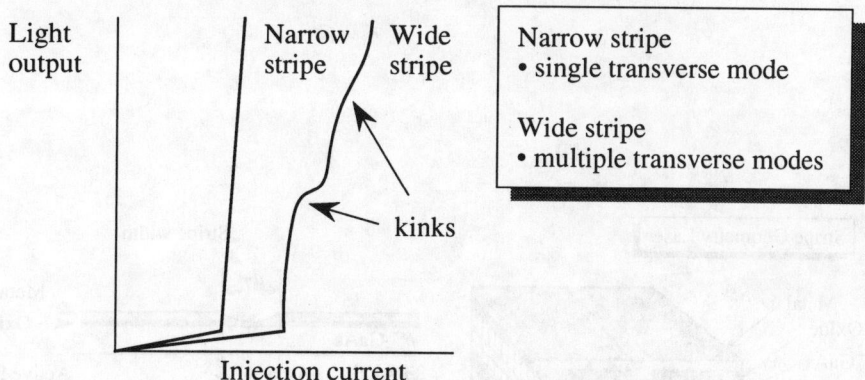

Figure 8.14: The shift in transverse modes participating in the optical output of a laser produces kinks in the light output-current curves.

The buried heterostructure laser, if fabricated correctly, does not suffer from kinks in the light-current curve. The output is single mode and the threshold current is very small. Of course, to take full advantage of the electronic properties, the active region must contain quantum well structures.

## 8.7.2   The Distributed Feedback Lasers

$\mathcal{R}$  The Fabry-Perot cavity laser, although easy to fabricate, suffers from a number of drawbacks. Since a simple mirror is used to create the stationary states, there is no special preference given to a particular optical mode as far as the optical cavity is concerned. The determination as to which modes will dominate is left entirely up to the gain spectra determined by the electronic properties of the active region. Since the mode spacing is only 4-5 Å, and the gain spectrum is rather flat on this scale, several modes end up lasing in the Fabry-Perot cavity. Of course, as discussed earlier, at high powers the side modes are suppressed relative to the peak mode. Nevertheless, at most operating conditions the laser linewidth is ∼ 20 Å for Fabry-Perot cavity, even though each mode is extremely narrow. The question naturally arises: Can the cavity itself provide mode selection? After all, in electronic circuits, resonant cavities can be designed with tremendous mode selectivity. Indeed, where would the microwave field be without resonant cavities? While, in semiconductor optoelectronics, one cannot design mode selective optical cavities with as much ease as in electronics, there are some solutions.

An important approach toward design of a mode selective optical cavity is the distributed feedback (DFB) structure. The DFB approach to create strong mode selectivity is based upon the propagation of waves in periodic structures. We know

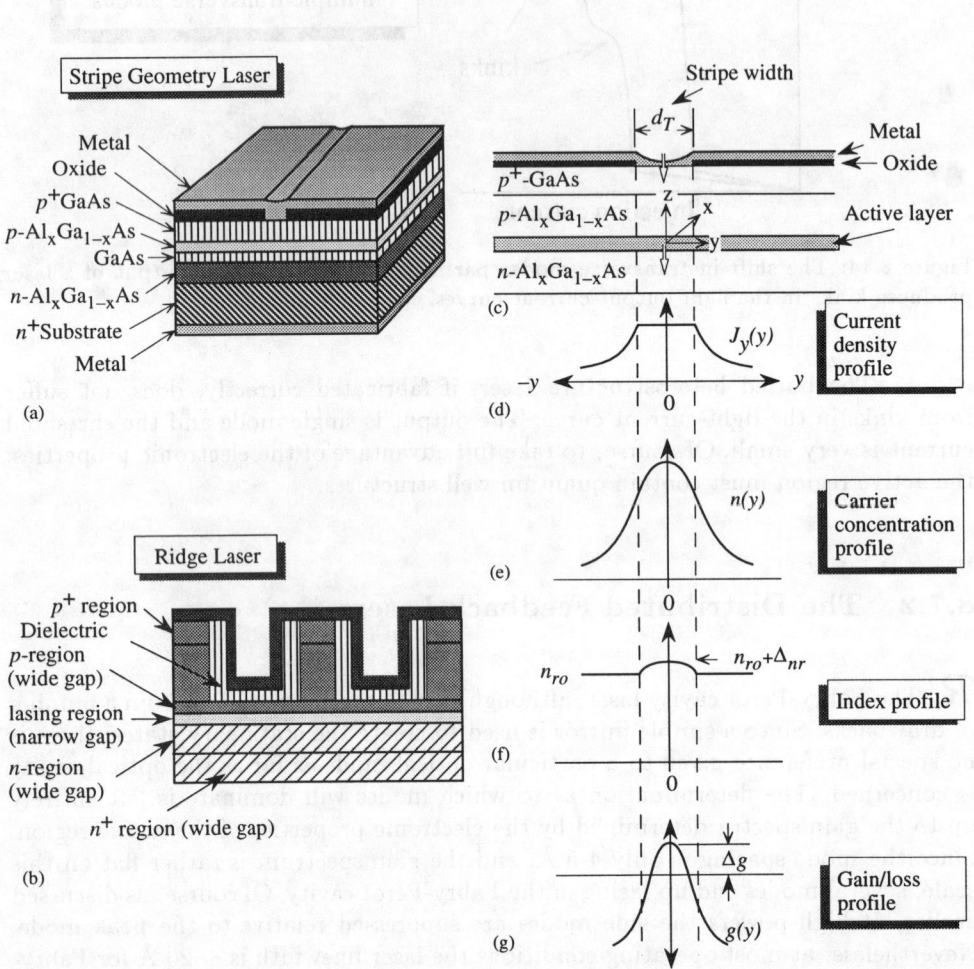

Figure 8.15: (a) The stripe geometry laser, (b) the ridge laser, (c) the current injection into the laser, (d) the current density profile, (e) the electron (hole) density profile in the active region, (f) The refractive index profile, and (g) the gain and loss profile.

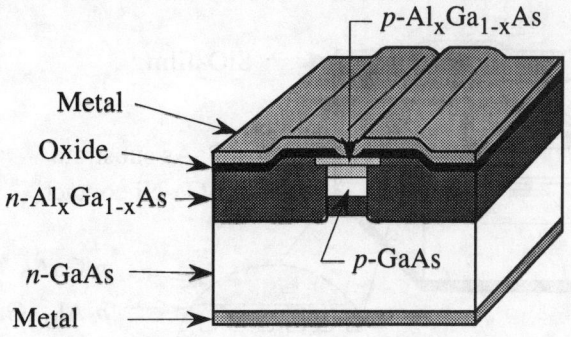

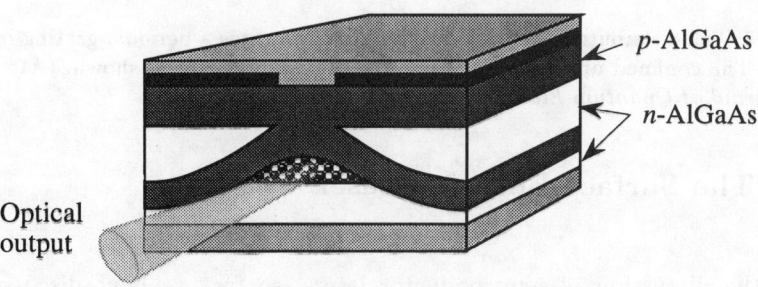

Figure 8.16: Index guided laser cavities. Etching and regrowth techniques are employed to produce buried active regions.

that in periodic structures, special effects occur when the wavelength of the wave approaches the wavelength of the periodic structure. In semiconductor crystals, this leads to bandgaps and Bragg reflections. Similar effects occur for optical waves.

In the DFB structure, a periodic grating is incorporated into the laser structure as shown in Fig. 8.17. The fabrication process is by no means trivial and involves growing the basic laser structure, etching a periodic structure and regrowing the top layer. The grating should be as close as possible to the active region so that the optical wave interacts strongly with the grating. However, since the placement of the grating creates defects, the grating cannot be too close to the active region. The difficulty in the fabrication of the DFB lasers is manifested in its cost, which is a thousand times greater than that of a Fabry-Perot laser.

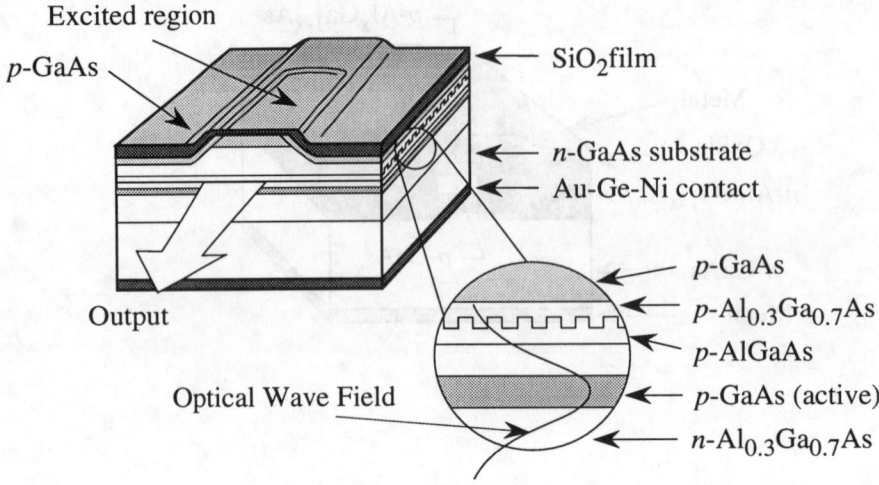

Figure 8.17: The distributed feedback structure incorporates a periodic grating in the laser structure. The confined optical wave senses the periodic grating as shown. (After K. Aiki, *IEEE Journal of Quantum Electronics*, **QE**-12, 601 (1976).)

## 8.7.3    The Surface-Emitting Laser

$\mathcal{R}$    In the discussion of semiconductor lasers, so far, we have discussed "edge emitting" lasers where the laser output comes out from the edge and is parallel to the substrate on which the device is grown. As noted earlier the lasing condition has the form

$$(\Gamma g_{th} - \alpha_c) > -\frac{1}{L}\ell n R \qquad (8.59)$$

where $L$ is the cavity length. Usually $R$ has a value of $\sim 0.3$, so that to ensure a reasonably low threshold, the laser length $L$ has to be $\gtrsim 100$ $\mu$m. Thus the edge emitting laser is a fairly large device on the scale of other microelectronic devices such as transistors. Additionally, it is difficult to produce a high density array of edge-emitting lasers on a single chip. While this may not be a serious problem for the usual optical communication applications, it places a severe limitation on the use of lasers for high density optical interconnects for computer chips.

The surface-emitting laser (SEL) or the vertical cavity surface emitting laser (VCSEL) overcomes the problems outlined above. These lasers can be produced in very high density on a single wafer and the light output emerges perpendicular to the wafer. The devices can also be made very small so that threshold currents can be in the range of sub-milliamperes.

In the SEL, the mirrors are placed not at the edges of the laser, but at the top and the bottom of the device. However, since it is difficult to epitaxially grow very thick layers, the thickness of the cavity is limited to $\sim 10~\mu$m. If the reflectivity is maintained at around 0.3, the reflection losses will approach $10^3$ cm$^{-1}$ or higher. This would cause the threshold current to be extremely high. This problem is overcome by developing mirrors with very high reflection coefficients. For example, if the reflectivity of the mirrors can be made to be 0.99, the mirror loss becomes only 10 cm$^{-1}$ which is quite acceptable for low threshold current density lasers. Thus the success of the SEL depends upon incorporating very high reflectivity mirrors in the structure.

In the previous section we have briefly mentioned how a periodic structure can be used to create an effective reflectivity for light with a wavelength that matches the spatial periodicity. This concept of distributed Bragg reflectors (DBRs) is used in the SEL to produce high quality mirrors. The DBR provides the very high reflectivity needed to compensate for the short cavity length. A typical DBR structure is shown in Fig. 8.18a. The active region of the SEL is essentially similar to that of a Fabry Perot cavity. However, the cladding layers are chosen as DBRs which consist of a periodic arrangement of layers with thicknesses $d_1$ and $d_2$ and refractive indices $n_{r1}$ and $n_{r2}$ as shown in Fig. 8.18a. The periodicity is chosen so that

$$n_{r1}d_1 + n_{r2}d_2 = \lambda_0/2 \tag{8.60}$$

which corresponds to the Bragg reflection condition for an optical beam with free space wavelength $\lambda_o$. The wavelength is chosen so that the photon energy coincides with the peak in the gain curve.

The DBR structure can be formed from crystalline materials like GaAs/AlAs or from amorphous materials such as Si/SiO$_2$ as shown in Figs. 8.18b and 8.18c. The reflectivity of a DBR structure is highly dispersive. The reflectivity is very high at the Bragg condition and drops off strongly when the wavelength is different from the Bragg condition wavelength. A schematic of the reflectivity is shown in Fig. 8.18a.

To carry out an accurate analysis of the SEL, the reflectivity of the DBRs has to be carefully calculated. Numerous methods have been developed to study this problem. For wide SEL, with device lateral diameters greater than $\sim 10~\lambda$, the reflectivity is given by

$$R = \left( \frac{1 - \left(\frac{n_{r1}}{n_{r2}}\right)^{2N}}{1 + \left(\frac{n_{r1}}{n_{r2}}\right)^{2N}} \right)^2 \tag{8.61}$$

where $n_{r1}$ is smaller than $n_{r2}$ and we have chosen $d_1 = d_2$. Here $N$ is the number of periods used in the DBR. The reflectivity can easily reach 99% if a proper combination of $N$ and $n_{r1}/n_{r2}$ is used. The reflectivity is somewhat lower if the diameter of the cavity starts to approach a few times $\lambda$.

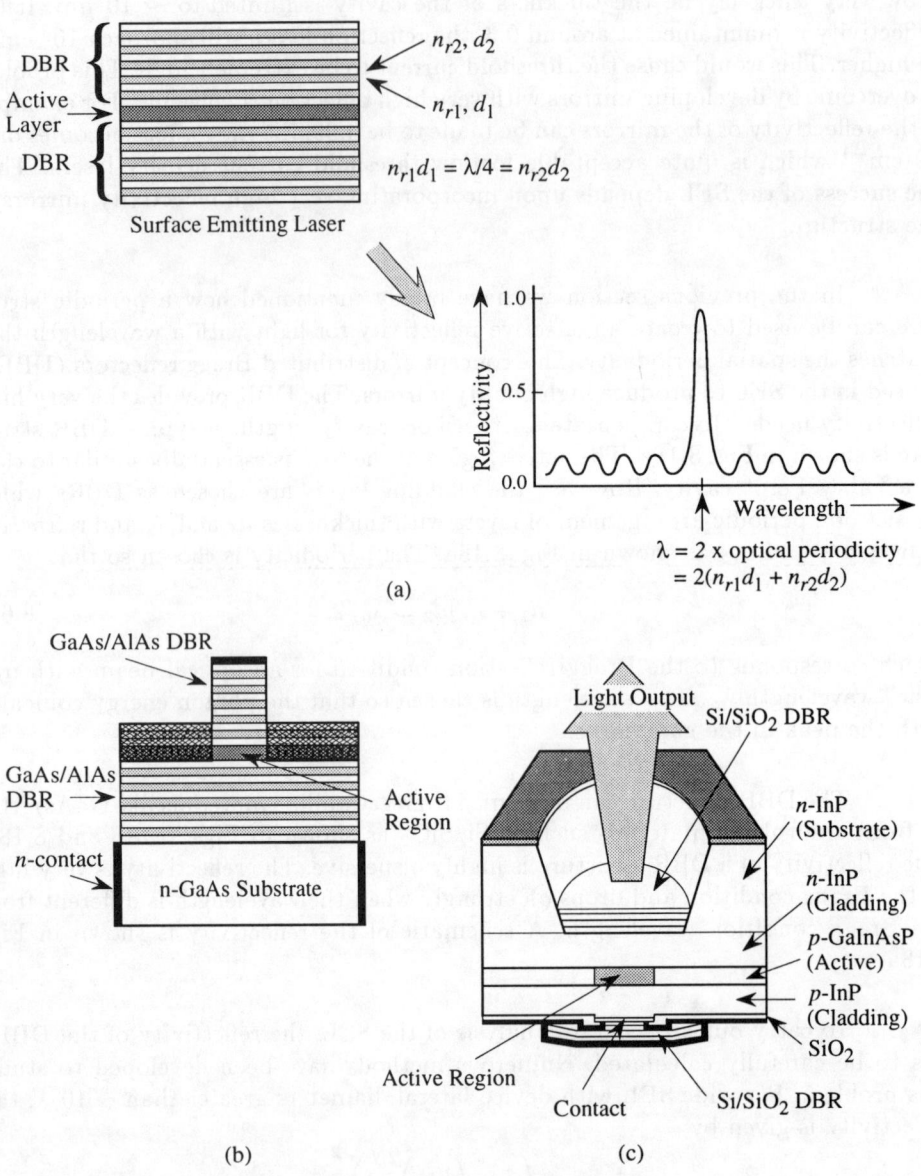

Figure 8.18: (a) A schematic of the placement of DBR reflectors used in a surface-emitting laser. A schematic of the reflectivity of a DBR stack is shown. (b) A typical structure using crystalline (GaAs/AlAs) DBR. (c) A structure using amorphous DBR stacks. (Figures such as ones shown can be seen in the special issue of *IEEE J. Quantum Electronics*, **QE**-29 (1993).)

Laser arrays can be fabricated using the SEL concept and such arrays have been shown to have output optical power approaching 1 watt. However, the SEL structure does face some difficulties, especially when compared to the edge-emitting device. These difficulties arise from the following considerations:

i) *Charge injection*: The presence of the DBR mirrors causes considerable difficulty in charge injection into the SEL. Great care has to be taken in developing a low resistance path to the active region. If the charge is to be injected through the DBR, this structure is to be doped. The thick DBR offers significant resistance to the current flow. If ring contacts are to be used to inject charge directly into the active region, by passing the DBR, current crowding effects become significant. In this case the edges of the device carry the current while the center has very low current flow. Improvements in SEL technology are likely to come from innovative techniques for DBR fabrication and charge injection.

ii) *Device heating*: The high resistance in the SEL current path causes considerable device heating. As a result, it becomes difficult to operate the device at high drive current. At high drive current, device heating causes loss in the device efficiency due to current leakage and Auger recombination (for low bandgap materials) and the optical output tends to saturate.

We have mentioned above the use of SELs in laser arrays. The arrays are important not only for inter-chip interconnects, but for high power optical devices. It is difficult to make a single laser structure with large enough area to give, say, 1 watt of optical power. Structural non-uniformities and technology limitations make it difficult for a single laser to emit more than $\sim 10$ mW of power. However, with laser arrays the power level can approach several watts. While the output power of laser arrays can be high, it is difficult to control the relative phase of the different lasers on an array. Considerable research is still ongoing to develop techniques to control the phase of the individual optical emitters.

## 8.7.4 The DBR Laser

$\mathcal{R}$ In the periodic structure based cavities discussed so far, the periodicity of the structure is fixed. As a result, the wavelength of emission that is given selective preference is also fixed. *In the DBR laser, the periodic reflecting stack is placed outside the active lasing region. Also, the refractive index of the stack is alterable by, say, current injection.* In Fig. 8.19 we show a typical DBR laser. The DBR is responsible for providing a reflection to the light emitted from the active region. The wavelengths that get the highest feedback must satisfy

$$\lambda_B = 2qa \qquad (8.62)$$

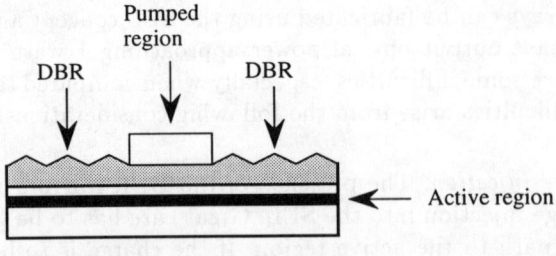

Distributed Bragg reflector structure

Figure 8.19: The DBR laser, where a Bragg reflector is placed outside the active region. The optical periodicity of the DBR ($n_{r1}$d) can be altered electronically.

where $q$ is a positive integer and $a$ is the optical periodicity of the structure

$$a = n_{r1}d_1 + n_{r2}d_2 \qquad (8.63)$$

The values of $n_{r1}$ and $n_{r2}$ can be altered electronically and, therefore, the laser can have a certain degree of wavelength tunability (say $\sim 30$ Å).

## 8.8   TEMPERATURE DEPENDENCE OF LASER OUTPUT

$\longrightarrow$   As in the case of the LED, the temperature dependence of the laser diode output is extremely important for most applications. We have seen in Section 8.5 that for high speed applications the laser has to be driven at a very high input power. As a result, the device gets heated even with good heat sinking. Issues of importance in the study of the temperature dependence are: i) Effect of temperature on the threshold current and the optical intensity; ii) Effect of temperature on the frequency of emission. We will briefly discuss these issues.

### 8.8.1   Temperature Dependence of the Threshold Current

$\longrightarrow$   As the temperature of the laser diode increases, the threshold current increases and at a given injection, the photon output decreases. One can ascribe three reasons for this to occur:

i) The increase in temperature causes the quasi-Fermi functions $f_e$ and $f_h$ to "smear" out more. As a result, the condition for inversion $f_e + f_h > 1$ requires a higher injection carrier density to be fulfilled. This increases the threshold current. This effect arises in all semiconductor lasers.

ii) The increase in the temperature causes the electrons and holes distribution to be spread out into higher energies. As a result, a greater fraction of the injected charge can cross over the active region and end up in the cladding or contact region. This leakage current was described for the LED in the previous chapter (Section 7.6.4). The leakage current depends upon the design of the laser and can be suppressed by using a wide active region or a graded index structure for quantum well lasers.

iii) Increase in temperature causes more electrons and holes to possess energies greater than the threshold energy needed for Auger recombination. This, coupled with the increase in the threshold carrier density, causes the Auger recombination to increase exponentially with temperature. The Auger processes are particularly important for narrow gap materials.

As a result of the three effects discussed above the threshold current density of a laser can, in general, be described by the form

$$J_{th}(T) = J_{th}^o exp(T/T_o) \qquad (8.64)$$

A large value of $T_o$ is desirable. Typical values in GaAs lasers are $\sim$ 120 K. For long wavelength lasers ($\lambda = 1.55 \ \mu$m), $T_o$ values are smaller ($\sim$ 50 K).

## 8.8.2 Temperature Dependence of the Emission Frequency

$\longrightarrow$ For many applications it is important that the laser frequency remain stable over the operating conditions. If the temperature changes, the emission frequency shifts. There are two effects that control the shift of the laser frequency:

i) Change in the semiconductor bandgap causes the entire gain spectrum to shift to lower energies as the temperature increases. The bandgap change in most semiconductors is about -0.5 meV/K. This would cause the gain spectrum to change by $\sim$ 3 to 4Å/K if there were no additional factors as shown in Fig. 8.20a. However, in the laser the emission depends not just on the position of the gain peak, but on the Fabry-Perot mode that is closest to the gain peak. This brings us to the second effect.

ii) As temperature changes, the thermal expansion of the laser cavity and the change in the refractive index alters the position of the resonant modes. The

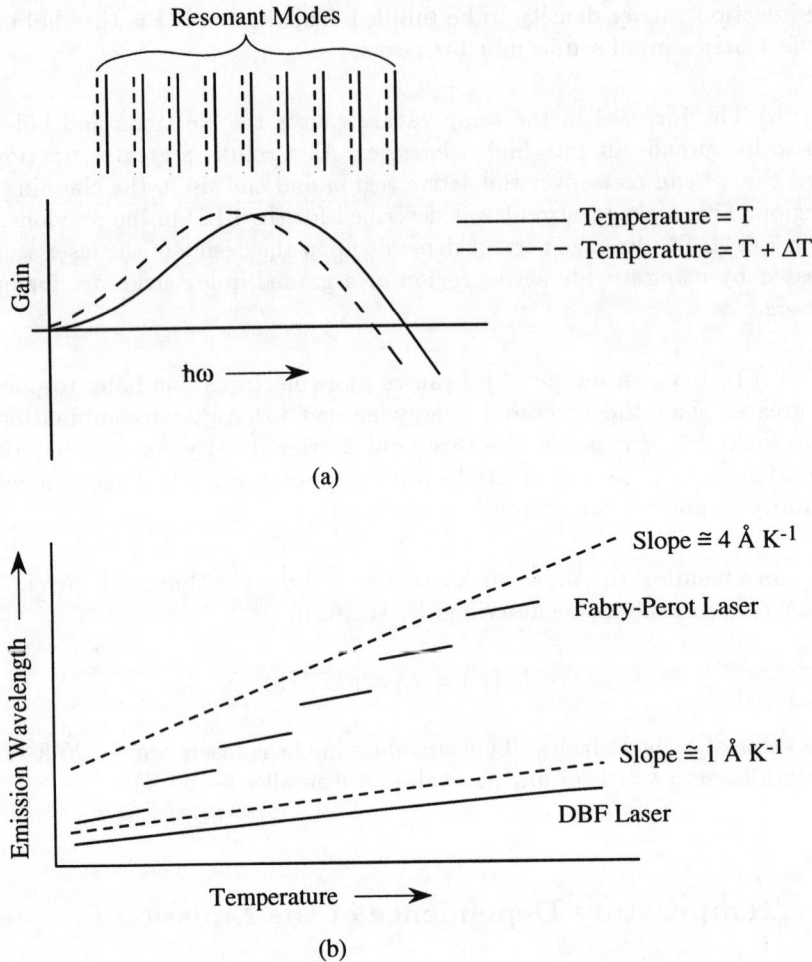

Figure 8.20: (a) The shift of the gain spectra and the resonant modes of a cavity with temperature. (b) Shift in the emission wavelength of a laser with temperature.

resonant modes are given by ($q$ is an integer)

$$q\lambda_q = 2L; \lambda_q = \frac{\lambda_{qo}}{n_r} \tag{8.65}$$

If the effective cavity length increases due to the temperature, the positions of the modes will shift relative to the gain spectrum which is itself shifting with temperature. This is shown schematically in Fig. 8.20b. For most semiconductors, the overall effect is to produce a shift in the resonant wavelengths of about 1 Å $K^{-1}$.

As a result of the two effects, the emission wavelength of a Fabry-Perot laser has a behavior schematically shown in Fig. 8.20b. The emission wavelength

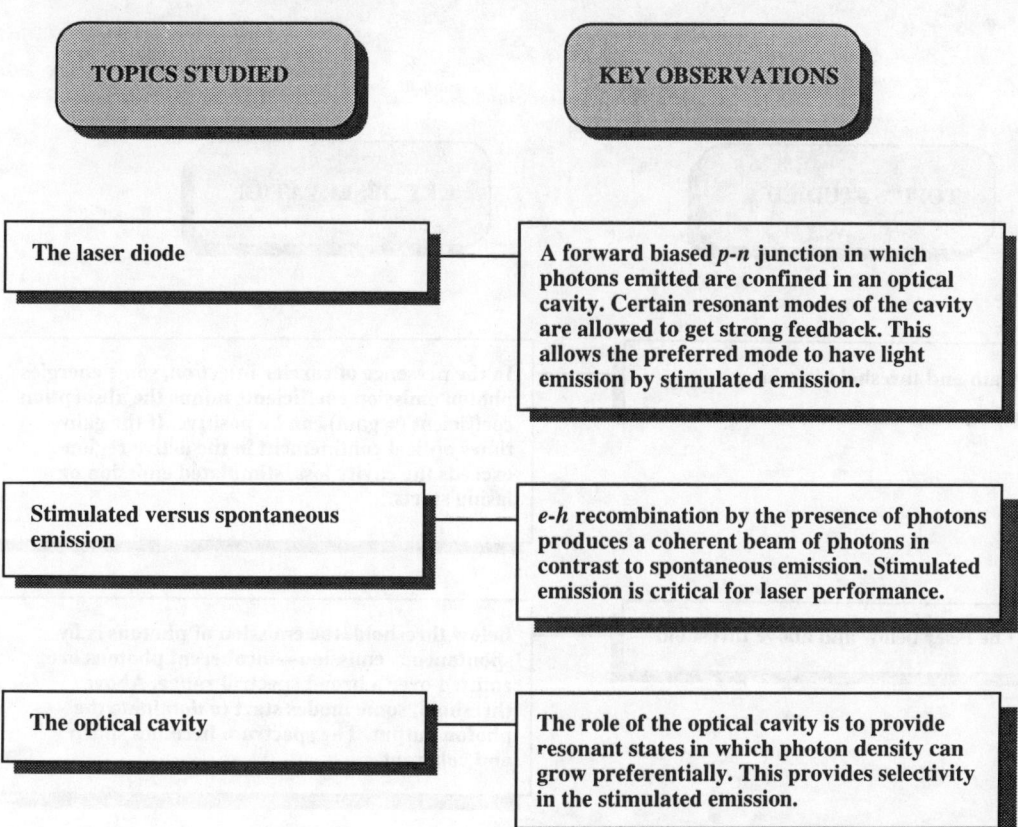

Table 8.1: Summary table.

shifts by $\sim 1$ Å $K^{-1}$ until an adjacent mode becomes closer to the gain peak. Thus mode hopping occurs. However, in the case of a DFB laser where the lasing mode is not selected by the gain peak but by the spacing of the grating, the mode hopping does not occur and the lasing wavelength shifts by only about $1$ Å $K^{-1}$ over a wide temperature range as shown in Fig. 8.20b.

## 8.9  CHAPTER SUMMARY

$\mathcal{R}$  In this chapter we have discussed laser diodes using semiconductors. The summary tables (Tables 8.1 and 8.2) highlight the important areas discussed in this chapter.

| TOPICS STUDIED | KEY OBSERVATIONS |
|---|---|
| Gain and threshold in a laser | In the presence of carrier injection, some energies photon emission coefficients minus the absorption coefficient (= gain) can be positive. If the gain times optical confinement in the active region exceeds the cavity loss, stimulated emission or lasing starts. |
| The laser below and above threshold | Below threshold, the emission of photons is by spontaneous emission—incoherent photons are emitted over a broad spectral range. Above threshold, some modes start to dominate the photon output. The spectrum becomes sharp and coherent. |
| Laser structures for superior performance: electronic tailoring | Laser threshold current is reduced by using narrow active regions, quantum wells and strained quantum wells. Essentially, the improvements results from improvements in inversion due to the density of states. |
| Laser structures for superior performance: optical cavity tailoring | Laser performance is improved by going from gain guided cavities to index guided cavities. The use of DFB lasers allows suppression of side modes. Surface emitting lasers are useful for vertical emission. DBR lasers have excellent tunable output. |

Table 8.2: Summary table.

## 8.10 PROBLEMS

### Section 8.3

**8.1** Consider a GaAs laser with a cavity length of 75 $\mu$m. Calculate the number of allowed longitudinal modes in an energy width $E_g \pm \frac{1}{2}k_B T$ where $E_g = 1.43$ eV and $T = 300$ K.

**8.2** Consider two optical cavities, one a Fabry-Perot cavity with a length of 100 $\mu$m and a mirror reflectivity of 0.33 and another with a length 10 $\mu$m. What should be the mirror reflectivity of the second cavity to ensure that a photon emitted inside spends the same time in the two cavities before escaping? Cavity loss due to absorption is 10 cm$^{-1}$.

**8.3** Consider a Fabry-Perot cavity of length 100 $\mu$m. The mirror reflectivity is 0.33 and the absorption loss in the cavity is 20 cm$^{-1}$. Calculate the photon lifetime $\tau_{ph}$. The refractive index in the cavity is 3.6.

### Sections 8.4 and 8.6

**8.4** Consider the semiconductor alloy InGaAsP with a bandgap of 0.8 eV. The electron and hole masses are 0.04 $m_0$ and 0.35 $m_0$, respectively. Calculate the injected electron and hole densities needed at 300 K to cause inversion for the electrons and holes at the bandedge energies. How does the injected density change if the temperature is 77 K? Use the Joyce-Dixon approximation.

**8.5** Consider a GaAs based laser at 300 K. Calculate the injection density required at which the inversion condition is satisfied at i) the bandedges; ii) at an energy of $\hbar\omega = E_g + k_B T$. Use the Joyce-Dixon approximation.

**8.6** Consider a GaAs based laser at 300 K. A gain of 30 cm$^{-1}$ is needed to overcome cavity losses at an energy of $\hbar\omega = E_g + 0.026$ eV. Calculate the injection density required. Also, calculate the injection density if the laser is to operate at 400 K.

**8.7** Consider the laser of the previous problem. If the time for *e-h* recombination is 2.0 ns at threshold, calculate the threshold current density at 300 K and 400 K. The active layer thickness is 2.0 $\mu$m and the optical confinement is unity.

**8.8** Two GaAs/AlGaAs double heterostructure lasers are fabricated with active region thicknesses of 2.0 $\mu$m and 0.5 $\mu$m. The optical confinement factors are 1.0 and 0.8, respectively. The carrier injection density needed to cause lasing is $1.0 \times 10^{18}$ cm$^{-3}$ in the first laser and $1.1 \times 10^{18}$ cm$^{-3}$ in the second one. The radiative recombination times are 1.5 ns. Calculate the threshold current densities for the two lasers.

**8.9** Consider a laser in which the carrier masses could be tuned. Assume that the hole density of states mass is 0.5 $m_0$ while the electron density of states mass changes from 0.02 $m_0$ to 0.2 $m_0$. Calculate and plot the transparency density needed at an energy of $E_g + k_B T$ where $E_g$ is 1.4 eV. The temperature is 300 K.

**8.10** In the previous problem, the electron mass was tuned. Consider now the case where the electron mass is fixed at 0.067 $m_0$ and the hole mass varies between

$0.1 \, m_0$ and $0.5 \, m_0$. Calculate the corresponding transparency density.

**8.11** Calculate the temperature dependence of the transparency condition injection density from 77 K to 400 K. Use the energy $\hbar\omega = E_g + k_B T$ to define the point where $f^e + f^h = 1$. The electron mass is $0.067 \, m_0$ and the hole mass is $0.45 \, m_0$.

**8.12** Consider a GaAs/Al$_{0.3}$Ga$_{0.7}$As laser with an active region thickness of $0.1 \, \mu$m at 300 K. The active region is doped $p$-type at $1.0 \times 10^{18}$ cm$^{-3}$. The cavity loss is 50 cm$^{-1}$ and $\Gamma = 0.12$. Calculate the transparency density for the bandgap energy and the threshold density for this device.

**8.13** Consider the device of the previous problem. Calculate the threshold current density if the radiative lifetime at threshold is 2.4 ns. What would the threshold current density be if the device active region was undoped and the lifetime was unchanged? Consider only the radiative current.

**8.14** Consider a GaAs/Al$_{0.3}$Ga$_{0.7}$As laser with an active region thickness of $0.1 \, \mu$m at 300 K. Calculate the gain curves versus energy for injection densities of 0; $8.0 \times 10^{17}$ cm$^{-3}$; $1.0 \times 10^{18}$ cm$^{-3}$ and $1.30 \times 10^{18}$ cm$^{-3}$.

**8.15** Consider a laser with an active region bandgap of 0.8 eV and a thickness of $0.1 \, \mu$m with $\Gamma = 0.1$. Calculate the threshold current for the device at 300 K. The device is described by the following parameters:

| | |
|---|---|
| Spontaneous emission lifetime at threshold | $= 3.0 \; ns$ |
| Laser length | $= 300 \; \mu m$ |
| Laser width | $= 10 \; \mu m$ |
| electron mass | $m_e^* = 0.04 \, m_0$ |
| hole mass | $m_h^* = 0.4 \, m_0$ |
| cavity loss | $\alpha_c = 40 \; cm^{-1}$ |

**8.16** Consider a laser emitting at $1.3 \, \mu$m with a threshold current of 1 mA. The laser is biased at 10 mA so that you can assume that all the photons are being emitted in a single mode. Calculate the photon density in the peak mode in the cavity which has a dimension $5 \, \mu$m$\times 150 \, \mu$m and a loss $\alpha_c = 40$ cm$^{-1}$. What is the output power of the device at this bias? The cavity refractive index is 3.3.

## Section 8.5

**8.17** A GaAs/AlGaAs laser has a threshold current density of 200 A/cm$^{-2}$. In a large signal switching, the laser is switched from zero current to 4 times the threshold current. Calculate the time delay before photon emission if the carrier lifetime is 2 ns.

**8.18** Discuss the effect of decreasing the cavity length of a Fabry-Perot laser on the modulation speed of the laser.

**8.19** Consider a GaAs/AlGaAs laser at 77 K and 300 K with the optical confinement factor $\Gamma = 1.0$. Calculate $dg/dn$ for the laser near threshold if the cavity loss is 100 cm$^{-1}$. What can you say about the effect of temperature on the small signal modulation?

## 8.11  REFERENCES

• **General**

– H. Kressel and J. K. Butler, *Semiconductor Lasers and Heterojunction LEDs*, Academic Press, New York (1977).

– R. Baets, "Heterostructures in III-V Optoelectronic Devices," *Solid State Electronics*, 30, 1175 (1987).

– J. Singh, *Semiconductor Optoelectronics: Physics and Technology*, McGraw-Hill, New York (1995).

# CHAPTER
# 9

# MODULATION AND
# DISPLAY DEVICES

## 9.1   INTRODUCTION

A nineteenth-century invention—the cathode ray tube or CRT—continues to play a central role in our lives. The television based on the CRT gobbles up our time to the great chagrin of parents and school teachers. Computer jocks spend hours peering into their computer displays, also based on the CRT. How has such an "old" device come to dominate technology or, more appropriately, become an enabling technology for so long? Certainly the CRT is not compatible with the other tools of the information age—the computer chips. The CRT provides a display system which allows us to ingest data that otherwise might completely overload our brains. The way our brain is configured, a picture is indeed worth a thousand words—or more appropriately these days, a billion bits! And the display system is becoming increasingly important, thanks to the complexity of the information being churned out by the computers. Compared to the IC, the CRT is a bulky, energy-guzzling device. These days an incredible amount of computing power can be carried in the backpack of a kindergartner going off to school. But if you want to view the results of the calculations, the display system has to be carried by a football linebacker. Fortunately, this embarrassing situation is beginning to change. The device that is rapidly evolving and will certainly replace the CRT is the liquid crystal display (LCD). After making its mark in wristwatches and pocket calculators, LCD technology is now an exploding technology, thanks to the laptop computer. Images appearing on ten-inch liquid crystal screens now approach the quality of CRTs.

LCD technology is a truly enabling technology which has given birth to the laptop and notebook computer industry. Display screens on airplanes, automobiles, etc., are also rapidly changing over to LCDs. Indeed, the LCD market is the biggest chunk of the optoelectronics market—projected to reach 100 billion dollars by the year 2010 from its 1994 value of $\sim$ 10 billion dollars. At this pace it will become as important as the IC market, a fact that is being rapidly realized by industries around the world.

In Chapter 3 we have discussed the basic physics behind the operation of the liquid crystal cell. We will review this briefly in the next section. The transition from a single liquid crystal cell to a full-scale display is a difficult challenge, comparable in complexity to the fabrication of complicated ICs such as DRAMS.

The LCD has been classified into two categories—the passive matrix LCD and the active matrix LCD. We will discuss the operation of these two kinds of displays.

The liquid crystal cell is essentially a light modulation device. However, it cannot be used in high speed modulation applications. The switching speeds of LCD cells are anywhere from milliseconds to microseconds. At these slow speeds these devices cannot be used for optical communications where modulation speeds of gigahertz or more are needed. For such applications, electro-optic devices based on materials like lithium niobate and gallium arsenide are needed. In such devices, switching times can approach 10 to 100 picoseconds. We will also examine such electro-optic devices in this chapter.

## 9.2   LIQUID CRYSTAL CELLS: THE UNDERLYING PRINCIPLES

$\longrightarrow$   While liquid crystals have been known for over a century, the first demonstration of their display potential was reported in 1968 by G. H. Heilmeier, L. A. Zanoni, and L. A. Barton. The physical principle used in this earlier demonstration is different from what is used in today's LCDs. This earlier effect was based on "dynamic scattering" of light when an electric current passed through the liquid crystal. In the absence of the electrical signal, long-range order is preserved in the crystal allowing light to transmit. The passage of current causes the long-range order to collapse. Instead, small domains of ordered molecules are created. These domains are randomly oriented and, as a result, scatter the light, causing the crystal to be more reflective or cloudy.

In 1971 the twisted nematic crystals were described by M. Schadt and W. Helfrich. Devices based on the twisted nematic depend upon the use of an electric field to alter the orientation of the optic axis of the liquid crystal. The reader should review Section 2.5 where we have discussed propagation of light in anisotropic media. Liquid crystals have remarkable properties that have allowed them to become the material of choice for flat panel displays. The remarkable progress in technology has allowed the fabrication of flat panel displays which, at sizes of $\sim 20$ cm $\times$ 20 cm, perform as well as the bulkier cathode ray tube based displays. As technology advances, the size of these displays will continue to increase providing important new markets in areas such as television, desktop computers (in addition to portable computers), cockpits, automobiles, etc. Here we will examine some of the underlying physics behind the operation of liquid crystal displays. The reader is urged to revisit Chapter 2, Sections 2.5 and 2.6, where the physics of light propagation in crystals has been discussed. In Chapter 2 we have discussed the anisotropy of the optical properties of crystals, in particular, the concepts of ordinary and extraordinary indices and optic axis. In this chapter we will discuss devices that exploit the fact that the optical properties of crystals can be altered by the application of electric fields.

In solid crystals, the effect of the electric field is to alter the anisotropy between $n_{re}$ and $n_{ro}$ so that the phase difference and hence polarization of the optical signal can be altered. The electric field causes this change by slightly altering the electron distribution at each atom on the crystal. There is no physical distortion or reorientation of the atoms since the force created by the electric field is too small to cause movement of atoms. Devices based on this effect will be discussed later in this chapter. Unlike the solid crystals, the liquid crystals are not very rigid (as discussed in Chapter 1, Section 1.7). *The liquid crystals are characterized by force constants that are quite small and, as a result, a relatively low electric field can cause realignment of the atoms. This allows the optic axis of the liquid crystal to be altered.* This is the basis of all modern LCDs.

There are three main types of distortions that can be produced in a nematic liquid crystal. As shown in Fig. 9.1, these are characterized by i) splay, where a force causes the rod-like molecules to distort as shown in Fig. 9.1a; ii) twist, which is produced by causing a rotation in the alignment of the molecules; and iii) bends, where the crystal is distorted so that a bend is produced in the rod like molecules. The elastic constants defining the energy per unit length to create these distortions are denoted by $K_1$, $K_2$ and $K_3$, respectively. Typical values of these elastic constants are in the range of $10^{-5}$ to $10^{-7}$ dyne.

When an electric field is applied to a liquid crystal, a force acts on the crystal which in turn causes a distortion in the orientation of the optic axis. This can be exploited in liquid crystal cells to cause a modulation of light passing through the cell. Let us remind ourselves that the liquid crystal is characterized by the dielectric constants $\epsilon_\parallel$ and $\epsilon_\perp$ in directions parallel and transverse to the optic axis (known as the director). Also, it is important to note that the dielectric constant is directly related to the polarizability of the material. The higher the polarizability (i.e., dipole moment induced due to an external field) the higher the dielectric constant. We can define the dielectric anisotropy as

$$\Delta\epsilon = \epsilon_\parallel - \epsilon_\perp \qquad (9.1)$$

When an electric field is applied, dipoles are induced in the direction of the optic axis as well as perpendicular to it. If $\Delta\epsilon$ is positive, the molecules will tend to have a lower energy if the optic axis is aligned parallel to the field. If $\Delta\epsilon$ is negative, the lowest energy configuration is one where the optic axis is perpendicular to the field. The torque that is produced by the electric field is given by

$$Torque = \alpha\Delta\epsilon E^2 \qquad (9.2)$$

where $\alpha$ is a geometric factor that depends upon the details of the glass plates of the cell and the placement of the liquid crystal. The distortion torque has to be balanced against the force constants ($K_1$, $K_2$ and $K_3$) discussed above.

Several possible orientations can be used to produce a liquid crystal cell. Some examples for the nematic crystal are shown in Fig. 9.2. We will summarize some important results for these configurations. It may be noted that the twisted nematic configuration is used widely in most consumer electronics products.

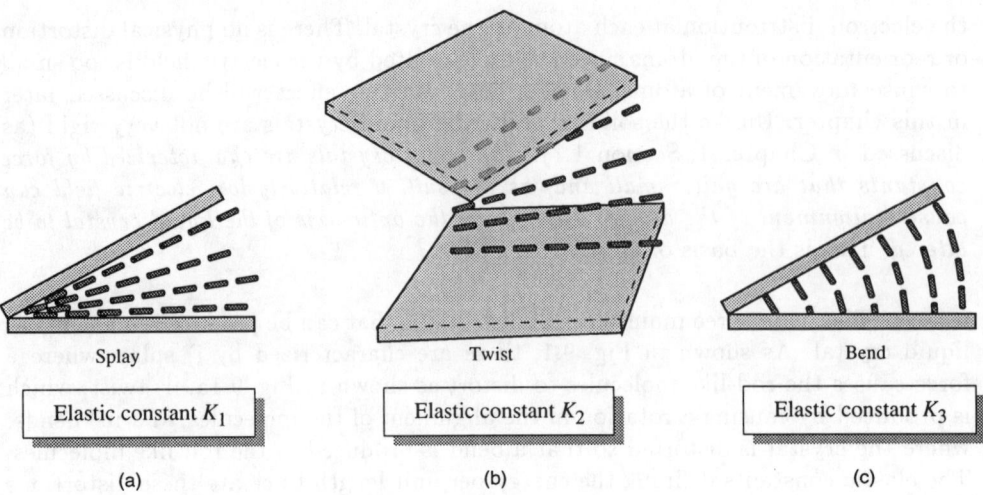

| Splay | Twist | Bend |
|---|---|---|
| Elastic constant $K_1$ | Elastic constant $K_2$ | Elastic constant $K_3$ |
| (a) | (b) | (c) |

Figure 9.1: The three distortions occuring in nematic liquid crystals. The distortions can be created by appropriate forces applied to two glass walls.

To exploit the ability of the field to alter the optic axis, several possible configurations of the liquid crystal cells can be used. The operation of these cells can be understood if it is recalled (from Chapter 2, Sections 2.5 and 2.6) that:
i) when light is travelling along the optic axis, there is no change in the polarization due to the changes in $n_{re}$ and $n_{ro}$ since for this propagation the two are equal;
ii) when light is propagating in a direction perpendicular to the optic axis, the difference in $n_{re}$ and $n_{ro}$ can alter the polarization of light as it travels. In particular, the polarization can change by 90°, if the cell thickness is chosen appropriately;
iii) *when light is propagating in a liquid crystal whose optic axis is slowly twisting, the polarization follows the twist in the crystal.*

Using the facts mentioned above, one uses two cross polarizers to modulate light. The first polarizer prepares the polarization of light as it enters the liquid crystal cell. As light passes through the cell the polarization may alter depending upon the status of the crystal. The second polarizer allows only that component of light to pass whose polarization is aligned with its own. In principle, three types of cells can be designed. We will discuss these three configurations and provide equations without derivation for the threshold voltages needed to cause the change in the optic axis.

### Parallel Orientation

This configuration can be used if $\Delta\epsilon$ is positive. In absence of the electric field the glass plates are prepared so that the optic axis of the liquid crystal lies parallel to the plates as shown in Fig. 9.2a. When a field is applied, the molecules tend to orient themselves so that the optic axis is parallel to the field. A threshold voltage can be defined above which the torque due to the electric field is large enough to overcome the restoring elastic torque. The threshold voltage is given by

$$V_{th} = \sqrt{\frac{\pi K_1}{4\Delta\epsilon}} \qquad (9.3)$$

Typically, the values for the threshold voltage are in the range of 2 to 6 volts.

**Perpendicular Orientation**
This orientation is used if $\Delta\epsilon$ is negative as shown in Fig. 9.2b. The threshold voltage in this case to reorient the optic axis of the liquid crystal is given by

$$V_{th} = \sqrt{\frac{\pi K_3}{4|\Delta\epsilon|}} \qquad (9.4)$$

**Twisted Orientation**
This is the most popular orientation used in display technology. We have already discussed how the twist geometry is used to alter the polarization of the incoming optical signal. The twist is created by rubbing two glass plates in a particular direction, as discussed in Section 1.7. This twist is destroyed when an electric field is applied. The threshold voltage for removing the twist is given by

$$V_{th} = \frac{\pi}{4} \left( \frac{4K_1 + K_3 - 2K_2}{\pi\Delta\epsilon} \right)^{1/2}$$

   In modern displays based on the liquid crystals, the configuration used is the one involving the twisted crystal. A typical liquid crystal cell based on the twisted nematic is shown in Fig. 9.3. The twist in the crystal is produced by preparing two glass plates that have been rubbed by a cloth to establish the orientation directions along which the liquid crystal molecules align themselves. The two plates are then oriented with the rubbing directions perpendicular to each other and with a small space between them. Typical space sizes between the plates are 4 to 10 microns. When liquid crystal is poured into the space, the molecules at near the glass plates orient themselves along the rubbing direction, and the desired twist is produced.

   The threshold voltages discussed above, do not produce an abrupt change in the optic axis from one state to another. The change is non-linear but not entirely abrupt. *Also, it must be kept in mind that even in the transparent state, there is considerable absorption in the liquid crystal.* Focusing on the most widely used configuration of the twisted crystal cell, in an ideal liquid crystal, the transmittance of the liquid crystal cell should change abruptly at $V_{th}$ as shown by the broken curve in Fig. 9.4. However, in real crystals this is not the case, since the crystal twist is relieved gradually and the transmittance change is, therefore, soft as shown. This

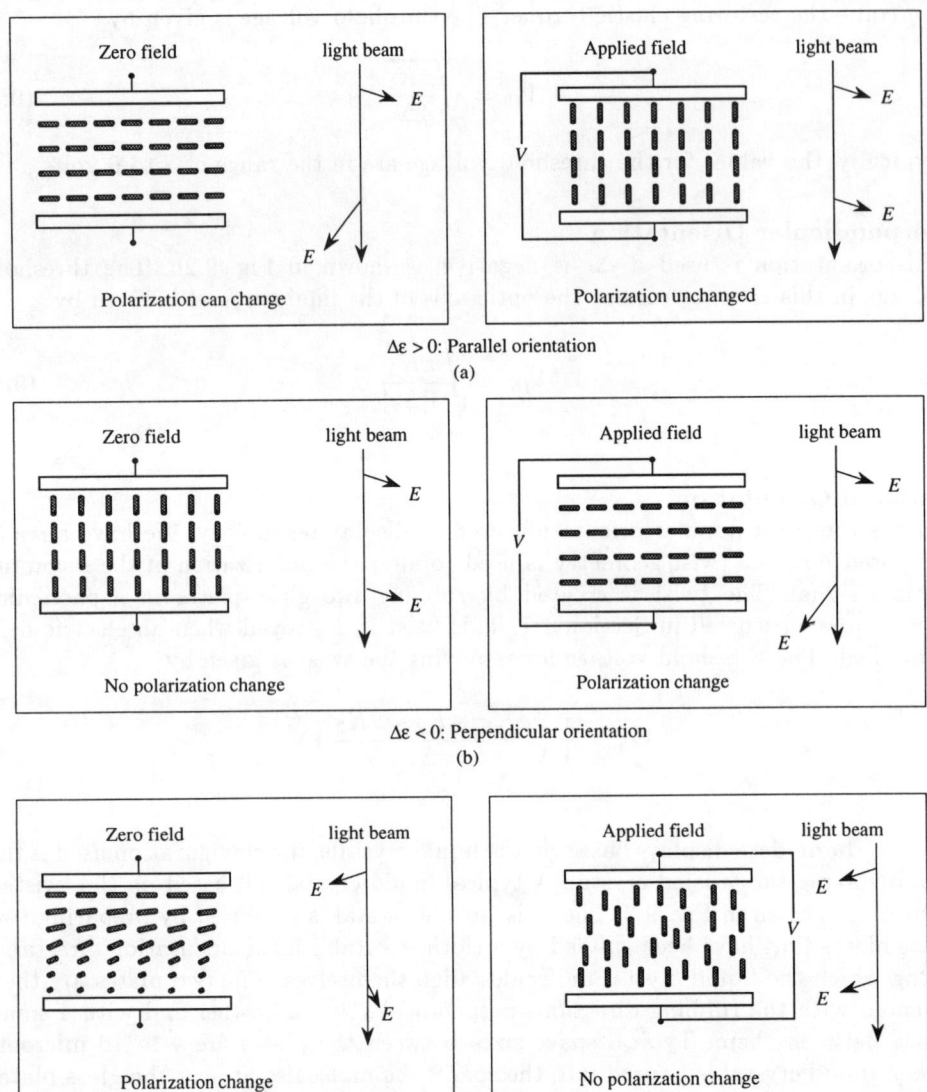

Figure 9.2: Various possible orientations capable of producing "light valves" based on liquid crystals. (a) In the parallel orientation, the optic axis is parallel to the glass plates under zero bias conditions and parallel to the applied field when the field is on. (b) In the perpendicular orientation, the optic axis is initially perpendicular to the glass plates and upon application of a bias, is parallel to them. (c) In the twisted configuration, the twist present in the absence of an applied bias is removed by the field.

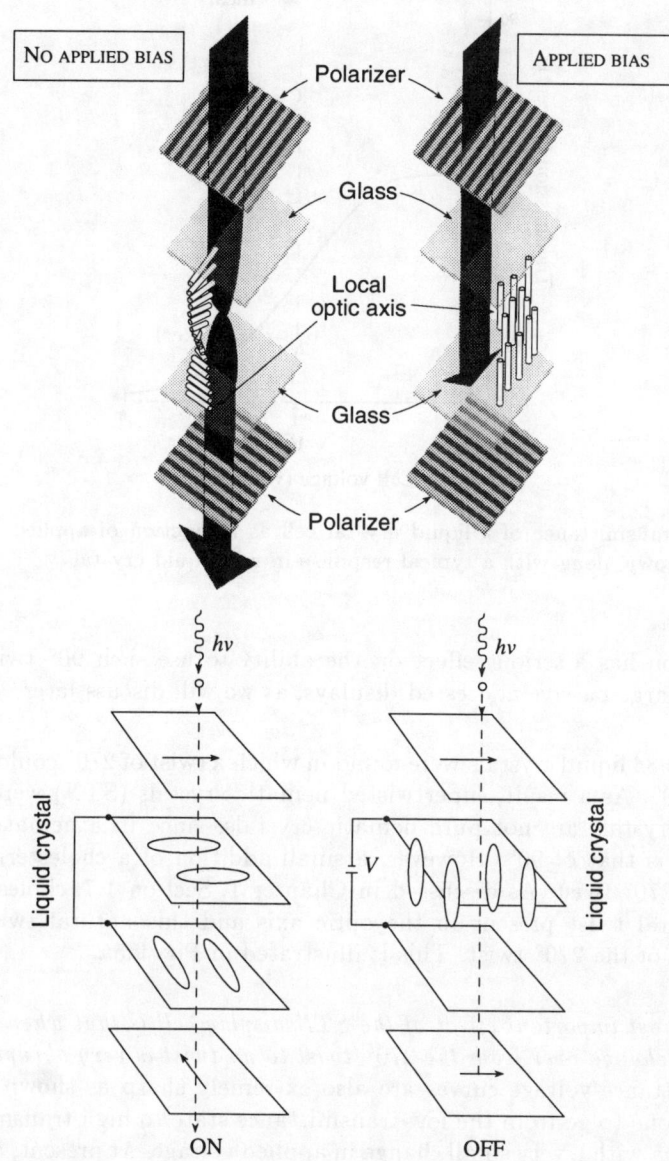

Figure 9.3: A schematic of the operation of a twisted nematic liquid crystal cell. In the upper figure, the arrangement of the various components of the cell are shown. In the lower figure, a schematic of the polarization changes in the cell are shown.

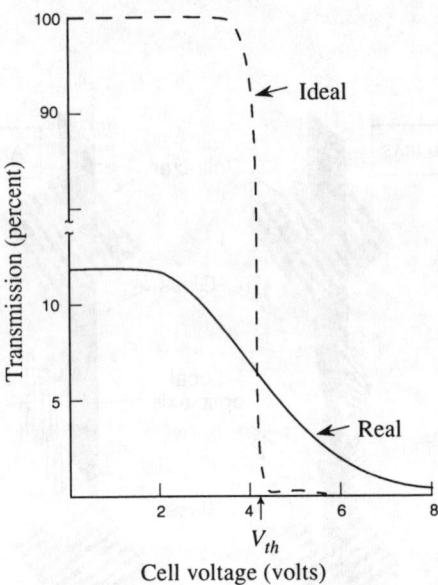

Figure 9.4: Transmittance of a liquid crystal cell as a function of applied bias. An ideal response is shown along with a typical response in real liquid crystals.

soft transition has a serious effect on the ability to use such 90° twisted nematic crystals for large passive addressed displays, as we will discuss later.

In 1984 liquid crystals were found in which a twist of 270° could be produced instead of 90°. As a result, supertwisted nematic crystals (STN) were possible. In fact, these crystals are not pure nematic crystals, since in a nematic crystal the stable twist is that of 90°. However, a small addition of a cholesteric component stabilizes a 270° twist. As discussed in Chapter 1, Section 1.7, cholesteric crystals have a natural twist present in the optic axis and this natural twist allows the stabilization of the 270° twist. This is illustrated in Fig. 9.5a.

*A most important effect of the STN display cell is that when a potential is applied, the change over from the 270° twist to no twist is very abrupt.* As a result, the transmittance-voltage curves are also extremely sharp as shown in Fig. 9.5b. This allows one to go from the low transmittance state to high transmittance state and vice versa with a very small change in applied voltage. At present, a wide variety of STN crystals are in use with twists ranging from 180° to 270°. The key attraction of all these structures is the extremely sharp transmittance—voltage curve.

While the operation of a single liquid crystal cell is quite simple, the scaling process to produce a display with a million cells is an enormous challenge. In the next section we will discuss these challenges and how they are being resolved.

**EXAMPLE 9.1** A nematic crystal is used in the parallel geometry. It has a splay elastic constant of $K_1 = 10^{-10}$ Newton and the dielectric anisotropy is $\Delta \epsilon = 5\epsilon_o$. Calculate the threshold voltage for the device.

The threshold voltage is given by

$$V_{th} = \left[ \frac{\pi (10^{-10}\ N)}{4(5 \times 8.85 \times 10^{-12}\ F/M)} \right]^{1/2}$$

$$= 1.33\ V$$

**EXAMPLE 9.2** A twist cell is made from a liquid crystal with the following properties:

| | |
|---|---|
| Dielectric anisotropy | $\Delta \epsilon = 4\epsilon_o$ |
| Splay elastic constant | $K_1 = 5 \times 10^{-11}\ N$ |
| Twist elastic constant | $K_2 = 8 \times 10^{-11}\ N$ |
| Bend elastic constant | $K_3 = 10^{-9}\ N$ |

Calculate the threshold voltage.

The threshold voltage is

$$V_{th} = \frac{\pi}{4} \left[ \frac{(4 \times 5 \times 10^{-11}) + (10^{-9}) - (2 \times 8 \times 10^{-11})\ N}{\pi (4 \times 8.85 \times 10^{-12}\ F/M)} \right]^{1/2}$$

$$= 2.4\ V$$

## 9.3  CHALLENGES IN SCALING TO A DISPLAY SCREEN

$\longrightarrow$  The chief attraction for liquid crystal devices is their potential application in large displays where one may have up to a million liquid crystal cells. The area of the display may be 25 centimeters by 25 centimeters and more. This offers several important challenges which are both physics related and processing related. In the next subsection we will discuss one important challenge that arises in a large matrix array—the challenge of addressing the individual pixels.

### 9.3.1  The Pixel Addressing Challenge

If LCD technology is to effectively compete with CRT technology, it has to offer comparable resolution and picture quality. It should offer the capability of color

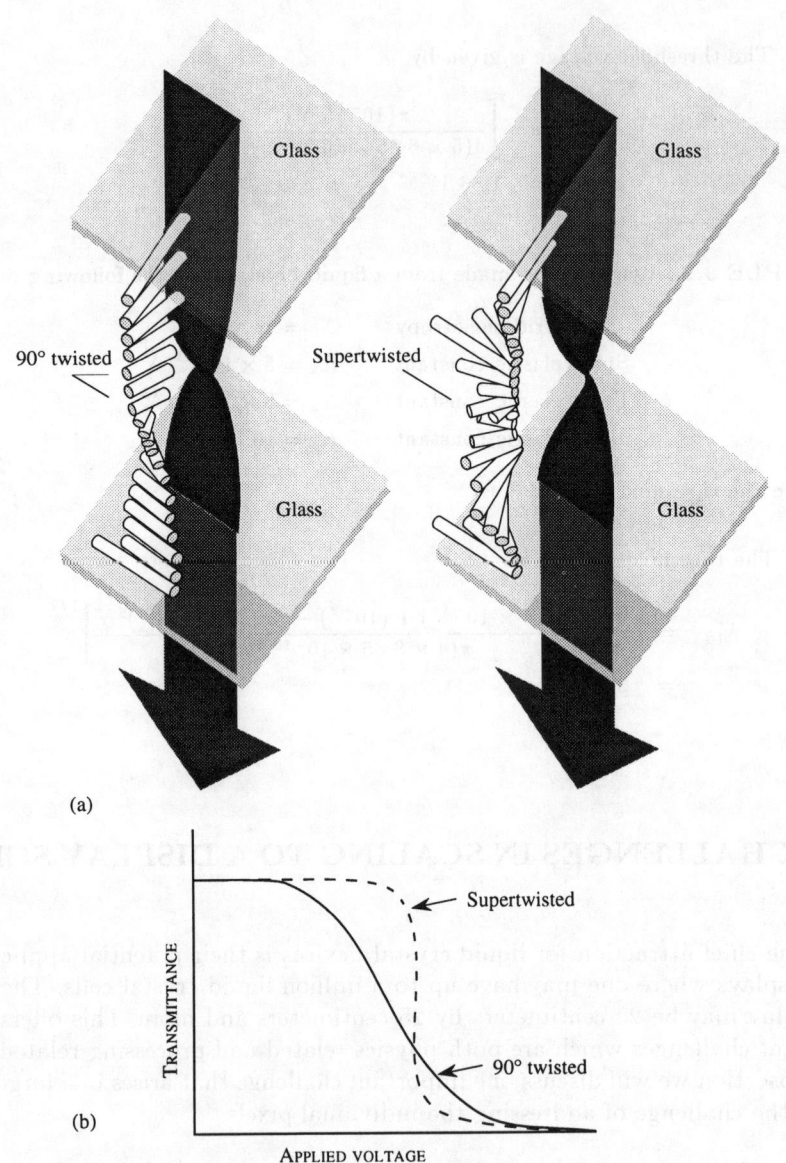

(a)

(b)

Figure 9.5: (a) A comparison of a 90° and 270° twist in a liquid crystal. (b) The transmittance-voltage relation in a supertwisted nematic crystal cell.

display as well as gray scale display. This requires one to have as many as a million pixels (picture elements) on the display. How does one address such a large number of elements so as to present a flicker-free image to a human eye? In some sense, the complexity of the problem is similar to that encountered in dynamic random memory (DRAM) where one has a similar number of elements that have to be addressed. However, the display problem is considerably more cumbersome because of the nature of the transmission- voltage (T-V) characteristics which are not sharp for the 90° twisted liquid crystals. Also, unlike the DRAM, one does not simply have a 0 or 1 state since gray scale may need to be incorporated along with color. If red, green, blue (RGB) display is needed, three pixels are placed together with appropriate dyes. All these elements must be addressed and refreshed, say, 30 times a second, to present a continuous image to the eye. *An important point to keep in mind in the liquid crystal display is that a voltage level has to be maintained between the two plates enclosing the liquid crystal.*

A brute force approach to addressing an array of pixels could be to individually access each pixel. However, this is impossible when the array size increases beyond a few hundred pixels. The approach used is then to place the elements on a matrix grid as shown in Fig. 9.6 and address each pixel one by one by applying appropriate voltage sequences to the rows and columns. To discuss the limitations imposed by such an addressing system, let us describe the T-V curve by a simple form shown in Fig. 9.7.

The T-V curve is characterized, for simplicity, by a voltage $V_{th}$, below which the device is "off" and a voltage level $V_{th}+\Delta$, above which the device is "on." A most important point to keep in mind is that this is the *static* picture, i.e., the applied voltage and transmittance corresponds to the dc biasing. In a display screen which is to be addressed at a rapid rate, it is not possible to hold a voltage level for a pixel to satisfy dc conditions. Let us consider this very important point in more detail.

We consider a matrix of size N × N which is to be addressed by applying sequential voltage pulses to $N$ horizontal and $N$ vertical lines. The row signals are applied to one plate surrounding the liquid crystal while the column signals are applied to the other plate. Let us say that the entire display has to be refreshed in a time $T$ (which is typically about 20 milliseconds). As a result, each pixel receives a voltage pulse only for a time $T/N$ as shown in Fig. 9.8. During the remaining period of time $(T - T/N)$ the pixel has a constant potential applied to it as shown. A strobe signal of value $V_S$ is applied to the rows while the columns receive the display information signal which is chosen to be $+V_D$ (for OFF state) or $-V_D$ (for ON state). The unselected pixels are maintained at $V_D$. The OFF state has a potential $V_S - V_D$ applied to the cell, while in the ON state, the voltage is $V_S + V_D$. *We note that the ON state and OFF state are different only over a time $T/N$ as shown in Fig. 9.8.* Obviously, in such a scheme, as the size of the matrix increases, the difference between the states starts to decrease. It is then intuitively obvious

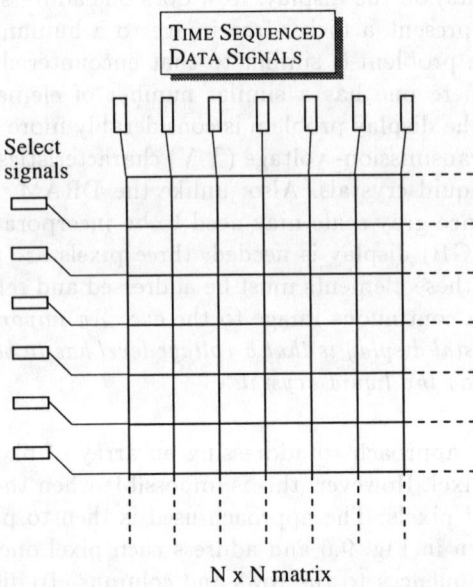

Figure 9.6: A matrix addressing approach used to address individual liquid crystal cells.

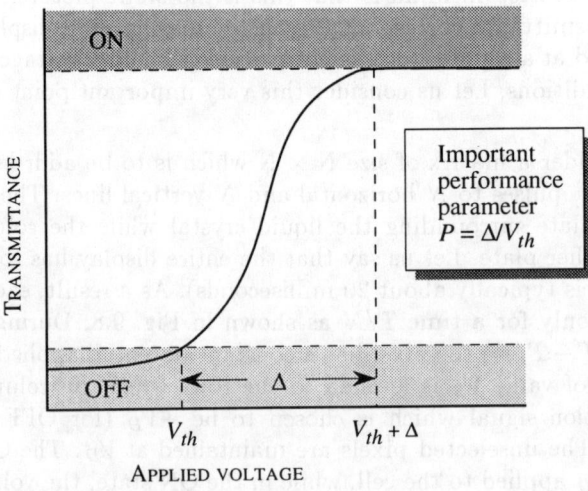

Figure 9.7: Static transmission-voltage curves (T-V) for a typical liquid crystal cell.

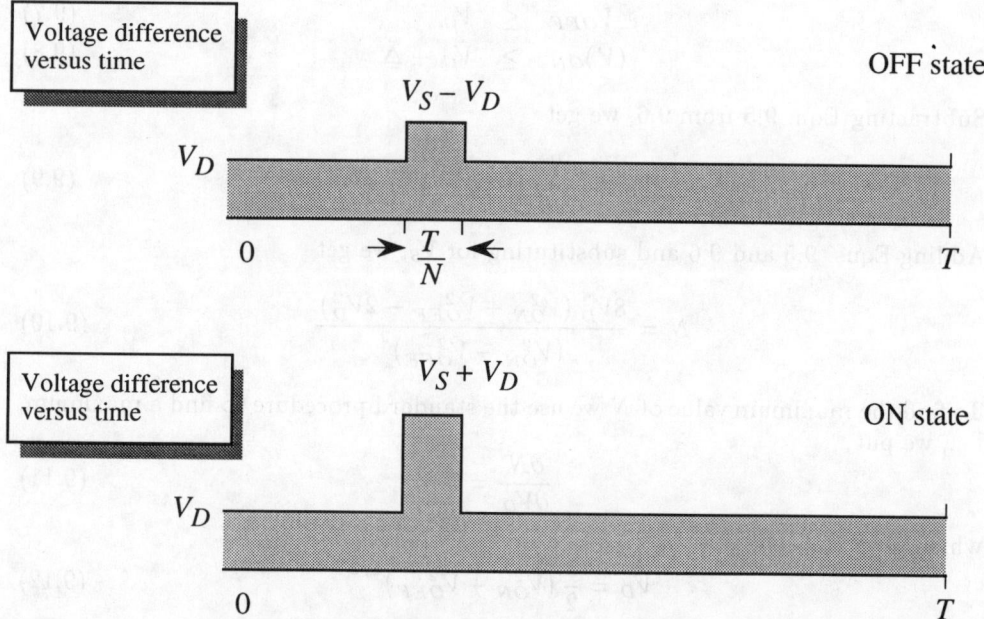

N x N MATRIX ADDRESSING BY N ROWS AND N COLUMNS

• Pixel transmittance responds to the difference between row and column voltages.
• A strobe signal $V_S$ is applied to the rows.
• Information signal $+V_D$ (for OFF) or $-V_D$ (for ON) is applied to the columns.
• Unselected pixels are maintained at a voltage difference of $V_D$.

Voltage difference versus time

OFF state

$V_S - V_D$

$V_D$

0    $\dfrac{T}{N}$    $T$

Voltage difference versus time

ON state

$V_S + V_D$

$V_D$

0        $T$

Figure 9.8: Important issues in matrix addressing of a liquid crystal display. The voltage levels across an OFF and ON pixel are shown during one cycle.

that the parameter $\Delta$ of Fig. 9.7, i.e., the parameter describing the non-linear nature of the T-V curve has to be very small if a large sized matrix is to be addressed. This intuitive expectation is confirmed by simple mathematics.

Before we analyze mathematically the relation between the parameter $\Delta$ and the matrix size defined by $N$ we need to ask and answer an important question: If a voltage pulse having a shape as shown in Fig. 9.8 is applied to a liquid crystal cell, how does one determine the transmittance? Considerable experimental and theoretical effort has been expended to answer this question. A good approximation appears to be that the *device response is determined by the root mean square value of the voltage pulse.* Based upon this important finding, we are now ready to see how the parameters $\Delta$ and $N$ are related. Using a bar over a potential to indicate rms value, we have, for the voltage pulses over a time period $T$ for the OFF and ON states

$$\bar{V}_{OFF}^2 = \frac{1}{N}(V_S - V_D)^2 + V_D^2 - \frac{V_D^2}{N} \tag{9.5}$$

$$\bar{V}_{ON}^2 = \frac{1}{N}(V_S + V_D)^2 + V_D^2 - \frac{V_D^2}{N} \tag{9.6}$$

For the device to operate, we must have

$$\bar{V}_{OFF} \leq \bar{V}_{th} \tag{9.7}$$

$$(\bar{V})_{ON} \geq \bar{V}_{th} + \bar{\Delta} \tag{9.8}$$

Subtracting Eqn. 9.5 from 9.6, we get

$$V_S = \frac{N}{4V_D}\left[\bar{V}_{ON}^2 - \bar{V}_{OFF}^2\right] \tag{9.9}$$

Adding Eqns. 9.5 and 9.6 and substituting for $V_S$, we get

$$N = \frac{8V_D^2\left(\bar{V}_{ON}^2 + \bar{V}_{OFF}^2 - 2V_D^2\right)}{\left(\bar{V}_{ON}^2 - \bar{V}_{OFF}^2\right)^2} \tag{9.10}$$

To find the maximum value of $N$ we use the standard procedure to find a maximum, i.e., we put

$$\frac{\partial N}{\partial V_D} = 0 \tag{9.11}$$

which gives

$$V_D = \frac{1}{2}\left(\bar{V}_{ON}^2 + V_{OFF}^2\right)^{1/2} \tag{9.12}$$

This gives us the maximum number of rows or columns

$$N_{max} = \left[\frac{\bar{V}_{ON}^2 + \bar{V}_{OFF}^2}{\bar{V}_{ON}^2 - \bar{V}_{OFF}^2}\right]^{1/2} \tag{9.13}$$

Using the values of $\bar{V}_{ON} = \bar{V}_{th} + \bar{\Delta}$ and $V_{OFF} = \bar{V}_{th}$, and defining a parameter

$$P = \frac{\bar{\Delta}}{V_{th}} \tag{9.14}$$

we get, for the optimization case

$$\boxed{N_{max} = \left[\frac{(1+P)^2 + 1}{(1+P)^2 - 1}\right]^2} \tag{9.15}$$

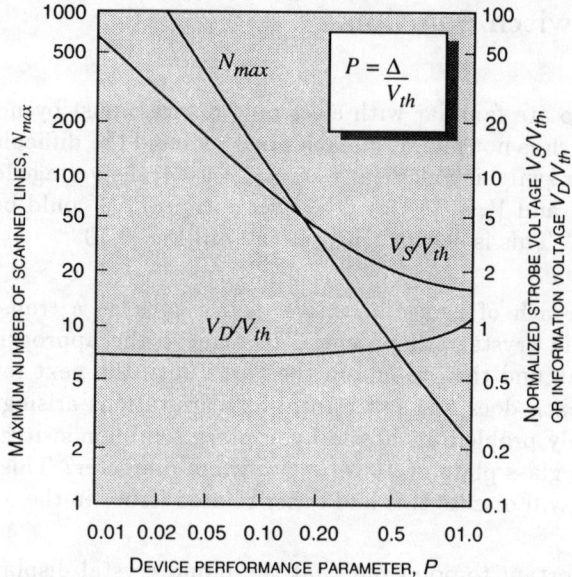

Figure 9.9: A plot of the maximum scanning capability and drive requirements in a liquid crystal display as a function of the device parameter $P$. (After P. M. Alt and P. Pleshko, *IEEE Trans. on Elect. Dev.*, Ed-21, 146 (1974).)

$$\frac{V_D}{V_{th}} = \frac{1}{2}\left[(1+P)^2 + 1\right]^{1/2} \tag{9.16}$$

$$\frac{V_S}{V_{th}} = \frac{1}{2}\left\{\frac{((1+P)^2 + 1)^3}{(1+P)^2 - 1}\right\}^{1/2} \tag{9.17}$$

Note that $V_D$ is very close to the threshold voltage $V_{th}$, but as $N$ increases, the value of $V_S$ increases rapidly. The value of $N_{max}$ depends critically on the rate given by P. Obviously, for a large value of $N_{max}$ one needs a small $\Delta/V_{th}$ value. In Figs. 9.9 we show a general plot of the dependence of the device performance on P. This figure is adapted from P. M. Alt and and P. Pleshko who first provided a quantitative analysis of this important issue for LCDs.

From Fig. 9.9 it is clear that it is quite difficult to increase the value of $N_{max}$ if $\Delta/V_{th}$ has a value of say, 0.1, which is typical of twisted nematic crystals with a 90° twist. However, as discussed earlier supertwisted nematics have a very small value of $P$ and it is possible to increase $N$ to approach several hundred.

## 9.3.2 The Switch Solution

$\longrightarrow$ Readers who are familiar with electronic devices must by now be wondering why a simple switch is not placed at each pixel to avoid the difficulty outline above. Indeed, if a proper switching element were to be used, the voltage level applied (i.e., $V_S - V_D$ for OFF and $V_S + V_D$ for ON) over a time $T/N$ could be maintained for the entire cycle $T$. This is shown schematically in Fig. 9.10.

The approach of using an active device such as a transistor or a diode to allow the liquid crystal cell potential to achieve the appropriate value of the information signal, and then maintain the value until the next refresh cycle. As a result, this approach does not suffer from any limitations arising from the values of $\Delta/V_{th}$. The only problem is, how do you place a million switches—one for each pixel—on a large glass plate of 10 inches or more diameter? This is an important challenge and we will discuss this and other related issues in the following sections.

It is important to point out that the liquid crystal display where switches at each pixel are not employed are called passive liquid crystal displays. Displays where a switch is incorporated at each pixel are called active matrix liquid crystal displays (AMLCDs). The reader may have seen these words in glossy ads for laptop computers with the AMLCD based laptops costing a thousand dollars more than the ones with a passive display.

**EXAMPLE 9.3** In a particular twisted nematic liquid crystal cell, the value of $P = \Delta/V_{th}$ is 0.2 while in sample of STN it is 0.05. Calculate the maximum size of the matrix that can be addressed if the passive addressing scheme is to be used.

The maximum number of lines that can be used is

$$N_{max} = \left[ \frac{(1+P)^2 + 1}{(1+P)^2 - 1} \right]^2$$

For the 90° twist case, we have

$$N_{max}(90°) = \left[ \frac{1 + 0.2)^2 + 1}{(1 + 0.2)^2 - 1} \right]^2$$

$$= 30.75$$

Thus the display size has to be less than 30 × 30. For the STN case, we have

$$N_{max}(STN) = 420.75$$

This size could reach 420 × 420 or 0.177 million pixels.

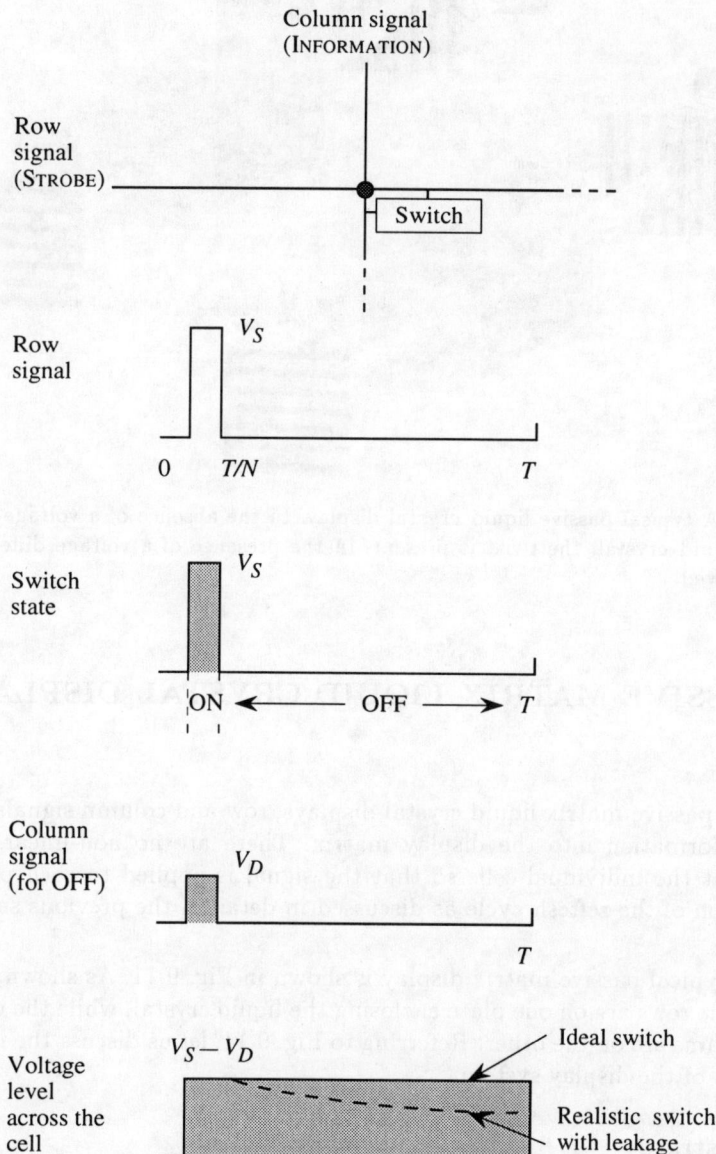

Figure 9.10: The input signal from the row and column voltages to a pixel and the voltage level on the pixel in the presence of a switch.

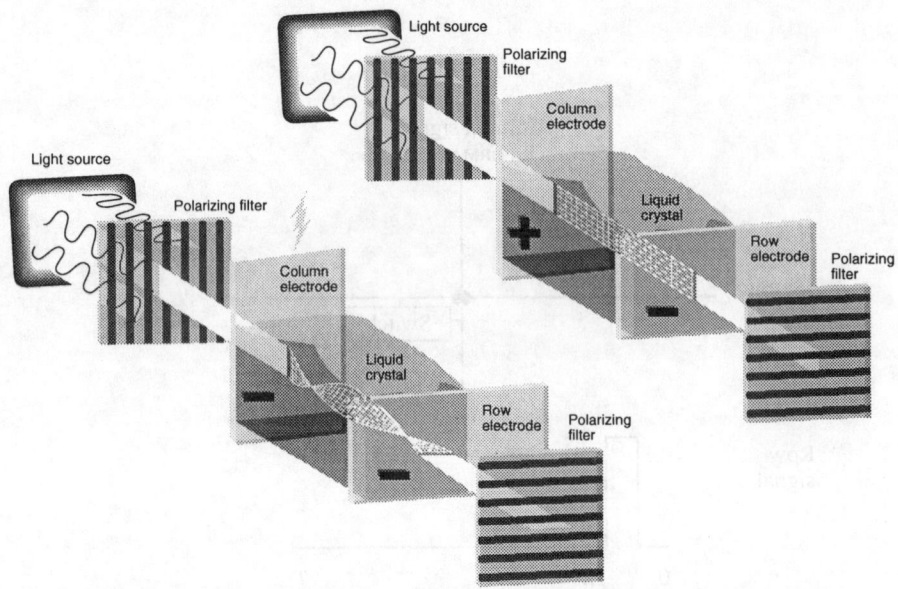

Figure 9.11: A typical passive liquid crystal display. In the absence of a voltage difference across the liquid crystal, the twist is present. In the presence of a voltage difference, the twist is removed.

## 9.4   PASSIVE MATRIX LIQUID CRYSTAL DISPLAY

$\longrightarrow$   In the passive matrix liquid crystal displays, row and column signals are used to input information into the display matrix. There are no non-linear elements (switches) at the individual cells so that the signal is applied to each pixel for a small fraction of the refresh cycle as discussed in detail in the previous section.

A typical passive matrix display is shown in Fig. 9.11. As shown, the electrodes for the rows are on one plate enclosing the liquid crystal, while the electrodes for the columns are on the other. Referring to Fig. 9.11, let us discuss the important components of the display system.

### Glass Substrate
The glass substrates that are used for the display have to be highly polished and clean. The topography of the substrates should be extremely flat, especially for the passive matrix displays. This is because the STN crystals have a high twist angle and are extremely sensitive to tiny bumps on the substrate. These bumps could cause the twist to unwind. The substrates are rubbed with a velvet cloth to cause the liquid crystal to align along the preferential (rubbing) direction.

### Transparent Electrodes
For obvious reasons, the conductors across which the voltages are to be applied on the cell have to be transparent. This technology is well developed and is often used in semiconductor detectors. Indium tin oxide (ITO) is the material of choice.

### STN Liquid Crystal
The supertwisted nematic liquid crystal is placed between the substrates after the edges have been sealed by special epoxy. Tiny spacer balls are used to create the proper gap between the plates. Typical gaps are of the order of 4-6 $\mu$m. The gap alignment is very critical when STN crystals are used, for reasons discussed above. Tolerances of $< 0.1\mu$m are needed.

### The Backlight
Unlike the semiconductor LED or laser diode, the liquid crystal cell does not produce any light; it simply modulates the light. Therefore, a backlight is provided to flood the backplane of the display. The back polarizer allows the light polarized along the allowed direction to pass through. The rest of the light is wasted. The backlight is one of the biggest power consumption sources in a liquid crystal display.

### Color Filters
If the display is to be a color display, it is necessary to include color filters in front of each pixel. Red, green and blue color filters are used. The three pixels for each color are individually controlled and are in close proximity so that, to the eye, a single color appears. The three colors can produce any color that is desired.

The passive matrix liquid crystal displays are widely used for a variety of display applications. As noted earlier, by using STN crystals the display size $N$ can approach several hundred lines. This is adequate for most applications (a computer display requires 640 × 480 pixels).

## 9.5  ACTIVE MATRIX LIQUID CRYSTAL DISPLAYS

$\longrightarrow$  As we have discussed previously, the difference between the active and passive display is the use of a switching device at each pixel. This switch allows the signal voltage to be applied to the liquid crystal cell for the entire cycle time between refreshes. As discussed earlier, this leads to better overall performance and most importantly allows one to use a 90° twist in the liquid crystal.

A schematic of the AMLCD is shown in Fig. 9.12. Unlike the passive matrix display, the row and column lines are placed on the same substrate. This creates considerable simplification in the manufacturing process. A detailed sketch of the

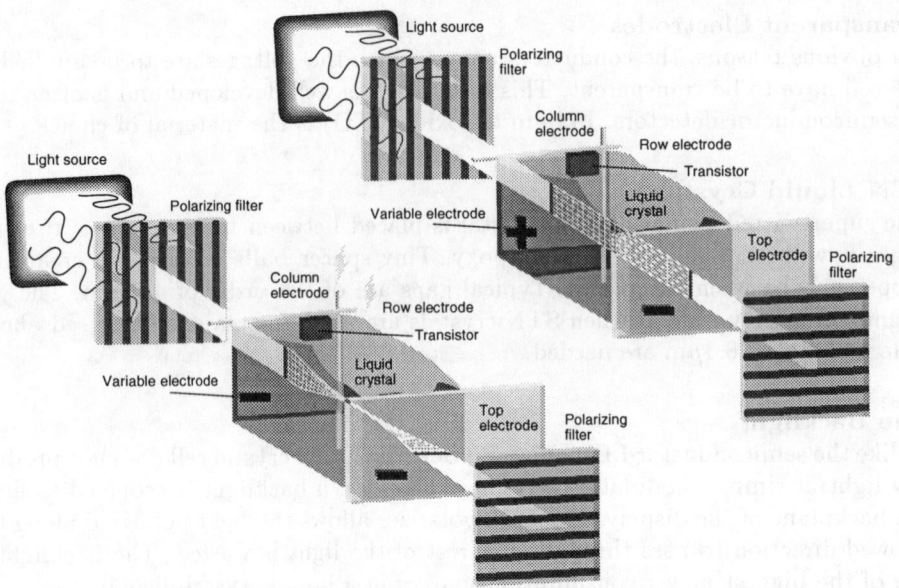

Figure 9.12: A schematic of an active matrix liquid crystal display.

individual pixel along with a thin film transistor (TFT) is shown in Fig. 9.13. A sample pulse is applied to the gate of a transistor pulling the device into inversion. A data voltage is applied via the column line to the drain (or the source) of the TFT. The source (or the drain) is connected to a storage capacitor which holds the applied voltage once the gate pulse is removed.

The TFT must have the properties that its resistance is very low when the gate bias is ON and very high when the gate bias is OFF. This allows the storage capacitor to charge to the applied potential during the time the signal is on. Also, the capacitances $C_{gs}$ (the gate source), $C_{gd}$ (gate drain), $C_S$ (storage), $C_{LC}$ (liquid crystal cell) and $C_{parasite}$ (with adjacent lines) should be such as to allow minimum charge leakage during a cycle time. Since the success of the AMLCD depends critically upon the TFT, it is important to examine this device in some detail.

## 9.5.1   The Thin Film Transistor

The key element in the AMLCD is the thin film transistor (TFT). Up to a million TFTs are required for a full color display, and in this sense the complexity of the system is comparable to the DRAM. However, in reality the complexities are greater because unlike the DRAM, the devices are not fabricated on a crystalline material.

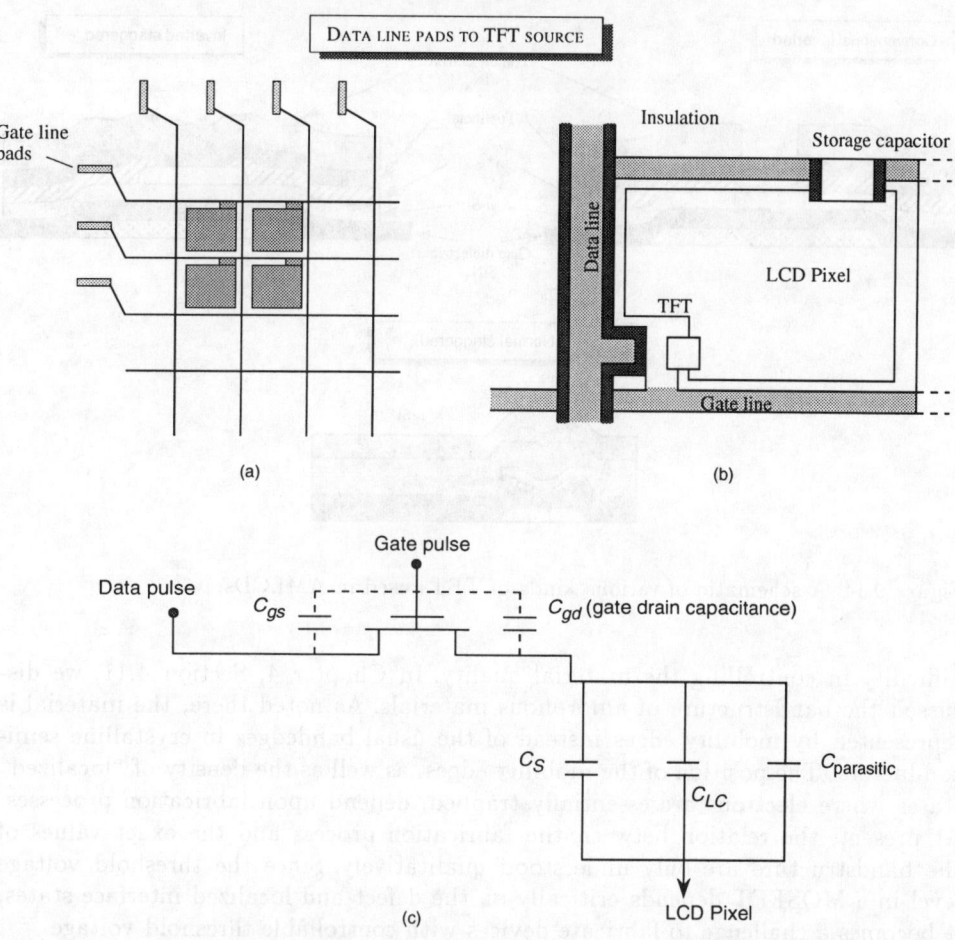

Figure 9.13: (a) A schematic of a layout for a AMLCD pixel array. (b) A detailed view of the pixel. (c) An equivalent circuit of the TFT and the cell storage capacitor.

Either polysilicon or amorphous silicon (a-Si) is used to fabricate the device which is basically an insulated gate field effect transistor. Polycrystalline Si devices have better performance because of superior mobilities of the electrons. However, the processing of the devices requires high temperatures that are not compatible with a glass substrate. As a result, a more expensive quartz substrate is used for polysilicon FETs. Small displays such as viewfinders do use AMLCDs based on polysilicon devices. However, larger area displays rely on glass substrates and on the use of a-Si TFTs.

Typical structures used for TFTs are shown in Fig. 9.14. In the inverted structures, the gate is deposited first, while in the normal structure the semiconductor is deposited first. The TFT offers a tremendous challenge because of the

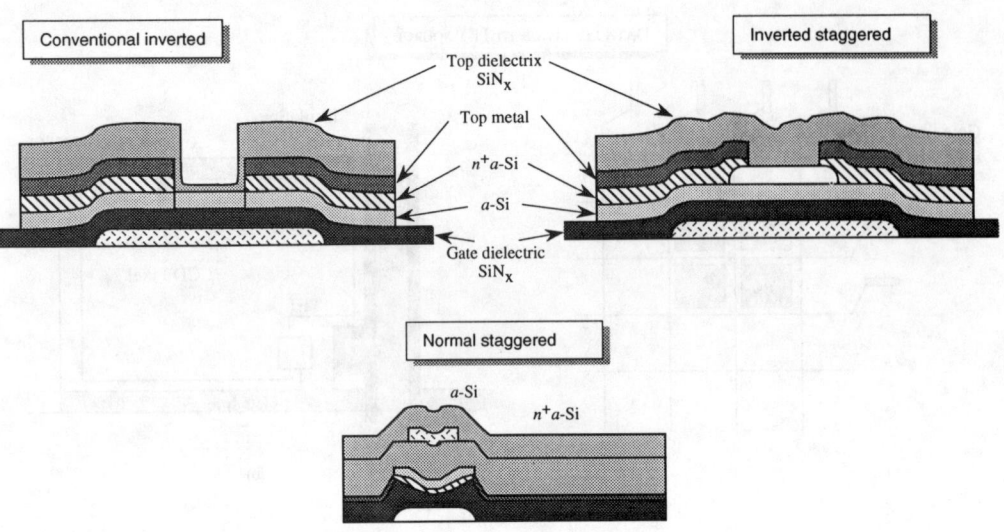

Figure 9.14: A schematic of various kinds of TFTs used in AMLCDs.

difficulty in controlling the material quality. In Chapter 4, Section 4.11, we discussed the bandstructure of amorphous materials. As noted there, the material is represented by mobility edges instead of the usual bandedges in crystalline semiconductors. The position of the mobility edges, as well as the density of "localized" states where electrons are essentially trapped, depend upon fabrication processes. At present, the relation between the fabrication process and the exact values of the bandstructure are only understood qualitatively. Since the threshold voltage level in a MOSFET depends critically on the defect and localized interface states, it becomes a challenge to fabricate devices with controllable threshold voltage.

Another important challenge for the TFT is the low mobility of electrons in the transistor. In crystalline MOSFETs, the mobility of electrons in the channel is $\sim 600$ cm$^2$ V$^{-1}$ s$^{-1}$ (in pure silicon it is $\sim 1100$ cm$^2$ V$^{-1}$ s$^{-1}$), but in a-Si TFTs, it is only 1 - 10 cm$^2$ V$^{-1}$ s$^{-1}$. Moreover, the mobility has a strong dependence upon the carrier density in the channel. The low mobility results in a considerable voltage drop across the 3-5 $\mu$m TFT channel and, as a result, the full data line potential is not applied to the storage capacitor. This naturally results in higher power consumption. Considerable research is currently focussing on improving the TFT performance.

As can be judged from the discussion in this section, while many of the processing steps for the TFTs are quite similar to those in crystalline MOSFETs, numerous challenges remain. Most of the challenges arise from the poor electronic properties of the amorphous semiconductor. However, since for large displays one has to use either polycrystalline or amorphous materials, a number of research groups are addressing these important issues.

## 9.6   CHALLENGES FOR DISPLAY TECHNOLOGY

$\mathcal{R}$   The display technology based on liquid crystals (and other possible materials) holds the key to an explosively growing market which, according to some forecasts could be as big as the silicon integrated circuit market in the early part of the $21^{st}$ century. If you were to buy a laptop computer in 1995, the most expensive component would be the display—more expensive than the RAM, microprocessor, the disc drive, or the keyboard. And, of course, without the display screen, there would be no laptop market. Portable and programmable displays will create new products as time passes and as the quality of the displays improve. Challenges faced in this technology are: i) high power consumption to drive the display; ii) poor transmission of light in the ON state with the result that the display is either very dim or the backlight has to be very strong; iii) poor viewing angle of the display; iv) size limitations on the display screen. All these challenges require thrusts in new materials, improved devices and improved processing.

In Fig. 9.15 we show a summary of the areas that need to be addressed to improve the displays to a level that they will eventually replace the venerable CRT. Let us take a brief look at approaches needed to advance the technology. As shown in Fig. 9.15, we will characterize the approaches as material and device related and processing related.

**Material and Device Related Challenges**
From the point of view of improving the display performance, several issues need to be addressed to improve the material and device response. We will briefly discuss these:

• *Novel Liquid Crystals*: The performance of the liquid crystals is obviously critical in the performance of the overall display. Materials which have a low threshold voltage $V_{th}$ and a strong non-linearity (small $\Delta/V_{th}$) are needed. A considerable amount of work is being devoted to develop new kinds of liquid crystals which are highly stable but have these properties. In addition to low $\Delta/V_{th}$ value, the crystals should also not absorb light themselves. The absorbed light is wasted and requires the backlight to be stronger.

An important issue in choosing the liquid crystal is resistivity of the material. For AMLCDs, the liquid crystal cell should have a large RC time constant so that the voltage level applied can last until the next refresh cycle ($\sim 20$ ns).

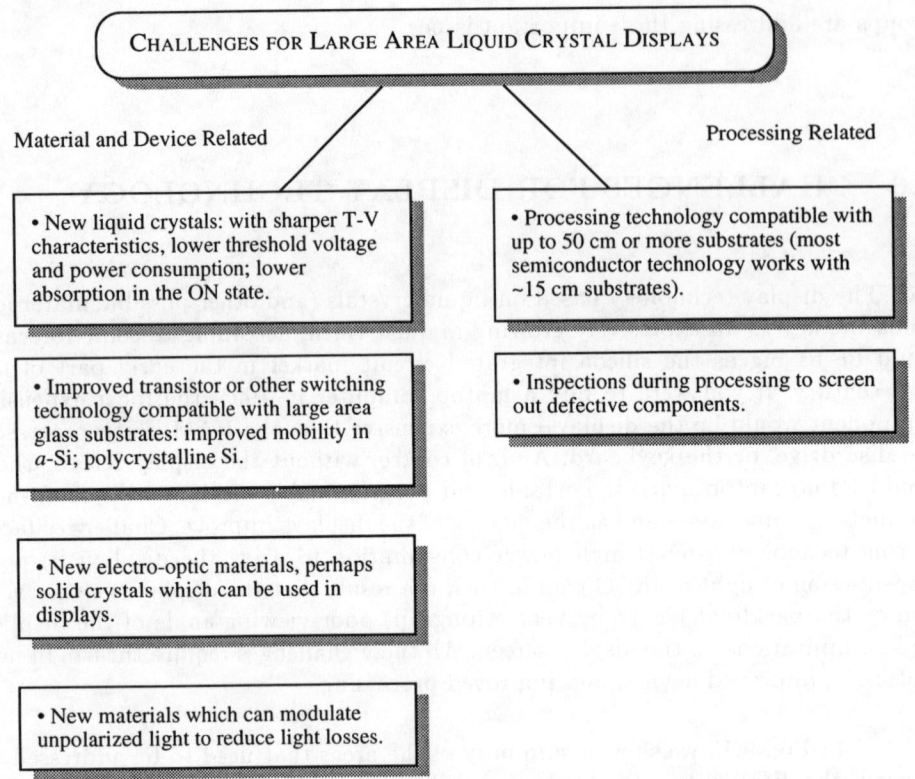

Figure 9.15: Challenges faced by the display technology.

- *Improved Amorphous and Polycrystalline Films*: As we have discussed in the previous section, the performance of the TFT is quite important in overall performance of the display. Issues that are currently being addressed are threshold voltage control and improved mobility in the channel.

- *New Materials for Displays*: The liquid crystals that are currently being used in displays have to be poured into the two glass plates. If, on the other hand, solid films could be developed that could be simply evaporated onto the substrate, the manufacturing process would benefit considerably. At present, however, there are no such materials that can compete in cost and power requirements with liquid crystals.

- *Novel Materials That Have Polarization Independent Response*: The backlight used in the display system emits light which is unpolarized. Since the LCD depends upon selecting a particular polarization, a large fraction of light is wasted. Also, this selection process requires front and back polarizers. If materials can be found that can modulate the optical intensity regardless of the polarization of light, the overall efficiency of the display system can greatly improve.

**Processing Related Challenges**

The processing steps in a display structure are quite similar to those in modern electronic ICs. The key difference is the size of the substrate. In Si ICs, the substrate size rarely goes above 15 centimeters. However, for displays, the size can be 30 centimeters and more. Certainly, if LCDs are to be used in home TVs, desktop computers, etc., the size of the glass substrate would have to be 50 centimeters and higher. This necessitates the development of all the processing tools (i.e., film deposition systems, steppers, etchers, etc.) that can operate on such a large scale. Step and repeat patterning techniques are used at present, but these are expensive.

It is well known that, as the substrate size increases, the yield of a process starts to decrease. For example, while a defect density of $10^{-2}$ cm$^{-2}$ can provide a yield of 80% for a 10 centimeter display, the yield falls to 10% for a 30 centimeter display. This places great importance on being able to test the structure during processing so that if an irreparable defect develops, one can abandon the process before going to completion.

## 9.6.1   Field Emission Displays

$\mathcal{R}$ Liquid crystal based flat panel display technology, while being commercially the most successful, suffers from a number of problems that have been already discussed. An important drawback is that it lacks the brightness of a CRT, since there is no light emission at the pixel level. There is considerable research effort to produce flat panel displays based on electron emitters and phosphors. These displays promise high contrast and excellent viewing angles. In this subsection we will take a brief look at this technology which is, perhaps, an important future technology. We will also mention the more futuristic technology based on novel polymer materials which can now be tailored to emit light. Advances in electronic devices based on polymers are allowing the possibility of addressable displays based on the polymer technology.

In Chapter 5, Section 5.7.1, we have discussed properties of phosphors that are widely used in display technologies based on the CRT. While the phosphors are, of course, quite compatible with flat panel display requirements, the electron gun that is used to excite the phosphor pixels is rather bulky. In a conventional CRT, electrons are produced from a cathode and after passing through focusing and deflecting regions, impinge on the phosphor screen. In CRTs, one has either a single electron gun (for monochrome) or three guns (for color) which requires the cathode to be placed at a certain distance from the screen. The system has to be housed in a high vacuum to avoid electron scattering. This, in turn, requires the use of strong and heavy casing for the display screen.

In the field emission display (FED), each phosphor pixel has its own electron source. Electron emission is produced by a potential applied between the gate and the base plate. The drive circuitry considerations are similar for this display as in the LCD displays.

The emission of the electrons results from the extraction of the electrons inside a material to the "vacuum" or free state. Various kinds of emitter tips are being investigated, including metal micro tips, silicon tips, and diamond tips. The electron current is due to the tunneling of the electrons from the emitter tip into the vacuum level. The emitter is formed into a sharp tip to ensure a high emission efficiency. A large number of tips would be used for exciting a single pixel.

While the FED is a very attractive alternative to AMLCDs, serious challenges still remain to be solved in this technology. These include high efficiency tips which can last over a long period of time and consume low power. The fabrication of the tips and integrating the display elements also is a serious challenge. Nevertheless, several companies have demonstrated small scale displays based on electron emission.

## 9.7   NEEDS FOR HIGH SPEED LIGHT MODULATION

$\mathcal{R}$   The liquid crystal cell discussed in the sections above is essentially a light modulator or a light valve. However, while this device is quite adequate for display applications, it cannot be used for optical communication applications. In displays which are seen by humans, the demands of the device are set by our sensory responses. In particular our own response time to external stimuli is 20 to 30 milliseconds. This fact is exploited for motion pictures and television displays. The liquid crystal cell which has switching times of microseconds to milliseconds is therefore quite adequate. However, in communications the device response should be as fast as possible. The reason is that one may be sending an enormous amount of data from one computer to another or millions of voice streams may be sent over the same line. Also, while our own response is slow, we are able to somehow process information in parallel very effectively. Thus we can follow a video film quite easily even though a typical one hour video film may have several gigabits of data. In optical communication the fiber has the potential to carry data at speeds of terabits per second. To fully utilize this enormous bandwidth, one must have high speed modulation devices.

In the last several chapters we have discussed semiconductor devices that can detect and generate light. While these devices are of obvious importance in the information processing age we live in, they are not "intelligent" devices. Intelligence comes from manipulating information—switching, modulating and carrying out arithmetic and logic operations. In the electronic domain, the flip-flops, registers, and logic gates, all based on transistor technology, provide the intelligent devices that run today's computers, microprocessors and even optoelectronic systems.

Consider an optical signal as it carries a data bit from New York and heads to Tokyo. There is obviously no single one to one optical fiber connection between the two end points. The data bit has to go through several repeaters and exchanges as it goes along. The repeaters are there to compensate for the optical power loss suffered in transmission. At present the repeaters convert the optical pulse to an electronic pulse, amplify it and reconvert it to a boosted optical pulse. The same is done in the switching exchanges. Thus all the data manipulation is done by the same electronic devices that run our personal computers (PC's). All this seems rather embarrassing to the proponents of optics and "optical computing," but at present there are few optoelectronic devices that can compete with the transistor in data manipulation. However, intelligent optoelectronic devices are improving rapidly and are beginning to make a serious impact on technology. It may still be a long time before these devices find their way into our PC's, but they will impact high speed communication systems in the near future.

In optical communication, information content is coded in optoelectronic devices as a variation of some aspect of the optical output from an LED or a laser diode (LD). The property of light that is modulated depends upon the particular application and technology available, and could involve the amplitude of the optical signal, the phase of the signal, widths of pulses being sent, etc. Regardless of the coding scheme used, it is clear that modulation of light is a critical ingredient of optoelectronic technology. In this regard the optoelectronic devices are somewhat lacking when compared to electronic devices.

In the electronic devices (MOSFET, MESFET, MODFET, BJT, HBT, etc.), the electrical signal is readily modulated by a gate signal or a base signal. The gate or the base is an integral part of the device. The device dimensions are quite small and allow modulation frequencies of up to 20 GHz in advanced silicon technology and up to several hundred gigahertz in advanced heterostructure technologies. When compared with electronic devices, the size of optoelectronic sources is rather huge. We have already discussed the limitations for LEDs in high speed applications. The laser diode can, in principle, overcome some of the problems but, as discussed in Chapter 8, is still limited to 20-30 GHz. Additionally, the laser diode does not have a simple "gate" from which one can control the light output.

There are two schemes used to modulate the optical signals in LEDs or LDs. The first one is direct modulation in which an electronic circuit is designed to simply modulate the current injected into the device. Since the light output is controlled by the injected current, one has the desired modulation. The "driver" for this direct modulation may be an FET or an HBT. Because of the different structural nature of the LD and the electronic device, it is not a trivial matter to fabricate these devices on the same chip. Thus such circuits are usually based on hybrid technology. The goal to build the driver and the source on the same substrate and develop an OEIC technology for the transmitter is being pursued actively in a number of labs.

Direct modulation of the laser is simple, which is the biggest attraction of this scheme as shown in Fig. 9.16. Note that, compared to electronic devices, the circuit is by no means simple or easily "manufacturable." It is simple when compared to other modulation approaches we will discuss later. However, direct modulation has several problems including the upper modulation frequencies (which are $\sim 40$ GHz) and shift in emission frequency.

External modulation of lasers offers the advantages and disadvantages listed in Fig. 9.16. The key disadvantage is that the modulator is usually large on the scale of microelectronic devices and is usually not a part of a simple integrated circuit. Not only that, it is usually fabricated from materials that are not compatible with semiconductor technology. Recently, however, this is changing with the use of quantum well systems based upon the same semiconductors that are used in laser diodes.

In the external modulation scheme, the light output passes through a material whose optical properties can be modified by an external means. Depending upon the means used, one can have electro-optic, acousto-optic, or magneto-optic modulators. The electro-optic effect is most widely used for high speed applications and is most compatible with modern electronics.

## 9.7.1   Figures of Merit for Modulators

Optical modulators are used to code information onto an optical signal by altering the amplitude, phase, or intensity of the signal (see the next chapter, Section 10.4). Depending upon the application, several figures of merit have been identified for modulators. These are broadly characterized as: modulation efficiency, modulation bandwidth, insertion loss, power consumption of the modulator and isolation between different channels. We will briefly discuss these figures of merit.

• *Modulation Efficiency*: To define the modulation efficiency, $\eta$, one needs to first identify the nature of modulation being carried out. For intensity modulation where the output light intensity varies between the maximum intensity $I_{max}$ and minimum intensity $I_{min}$, the efficiency is defined as

$$\eta = \frac{I_{max} - I_{min}}{I_{max}} \; (\times 100\%) \qquad (9.18)$$

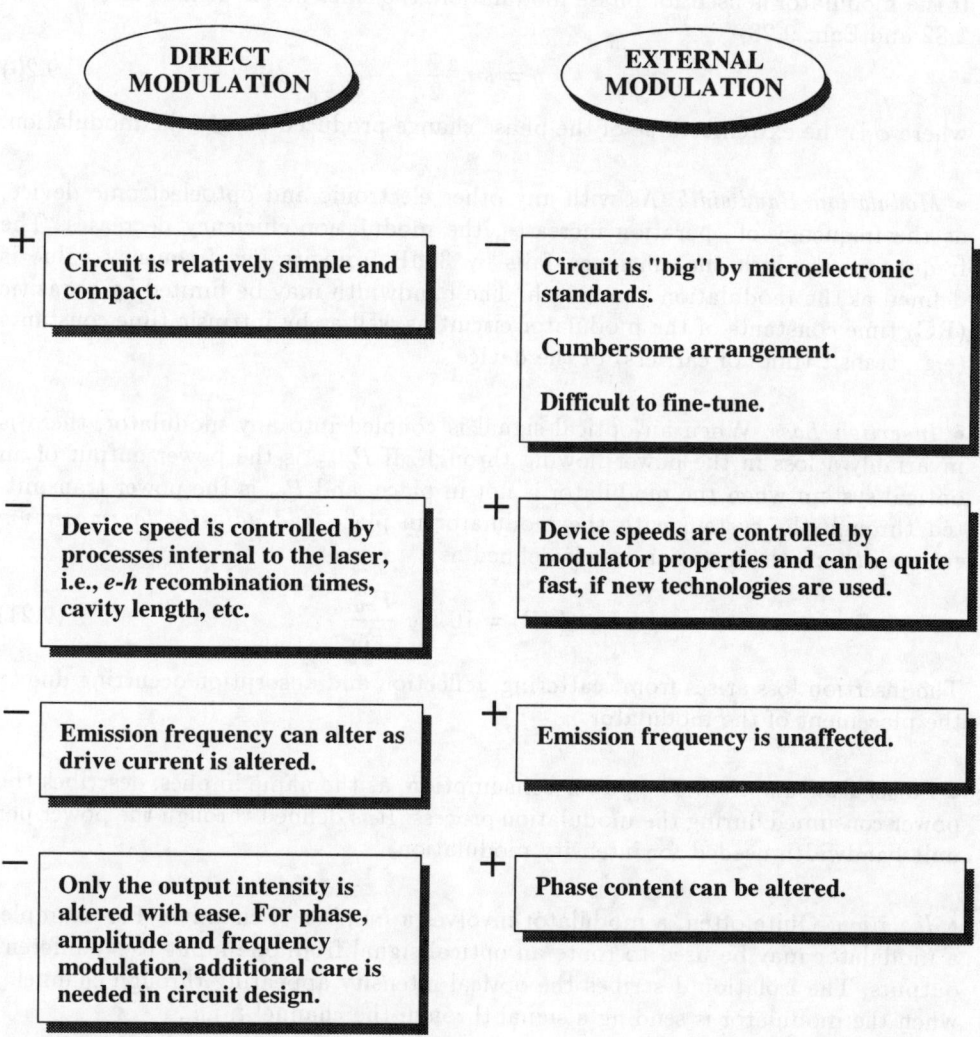

Figure 9.16: A comparison of the advantages (+) and disadvantages (−) of internal (or direct) and external modulation of laser diodes.

The modulation depth is also defined in decibels as

$$\text{contrast rate} = 10\ log\ \frac{I_{max}}{I_{min}} \qquad (9.19)$$

If the modulator is used for phase modulation, the efficiency is defined as (see Eqn. 2.82 and Eqn. 9.28)

$$\eta = sin^2 \frac{\phi}{2} \qquad (9.20)$$

where $\phi$ is the extreme value of the phase change produced due to the modulation.

• *Modulation Bandwidth*: As with any other electronic and optoelectronic device, as the frequency of operation increases, the modulation efficiency decreases. The frequency at which the efficiency falls by 3 dB from its low frequency value is defined as the modulation bandwidth. The bandwidth may be limited by parasitic (RC) time constants of the modulator circuit as well as by intrinsic time constants (e.g., transit times of carriers) of the device.

• *Insertion Loss*: When an optical signal is coupled into any modulator, there is invariably a loss in the power flowing through. If $P_{max}$ is the power output of an optical system when the modulator is not in place, and $P_{in}$ is the power transmitted through the system with the modulator in place *and adjusted for maximum transmittance*, the insertion loss is defined as

$$\text{Loss}(dB) = 10\ log\ \frac{P_{out}}{P_{in}} \qquad (9.21)$$

The insertion loss arises from scattering, reflection and absorption occurring due to the placement of the modulator.

• *Power Consumption*: The power consumption, as the name implies, describes the power consumed during the modulation process. It is defined through the power per unit bandwidth needed for intensity modulation.

• *Isolation*: Quite often, a modulator involves a number of channels. For example, a modulator may be used to route an optical signal from one input to $N$ different outputs. The isolation describes the optical intensity appearing through channel $j$ when the modulator is sending a signal through the channel $i$, i.e.,

$$\text{Isolation}(dB) = 10\ log\ \frac{I_i}{I_j} \qquad (9.22)$$

In most optical networks, an isolation of $\sim$ 40 dB is required.

Having discussed the figures of merit of modulators, we will now examine some important modulator configurations and the basic physical effects upon which they are based.

## 9.8   ELECTRO-OPTIC MODULATORS

Electro-optic modulators can produce amplitude, frequency or phase modulation in an optical signal by exploiting the electro-optic effect in which the optical properties of a crystal can be altered by an electric field. A number of crystals exist which have desirable response behavior. These include potassium dihydrogen phosphate (KDP), ferroelectric peroskites such as $LiNbO_3$ and $LiTaO_3$ as well as semiconductors such as GaAs and CdTe. We have discussed the basis of the electro-optic effect in Section 5.10.

In general, the refractive index of a crystal is not isotropic and is described in terms of an index ellipsoid (or indicatrix). The equation for the indicatrix, along the principle axis $x_i$, is (see Chapter 2, Section 2.5)

$$\sum_{i=1}^{3} \frac{x_i^2}{n_{ri}^2} = 1 \tag{9.23}$$

where $n_{ri}$ are the principle refractive indices. When an electric field is applied along a particular direction, the refractive indices are affected. However, because of anisotropy in the crystal, different indices are affected differently. In uniaxial crystals, there are two axis, say $x_1$ and $x_2$, along which the refractive index is the same, say $n_{ro}$, and one along which the index is $n_{re}$. These are the *ordinary* and *extraordinary* indices, respectively (as discussed in Section 2.5).

When a linearly polarized light wave enters the modulator as shown in Fig. 9.17, it resolves into two components. In general, as discussed in Section 2.5, the two directions have a different refractive index and a phase $\phi$ develops between them after they propagate a distance $L$. Consider an input signal that is linearly polarized and given by

$$E_x = \frac{E_o}{\sqrt{2}} \, exp(i\omega t) \tag{9.24}$$

$$E_y = \frac{E_o}{\sqrt{2}} \, exp(i\omega t) \tag{9.25}$$

After transmission through the modulator, the wave emerges with a general polarization given by

$$E_x = \frac{E_o}{\sqrt{2}} \, exp(i\omega t + i\theta_1) \tag{9.26}$$

$$E_y = \frac{E_o}{\sqrt{2}} \, exp(i\omega t + i\theta_2) \tag{9.27}$$

with the phase difference given by $\phi = \theta_2 - \theta_1$. If $\phi$ is $\pi/2$, the output beam is circularly polarized and if it is $\pi$, it is linearly polarized with polarization 90° with

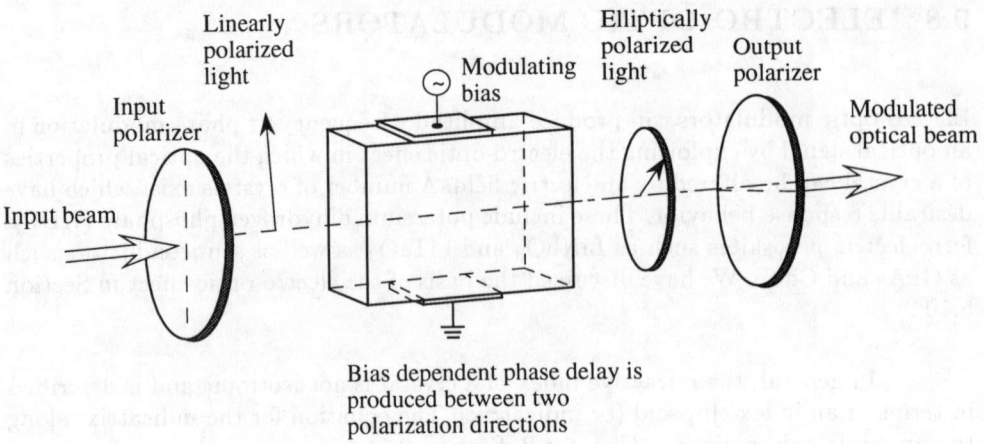

Figure 9.17: The use of an electro-optic device to modulate an optical signal. The applied bias introduces a phase change between light travelling in two polarization directions and the output light is modulated.

respect to the input beam. If the output beam passes through a polarizer at 90° with respect to the input beam polarizer as shown in Fig. 9.17, the modulation ratio is given by

$$\frac{I_{out}}{I_{in}} = sin^2 \frac{\phi}{2} \tag{9.28}$$

Thus if $\phi$ can be controlled by an electric field, the intensity can be modulated. For GaAs, the electric field dependent phase is given by (see Section 5.10)

$$\phi = \frac{2\pi}{\lambda} L n_{ro}^3 r_{41} \frac{V}{d} \tag{9.29}$$

where $\lambda$ is the wavelength of light, $L$ the device length, $n_{ro}$ the GaAs refractive index, $r_{41}$ the electro-optic coefficient for GaAs, $V$ the transverse applied bias, and $d$ the thickness of the modulator. A similar analysis for materials like KDP shows that the phase change between the two polarized waves at the output is given by

$$\phi = \frac{2\pi n_{ro}^3 r_{63} E L}{\lambda_o} \tag{9.30}$$

where $r_{63}$ is the electro-optic coefficient and $E$ is the electric field ($= V/d$). If cross-polarized polarizers are used, the maximum transmittance occurs when the phase change is $\pi$.

It may be noted that the electric field can be applied in a transverse or longitudinal way to the modulator. When a transverse field is applied, the effect is called Kerr effect. On the other hand, when the field is longitudinal the effect is called Pockel effect.

| Material | $\lambda(\mu m)$ | $r_{ij}$ ($10^{-12}$m/V) |
|----------|-----------|-------------------------|
| GaAs | 0.8 -1.0 | $r_{41} = 1.2$ |
| Quartz | 0.6 | $r_{11} = -0.47$ <br> $r_{41} = -0.2$ |
| LiNbO$_3$ | 0.5 | $r_{13} = 9.0$; $r_{22} = 6.6$ <br> $r_{42} = 30$ |
| KDP | 0.5 | $r_{41} = 8.6$; $r_{63} = 9.5$ |

Table 9.1: Electro-optic coefficients for some materials. (After P.K. Cheo, *Fiber Optics, Devices and Systems*, Prentice-Hall, New Jersey (1985).)

It is clear from Eqns. 9.29 and 9.30 that a high electro-optic coefficient can allow one to achieve a modulation using a smaller interaction length for the same applied field. However, the electro-optic coefficients of most materials are rather small ($\sim 10^{-12}$ m/V) as can be seen from Table 9.1, so that for realistic bias values, the length required is quite long (millimeters or more).

As noted above, the small values of the electro-optic coefficients force one to make external modulators from long devices that are incompatible with modern microelectronics. Thus there is an intense search for materials with improved physical properties. An important advance in this direction is due to quantum well structures.

### Electro-Optic Materials and Image Recording

The electro-optic effect discussed in this section is useful not only for optical modulation but also for optical image recording. Ferrroelectric materials which have a strong electro-optic effect find important uses in applications involving image recording, image contrasting etc. The material most widely used for this application is the polycrystalline ceramic PLZT discussed in Chapter 1, Section 1.5. Being polycrystalline, this material can be fabricated in large dimensions at low cost.

We recall from Chapter 1 that ferroelectric materials have a non-zero spontaneous polarization. When an external electric field is applied to them, the polarization can be altered. In particular the polarization can be reduced to zero by applying a field called the coercive field. The coercive field that is needed can be altered if a built-in field can be created by photo-generated carriers. This is the principle behind optical image recording.

The photoferroelectric (PFE) imaging device consists of a thin (0.1-0.3 mm) plate of PLZT ceramic with transparent electrodes applied to the major faces. The image to be stored is made to illuminate the face of the plate using near-ultraviolet illumination. Simultaneously a voltage is applied to the device. When light shines on the photo-sensitive PLZT, photo-generated carriers are produced with a local concentration proportional to the local image intensity. The carriers (electrons and holes) are separated by the applied field and trapped at defect sites. A local field is superimposed on the external field changing the coercive field and therefore the local polarization of the material. This in turn results in local strain variations on the PLZT plate. The image is thus faithfully recorded on the plate and can be read through a projection device.

By using proper fabrication techniques, PLZT ceramics can be made with grain sizes of $\sim 2$ micrometers. Such plates can store images with resolution of up to 100 lines per centimeter. The stored images can be erased by shining a uniform beam of near-ultraviolet light on the PLZT plate and simultaneously applying a voltage pulse to switch the ferroelectric polarization to its initial remnant state.

**EXAMPLE 9.4** A bulk GaAs device is used as an electro-optic modulator. The device dimension is 1 mm and a phase change of 90° is obtained between light polarized along $< 01\bar{1} >$ and $< 011 >$. The wavelength of the light is 1.5 $\mu$m. Calculate the electric field needed.

The phase change produced is ($\xi = 1$)

$$\Delta\phi = \frac{2\pi}{\lambda} n_{ro}^3 r_{41} EL = \frac{\pi}{4}$$

$$E = \frac{\lambda}{8n_{ro}^3 r_{41} L}$$

$$= \frac{(1.5 \times 10^{-6} \ m)}{8(3.3)^3 (1.2 \times 10^{-12} \ m/V)(10^{-3} \ m)}$$

$$= 4.35 \times 10^6 \ V/m$$

**EXAMPLE 9.5** The crystal KD*P (potassium dideuterium phosphate) is an important material for optoelectronics. Calculate the voltage needed to produce a phase change of $\pi$ in a KDP device. This voltage is called the half-wave voltage. The wavelength of light is 1.064 $\mu$m. The refractive index is 1.52.

The half-wave voltage is ($r_{63} = 26.4 \times 10^{-12} \ m/V$)

$$V(\lambda/2) = EL = \frac{(1.064 \times 10^{-6} \ m)\pi}{2(1.52)^3 (26.4 \times 10^{-12} \ m/V)}$$

$$= 5.74 \ kV$$

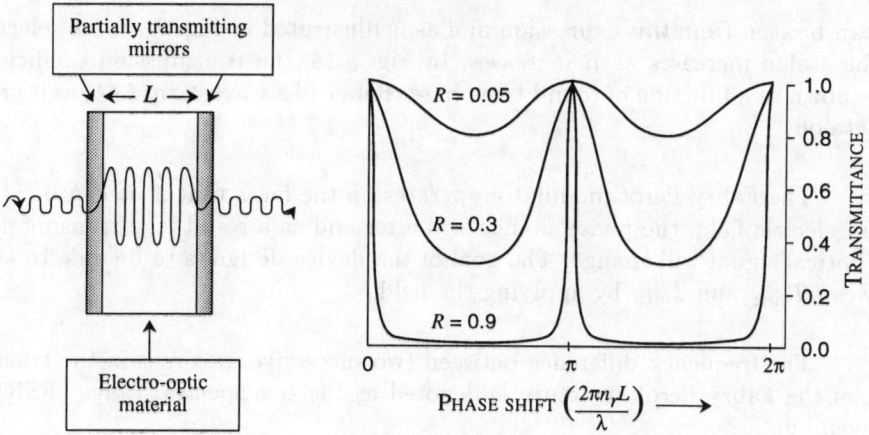

Figure 9.18: A schematic of a Fabry-Perot modulator. In epitaxially grown devices, the partially transmitting structures are distributed Bragg reflectors. The transmission coefficient of a Fabry-Perot structure as a function of the phase shift is shown at the right of the figure. $R$ is the mirror reflectivity.

## 9.9 INTERFEROMETRIC MODULATORS

$\longrightarrow$ In the previous section, we have seen how the phase change produced by the electro-optic effect can be used to modulate a signal. However, the device configuration shown in Fig. 9.17 is not the only configuration that is used to design modulators. There are a number of other configurations that do not require polarizers, but modulate a signal through interferometric effects.

### 9.9.1 Fabry-Perot Modulators

The Fabry-Perot modulator (often called the *etalon*) consists of two partially transmitting mirrors enclosing an electro-optic material as shown in Fig. 9.18. If $n_r$ is the refractive index of the electro-optic material and $L$ is its length, the transmission through the etalon is maximum when ($\lambda$ is the free space wavelength)

$$L = \frac{m\lambda}{2n_r} \qquad (9.31)$$

The transmission coefficient for an etalon with mirror reflectivity $R$ is given by

$$T = \frac{1}{1 + \frac{4R}{(1-R)^2} sin^2 \left(\frac{2\pi n_r L}{\lambda}\right)} \qquad (9.32)$$

As can be seen from this expression and as is illustrated in Fig. 9.18, the selectivity of the etalon increases as $R$ increases. In Fig. 9.18, the transmission coefficient is also shown as a function of round trip phase change of a wave, $2\pi n_r L/\lambda$, as it crosses the etalon.

The Fabry-Perot modulator operates on the basis that if $n_r$ can be altered by an electric field, the phase change will alter and as a result, the transmission of the optical signal will change. The goal of the device design is to be able to switch between $T_{max}$ and $T_{min}$ by applying the field.

The frequency difference between two successive maximas in the transmission of the Fabry-Perot structure is denoted as the free spectral range (FSR) and is given by

$$FSR = \frac{c}{2n_r L} \qquad (9.33)$$

An important parameter for an etalon is its finesse which gives the ratio of the FSR and the full width at half maximum of any transmission peak. It has a value

$$F = \frac{\pi (R_1 R_2)^{1/4}}{1 - (R_1 R_2)^{1/2}} \qquad (9.34)$$

where $R_1$ and $R_2$ are the reflection coefficients of the front and back mirrors of the etalon. Fabry-Perot modulators have been demonstrated to operate up to 10 GHz with contrast ratios up to 10 dB.

## 9.9.2   Mach-Zender Modulators

The phase modulation produced by electro-optic effect can be used to create intensity modulation in a Mach-Zender interferometer. In Fig. 9.19, we show a schematic of the modulator. An optical signal coming form a single-mode waveguide is split by a 3 dB coupler. The two split beams travel through two different guides (in general of different length) and then recombine to produce the output.

If the optical paths in the two arms of the interferometer is an integral number of optical wavelengths, the two waves will arrive at the second coupler in phase and interface constructively to produce a high intensity. If an electric field is now used to create a relative phase difference between the two waves, the intensity can be reduced. If the overall phase difference between beams traveling in the two paths is $\pi$, a minimum intensity will be produced.

Due to the loss in optical signal suffered as a result of the couplers, the Mach-Zender modulators do not have very high efficiencies.

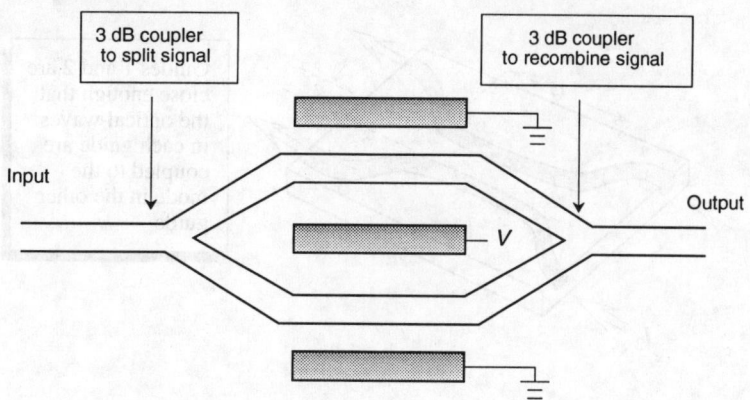

Figure 9.19: A schematic of the Mach-Zender interferometer used as a modulator. Two 3 dB couplers are used to split and recombine an incoming signal.

## 9.10   THE DIRECTIONAL COUPLER

⤳ In the previous section we have seen how the electro-optic effect can be used to modulate an optical signal. By changing the refractive index of a material, one can introduce a phase modulation, polarization modulation or, through interference of two beams, an intensity modulation. It is straightforward to use the same concept for switching applications. An important application in switching is satisfied by the directional coupler. We have discussed the basic theory behind the operation of the directional coupler in section 3.8. Here we will see how it can be used to modulate or switch light.

The directional coupler in its basic form couples two optical inputs $I_1$ and $I_2$ to two outputs $O_1$ and $O_2$ as shown in Fig. 9.20a. The device should be able to connect an entering optical signal to either $O_1$ or $O_2$. Such a device is extremely important in communication applications, where it can be used to route signals.

The basic operation of how a directional coupler operates has been discussed in Section 3.9.1. We will use the results obtained in that section. Fig. 9.20b illustrates that the optical power of a signal entering one of the guides sloshes back and forth between the two guides, as the signal progresses through the device.

A complete transfer of power occurs when the signal propagates a distance

$$L = \frac{\pi}{2K} \tag{9.35}$$

where $K$ is the coupling coefficient due to the overlap in the modes of the two guides.

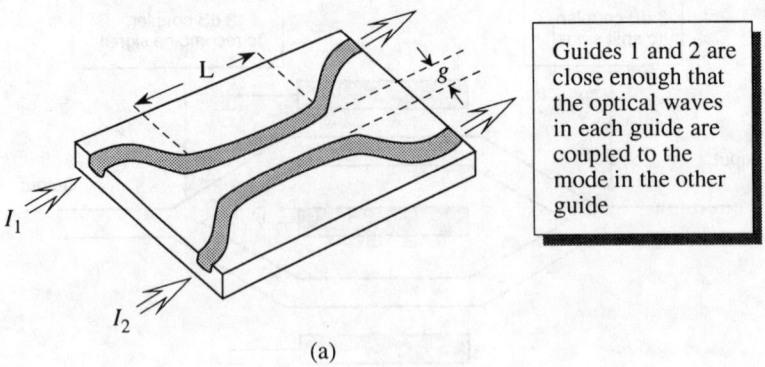

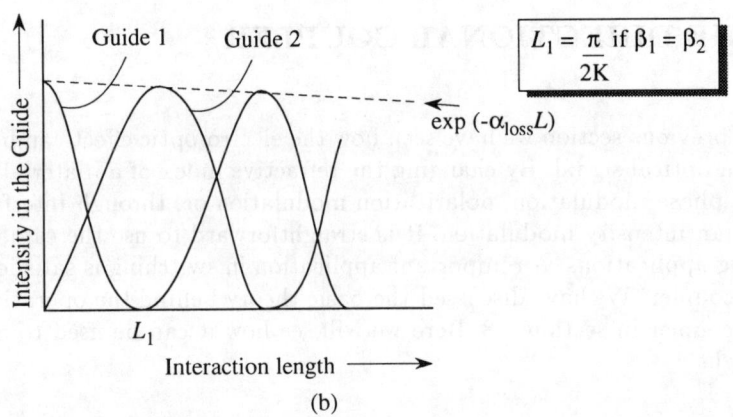

Figure 9.20: (a) A schematic of the directional coupler where two waveguides are separated by a gap $g$ over a length $L$. (b) The transfer of optical energy from one guide to another when initially the light signal enters Guide 1.

If the guides are not phase matched ($\beta_1 \neq \beta_2$), the entire light is not coupled from one guide to another. If we write

$$2\Delta = |\beta_1 - \beta_2| \neq 0 \tag{9.36}$$

It can be shown that if $A_1(0) = 1; A_2(0) = 0$ ($A_1$ and $A_2$ are the fields in the two guides), one gets

$$\frac{P_2(z)}{P_1(0)} = \frac{K^2}{K^2 + \Delta^2} \, sin^2 \left[(K^2 + \Delta^2)^{1/2} z\right] \tag{9.37}$$

Clearly, if $\Delta^2 >> K^2$, there is no coupling of light.

The working of the directional coupler depends upon altering the value of $\Delta$ by an electric field. Note that if $\Delta = 0$, the length over which the first transfer of energy occurs is

$$L_1 = \frac{\pi}{2K}$$

(9.38)

Now if one applied a voltage so that a change $\Delta_{sw}$ occurs so that

$$(K^2 + \Delta_{sw}^2)^{1/2} L_1 = \pi$$

(9.39)

then there will be no light emerging from guide 2 and the light will re-emerge from the guide 1 as shown in Fig. 9.21a.

The switching is controlled by the coupling coefficient $K$ and the strength of the electro- optic effect and the bias applied (through $\Delta$). In order that the voltage level needed is small, it is important that the material have a large electro-optic effect. The coupling coefficient depends exponentially on the separation $g$ of the two waveguides, since the coupling is due to the evanescent waves in each guide.

In order to require a smaller total applied potential, it is common to use a "push-pull" electrode arrangement as shown in Fig. 9.21b. The total shift $\Delta$ is then doubled. The shift $\Delta$ has been discussed in Section 5.10 and is given by the expression

$$\Delta = \frac{\pi n_{ro}^3 r_{eff}(2E_y)}{2\lambda}$$

(9.40)

where $r_{eff}$ is the effective electro-optic coefficient. For GaAs the value to be used is of $r_{41}$. The factor $2E_y$ is used for the push-pull electrode arrangement.

For directional coupler devices based on bulk materials ($L_i NbO_3$, GaAs, $L_i TaO_3$, etc.,) the size of the device is fairly large, since the electro-optic coefficient is not so large. Typically the devices require an interaction length of $\sim$ 1 cm and a bias voltage of $\sim$ 5-10 V. The use of quantum well structures are allowing smaller device structures due to the improved electro-optic effect related to the excitonic features as discussed in the next section.

## 9.11 ADVANCED MODULATION AND SWITCHING DEVICES

$\mathcal{R}$  In the previous section we have mentioned that the electro-optic coefficient in most materials is quite small, resulting in devices that are either quite long or require high voltages. An important new development in improving this situation is the use of quantum well structures. In Chapter 4, Section 4.7, we have discussed the

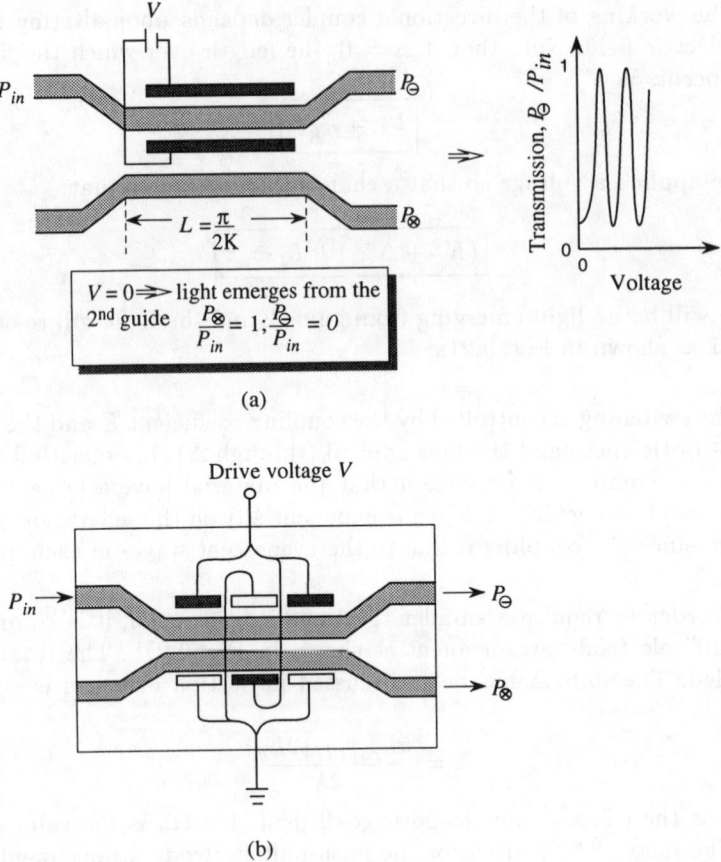

Figure 9.21: (a) A directional coupler where the length is chosen so that complete transfer occurs at zero applied bias. When the bias is changed, the light coupling can be changed. (b) A "push-pull" electrode arrangement for a directional coupler to double the phase change between the guides.

quantum well structures produced by sandwiching a narrow bandgap semiconductor between two enclosing regions of wider bandgap. We will briefly describe some of the issues in quantum-well-based devices for optical modulation.

### 9.11.1   Motivations For Quantum Well Devices

With the ability to fabricate quantum well devices (around the mid-1970s), an enormous number of new device concepts have been examined and demonstrated. Many of these concepts have found applications in high performance electronic and optoelectronic devices. Let us examine some motivations for quantum well structures in the area of optical modulators.

### Density of States Modification

In Chapter 4, Section 4.7, and in Appendix C, we have discussed that the density of states in a 2D system (a quantum well) has a step function-like behavior. Within the conduction and valence bands, one has subbands that lead to step-like density of states. This is to be contrasted to 3D (bulk) density of states which has a mono- tonically increasing form at the bandedges, as shown in Fig. 9.22. Since optical processes such as absorption coefficient depend upon the density of states, these processes are modified in quantum wells.

### Exciton Effects

In our discussions on optical absorption we talked about how e-h pairs are generated by a photon. The reader might think of the following question: electrons and holes are oppositely charged particles. Does the Coulombic attraction between them alter the optical absorption process? Indeed the Coulombic effect does play a very impor- tant role in the light-semiconductor interactions. To bring out the intricacies of this e-h interaction, let us invoke our understanding of optical spectra in an atom. In an atom, there is an electron-proton interaction which leads to sharp levels (denoted by symbols 1s, 2s, 2p, etc.). In optical absorption these levels result in sharp peaks in the absorption versus energy relations.

A similar phenomenon occurs for the e-h pairs. The e-h pairs attracted by their Coulombic interaction are called excitons. These excitons cause peaks in the absorption spectra, as shown schematically in Fig. 9.22. In quantum wells, the strength of these peaks is much stronger than in bulk materials, because of the stronger overlap between the electron and hole wavefunctions.

### Modification of Optical Properties by Fields

Another extremely important outcome of quantum well structures is that in the presence of an electric field, the optical properties (absorption coefficient, refractive index) of the material can change dramatically. This is shown schematically in Fig. 9.22. The large change in the absorption coefficient shown can be exploited for modulation of light.

## 9.11.2   Electro-Absorption Modulators

When an electric field is applied to a quantum well structure, the absorption spec- trum shifts toward a lower energy as shown in Fig. 9.23a. As a result, if a light beam with frequency $\hbar\omega$ is impinging on the device, the light transmitted can be modulated by a simple bias. The phenomenon of the shift of the absorption spec-

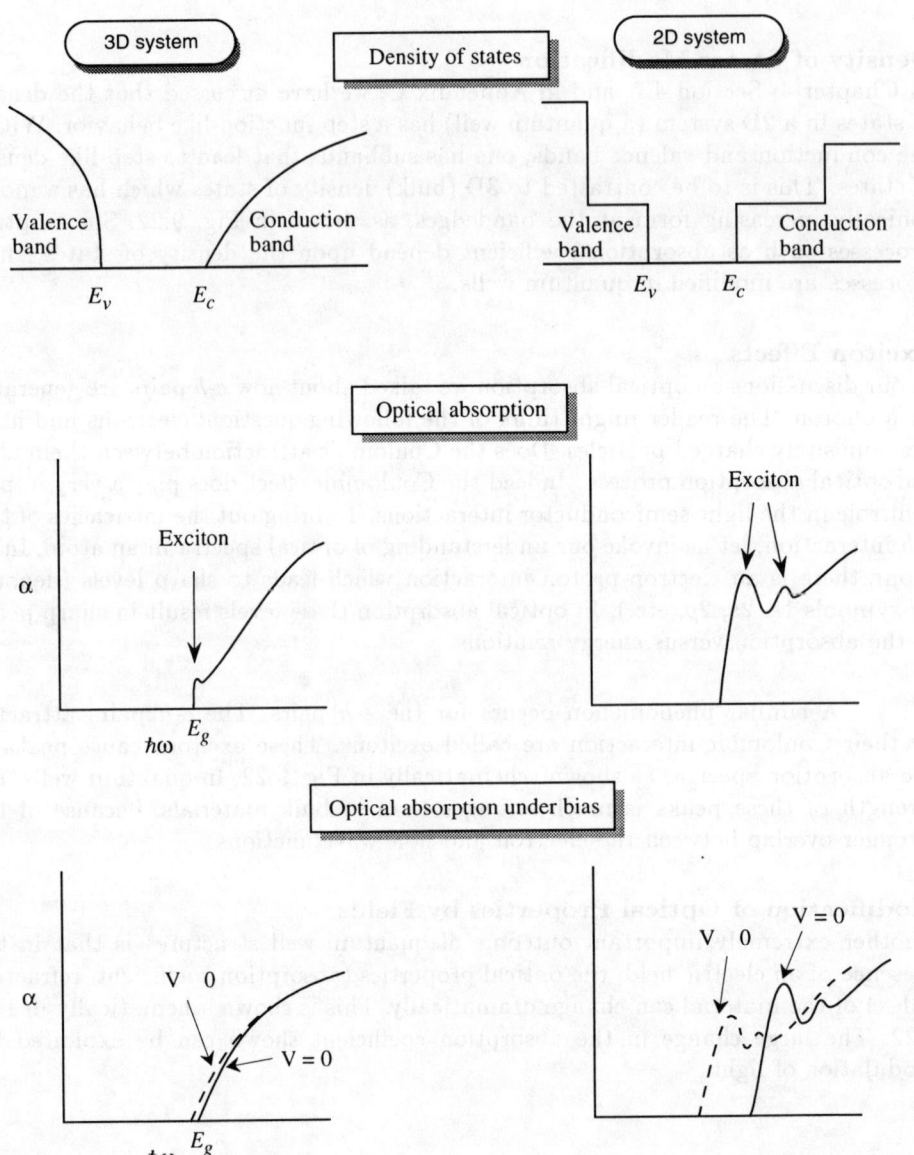

Figure 9.22: The difference between 3-dimensional (3D) and 2-dimensional (2D) systems that are exploited to design superior optical modulators. The excitonic effects are very strong in quantum wells, and the absorption spectra can be modified strongly by an applied field.

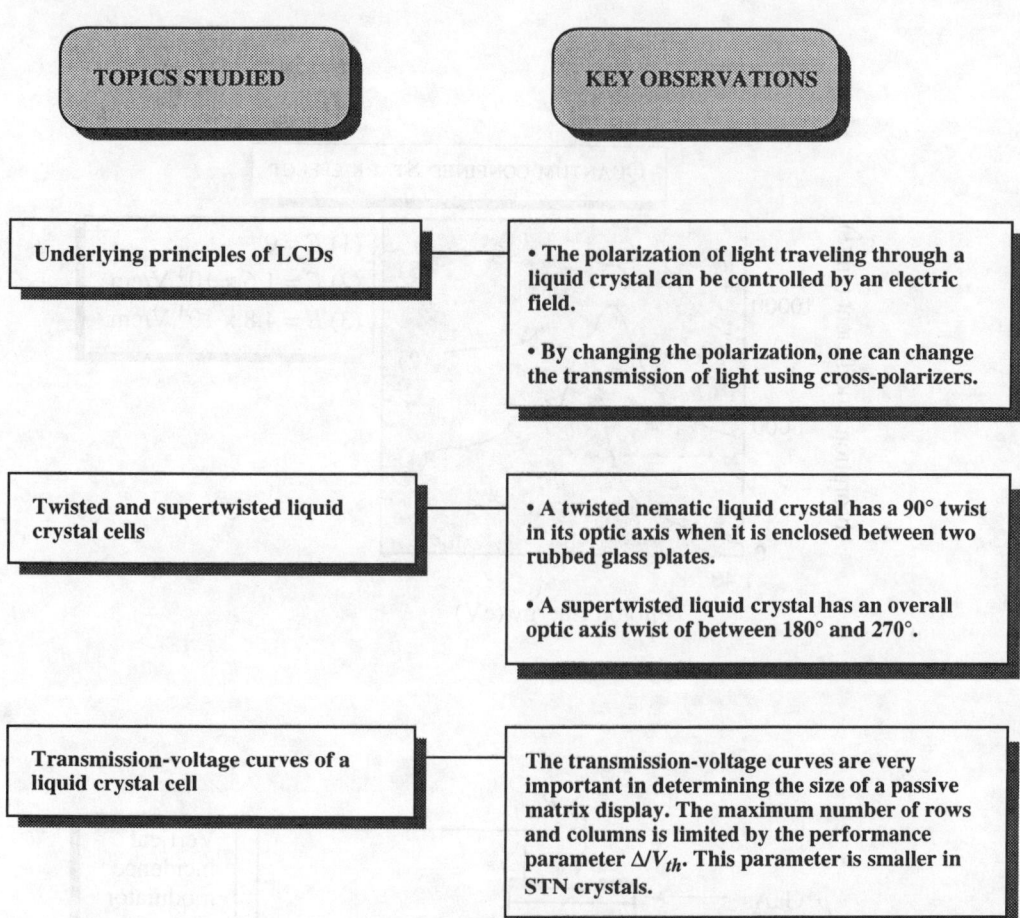

| TOPICS STUDIED | KEY OBSERVATIONS |
|---|---|
| Underlying principles of LCDs | • The polarization of light traveling through a liquid crystal can be controlled by an electric field. <br><br> • By changing the polarization, one can change the transmission of light using cross-polarizers. |
| Twisted and supertwisted liquid crystal cells | • A twisted nematic liquid crystal has a 90° twist in its optic axis when it is enclosed between two rubbed glass plates. <br><br> • A supertwisted liquid crystal has an overall optic axis twist of between 180° and 270°. |
| Transmission-voltage curves of a liquid crystal cell | The transmission-voltage curves are very important in determining the size of a passive matrix display. The maximum number of rows and columns is limited by the performance parameter $\Delta/V_{th}$. This parameter is smaller in STN crystals. |

Table 9.2: Summary table.

trum is called the quantum confined stark effect (QCSE). A number of novel devices and switches have been demonstrated based on QCSE. The QCSE device response time can be very fast (of the order of 10 ps) and thus, very high speed modulation devices can be produced. In Fig. 9.23b we show a simple *p-i-*(multiple quantum well)*n* diode which can be used as an electro-absorption modulator.

## 9.12 CHAPTER SUMMARY

In this chapter we have examined the modulation and display devices. Tables 9.2 to 9.4 summarize the findings of this chapter.

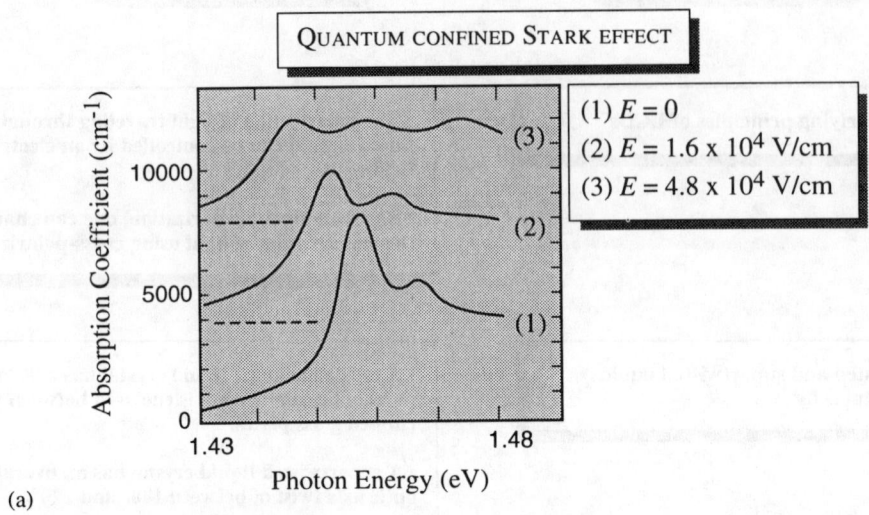

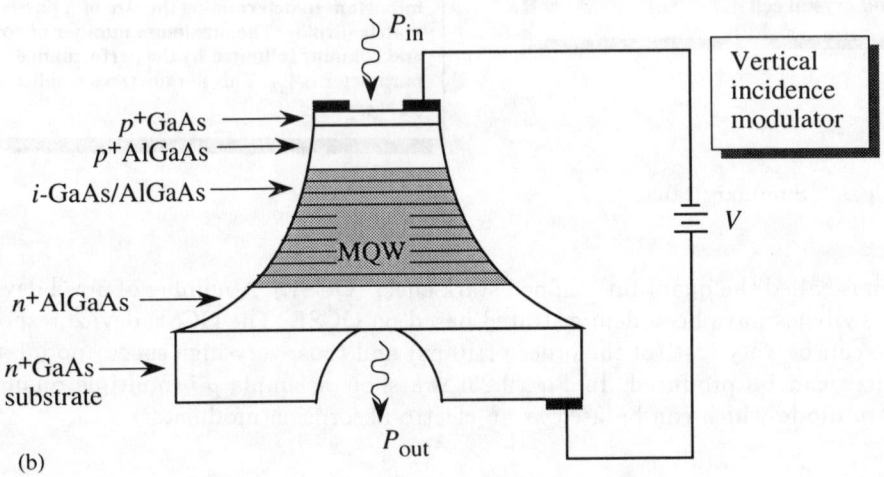

Figure 9.23: (a) The shift in the optical absorption with applied field. (b) A quantum confined stark effect modulator. By applying a bias, the output light intensity can be modulated.

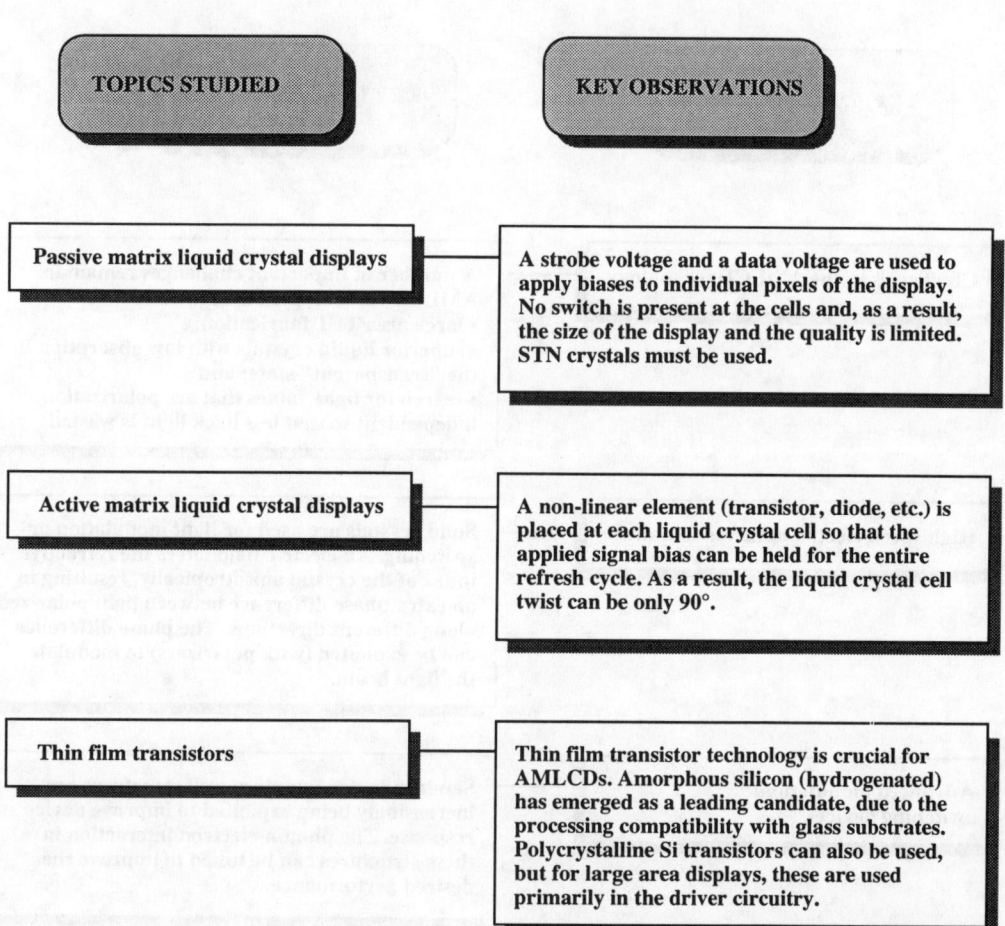

**TOPICS STUDIED**

**KEY OBSERVATIONS**

**Passive matrix liquid crystal displays**

A strobe voltage and a data voltage are used to apply biases to individual pixels of the display. No switch is present at the cells and, as a result, the size of the display and the quality is limited. STN crystals must be used.

**Active matrix liquid crystal displays**

A non-linear element (transistor, diode, etc.) is placed at each liquid crystal cell so that the applied signal bias can be held for the entire refresh cycle. As a result, the liquid crystal cell twist can be only 90°.

**Thin film transistors**

Thin film transistor technology is crucial for AMLCDs. Amorphous silicon (hydrogenated) has emerged as a leading candidate, due to the processing compatibility with glass substrates. Polycrystalline Si transistors can also be used, but for large area displays, these are used primarily in the driver circuitry.

Table 9.3: Summary table.

| TOPICS STUDIED | KEY OBSERVATIONS |
|---|---|
| Challenges in the AMLCD technology | A number of important challenges remain in AMLCD technology. These include:<br>• large area TFT fabrication;<br>• superior liquid crystals with low absorption in the "transparent" state; and<br>• search for light values that are polarization independent so that less back light is wasted. |
| High speed electro-optic modulators | Solid crystals are used for light modulation or switching. An electric field alters the refractive index of the crystal anisotropically, resulting in an extra phase difference between light polarized along different directions. The phase difference can be exploited (with polarizers) to modulate the light beam. |
| Advanced modulation/ switching devices | Semiconductor quantum well structures are increasingly being exploited to improve device response. The photon-electron interaction in these structures can be tuned to improve the desired performance. |

Table 9.4: Summary table.

## 9.13 PROBLEMS

### Section 9.2

**9.1** Explain qualitatively why $\epsilon_\parallel$ and $\epsilon_\perp$ are different in a liquid crystal. Note that the dielectric constant in a material is determined by the response of the material to an external field.

**9.2** Consider the parallel configuration shown in Fig. 9.2a. Under zero bias, the cell is to be designed so that the phase shift of an incoming light beam is 90° when the beam passes through the cell. Calculate the minimum liquid crystal thickness needed. Assume that $n_{re} = 3.1; n_{ro} = 2.8; \lambda = 1.0\mu$m.

**9.3** Consider a liquid crystal cell using the parallel configuration (Fig. 9.2a). The splay elastic constant is found to be $K_1 = 2 \times 10^{10}$ $N$ for a wide range of liquid crystals. If the maximum voltage available is 2.0 V, calculate the minimum dielectric anisotropy needed to use the cell in a display application.

**9.4** In a twisted orientation liquid crystal cell (Fig. 9.2c), what should be the polarization of the incoming light for optimum performance?

**9.5** A twisted liquid crystal cell is made from a crystal with force constants $K_1 = 5 \times 10^{-11}$ $N; K_2 = 10^{-10}$ $N; K_3 = 10^{-9}$ $N$. Calculate the minimum dielectric anisotropy needed to ensure that a 3 volt supply will be sufficient for this device.

**9.6** In a twisted cell, trace the polarization of an arbitrarily polarized light as it passes through the cell. Consider both the zero bias case and the case where the applied bias is greater than the threshold voltage.

**9.7** Discuss the possible reasons why the transmission-voltage curve of a real liquid crystal is different form an ideal case.

**9.8** Discuss why a nematic crystal cannot be used with a 270° twist, while a cholesteric crystal can be.

### Section 9.3

**9.9** Calculate approximately the maximum switching time acceptable in a liquid crystal cell that is to be used in a 500 × 500 display.

**9.10** Discuss the difficulties of individually addressing an N × N matrix of liquid crystal cells.

**9.11** Derive Eqns. 9.5, 9.6, and 9.13.

**9.12** A simple nematic crystal cell with a 90° twist and an STN cell with a 270° twist have $\Delta/V_{th}$ values of 0.2 and 0.02, respectively. Calculate the maximum values of rows (and columns) that are possible in a passive matrix display made from these two devices.

**9.13** If you examine a passive matrix based laptop computer, you will find a "shadowing" effect, i.e., a letter $A$ may cast a shadow around it along the rows or columns. Explain why this occurs.

**9.14** Consider watching the display screen of a laptop computer with polarizer glasses on. What do you expect when you turn your head? Should a similar effect occur when you view your TV screen?

**9.15** Consider a passive matrix liquid crystal display with $P = \Delta/V_{th} = 0.02$ and $V_{th} = 0.3\ V$. Calculate the maximum size of the display matrix one can build with this crystal. What kind of voltage supply values would you need?

**9.16** An active matrix display is designed with a transistor switch and a cell capacitor which can hold its voltage value for 20 ms. The refresh cycle is 30 ms for the display. Calculate the maximum display matrix possible if $P = \Delta/V_{th}$ is 0.4.

**9.17** An active matrix display is designed with a switch and cell capacitor which holds the voltage level for a time $t_h$ while the refresh cycle has a time period $t_r$. Calculate the maximum rows (columns) possible in the display with a given value of $\Delta/V_{th}$.

**9.18** A non-linear element (a switch capacitor combination) is developed for an AMLCD which switches in a time of 10 ms, but can hold its charge for a time of 100 ms. The refresh cycle time for the display is to be 30 ms and a 500 row (column) display is to be designed. Calculate the value of $\Delta/V_{th}$ that is needed.

## Section 9.5

**9.19** An a-Si:H TFT is developed with a channel mobility of 1 cm$^2$/V-s and a channel length of 5 $\mu$m. If the field in the channel (the source to drain field) is 10 kV/cm, calculate the switching time of the TFT if this time is limited by carrier transit time.

**9.20** Consider an a-Si:H TFT in which the channel mobility is 1 cm$^2$/V-s. Assume that the channel charge density goes from $10^{16}$ cm$^{-7}$ to $10^{18}$ cm$^{-3}$ during switching. Calculate the channel conductivity in the ON and OFF states.

**9.21** Discuss the importance of channel mobility and channel charge density in the ON and OFF states of a TFT for an AMLCD.

**9.22** Discuss some of the differences in design of a DRAM and AMLCD which make the AMLCD manufacturing a greater challenge.

**9.23** The absorption coefficient for a liquid crystal cell of thickness 5 $\mu$m is 1000 cm$^{-1}$ in the ON (transparent) state. Calculate the fraction of the back light that passes through a display matrix element in the ON state, accounting also for the polarization losses. The back light is unpolarized.

## Sections 9.8–9.10

**9.24** Discuss why a liquid crystal cell is not suitable for signal coding in an optical communication system.

**9.25** A GaAs electro-optic modulator is needed in an optical communication system. The maximum voltage available is 10 V. A device of length no more than 1 mm is needed. Calculate the thickness of the device. How long does it take for an optical signal to pass the device? Use data in Table 9.1 with $n_r = 3.6$ and $\lambda = 1.0\ \mu$m.

**9.26** A lithium niobate modulator is to be designed for a 1.06 $\mu$m system. The device length is 1 mm and the thickness is 10 $\mu$m. Calculate the voltage needed for the modulator.

**9.27** Discuss the incompatibility of modern microelectronic devices with bulk electro-optic modulator. Note that the dimensions of most microelectronic devices are $\sim 1\ \mu$m.

**9.27** Show that the power required to produce a modulated digital optical signal in an ideal (no ohmic losses) phase modulator is approximately

$$P = \frac{\epsilon}{2} V E_0^2 \Delta \nu$$

where $V$ is the volume of the modulation region, $E_0$ is the electric field swing needed for modulation, and $\Delta \nu$ is the bandwidth of the signal.

## 9.14  REFERENCES

- **Liquid Crystal Displays**

  - A very good overview of the field is presented in articles in *Solid State Electronics*: December (1991), January (1992), February (1992), April (1992), May (1992), June (1992), August (1994), and September (1992).

  - A special issue of IBM *Journal of Research and Development*, 36 (1992) is devoted to the area of flat panel displays.

- **Electro-optic Effect and the Directional Coupler**

  - A. B. Buckman, *Guided Wave Photonics*, Saunders College Publishing, Harcourt Brace Javanovich College Publisher, Orlando, FL (1992).

  - J. M. Hammer, "Modulation and Switching in Dielectric Waveguides," *Integrated Optics*, ed. T. Tamir, Springer-Verlag, New York (1982). Also see other articles in this book.

  - C. Pollack, *Fundamentals of Optoelectronics*, Irwin, Chicago (1995).

  - M. Wegener, T. Y. Chang, I. Bar-Joseph, J. M. Kuo, and D. S. Chemla, *Applied Physics Letters*, 55, 583 (1989).

  - J. E. Zucker, I. Bar-Joseph, G. Sucha, U. Koren, B. I. Miller, and D. S. Chemla, *Electronic Letters*, 24, 458 (1988).

# CHAPTER
# 10

# OPTICAL
# COMMUNICATION
# SYSTEMS:
# DEVICE NEEDS

## 10.1   INTRODUCTION

In Chapters 6, 7, 8 and 9 we have discussed some important devices that make up the arsenal of optoelectronics. Are these devices capable of taking on the well established electronic devices and contributing to the age of information processing? While the vision of all "optical computers" which carry out calculations at the "speed of light" (whatever that means) has not come true, there are a number of areas where optoelectronic devices have matched and surpassed the performance of electronic devices. A shining example is the area of optical communication. Optical communication networks have given a major boost to the compound semiconductor technology and have already established themselves as the high performance networks of choice.

The use of optics in communicating information probably started when human beings discovered fire. Light has been used to convey important battlefield strategies for thousands of years. In the modern information age, the era of optical communication was launched with the invention of the LED and the laser and the availability of low loss optical fibers. Without the optical fiber, there would be no optical communication. Because of the central role of optical fibers in the communication system, it is essential that the optoelectronic devices be compatible with the demands placed by the fiber on light sources and detectors. It is fair to say that a large fraction of research in semiconductor optoelectronics is motivated by the desire to optimize and exploit the potential of the optical fiber. While optoelectronic devices are used in a variety of applications, the demands of a high performance communication system are extremely stringent.

In Chapters 6, 7 and 8 we discussed some important performance parameters of detectors and light sources. An important question in the reader's mind must be : What is an adequate performance for an optoelectronic device? Is an LED emitting a 200  Å spectral width optical beam at 8800 Å with a 100 $\mu$W power adequate for a 1 Gbps transmission system? Why does one need a laser source with a 3 Å spectral width? What should be the detectivity of a detector to qualify for an optical communication system? These questions cannot be answered without understanding the fundamentals of a communication system in general, and the optical fiber-based system in particular.

## 10.2   A CONCEPTUAL PICTURE OF THE OPTICAL
## COMMUNICATION SYSTEM

$\longrightarrow$ Optical communication is one of the fastest growing segments of optoelectronics. This is in spite of the fact that the ideal optoelectronic devices that can exploit

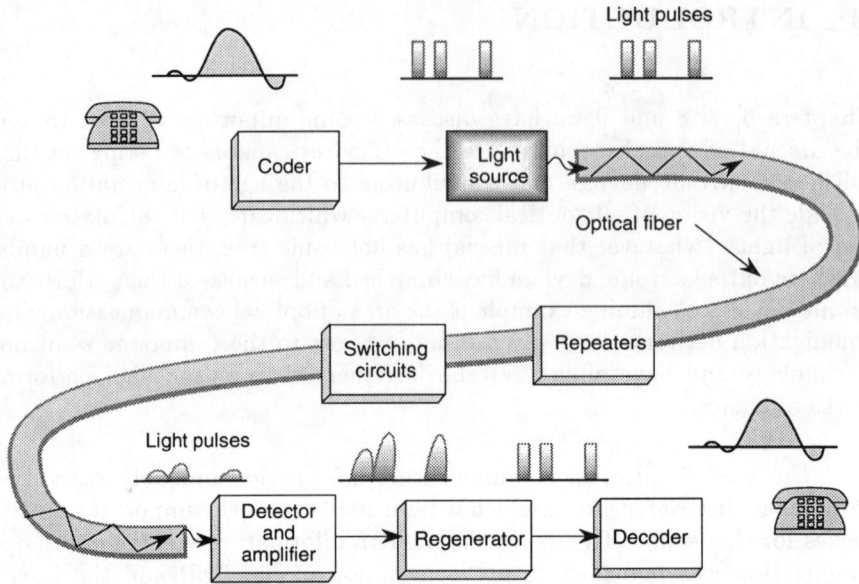

Figure 10.1: Components of an optoelectronic communication system.

the full potential of optical communication still do not exist. Consequently, a very small capacity of the optical system is used at present. This, of course, suggests that the future of optical communication will be even brighter.

The chief advantage of the optical communication system comes from the properties of the optical fiber which is the medium used to convey information. In Fig. 10.1 we show a typical optoelectronic communication system layout. The information to be transmitted (data, voice, etc.) is first coded onto an optical signal. This requires a proper driver or modulator and an optical source. The signal is next coupled to an optical fiber, which is a key component which has made optical communication possible. As the signal passes along the fiber, at some state it may need to be switched to other channels. This requires appropriate switching elements. Once the data have reached the desired point, they are detected by an optical detector. The generated signal is amplified and received.

In real systems, the optical signal attenuates as it propagates. This requires the placement of "repeaters" to regenerate the optical signal or amplifiers which amplify the signal. In addition, the optical data may need to be switched from one channel to another. The switching circuitry can be quite complicated requiring lasers and detectors as well as silicon based decision making circuitry, etc. The basic devices involved in the communication system have already been discussed in the previous chapters. Here we are more interested in identifying the requirements for these devices.

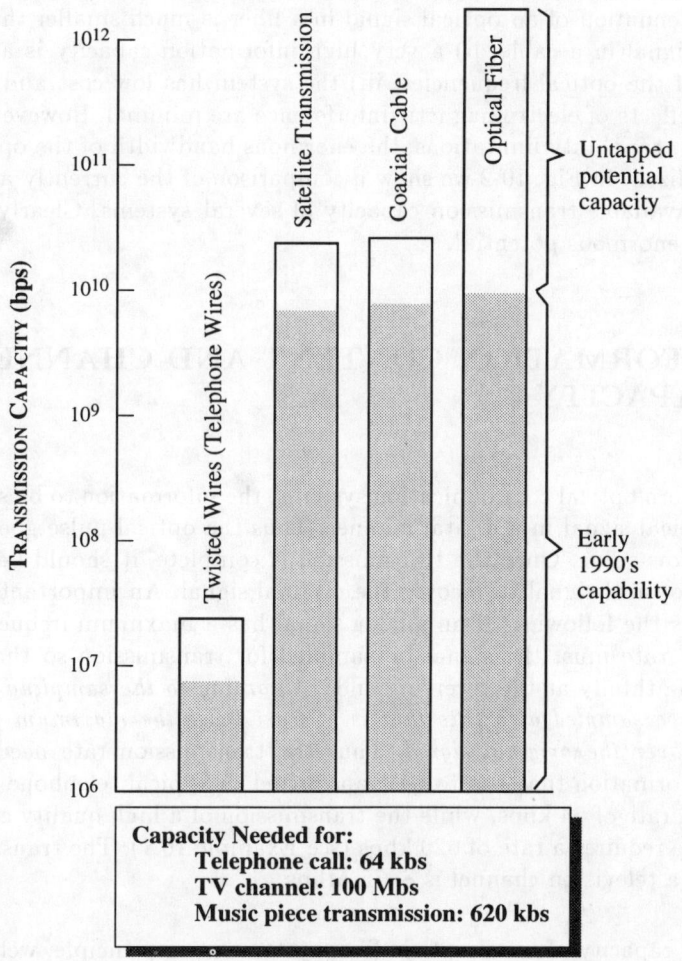

Figure 10.2: A comparison of the transmission capacity of some important transmission media. The current and potential capacities are shown. Transmission rates needed for some day to day applications are also shown. Data compression techniques can significantly reduce the bandwidth needed. (The capacities shown are an updated version of a figure in *Computer Engineering*, March (1987).)

The optical communication system is quite similar in concept to the electronic (microwave based) communication system. The key differences are in the wavelength of the radiation used and the use of an optical fiber instead of a metallic cable. The use of fibers provides four key areas of advantage for the optical system over the microwave based system: i) far greater repeater spacings are possible since the attenuation of an optical signal in a fiber is much smaller than that of a microwave signal in a cable; ii) a very high information capacity is available due to the use of the optical frequencies; iii) the system has low cost and low weight; and iv) the effects of electromagnetic interference are minimal. However, at present due to device capability limitations, the enormous bandwidth of the optical fiber is not fully utilized. In Fig. 10.2 we show a comparison of the currently available and potentially available transmission capacity of several systems. Clearly the optical fiber has an enormous potential.

## 10.3   INFORMATION CONTENT AND CHANNEL CAPACITY

$\longrightarrow$ In modern optical communication systems, the information to be sent is coded onto the optical signal in a digital manner. Thus the optical pulse goes between a high and a low value. Once the transmission is complete, it should be possible to convert the optical signal to recover the original signal. An important question in this regard is the following: If an analog signal has a maximum frequency content $f_m$, at what rate must the signal be sampled for transmission so that it can be reproduced faithfully at the receiving end? *According to the sampling theorem the signal must be sampled at a rate that is at least twice the maximum frequency in order to recover the original signal.* Thus the transmission rate needed depends upon the information that has to be transmitted. A typical telephone call needs a transmission rate of 64 kbps, while the transmission of a high quality compact disc musical piece requires a rate of 620 kbps (see Example 10.1). The transmission rate required for a television channel is $\sim$ 75 Mbps.

The capacity of usual optical fiber systems is, in principle, well over terra-bit/s ($\sim$ 10% of the optical frequency), so that to maximize the system performance the information from different sources is multiplexed as shown schematically in Fig. 10.3.

At present there are no practical optical systems which can use frequency division multiplexing (FDM) in which a number of frequency channels are used as carrier frequencies and the information is coded as to these frequencies. This would be the optimum system but at present it is difficult to develop detection systems that can carry out "coherent" detection to selectively decode the information on the carrier frequencies. Research in the area of narrow linewidth stable laser sources is

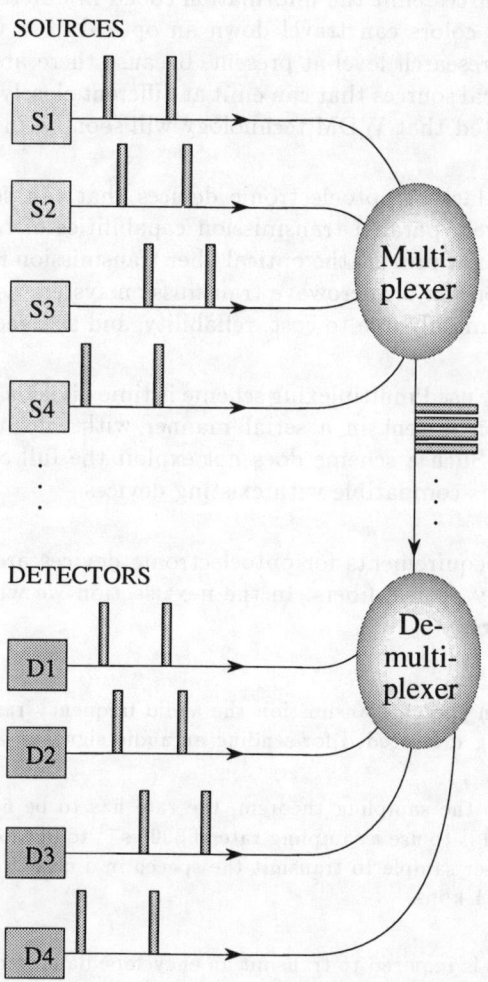

Figure 10.3: A schematic of a multiplexing (demultiplexing) scheme used to exploit the large capacity of a communication channel. At present the optical communication cannot exploit multiplexing of channels with different frequencies or wavelengths. As a result the multiplexing is done "serially."

to a large part driven by the desire for coherent (i.e., systems that can detect the phase or frequency of a signal) optical systems.

Another related multiplexing scheme is wavelength division multiplexing (WDM) in which sources emit the information coded in different "color" or "wavelength" light. These colors can travel down an optical fiber without interference. This scheme is at a research level at present, because there are many challenges in reliable tunable optical sources that can emit at different closely spaced wavelengths. However, it is expected that WDM technology will soon be in the marketplace.

Due to the lack of optoelectronic devices that can serve FDM or WDM schemes, the massively parallel transmission capabilities of optical fibers are not fully exploited. Thus, at present the optical fiber transmission rates are not superior to what could be done in a microwave transmission system as shown in Fig. 10.2. The advantages are mainly due to cost, reliability, and size reductions.

A commonly used multiplexing scheme is time division multiplexing (TDM) in which information is sent in a serial manner with information being sent in different time slots. Such a scheme does not exploit the full capabilities of optical communication but is compatible with existing devices.

Important requirements for optoelectronic devices are based upon the requirements placed by optical fibers. In the next section we will examine the properties of optical fibers.

**EXAMPLE 10.1**  In speech transmission the audio frequency range is 300 to 3400 Hz. Calculate the sampling rate needed for sending an audio signal in a telephone line.

According to the sampling theorem, the rate has to be $6800 \ s^{-1}$. However, the international standard is to use a sampling rate of $8000 \ s^{-1}$ to improve the margin of error. Also one uses 8 bits per sample to transmit the speech in a clear manner. This requires a transmission rate of 64 kbps.

**EXAMPLE 10.2**  It is required to transmit an encyclopedia stored in a computer having 10,000 pages of written text. Estimate the information content of the encyclopedia. A communication channel is available with a transmission capacity of 500 Mbps. Calculate the time taken for transmission.

Let us assume that on an average a page has 45 lines, each with 12 words and each word has 6 letters. In the ASCII code, each letter is represented by 7 bits. The total information content of the encyclopedia is then

$$(10^4)(45)(12)(6)(7) = 2.268 \times 10^8 \, bits$$

Using the transmission capability of the channel, the time of transmission is

$$t = \frac{2.268 \times 10^8 \, \text{bits}}{500 \times 10^6 \, \text{bps}} = 0.4536s$$

## 10.4 MODULATION AND DETECTION SCHEMES

$\mathcal{R}$ Optoelectronic detectors are used for a variety of applications, one of the most important being in the area of communications. The information to be transmitted is coded into a carrier beam (an optical beam for optical communication) which is then transmitted over an appropriate medium. The detector system is responsible for decoding the information sent. The design and performance of the detection system is intimately tied to the coding scheme used. At present, limitations placed by the performance of semiconductor optoelectronic sources and detectors do not allow one to use the full range of coding schemes available. Nevertheless, we will examine these schemes briefly since eventually, as devices advance, the system designer will be offered these choices.

The optical beam that carries the information is characterized by its amplitude, frequency, or wavelength and phase as well as its intensity (which is determined by the amplitude). All of these parameters can be modulated to code information provided that devices exist to code and decode the parameters.

### 10.4.1 Amplitude Modulation

$\mathcal{R}$ Amplitude modulation (AM) is an important modulation scheme widely used to send information using microwaves. The field of the carrier wave is represented by

$$F_c = F_{co} sin\omega_c t \tag{10.1}$$

and that of the modulation signal by

$$F_m = F_{mo} sin\omega_m t \tag{10.2}$$

In amplitude modulation, the maximum amplitude $F_{co}$ of the carrier wave is made proportional to the instantaneous modulating field $F_{mo} sin\omega_m t$. The modulation index is defined as

$$m = \frac{F_{mo}}{F_{co}} \tag{10.3}$$

The amplitude of the amplitude modulated carrier becomes (see Fig. 10.4)

$$
\begin{aligned}
A &= F_{co} + F_m = F_{co} + F_{mo} sin\omega_m t \\
&= F_{co} + mF_{co} sin\omega_m t \\
&= F_{co}(1 + m sin\omega_m t)
\end{aligned}
\tag{10.4}
$$

The field associated with the modulated carrier wave is now

$$
\begin{aligned}
F &= A sin\omega_c t = F_{co}(1 + m sin\omega_m t) sin\omega_c t \\
&= F_{co} sin\omega_c t + \frac{mF_{co}}{2} cos(\omega_c - \omega_m)t - \frac{mF_{co}}{2} cos(\omega_c + \omega_m)t
\end{aligned}
\tag{10.5}
$$

The amplitude modulated carrier contains three terms : i) the unmodulated carrier term (see Fig. 10.4); ii) an upper side band (USB) with frequency $\omega_c + \omega_m$; iii) a lower side band (LSB) with frequency $\omega_c - \omega_m$.

It must be kept in mind that the optical carrier frequencies are in the range of $10^{14} - 10^{15}$ Hz while the modulating frequencies are (limited by the electronics and the optical transmitter (lasers)) $\sim 10^{10}$ Hz. In order to be able to decode the information being sent, it should be possible to have a detection system that can isolate the USB or the LSB. It is important to note that while in the microwave domain it is possible to do such detection, in the optical regime severe challenges still remain.

## 10.4.2   Frequency Modulation

$\mathcal{R}$   Another important modulation scheme used widely in transmission of signals is the frequency modulation (FM) approach. As the name implies, the carrier signal frequency is modulated in this approach. Once again the unmodulated carrier wave can be written as

$$
F_c = F_{co} sin(\omega_c t + \phi)
\tag{10.6}
$$

where we have included a phase term for completeness. If the modulating signal is (we choose a cosine term for simplicity)

$$
F_m = F_{mo} cos\omega_m t
\tag{10.7}
$$

the *frequency of the modulated signal is*

$$
f = f_c(1 + k \, F_{mo} cos\omega_m t)
\tag{10.8}
$$

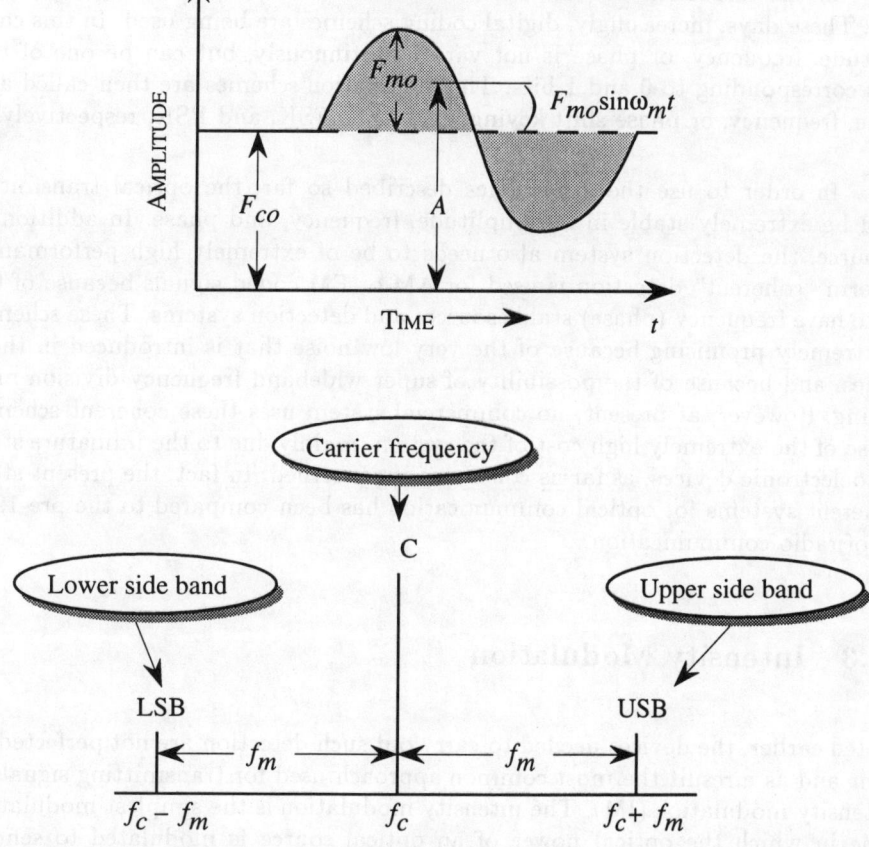

Figure 10.4: The amplitude of the amplitude modulated carrier wave and the frequency components of the carrier.

where $k$ is the proportionality constant which is dependent upon the details of how the modulation is done. The extremes in the frequency of the modulated carrier occur at value

$$f = f_c(1 \pm k \, F_{mo}) \tag{10.9}$$

As a result, the signal has not only the carrier frequency $\omega_c$ but a number of other frequency terms. The FM signal requires a larger spectral width for transmission. The higher bandwidth needed is, however, compensated by the lower noise that is possible in FM detection systems.

Closely related to the FM scheme is the phase modulation scheme where the modulating signal modulates the phase of a carrier signal.

In the discussion above, we have discussed the modulation by an analog signal. These days, increasingly, digital coding schemes are being used. In this case, amplitude, frequency, or phase is not varied continuously, but can be one of two values corresponding to 0 and 1 bits. The modulation schemes are then called amplitude, frequency, or phase shift keying, i.e., ASK, FSK, and PSK, respectively.

In order to use the approaches described so far, the optical transmitter should be extremely stable in its amplitude, frequency, and phase. In addition to the source, the detection system also needs to be of extremely high performance. The term "coherent" detection is used for AM or FM coded signals because of the need to have frequency (phase) stable sources and detection systems. These schemes are extremely promising because of the very low noise that is introduced in these schemes, and because of the possibility of super wideband frequency division multiplexing. However, at present, no commercial system uses these coherent schemes because of the extremely high cost of the system, mainly due to the immature state of optoelectronic devices, as far as coherence is concerned. In fact, the present state of coherent systems for optical communication has been compared to the pre-1930 state of radio communication.

## 10.4.3   Intensity Modulation

As noted earlier, the devices needed to carry out such detection are not perfected at present and as a result the most common approach used for transmitting signals is by intensity modulation (IM). The intensity modulation is the simplest modulation scheme in which the optical power of an optical source is modulated to send a signal. The signal is then directly detected by the detection. Such a modulation is compatible with detectors that have been discussed in the previous appendix. However, two important disadvantages exist in the intensity modulation scheme. The noise levels produced are quite high and the full bandwidth of the optical system cannot be used.

## 10.5   SOME PROPERTIES OF OPTICAL FIBERS

$\longrightarrow$   One of the most important motivations for the rapid advances in optoelectronic devices is the ability to transmit and receive information through an optical fiber. The superior properties of the fiber when compared to the metallic cables have been the most important driving forces for the use of optoelectronics in telecommunications, computer links, industrial automation, medical technology, and military applications. In Section 10.2 we saw that optical sources are an important component of the optical communication system. These sources are laser diodes and light emitting diodes (LEDs). The output of these sources must be coupled to the optical fiber for most applications.

In Chapter 3 we have discussed several important aspects of optical fibers. In Section 3.2, some of the structural properties were discussed; in Section 3.4, the acceptance angle and numerical aperture were discussed and finally, in Section 3.7, the modes in an optical fiber were discussed. The reader is urged to review that material. In this section we will discuss some additional important issues related to the optical communication system.

## 10.5.1   Fiber Losses

In an ideal fiber, the optical waves should propagate in the core region due to the confinement from the cladding region without any loss. However, in real fibers, there are losses which arise from various sources. Some of the important losses are due to the following sources:

*i) Absorption Loss*: The light can excite several transitions in the material making up the fibers. The absorption loss is simple $\exp(-\alpha L)$ where $\alpha$ is the absorption coefficients and $L$ is the optical path. The values of $\alpha$ are typically $\sim 0.02$ km$^{-1}$. The absorption loss is very important since it determines how much distance the optical signal can propagate before it needs to be regenerated by repeaters. In Fig. 10.5 we show a typical dependence of the loss on photon wavelength for silica fibers.

*ii) Scattering Loss*: In the fabrication process used for optical fibers, some irregularities are produced which can cause scattering losses. Improved techniques are reducing these losses.

*iii) Bending Losses*: If the optical fiber is bent too tightly (say, in circles of radius of a few millimeters), the light cannot bend along with the fiber and some of the lights can ooze out and be lost.

The consequences of attenuation on system performance will be examined later.

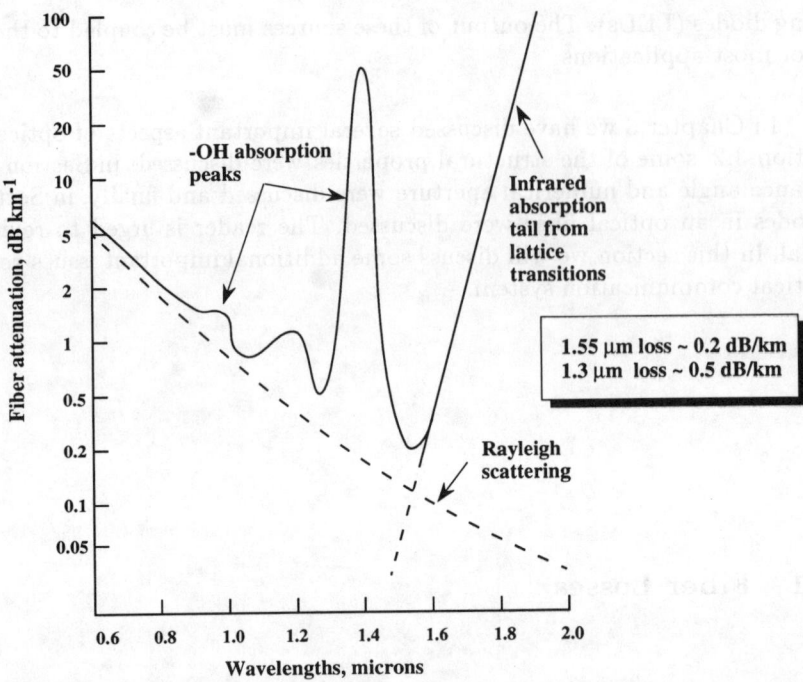

Figure 10.5: Attenuation in optical fibers as a function of wavelength. Low transmission losses occur at 1.3 $\mu$m and 1.55 $\mu$m. The loss figures are decreasing continuously as the technology improves.

## 10.5.2   Multipath Dispersion

A very important issue in optical fibers, especially in fibers with a wide core, is the problem of multipath dispersion. This problem arises from the fact that there are a number of paths that an optical beam can take to traverse a certain length of a fiber. Referring to Fig. 3.8 and the discussion of Section 3.4, let us assume that light signals come at angles between $\theta = 0$ (axial ray) to $\theta = \theta_A$ (extreme ray). The two extreme paths are defined by the axial ray and the most oblique ray which enters at an angle corresponding to $\phi_{1c}$. The rays travel with a velocity $v$ given by

$$v = c/n_{r1} \tag{10.10}$$

and take a time

$$t_a = \frac{\ell}{v} = \frac{n_{r1}\ell}{c} \tag{10.11}$$

for the axial ray to travel across a fiber of length $\ell$. The extreme ray effectively has to travel a distance $\ell/sin\phi_{1c}$. The time taken is (using the value of $sin\phi_{1c}$ from

Eqn. 3.16)

$$t_e = \frac{n_{r1}\ell}{c \sin\phi_{1c}} = \frac{n_{r1}^2 \ell}{n_{r2} c} \tag{10.12}$$

Thus, if a source sends in the two rays, the time difference between them when they emerge from the fiber is ($\Delta n_r = n_{r2} - n_{r1}$)

$$\tau_{mp} = t_a - t_e = \frac{n_{r1}\ell}{c}\left[1 - \frac{n_{r1}}{n_{r2}}\right] = \frac{n_{r1}\ell\Delta n_r}{n_{r2} c} \tag{10.13}$$

The multipath or model dispersion is given by the time spread per unit length,

$$\boxed{\frac{\tau_{mp}}{\ell} = \frac{n_{r1}\Delta n_r}{n_{r2} c}} \tag{10.14}$$

As shown in Fig. 10.6, the input pulse spreads in time as it passes through the fiber and emerges with a width $\tau_{mp}$. The multipath dispersion is closely related to the bandwidth the fiber can support over a given distance. To a good approximation the bandwidth $\Delta f$ limited by the multipath dispersion is simply

$$2\Delta f \sim \frac{1}{\tau_{mp}} \tag{10.15}$$

The bandwidth distance product is, therefore, (from Eqns. 10.14 and 10.15)

$$\boxed{\Delta f \ell = \frac{n_{r2} c}{2 n_{r1} \Delta n_r}} \tag{10.16}$$

For a high bandwidth distance product, the value of $\Delta n_r$ should be small. Since the numerical aperture is proportional to $\Delta n_r$, this means that for high frequency transmission, the numerical aperture should be small. However, it must be kept in mind that reducing $\tau_{mp}$ by reducing the numerical aperture can have a negative effect on the coupling of light into the fiber, especially if the light source is not highly collimated as is the case for an LED. Thus, to collect sufficient light, one needs a fiber with high numerical aperture and this in turn can reduce the bandwidth distance product. In semiconductor lasers, on the other hand, it is possible to get highly collimated beams and thus use fibers with high bandwidth-distance products.

To improve the bandwidth-distance product for a fiber, special approaches are taken. An important approach is the graded index fiber in which the index of the core is varied as shown in Fig. 3.2. The grading allows one to have a much higher bandwidth-distance product compared to the step index fiber.

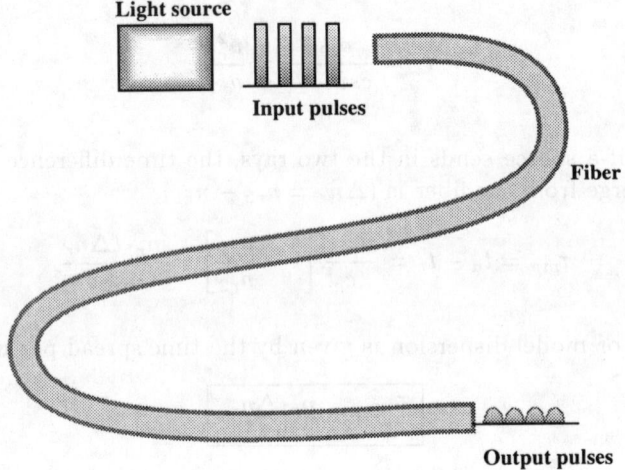

Figure 10.6: Due to multipath delays (and material dispersion delays) an optical pulse spreads as it passes through an optical fiber. The pulse spreading determines the maximum bandwidth the fiber can carry (in serial transmission) over a certain distance.

A fiber that is increasingly dominating advanced optical systems is the single mode fiber. As the diameter of the core of the fiber starts to become comparable to the wavelength of light, the simple ray picture developed above breaks down. One has to consider the proper waveguide theory where the Maxwell equations are solved in the fiber, as discussed in Section 3.7. *As the fiber dimensions shrink, one eventually reaches a stage where only one mode of radiation can propagate in the optical fiber.* Fibers having such parameters are called single mode fibers. The condition for a step index fiber of core diameter $a$ as shown in Fig. 3.2 is (see Eqn. 3.87)

$$\boxed{\frac{2\pi a \left(n_{r1}^2 - n_{r2}^2\right)^{1/2}}{\lambda} < 2.405} \qquad (10.17)$$

By choosing a small core diameter, a single mode fiber can be fabricated. The cladding diameter is kept at 60-100 $\mu$m to maintain the strength of the fiber. *Since one has a single mode propagating in a single mode fiber, there is no multipath dispersion.*

## 10.5.3  Material Dispersion

In addition to the multipath delay problem discussed above, another limitation on the capacity of the optical fiber is due to the material dispersion of the fiber.

The material dispersion arises from the variation of the refractive index of the optical fiber with wavelength. Because of the variation, radiation with different wavelength travels a different optical path and thus a time delay is introduced. Of course, if an optical signal has a single wavelength, one would not be concerned with material dispersion. However, even the purest optical sources have some spread in the wavelength which makes material dispersion a serious problem.

In a material in which the dielectric constant is dependent upon the wavelength of light, as discussed in Section 3.8, one can define a phase velocity $v_p$ which defines the phase change of a signal and a group velocity $v_g$ which describes the rate of energy transfer by the signal. We have

$$v_p = \frac{\omega}{k} \tag{10.18}$$

and

$$v_g = \frac{d\omega}{dk} \tag{10.19}$$

If the material has no dispersion the group velocity and the phase velocity are the same. We can also define the ordinary refractive index $n_r$ for the phase and the group refractive index $n_g$. These are related to the respective velocities by

$$n_r = \frac{c}{v_p} \tag{10.20}$$

$$n_g = \frac{c}{v_g} \tag{10.21}$$

In general $n_r$ and $n_g$ are a function of the wavelength $\lambda$. We can write (using Eqns. 10.18 through 10.21)

$$n_g = \frac{c}{v_g} = c\frac{dk}{d\omega} = c\frac{d}{d\omega}\left(\frac{\omega n_r}{c}\right) = n_r + \omega\frac{dn_r}{d\omega} \tag{10.22}$$

We also have

$$\frac{dn_r}{d\omega} = \frac{dn_r}{d\lambda} \cdot \frac{d\lambda}{d\omega} \tag{10.23}$$

Since,

$$\omega = \frac{2\pi c}{\lambda} \tag{10.24}$$

$$\frac{d\omega}{d\lambda} = -\frac{2\pi c}{\lambda^2} \tag{10.25}$$

and we get, from Eqns. 10.22 through 10.25,

$$n_g = n_r - \frac{2\pi c}{\lambda}\frac{dn_r}{d\lambda}\left(\frac{\lambda^2}{2\pi c}\right) = n_r - \lambda\frac{dn_r}{d\lambda} \tag{10.26}$$

This gives for the group velocity,

$$v_g = \frac{c}{n_g} = \frac{c}{n_r - \lambda dn_r/d\lambda} \tag{10.27}$$

An impulse of light at a fixed $\lambda$ travels a distance $\ell$ of the fiber in a time

$$t = \frac{\ell}{v_g} = \left[n_r - \lambda\frac{dn_r}{d\lambda}\right]\frac{\ell}{c} \tag{10.28}$$

If the optical signal is made up of a spread of wavelengths there will be a spread in the time intervals taken by the different wavelengths to traverse the fiber. If $\Delta\lambda$ is the wavelength spread, the time spread is given by differentiating Eqn. 10.28 to get

$$\frac{\Delta t}{\Delta\lambda} = -\frac{\ell}{c}\lambda\frac{d^2n_r}{d\lambda^2} \tag{10.29}$$

An important property of an optical signal is, therefore, the spread in wavelengths $\Delta\lambda$ over which the optical signal has, say, 50% or higher spectral power. The relative spectral width of the source is defined by

$$\gamma = \left|\frac{\Delta\lambda}{\lambda}\right| = \left|\frac{\Delta\omega}{\omega}\right| \tag{10.30}$$

An impulse, after travelling a distance $\ell$ through the fiber, will have a half power spread in time $\tau_{disp}$ given by $(\tau_{disp} \equiv \Delta t)$

$$\tau_{disp} = -\frac{\ell}{c}\lambda\Delta\lambda\frac{d^2n_r}{d\lambda^2} = -\frac{\ell}{c}\gamma\,\lambda^2\left(\frac{d^2n_r}{d\lambda^2}\right) \tag{10.31}$$

An important parameter is the spread in time per unit length,

$$\frac{\tau_{disp}}{\ell} = \frac{\gamma}{c}|Y_m| \tag{10.32}$$

where the quantity $Y_m$ represents the material dispersion

$$Y_m = \lambda^2\frac{d^2n_r}{d\lambda^2} \tag{10.33}$$

We can approximately define by $1/4\tau_{disp}$ the bandwidth of the signal the fiber system can support for a length $\ell$. The exact relation depends somewhat on the spectral shape of the light output. We then have

$$\boxed{(\Delta f)\ell = \frac{\ell}{4\tau_{disp}} = \frac{c}{4\gamma\,|Y_m|}} \tag{10.34}$$

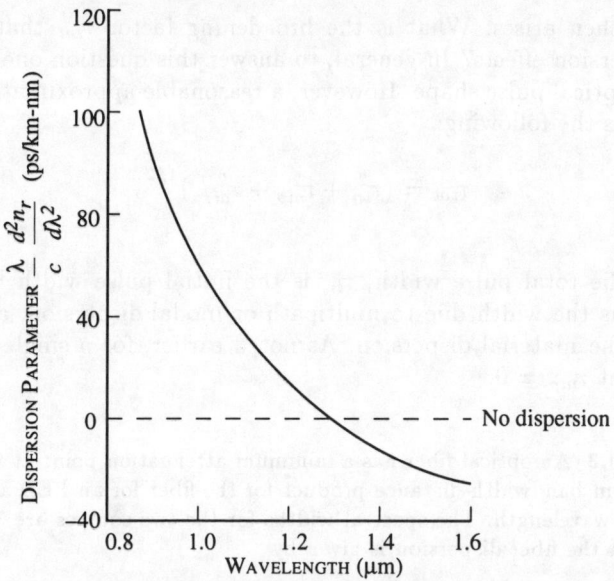

Figure 10.7: A typical dispersion curve versus wavelength for silica optical fibers. Notice that the dispersion goes to zero at $\sim 1.3$ $\mu$m. Note also from Fig. 10.5 that the attenuation is lowest in the silica fiber at 1.55 $\mu$m, although the attenuation is not too high at 1.3 $\mu$m.

*This relation emphasizes the importance of the optical source quality $\gamma$ and the fiber dispersion quality in determining the channel capacity and the distance up to which the signal can be carried.* A small spread in the optical source emission spectrum is critical and is a key reason for the use of semiconductor laser diodes instead of light emitting diodes in high performance optical communication systems. Typical dispersion values for silicon fibers are shown in Fig. 10.7. (Note that the values shown in Fig. 10.7 are not for $Y_m$, which can be obtained by multiplying the value given by $\lambda c$).

It is interesting to compare the capacity of the fiber-source system at the wavelength of no dispersion versus the wavelength of minimum attenuation (e.g., 1.3 $\mu$m vs. 1.55 $\mu$m transmission). The signal can be transmitted farther at 1.55 $\mu$m, but the transmission bandwidth is not as high as at 1.3 $\mu$m. The optimum system has to consider the properties of the detector and the modal dispersion of the fiber as well. In Example 10.4 we will carry out a comparison of various devices and emission wavelengths.

In this and the previous subsection we have examined two important sources that limit the transmission capacity of the fiber. The sources are modal dispersion and material dispersion. In general, both these limitations will be present in a fiber.

The question then arises: What is the broadening factor $\tau_{tot}$ that describes the combined dispersion effects? In general, to answer this question one needs to know details of the optical pulse shape. However, a reasonable approximation that can be easily applied is the following:

$$\tau_{tot} = \left(\tau_{in}^2 + \tau_{mp}^2 + \tau_{disp}^2\right)^{1/2} \tag{10.35}$$

where $\tau_{tot}$ is the total pulse width, $\tau_{in}$ is the initial pulse width before entering the fiber, $\tau_{mp}$ is the width due to multipath or modal dispersion, and $\tau_{disp}$ is the width due to the material dispersion. As noted earlier, for a single-mode fiber we can assume that $\tau_{mp} = 0$.

**EXAMPLE 10.3** An optical fiber has a minimum attenuation point at 1.55 $\mu$m. Calculate the maximum bandwidth-distance product for the fiber for an LED and a laser diode emitting at this wavelength. The spectral widths for the two sources are 300 Å and 30 Å, respectively, and the fiber dispersion is given by

$$Y_m = \lambda^2 \left.\frac{d^2 n_r}{d\lambda^2}\right|_{\lambda=1.55\mu m} = -0.01$$

The bandwidth-distance product is given by

$$\Delta f \cdot \ell = \frac{c}{4Y_m}\left(\frac{\lambda}{\Delta\lambda}\right)$$

This gives for the LED

$$\Delta f \cdot \ell = \frac{(3 \times 10^{10}\,cm/s)(1.55 \times 10^{-4}\,cm)}{4 \times (0.01) \times (300 \times 10^{-8}\,cm)} = 3.875 \times 10^{13}\,Hz \cdot cm$$

$$= 0.3875 GHz \cdot km$$

For the laser diode the value becomes ten times greater. Based upon this example, if this LED is to be used to send a signal over a 10-km fiber, the maximum bandwidth is 38.75 MHz.

## 10.5.4   Signal Attenuation and Detector Demands

An extremely important consideration in fiber optics is the attenuation suffered by light as it travels through the fiber. The success of the optical fiber communication system owes a great deal to the low attenuation suffered by the electromagnetic waves as they travel in the fiber. The original proposal to use fiber as a communication medium was made by Kao and Hockham, who estimated that an attenuation figure of 20 dB/km was needed if fibers were to be competitive. The attenuation in dB/km is given by

$$\text{Attenuation} = \frac{10 log_{10}(P_i/P_f)}{L} dB/km \qquad (10.36)$$

where $P_i$ is the initial optical power and $P_f$ is the power left after traveling a distance $L$ (in km). Note also that the absorption coefficient $\alpha$ also defines the ratio of $P_i$ and $P_f$ via the equation

$$\frac{P_f}{P_i} = exp\,(-\alpha L) \qquad (10.37)$$

Thus, for example, an attenuation of 20 dB/km corresponds to an absorption coefficient of 4.6 km$^{-1}$. Modern optical fibers are capable of attenuation lower than 0.2 dB/km at certain wavelengths. The attenuation curve for the most widely used silica fiber is shown in Fig. 10.5. It may be noted that fibers with high attenuation may be selected for certain special applications because they may have to be used in adverse environments, like under high radiation dosage. Under high radiation certain fibers do not degrade as much as other fibers and thus provide a better system.

The attenuation in the fiber is due to various scattering processes that occur causing light to be scattered or absorbed as it travels in the fiber. The improvements in the fiber attenuation have resulted from a detailed understanding of the absorption process. All fibers have minimas in the attenuation curve at some special wavelengths. The choice of the optical source and the detection system is greatly influenced by these special wavelengths. For example, as shown in Fig. 10.5, the most widely used silica fibers have an attenuation minimum at 1.55 $\mu$m. There is also a local minimum at 1.3 $\mu$m. The selection of an operating wavelength is affected by considerations such as fiber dispersion, attenuation, source spectral purity, as well as demands imposed upon the system. For example, if the communication system is for a local area network (LAN), a GaAs based source ($\lambda \sim 0.88$ $\mu$m) may be chosen due to the advanced level of GaAs based optoelectronics and electronics. However, for long-distance communication, the concerns of dispersion and attenuation are quite important. Example 10.4 brings out some very important features that influence choice of devices for communication systems.

## 10.5.5 Fiber Amplifier

In the previous sub-section, we have seen how an optical signal degrades as it moves through a fiber. One may ask the following question: Is it possible to have a fiber where there is positive gain present so that the optical signal grows as it propagates? Such fibers are now present and are called fiber amplifiers. An extremely important development in fiber amplifiers is the erbium doped fiber amplifier. The physics of the fiber amplifiers is qualitatively similar to that of the semiconductor laser amplifier. In the erbium doped (and other rare earth ion doped) fiber amplifiers (EDFAs), erbium ions are placed inside the fiber core during the fabrication process. The doped ions have associated with them energy bands at different wavelengths. In Fig. 10.8 we show the energy spectra of the erbium ions in a fiber. The levels that are exploited are the $4\ell_{15/2}$ and $4\ell_{13/2}$. Carriers are excited by a 1.48 $\mu$m laser semiconductor by optical pumping. The relaxing carriers can produce emission and gain over a broad spectrum ranging from 1.5 $\mu$m to 1.56 $\mu$m. Use of other rare earth ions can produce amplifiers at other spectral regions.

The development of EDFAs has revolutionized the optical communication systems. The loss of the optical signal as it travels can be compensated by these fiber amplifiers thus eliminating the need to convert to electrical signals and reconvert to optical signals.

**EXAMPLE 10.4**  Consider the specifications of the components of an optoelectronic communication system shown in Table 10.1. The detector sesitivity is such that the minimum power needed to detect a signal is given by 0.1 nW/Mbps.

Calculate the repeater spacing needed if the system is to be used for transmission of data at i) 100 Mbps; ii) 1.0 Gbps. Assume that the cable coupling losses and splicing losses are included in the attenuation specifications provided.

Given the system component specifications, the repeater spacing is determined by pulse spreading (dispersion) considerations or by attenuation considerations. The dispersion limit is determined by the source spectral purity and the dispersion in the fiber, while the attenuation limit is determined by the minimum power needed for detection. Let us consider the case of the LED first.

$$\lambda = 0.88\mu m; 0.1 \; Gbps \text{ transmission}; P_i = 5 \times 10^{-5} \; W$$

$$P_f = \text{Detector minimum power required} = (10^{-10} \; W/Mbps)(100 Mbps) = 10^{-8} \; W$$

$$\frac{P_i}{P_f} = \frac{50 \times 10^{-6} \; W}{10^{-8} \; W} = 5 \times 10^3$$

From Eqn. 10.36 we get (attenuation is 1.5 $dB/km$)

$$L \equiv l_{att} = \frac{10 log(5 \times 10^3)}{1.5} = 24.7 \; km$$

A material dispersion of 70 ps/km-nm corresponds to a value of the dispersion parameter $Y_m = 0.018$. Taking the bit rate $B$ as $2\Delta f$ we get for the single mode (SM)

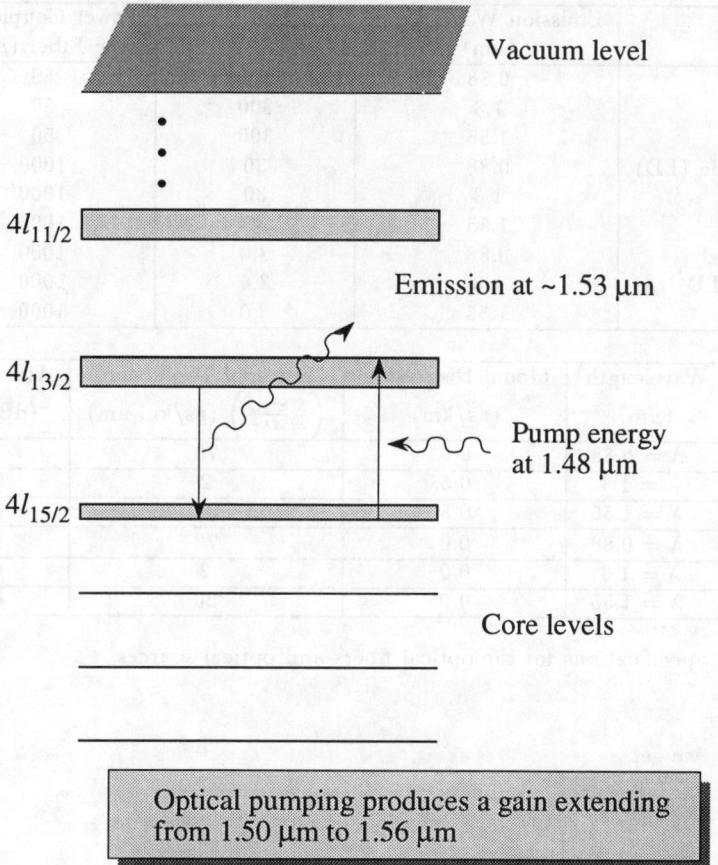

Vacuum level

$4l_{11/2}$

Emission at ~1.53 μm

$4l_{13/2}$

Pump energy
at 1.48 μm

$4l_{15/2}$

Core levels

Optical pumping produces a gain extending
from 1.50 μm to 1.56 μm

Figure 10.8: The energy diagram of erbium ions in a silica fiber. Gain is produced by optically pumping electrons from the $4\ell_{15/2}$ level to the broader $4\ell_{13/2}$ band.

fiber (see Eqn. 10.34)

$$L \equiv l_{disp}(SM) = \frac{c\lambda}{2B\Delta\lambda Y_m} = 2.4 \; km$$

which is smaller than the length obtained from attenuation considerations.

*Thus for the single mode fiber, the dispersion limit is the one which sets the repeater spacing of 2.4 km for the LED at 0.88 μm.* In the case of the graded index fiber, we must include the effect of the modal dispersion. The total broadening is

$$\frac{\tau_{tot}}{\ell} = \frac{(\tau_{mp}^2 + \tau_{disp}^2)^{1/2}}{\ell}$$

using $\tau_{mp}/\ell = 0.5$ ns/km and $\tau_{disp}/\ell = 2.07$ ns/km for the modal and material dispersion

| Source | Emission Wavelength ($\mu$m) | Spectral Width (Å) | Power Coupled to the Fiber ($\mu$W) |
|---|---|---|---|
| LED | 0.88 | 300 | 50 |
|  | 1.3 | 300 | 50 |
|  | 1.55 | 300 | 50 |
| Laser diode (LD) | 0.88 | 30 | 1000 |
|  | 1.3 | 30 | 1000 |
|  | 1.55 | 30 | 1000 |
| Distributed | 0.88 | 3.0 | 1000 |
| feedback LD | 1.3 | 3.0 | 1000 |
|  | 1.55 | 3.0 | 1000 |

| Fiber Type | Wavelength ($\mu$m) | Modal Dispersion (ns/km) | Material Dispersion $\left(\frac{\lambda}{c}\frac{d^2 n_r}{d\lambda^2}\right)$ (ps/km-nm) | Attenuation (dB/km) |
|---|---|---|---|---|
| Graded | $\lambda = 0.88$ | 0.5 | 70 | 1.5 |
| index | $\lambda = 1.3$ | 0.5 | 2 | 0.6 |
|  | $\lambda = 1.55$ | 0.5 | 20 | 0.2 |
| Single | $\lambda = 0.88$ | 0.0 | 70 | 1.5 |
| mode | $\lambda = 1.3$ | 0.0 | 2 | 0.6 |
|  | $\lambda = 1.55$ | 0.0 | 20 | 0.2 |

Table 10.1: Specifications for the optical fibers and optical sources.

broadening, we get

$$\frac{\tau_{tot}}{\ell} = 2.13 ns/km$$

This gives for the repeater distance for the graded-index (GI) fiber for bandwidth $B = 10^8$ bps

$$L(GI) = \frac{10^9}{2 \times 10^8 \times 2.13} = 2.35 km$$

Thus the repeater spacing is a little smaller in the graded-index fiber.

Similar calculations for the other device combinations give the repeater spacings listed in Table 10.2.

This example shows the complicated interplay between the various component specifications in deciding the performance of the total system. The reader is urged to carefully examine the repeater distances given in Table 10.2. The reader can look for the following comparisons: *i) the LED has a poor performance for 0.88 $\mu$m transmission, but can perform reasonably well if it is used at 1.33 $\mu$m or 1.55 $\mu$m; ii) the 1.33 $\mu$m source does not perform as well as the 1.55 $\mu$m source in a single mode fiber at low transmission rates, but performs better at a high transmission rate.*

| Source | Wavelength ($\mu$m) | Repeater Spacing for SM fiber (km) | | Repeater Spacing for GI Fiber (km) | |
|---|---|---|---|---|---|
| | | 0.1 Gbps | 1.0 Gbps | 0.1 Gbps | 1 Gbps |
| LED | 0.88 | 2.4 (D) | 0.24 (D) | 2.35 (D) | 0.24 (D) |
| | 1.33 | 61 (A) | 45 (A) | 10.0 (D) | 1.0 (D) |
| | 1.55 | 78 (D) | 7.8 (D) | 6.2 (D) | 0.6 (D) |
| LD | 0.88 | 24 (D) | 2.4 (D) | 10.0 (D) | 1.0 (D) |
| | 1.33 | 61 (A) | 45 (A) | 10.0 (D) | 1.0 (D) |
| | 1.55 | 78 (D) | 7.8 (D) | 10.0 (D) | 1.0 (D) |
| DFB-LD | 0.88 | 33 (A) | 24 (D) | 10.0 (D) | 1.0 (D) |
| | 1.33 | 61 (A) | 45 (A) | 10.0 (D) | 1.0 (D) |
| | 1.55 | 250 (A) | 200 (A) | 10.0 (D) | 1.0 (D) |

Table 10.2: Calculated results for repeater spacings for the design problem of Example 10.4. The symbol $D$ is for dispersion limited; $A$ is for attenuation limited.

## 10.6 SUMMARY OF DEVICE REQUIREMENTS

$\longrightarrow$ Optical communication places a number of requirements on optoelectronic devices such as sources, detectors, switches, etc. In this chapter we have examined some of the important requirements on source power, spectral purity, detector's detectivity, etc., and how these affect system performance. Some of these requirements are met by the devices. However, many of the requirements are not yet fulfilled. We will briefly summarize these issues.

**Sources**: The optical sources are still not at a stage where their phase and frequency are stable enough to be used in coherent detection. The coherent detection relies on the signal to maintain a phase coherence over a period of time. Due to this lack of phase (and frequency) coherence, the optical signals cannot use the tremendous potential of billions of optical signals being transmitted in parallel in an optical fiber.

The next level of requirements on sources is emission at a stable wavelength with very small spectral width. Even this requirement where coherence is not critical is difficult to satisfy by the modern laser. Laser linewidths are quite large and the wavelength is not very stable. Of course, progress is being made in this area, and it is currently possible to send up to 20 or so different "color" optical signals in parallel in a fiber.

The present optical sources are commercially being used for applications where only a single signal is sent down a fiber. For this application, the spectral purity of an LED is suitable only for LANs. For long distance communication, the sources of choice are laser diodes (preferably DFB laser diodes).

Since, at present, the parallel transmission potential of optical fiber is not being exploited, the data transmission rate can be pushed up only by faster optical sources. Currently, the LED modulation speeds are less than 5 GHz (with little possibility of further increase) while the laser diodes operate under 30 GHz. It is unlikely that laser diodes will reach the response times capable in electronic devices ($\geq 300$GHz). Thus, it is clear that the future of optical communication is closely tied to tunable stable laser sources which can first be used for WDM applications and eventually be used for FDM applications.

**Detectors**: The present detectors are not capable of wavelength- or frequency selective detection. Thus even if sources were available which could emit with a very closely spaced wavelength, the detectors would not be able to distinguish between these signals. The detectors are, however, quite adequate for the present single-signal transmission. By exploiting alloys and heterostructures, all the important detection wavelengths (0.88 $\mu$m for LANs, 1.3 $\mu$m and 1.55 $\mu$m for long-distance applications) are covered quite well.

**Intelligent Optoelectronic Devices**: An important aspect of the optical communication system involves the repeaters, switches, and routers. The repeaters involve devices to regenerate a signal once it has fallen to a low value. Most modern repeater systems convert the optical signal to an electronic signal which is amplified and used to drive laser diodes. An important area where a great deal of progress is being made is the area of "optical amplifiers," especially the fiber amplifiers.

The switches and routers are critical elements in a communication system. Currently, essentially all of these devices involve either electronic devices or use materials like LiNbO$_3$. However, advances are being made in using quantum well based devices for switches.

From the discussion of this section it is clear that there is a great deal of work to be done to develop semiconductor optoelectronic devices which will fully exploit the potential of optical communication. In this sense, the modern optoelectronic devices are still in their infancy. This, of course, makes the area of optoelectronics very exciting.

## 10.7   ADVANCED DEVICES: OPTOELECTRONIC INTEGRATED CIRCUITS (OEICs)

$\mathcal{R}$  The tremendous advantages of integrating devices on the same chip in electronics naturally suggest that the same be done with electronic and optoelectronic devices. The OEICs have, however, proven to be a difficult challenge. Integrating transistors and resistors and capacitors was not so difficult because of the compatibility of the fabrication process. However, the integration of the laser with its driver (a FET or a bipolar transistor) or a photodetector with an amplifier is proving to be quite a challenge because of the inherent incompatibilities in these devices. For example, the laser requires a *p-i-n* structure which is quite different from the structure of any transistor. This incompatibility requires that if the system is to remain planar, deep etching and regrowth must be carried out.

The area of regrowth is still in its infancy (as will be discussed in the next chapter). Also, it is clear that when a material is etched, certain defects are left on the surface so that, after regrowth, bandgap states are created which can trap electrons. A great deal of effort in OEIC technology is being devoted to develop less damaging etches and better regrowth techniques.

In OEICs, if regrowth is to be completely avoided, one can integrate devices on different levels (i.e., on different heights on the wafer). This, however, is not an optimum approach, although it does provide working circuits. A number of schemes have been reported and are being pursued to advance OEIC technology. Nearly all kinds of different electronic and optical devices have been integrated, and recent results have indicated performance levels approaching those in hybrid technologies. Of course, it is expected that as progress continues, the OEICs will achieve better performance and certainly better reliability than hybrid circuits. Advances in OEIC technology are essential to realize the full promise of optoelectronics.

## 10.8   CHAPTER SUMMARY

$\mathcal{R}$   In this chapter we have discussed some of the driving forces behind modern optoelectronic information processing systems. The optical communication system is one of the most important systems which is exploiting the best of optics and electronics. We discussed the device demands placed by the communication systems on light sources, detectors, switches, etc. We have noted that many of the device needs remain unfulfilled so that the full potential of the optical communication system has not yet been achieved. Tables 10.3 and 10.4 summarize the issues addressed in this chapter.

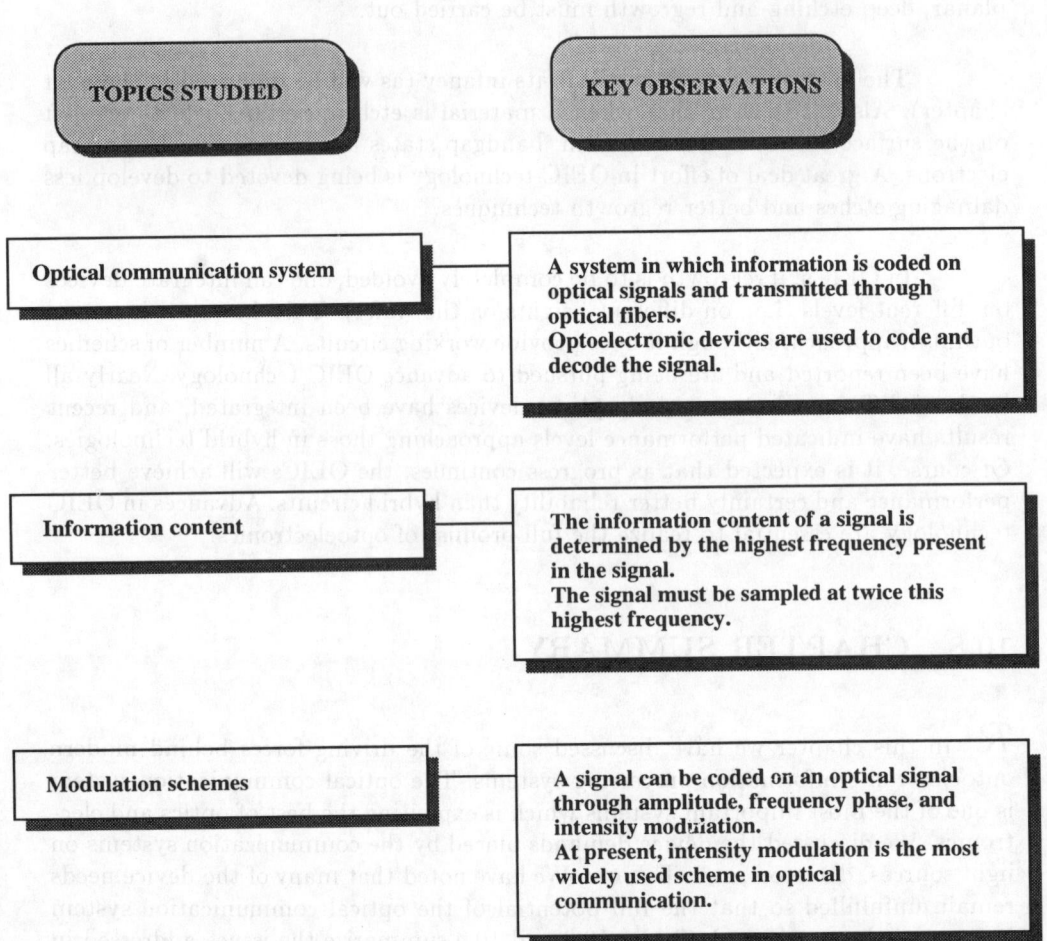

| TOPICS STUDIED | KEY OBSERVATIONS |
|---|---|
| Optical communication system | A system in which information is coded on optical signals and transmitted through optical fibers.<br>Optoelectronic devices are used to code and decode the signal. |
| Information content | The information content of a signal is determined by the highest frequency present in the signal.<br>The signal must be sampled at twice this highest frequency. |
| Modulation schemes | A signal can be coded on an optical signal through amplitude, frequency phase, and intensity modulation.<br>At present, intensity modulation is the most widely used scheme in optical communication. |

Table 10.3: Summary table.

| TOPICS STUDIED | KEY OBSERVATIONS |
|---|---|
| Signal dispersion in fibers | As optical signals propagate in multimode fibers, they broaden due to different speeds of propagation for various modes in the fiber. Also, signals suffer broadening due to the material dispersion in the fiber. |
| Signal attenuation in fibers | Optical signals suffer losses as they travel in a fiber. The loss is lowest at 1.55 μm for the widely used silica fibers. |
| Optical fiber amplifiers | Optical fibers doped with rare earth atoms can amplify optical signals. These fiber amplifiers are pumped by semiconductor lasers to provide positive gain. |

Table 10.4: Summary table.

## 10.9  PROBLEMS

**Section 10.3**

**10.1**  A compact disc is to be broadcast digitally over a communication channel. The disc has 60 minutes of music covering a bandwidth of 20 kHz. The recording has a dynamic range of 80 dB. Calculate the transmission rate and the total information content of the transmission. Dynamic range is given by

$$20log\frac{A_{signal}}{A_{noise}}$$

For a digital transmission, a 80 dB dynamic range means that $A_{signal}/A_{noise} = 10^4$. This requires one to have 14 digital bits assigned to each sampling ($2^{14} = 1.6 \times 10^4$).

**10.2**  A television channel is to be broadcast over a transmission system. The signal occupies a total of 5.5 MHz bandwidth and requires a dynamic range of 50 dB. What is the capacity needed if the signal is to be transmitted digitally? Recently, some telephone companies have suggested that by using data compression/decompression technologies they can send a TV signal over telephone wires. If the capacity of copper wires is $\sim$ 10 Mbps, what level of compression is needed to send a TV signal?

**10.3**  Using the information given in Fig. 10.2 on the present (early 1990's) capacity of various transmission systems, estimate the number of TV channels that can be sent simultaneously over the four transmission media mentioned in Fig. 10.2. Assume that each TV channel needs $\sim$ 100 Mbps capacity.

**Section 10.5**

**10.4**  Calculate the distance an optical signal will travel before it falls to a hundredth of its initial value in optical fibers with attenuations of i) 20 dB/km; ii) 2 dB/km; and iii) 0.2 dB/km.

**10.5**  Calculate the maximum acceptance angles $\theta_A$ for the following step index fibers:

i)    $n_{r1} = 1.470; n_{r2} = 1.45; n_{ro} = 1.0$

ii)   $n_{r1} = 1.46; \quad n_{r2} = 1.4; \; n_{ro} = 1.0$

**10.6**  Calculate the multipath dispersion parameter $\tau_{mp}/\ell$ for the two fibers of Problem 10.5. Also calculate the bandwidth-distance product.

**10.7**  In order to exploit the low loss, low dispersion window of silica fibers, an optical source emitting at 1.3 $\mu$m is selected. Why is it not possible to use silicon detectors for this technology?

**10.8**  An LED is used as an optical source for an LAN. The LED emits at 0.8 $\mu$m, and 15.0 $\mu$W of optical power is coupled into a fiber with attenuation of 2 dB/km. An InGaAs $p$-$i$-$n$ detector with detection limits defined by a minimum power of

1.0 nW/Mbps receives the signal. If the source-to-detector spacing is 1 km and the system performance is governed by signal attenuation, calculate the maximum transmission rate possible.

**10.9** Discuss the reasons why the simple detectors discussed in Chapter 10 cannot exploit the capacity of an optical fiber.

**10.10** A GaAs/AlGaAs laser emitting at 0.88 $\mu$m is to be used in an LAN. Assume that the material dispersion of the fiber sets the limit of the system performance. The parameter $Y_m$ describing the material dispersion has a value of 0.025. Calculate the maximum spectral width $\Delta\lambda$ of the laser if a 1 Gbps transmission is to occur with a repeater or detector spacing of 2 km.

## 10.10   REFERENCES

- General

  - T. Edwards. *Fiber Optic Systems, Network Applications*, John Wiley and Sons, New York (1989).

  - J. Gowar, *Optical Communication System*, Prentice-Hall, Englewood Cliffs, NJ (1984).

# CHAPTER
# 11

# FABRICATION AND PROCESSING OF DEVICES

## 11.1   INTRODUCTION

This textbook deals with the optoelectronic devices which are designed to provide the key components of the information age. The devices we have addressed are based on a wide range of materials, including semiconductors, glasses and liquid crystals. Semiconductors are currently the basis of most electronic devices used in information processing, and it makes sense to use the same materials for optoelectronic devices as much as possible. In this chapter we will discuss how semiconductors are manufactured and give an overview of the various techniques used to fabricate devices. We will also discuss some of the challenges that remain to be overcome in the device fabrication arena, particularly in regard to optoelectronic devices. In addition we will discuss the fabrication challenges in the area of optical fibers. A brief discussion of fiber splicing and coupling will also be given. Important issues in liquid crystal displays have been mentioned in Chapter 9.

## 11.2   SEMICONDUCTORS: BULK CRYSTAL GROWTH

$\mathcal{R}$   Bulk crystal growth techniques are used mainly to produce substrates on which devices are eventually fabricated. While for some semiconductors like Si and GaAs (to some extent for InP), the bulk crystal growth techniques are highly matured, for most other semiconductors it is difficult to obtain high quality, large area substrates. The aim of the bulk crystal growth techniques is to produce single crystal boules with as large a diameter as possible and with as few defects as possible. In Si the boule diameters have reached 30 cm with boule lengths approaching 100 cm. Large size substrates ensure low cost device production.

For the growth of boules from which substrates are obtained, one starts out with a purified form of the elements that are to make up the crystal. One important technique that is used is the Czochralski (CZ) technique. In the Czochralski technique shown in Fig. 11.1, the melt of the charge (i.e., the high quality polycrystalline material) is held in a vertical crucible. The top surface of the melt is just barely above the melting temperature. A seed crystal is then lowered into the melt and slowly withdrawn. As the heat from the melt flows up the seed, the melt surface cools and the crystal begins to grow. The seed is rotated about its axis to produce a roughly circular cross-section crystal. The rotation inhibits the natural tendency of the crystal to grow along certain orientations to produce a faceted crystal.

The CZ technique is widely employed for Si, GaAs, and InP and produces long ingots (boules) with very good circular cross-section. For Si up to 100 kg ingots can be obtained. In the case of GaAs and InP the CZ technique has to face problems arising from the very high pressures of As and P at the melting temperature of the

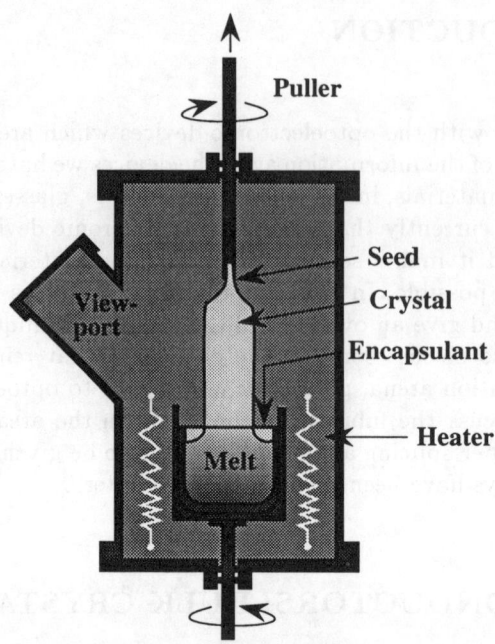

Figure 11.1: Schematic of Czocharlski-style crystal grower used to produce substrate ingots. The approach is widely used for Si, GaAs and InP.

compounds. Not only does the chamber have to withstand such pressures, also the As and P leave the melt and condense on the sidewalls. To avoid the second problem one seals the melt by covering it with a molten layer of a second material (e.g., boron oxide) which floats on the surface. The technique is then referred to as liquid encapsulated Czochralski, or the LEC technique.

A second bulk crystal growth technique involves a charge of material loaded in a quartz container. The charge may be composed of either high quality poly-crystalline material or carefully measured quantities of elements which make up a compound crystal. The container called a "boat" is heated till the charge melts and wets the seed crystal. The seed is then used to crystallize the melt by slowly lowering the boat temperature starting from the seed end. In the gradient-freeze approach the boat is pushed into a furnace (to melt the charge) and slowly pulled out. In the Bridgeman approach, the boat is kept stationary while the furnace temperature is temporally varied to form the crystal. The approaches are schematically shown in Fig. 11.2.

The easiest approach for the boat technique is to use a horizontal boat. However, the shape of the boule that is produced has a D-shaped form. To produce circular cross-sections vertical configurations have now been developed for GaAs and InP.

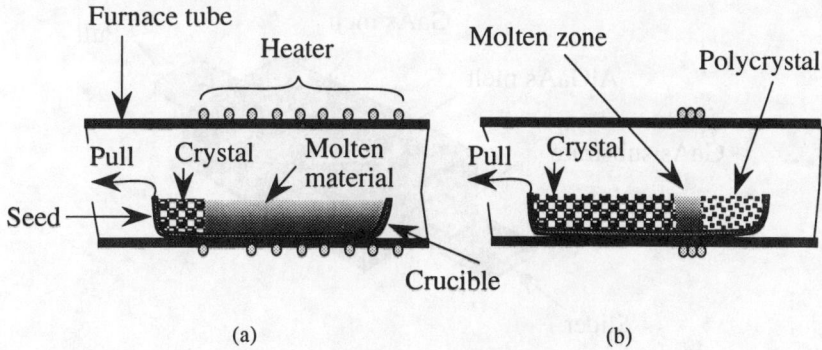

Figure 11.2: Crystal growing from the melt in a crucible: (a) solidification from one end of the melt (horizontal Bridgeman method); (b) melting and solidification in a moving zone.

In addition to producing high purity bulk crystals, the techniques discussed above are also responsible for producing crystals with specified electrical properties. This may involve high resistivity materials along with $n$- or $p$-type materials. In Si it is difficult to produce high resistivity substrates by bulk crystal growth and resistivities are usually $<10^4$ $\Omega$-cm. However, in compound semiconductors carrier trapping impurities such as chromium and iron can be used to produce material with resistivities of $\sim 10^8$ $\Omega$-cm. The high resistivity or semi-insulating (SI) substrates are extremely useful in device isolation and for high speed devices. For $n$- or $p$-type doping carefully measured dopants are added in the melt.

The availability of high quality substrates is essential to any device technology. Other than the three materials mentioned above (Si, GaAs, InP) the substrate fabrication of semiconductors is still in its infancy. Since epitaxial growth techniques used for devices require close lattice matching between the substrate and the overlayer, non- availability of substrates can seriously hinder the progress of a material technology. This is, for example, one of the reasons of slow progress in large bandgap semiconductor technology necessary for high power-high temperature electronic devices and short wavelength semiconductor lasers.

## 11.3 EPITAXIAL CRYSTAL GROWTH

$\longrightarrow$ The substrates that result once the bulk grown semiconductor boule is sliced and lapped are almost never used directly for devices. Invariably an epitaxial layer or epilayer is grown which may be a few microns in thickness. The word epitaxy comes from the Greek word "epi" (upon) and "taxis" (ordered) meaning the ordered continuation of the substrate crystal. The epitaxial growth techniques have a very slow growth rate (as low as a monolayer per second for some techniques), which

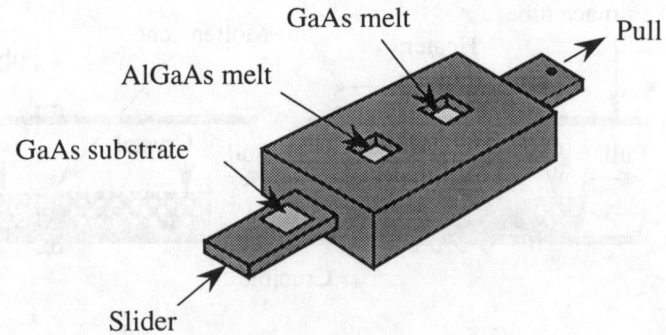

Figure 11.3: A schematic of the LPE growth of AlGaAs and GaAs. The slider moves the substrate, thus positioning itself to achieve contact with the different melts to grow heterostructures.

allows one to control very accurately the dimensions in the growth direction. In fact, in techniques like molecular beam epitaxy (MBE) and metal organic chemical vapor deposition (MOCVD), one can achieve monolayer ($\sim 3$ Å) control in the growth direction. This level of control is essential for the variety of heterostructure devices that are being used in optoelectronics. The epitaxial techniques are also very useful for precise doping profiles that can be achieved. In fact, it may be argued that without the advances in epitaxial techniques that have occurred over the last two decades, most of the developments in semiconductor physics would not have occurred. Table 11.1 gives a brief view of the various epitaxial techniques used along with some of the advantages and disadvantages.

### Liquid Phase Epitaxy (LPE)
LPE was an epitaxial growth technique of choice until the 70's when it gradually started to be replaced by other techniques. LPE is still used for growth of crystals such as HgCdTe for long wavelength detectors and AlGaAs for double heterostructure lasers. As shown in Table 11.1, LPE is a close to equilibrium technique in which the substrate is placed in a quartz or a graphite boat and covered by a liquid of the crystal to be grown (see Fig. 11.3). The liquid may also contain dopants that are to be introduced into the crystal. LPE is often used for alloy growth where the growth follows the equilibrium solid-liquid phase diagram. By precise control of the liquid composition and temperature, the alloy composition can be controlled. Because LPE is a very close to equilibrium growth technique, it is difficult to grow alloy systems which are not miscible or even grow heterostructures with atomically abrupt interfaces. Nevertheless heterostructures where interface is graded over 10-20 Å can be grown by LPE by sliding the boat over successive "puddles" of different semiconductors. For many applications such interfaces are adequate and since LPE is a relatively inexpensive growth technique, it is used widely in many commercial applications.

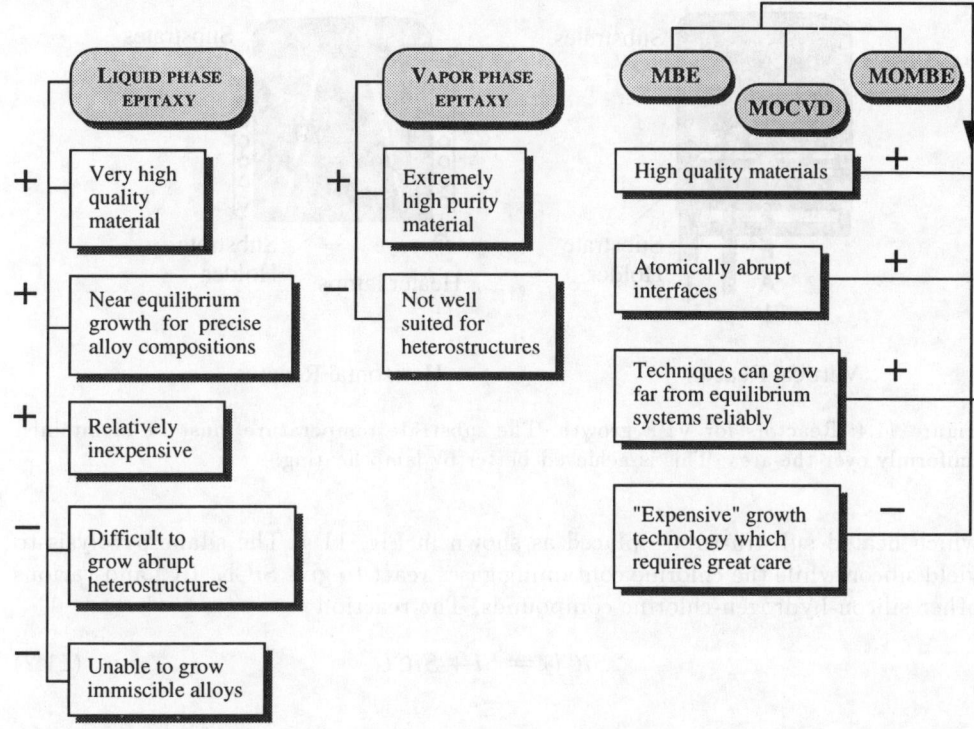

Table 11.1: A schematic of the various epitaxial crystal growth techniques and some of their positive and negative aspects.

## Vapor Phase Epitaxy (VPE)

A large class of epitaxial techniques relies on delivering the components that form the crystal from a gaseous environment. If one has molecular species in a gaseous form with partial pressure $P$, the rate at which molecules impinge upon a substrate is given by

$$F = \frac{P}{\sqrt{2\pi m k_B T}} \sim \frac{3.5 \times 10^{22} P(torr)}{\sqrt{m(g)T(K)}} \ mol./cm^2 - s \tag{11.1}$$

where $m$ is the molecular weight and $T$ the cell temperature. For most crystals the surface density of atoms is $\sim 7 \times 10^{14}$ cm$^{-2}$. If the atoms or molecules impinging from the vapor can be deposited on the substrate in an ordered manner, epitaxial crystal growth can take place.

The VPE technique is used mainly for homoepitaxy and does not have the additional apparatus present in techniques such as MOCVD for precise heteroepitaxy. As an example of the technique, consider the VPE of Si. The Si containing reactant silane (SiH$_4$) or dichlorosilane (SiH$_2$Cl$_2$) or trichlorosilane (SiHCl$_3$) or silicon tetrachloride (SiCl$_4$) is diluted in hydrogen and introduced into a reactor in

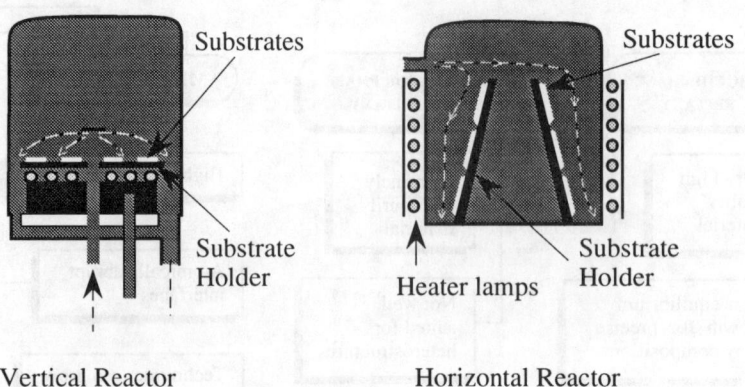

Vertical Reactor                    Horizontal Reactor

Figure 11.4: Reactors for VPE growth. The substrate temperature must be maintained uniformly over the area. This is achieved better by lamp heating.

which heated substrates are placed as shown in Fig. 11.4. The silane pyrolysis to yield silicon while the chlorine containing gases react to give $SiCl_2$, HCl and various other silicon-hydrogen-chlorine compounds. The reaction

$$2SiCl_2 \rightleftharpoons Si + SiCl_4 \qquad (11.2)$$

then yields Si. Since HCl is also produced in the reaction, conditions must be tailored so that no etching of Si occurs by the HCl. Doping can be carried out by adding appropriate hydrides (phosphine, arsine, etc.,) to the reactants.

An important consideration in VPE is safety related, since hydrogen, which is produced in the deposition, can explode in contact with any oxygen. Also, almost all the reactants are highly toxic.

VPE can be used for other semiconductors as well by choosing different appropriate reactant gases. The reactants used are quite similar to those employed in the MOCVD technique discussed later.

**Molecular Beam Epitaxy (MBE)**
Molecular beam epitaxy (MBE) is one of the most important epitaxial techniques as far as heterostructure physics and devices are concerned. Almost every semiconductor (other than a few very large bandgap semiconductors) has been grown by this technique. MBE is a high vacuum technique ($\sim 10^{-11}$ torr vacuum when fully pumped down) in which crucibles containing a variety of elemental charges are placed in the growth chamber (Fig. 11.5). The elements contained in the crucibles make up the components of the crystal to be grown as well as the dopants that may be used. When a crucible is heated, atoms or molecules of the charge are evaporated, and these travel in straight lines to impinge on a heated substrate.

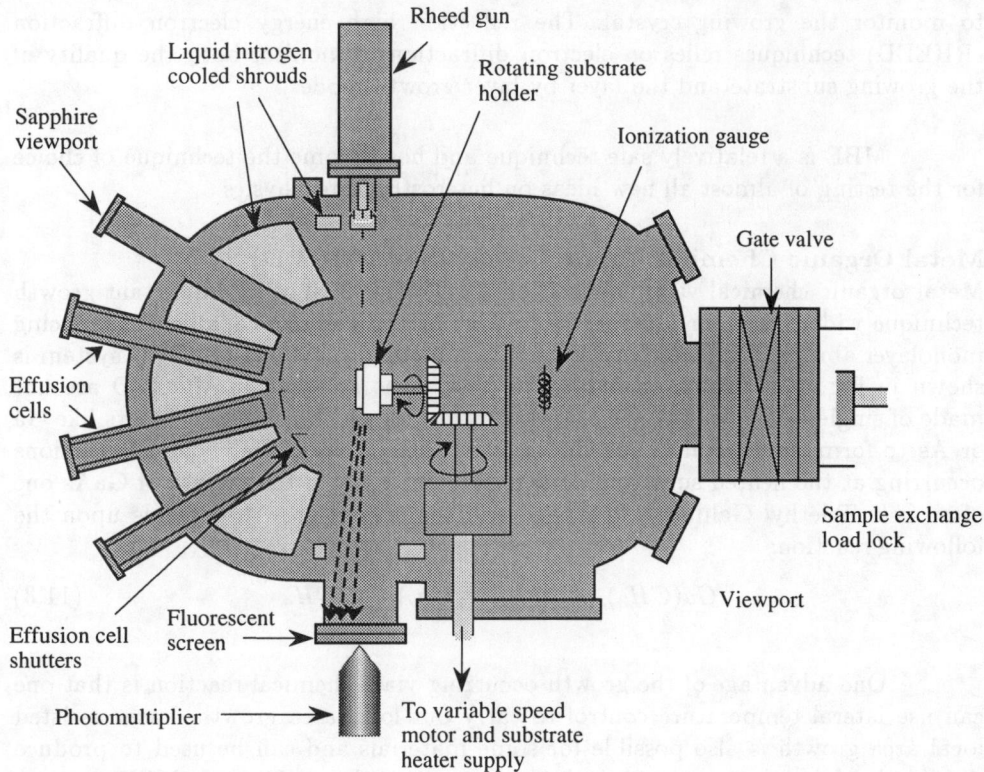

Figure 11.5: A schematic of the MBE growth system.

The growth rate in MBE is ~1.0 monolayer per second and this slow rate coupled with shutters placed in front of the crucibles allow one to switch the composition of the growing crystal with monolayer control. However, to do so, the growth conditions have to be adjusted so that growth occurs in the monolayer by monolayer mode rather than by 3-dimensional island formation. This requires that the atoms impinging on the substrate have enough kinetics to reach an atomically flat profile. Thus the substrate temperature has to be maintained at a point where it is high enough to provide enough surface migration to the incorporating atoms, but not so high as to cause entropy controlled defects.

Since no chemical reactions occur in MBE, the growth is the simplest of all epitaxial techniques and is quite controllable. However, since the growth involves high vacuum, leaks can be a major problem. The growth chamber walls are usually cooled by liquid $N_2$ to ensure high vacuum and to prevent atoms/molecules to come off from the chamber walls.

The low background pressure in MBE allows one to use electron beams to monitor the growing crystal. The reflection high energy electron diffraction (RHEED) techniques relies on electron diffraction to monitor both the quality of the growing substrate and the layer by layer growth mode.

MBE is a relatively safe technique and has become the technique of choice for the testing of almost all new ideas on heterostructure physics.

### Metal Organic Chemical Vapor Deposition (MOCVD)

Metal organic chemical vapor deposition (MOCVD) is another important growth technique widely used for heteroepitaxy. Like MBE, it is also capable of producing monolayer abrupt interfaces between semiconductors. A typical MOCVD system is shown in Fig. 11.6. Unlike in MBE, the gases that are used in MOCVD are not made of single elements, but are complex molecules which contain elements like Ga or As to form the crystal. Thus the growth depends upon the chemical reactions occurring at the heated substrate surface. For example, in the growth of GaAs one often uses Triethyl Gallium and Arsine and the crystal growth depends upon the following reaction:

$$Ga(CH_3)_3 + AsH_3 \rightleftharpoons GaAs + 3CH_4 \tag{11.3}$$

One advantage of the growth occurring via a chemical reaction is that one can use lateral temperature control to carry out local area growth. Laser assisted local area growth is also possible for some materials and can be used to produce new kinds of device structures. Such local area growth is difficult in MBE.

There are several varieties of MOCVD reactors. In the atmospheric MOCVD the growth chamber is essentially at atmospheric pressure. One needs a large amount of gases for growth in this case, although one does not have the problems associated with vacuum generation. In the low pressure MOCVD the growth chamber pressure is kept low. The growth rate is then slower as in the MBE case.

The use of the MOCVD equipment requires very serious safety precautions. The gases used are highly toxic and a great many safety features have to be incorporated to avoid any deadly accidents. Safety and environmental concerns are important issues in almost all semiconductor manufacturing since quite often one has to deal with toxic and hazardous materials.

In addition to MBE and MOCVD one has a hybrid epitaxial technique often called MOMBE (metal organic MBE) which tries to combine the best of MBE and MOCVD. In MBE one has to open the chamber to load the charge for the materials to be grown while this is avoided in MOCVD where gas bottles can be easily replaced from outside. Additionally, in MBE one has occasional spitting of material in which small clumps of atoms are evaporated off on to the substrate. This is avoided in MOCVD and MOMBE.

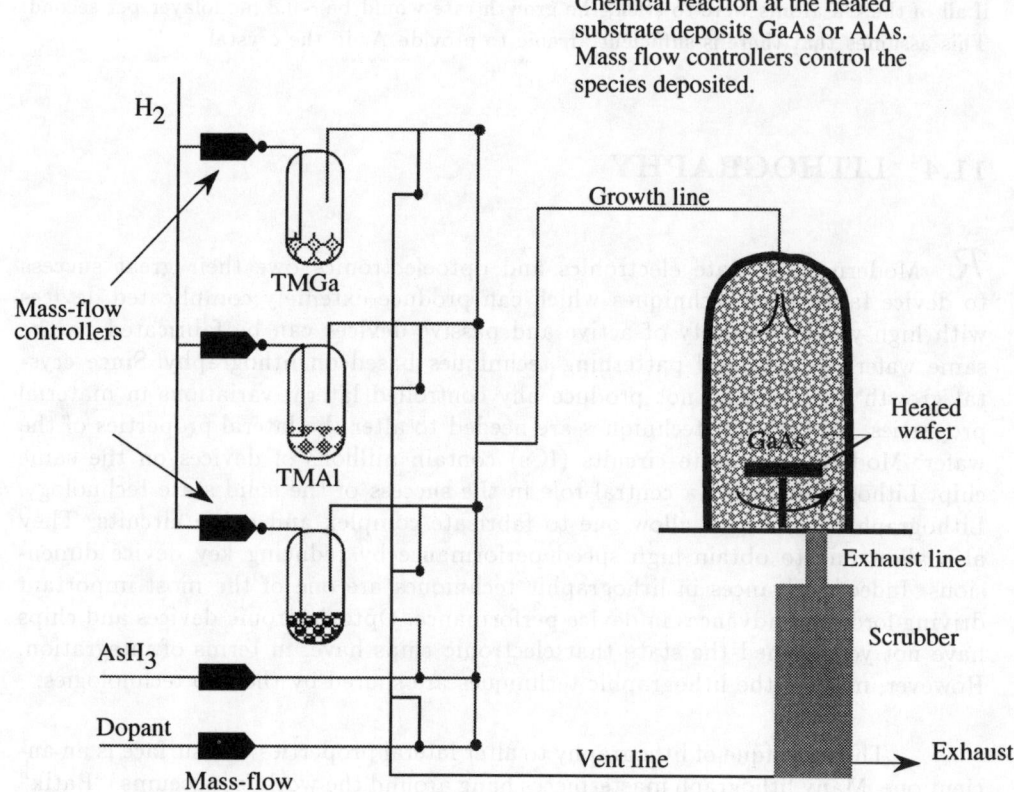

Chemical reaction at the heated substrate deposits GaAs or AlAs. Mass flow controllers control the species deposited.

TMGa : Gallium containing organic compound
TMAl : Aluminum containing organic compound
AsH$_3$ : Arsenic containing compound

Figure 11.6: Schematic diagram of an MOCVD system employing alkyds (trimethyl gallium (TMGa) and trimethyl aluminum (TMAl)) and metal hydride (arsine) material sources, with hydrogen as a carrier gas.

**EXAMPLE 11.1** Consider the growth of GaAs by MBE. The Ga partial pressure in the growth chamber is $10^{-5}$ Torr, and the Ga cell temperature is 900 K. Calculate the flux of Ga atoms on the substrate.

The mass of Ga atoms is 70 g/mole. The flux is (from Eqn. 11.1)

$$F = \frac{3.5 \times 10^{22} \times 10^{-5}}{\sqrt{70 \times 900}} = 5.27 \times 10^{14} \, \text{atoms/cm}^2$$

Note that the surface density of Ga atoms on GaAs is $\sim 6.3 \times 10^{14} \, \text{cm}^{-2}$. Thus, if all of the Ga atoms were to stick, the growth rate would be $\sim 0.8$ monolayer per second. This assumes that there is sufficient arsenic to provide As in the crystal.

## 11.4  LITHOGRAPHY

$\mathcal{R}$   Modern solid state electronics and optoelectronics owe their great success to device fabrication techniques which can produce extemely complicated devices with high yield. A variety of active and passive devices can be fabricated on the same wafer using lateral patterning techniques based on lithography. Since crystal growth processes do not produce any controlled lateral variations in material properties, lithographic techniques are needed to alter the lateral properties of the wafer. Modern solid state circuits (ICs) contain millions of devices on the same chip. Lithography plays a central role in the success of the solid state technology. Lithographic techniques allow one to fabricate complex and dense circuits. They also allow one to obtain high speed performance by reducing key device dimensions. Indeed, advances in lithographic techniques are one of the most important driving forces for advances in device performance. Optoelectronic devices and chips have not yet reached the state that electronic chips have, in terms of integration. However, most of the lithographic techniques are shared by the two technologies.

The technique of lithography to alter lateral properties of a surface is an ancient one. Many lithograph masterpieces hang around the world's museums. "Batik" paintings and t-shirts also rely on the concepts of lithography. The lithography process involves taking a certain design created in a computer or by an artist and transferring it onto the wafer. A number of steps are involved in this process. While new advances in lithographic techniques are introducing continual changes in the individual steps, the following discussion will provide an overview of the current state of the art.

### 11.4.1  Photoresist Coating

$\mathcal{R}$   The wafer in its virgin form has little irreversible sensitivity to optical or electron beams. Such a sensitivity is needed if a pattern of device and circuits is to be transferred to the wafer. To make the wafer (which is usually covered by a thin oxide film or some other dielectric passivation material) sensitive to an image, a photoresist is spread on the wafer by a process called spin coating. For the resist to be reliable it must satisfy three criteria: i) it must have good bonding to the substrate; ii) its thickness must be uniform; and iii) the thickness should be reliably controlled over different wafer runs.

Spin-coating has emerged as the most reliable technique for photoresist application. As shown in Fig. 11.7, a small puddle of the resist is applied to the center of the wafer which is held to a spindle by a vacuum chuck. The spindle is now spun at a rate of 2000-8000 rpm for 10 to 60 seconds. During the first couple of seconds of spinning, most of the resist is thrown off and carefully drained away. The remaining resist forms a thin layer whose thickness is controlled by the spin speed (thickness $\propto \frac{1}{\sqrt{\omega}}$, where $\omega$ is the spin frequency). An edge bead is formed which is several times the thickness of the film. A variety of details are introduced (e.g., variable spin-speed; applying the puddle to an already spinning wafer, etc.,) to obtain more uniform resists. The thickness of the resists are usually in the range of 0.7 to 1.0 $\mu$m. Once the resist is applied, it is soft baked at 90 to 100 C to improve adhesion to the oxide.

Once the resist is ready, it is exposed to an optical image through a mask (to be discussed next) for a certain exposure time as shown in Fig. 11.8. The resist is then developed by washing it in a solvent which dissolves away the regions of higher solubility. A resist which becomes more soluble when exposed to illumination is called positive. Its image is identical to the opaque image on the mask plate. A resist that loses solubility when illuminated is called negative.

The first materials which proved to have a high sensitivity and etchant resistance for successful microelectronic photoresists were negative resists. These were based on polyisoprene in the form of cyclized rubber. The resist consists of isoprene molecules (a hydrocarbon containing double carbon bonds and the basic building block of rubber) and a photo-sensitive $N_3$ group radical. When light falls on the resist, the radical causes formation of long chains from the individual isoprene hydrocarbons causing polymerization of the of the isoprene. The polymerized region is highly insoluble while the unexposed part is soluble.

The negative resist has the advantage that its solubility is almost digital. There is a tremendous contrast between exposed and unexposed regions. However, since it is based on chain formation, its resolution is limited to 2-3 $\mu$m.

The positive resist is not based on chain formation and can produce much sharper resolution. The resist relies upon molecules where $N_2$ is bonded to a hydrocarbon with benzene like structure. Exposure to light causes the dissociation of the $N_2$ group producing a carboxylic acid. This can be dissolved and washed away in an alkaline solution such as dilute sodium hydroxide.

Once a resist is in place, one needs to expose it with an optical image containing the pattern to be transferred to the wafer. This requires the fabrication and use of a mask.

**Spin coating of a resist on a wafer:  A photosensitive resist is "spun" onto the wafer.**

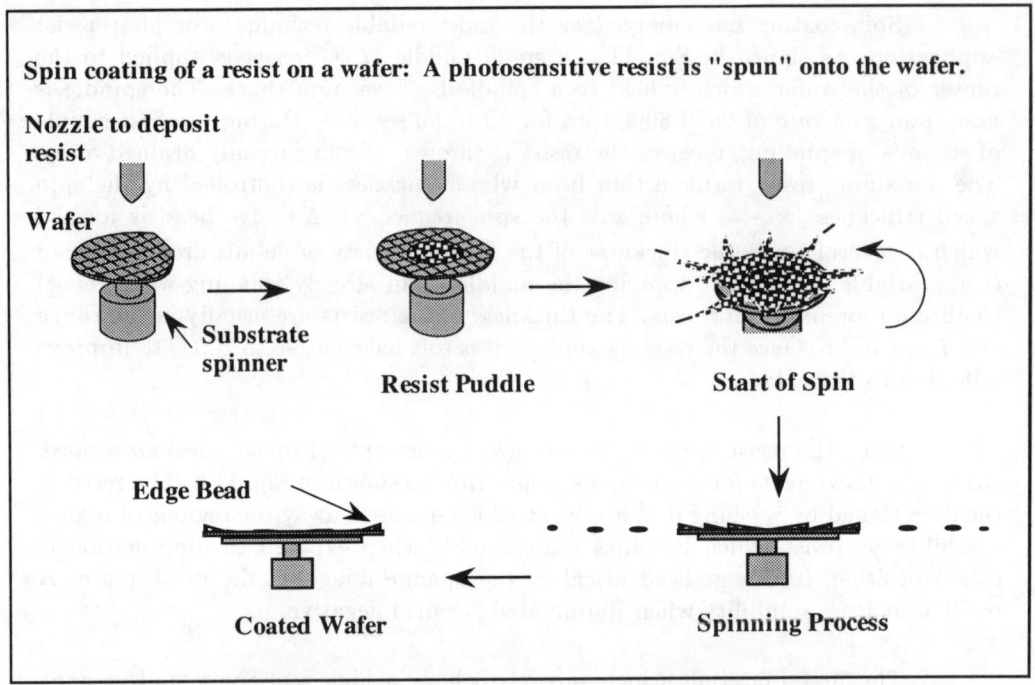

Figure 11.7: The process of spin coating a wafer starts the lithographic process.

**Transference of an image to the resist by using a mask and etching of the exposed regions.**

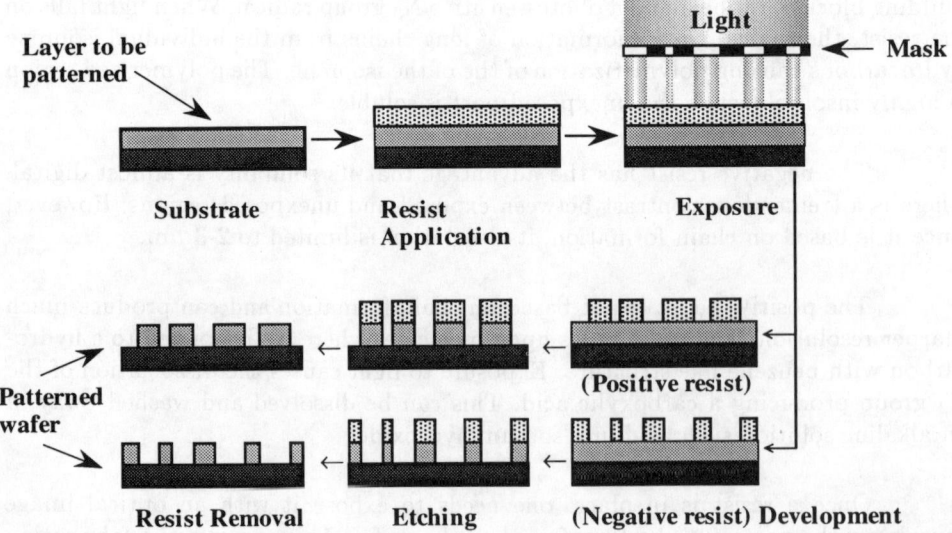

Figure 11.8: The exposure and etching process which allows one to transfer a pattern to the wafer. (After W.S. Ruska, *Microelectronic Processing, An Introduction to the Manufacture of Integrated Circuits*, McGraw-Hill, New York (1988).)

**EXAMPLE 11.2** A negative resist is made up of molecules with a mass of $10^5$ amu. The molecule is a straight chain of $CH_2$ units with a C-C spacing of 1.54 Å. The resolution of the resist is about the length of one molecule. Estimate this resolution. The atomic mass of C is 12 and of H is 1.

Each unit has a mass of 14. Thus, the number of units in the molecule is

$$n = \frac{10^5}{14} = 7143 \text{ units}$$

The total length is then

$$\ell = 7143 \times 1.54 \quad = \quad 1.1 \times 10^4 \text{ Å}$$
$$= \quad 1.1 \ \mu m$$

This resist is adequate for most technologies, but not for sub-micron technology.

## 11.4.2 Mask Generation and Image Transfer

$\mathcal{R}$ Like the negative in photography, the mask allows one to transfer a complicated circuit pattern on to the sensitive resist deposited on the wafer. These days, as shown in Fig. 11.9, computer aided design (CAD) software programs allow one to directly generate a design tape which is then analyzed and "written" on to a mask plate. The pattern on the final mask plate can be generated optically or by electron beam writing. The electron beam (e beam) produces much finer features and can be controlled with greater precision.

The mask plate which is later used to repeatedly transfer patterns to different wafers must be transparent (at proper positions) to the radiation used for final pattern transfer. The mask should also have good mechanical and thermal properties. Quartz or borosilicate glasses are often used for ultraviolet light lithography. The regions of the mask that are to be opaque are covered with metallic chrome or iron oxide.

Usually only the basic building block of the device or circuit is produced first on the mask. This single pattern is called a reticle, and a step and repeat camera (or stepper) is used to create the entire pattern from the reticle. The reticle can be used to generate a wafer size mask plate or, in some cases, may be directly used to generate the entire pattern on the wafer by the stepper, as shown in Fig. 11.9.

Once the mask plate is made and the resist is deposited, the next step is to transfer the pattern on the mask to the wafer. For feature sizes greater than 0.25 $\mu$m, one uses optical equipment for the pattern transfer. The limit of $\sim$0.25 $\mu$m is governed by the wavelength of the light available. Most materials become opaque to light once the wavelength goes below $\sim$1000 Å. If electromagnetic radiation is to be used, one must go down to x-ray lithography with $\lambda \sim$40-80 Å. Another way to decrease feature size is by use of e-beam lithography.

The patterns to be printed in microelectronics can be classified as windows and lines. The windows may be used to allow doping diffusion or to make connections with a conductor through an insulator. The conducting interconnects and resistors are usually formed into line patterns. The imaging process involves transferring an ideal image to the mask and then from the mask to the wafer. Once the image is transferred, the resist is developed and etching is carried out. In all these steps, the ideal pattern in the designers mind is gradually lost. This occurs even if the design considerations are well within the image transferring technology.

## 11.5   ETCHING

$\mathcal{R}$  Like the chisel and the drill for the carpenter and the sculptor, etchants are an important tool for the microelectronics processing engineer. The etchants allow one to remove material in a selective manner once the resist has been patterned. The choice and control of the etching process is crucial if the features in the resist film are to become a part of the substrate.

An ideal etching process must be able to remove a layer of material from the region where there is no resist. The etchant should not attack the resist, nor should it penetrate under the resist causing undercuts. It should also attack only one layer and should be self limiting (e.g., should etch $SiO_2$ and not Si, etc.). A number of techniques are developed to carry out etching. These are discussed below. Once etching is finished, one must remove the resist material by either other etchants or by the "lift off" technique. We will now briefly review the etching approaches.

### 11.5.1   Wet Chemical Etching

$\mathcal{R}$  The simplest and most commonly employed etching technique is the wet chemical etching in which the wafer is simply soaked in a liquid chemical which dissolves away the semiconductor. The etching process involves a chemical reaction in which the elements of the film to be etched react with the etching solution. The reaction products can then be rinsed away. The rate of the etching is proportional to the

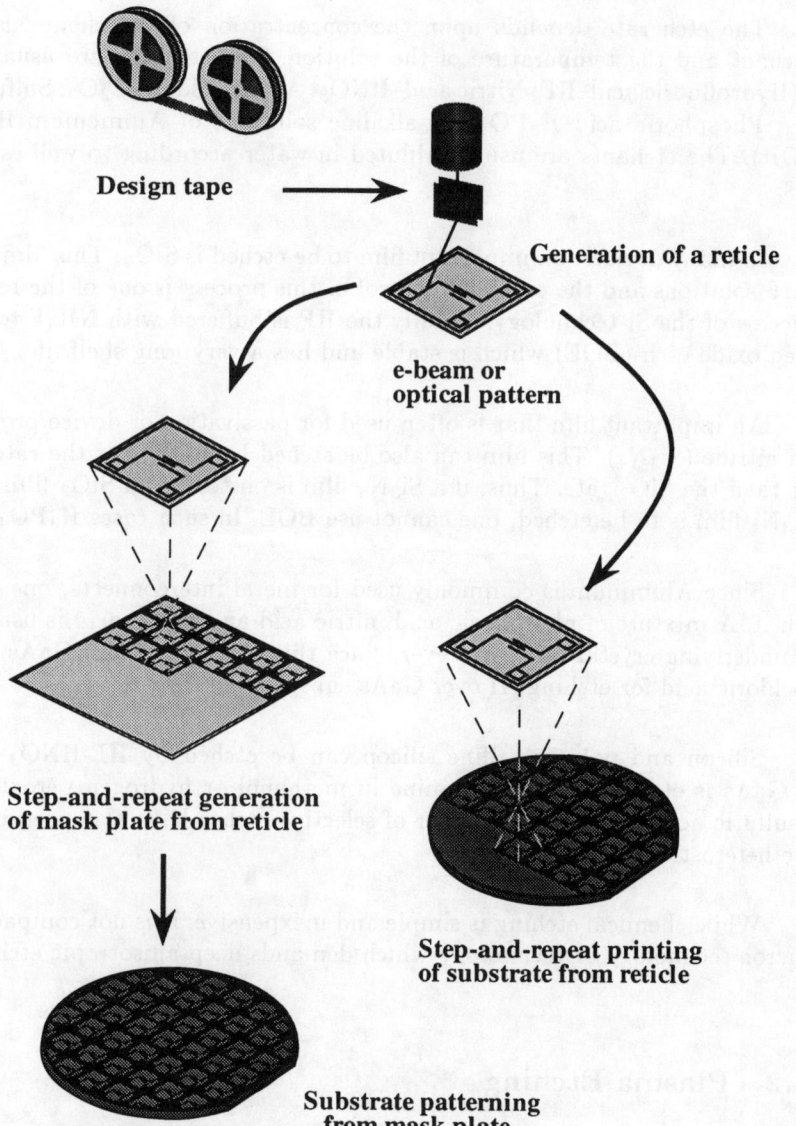

**Design tape**

**Generation of a reticle**

**e-beam or optical pattern**

**Step-and-repeat generation of mask plate from reticle**

**Step-and-repeat printing of substrate from reticle**

**Substrate patterning from mask plate**

Figure 11.9: The processes used in the generation of a mask for lithography. (After W.S. Ruska, *Microelectronic Processing, An Introduction to the Manufacture of Integrated Circuits*, McGraw-Hill, New York (1988).)

etch time and is usually isotropic. The isotropic nature of the wet etching is not very suitable for devices where very sharp sidewalls are to be produced.

The etch rate depends upon the concentration of the chemicals used in the etchant and the temperature of the solution. The etchants are usually either acids (Hydrofluoric acid–HF; Nitric acid–$HNO_3$; Acetic acid–$H_4C_2O_2$; Sulfuric acid–$H_2SO_4$; Phosphoric acid–$H_3PO_4$) or alkaline solutions of Ammonium Hydroxide ($NH_4OH$). The etchants are usually diluted in water according to well established recipes.

In many devices, an important film to be etched is $SiO_2$. This film is etched with HF solutions and the ease and control of this process is one of the reasons for the success of the Si-technology. Usually the HF is buffered with $NH_4F$ to produce buffered oxide etch (BOE) which is stable and has a very long shelf life.

An important film that is often used for passivation or device protection is silicon nitride ($Si_3N_4$). This film can also be etched by BOE, but the rate is much slower than the $SiO_2$ rate. Thus, if a $Si_3N_4$ film is on top of an $SiO_2$ film and only the $Si_3N_4$ film is to be etched, one cannot use BOE. In such cases $H_3PO_4$ is used.

Since Aluminum is commonly used for metal interconnects, one often has to etch it. A mixture of phosphoric acid, nitric acid and acetic acid is usually used if the underlying crystal is Si. However, since this etchant attacks GaAs, one uses hydrochloric acid for etching Al over GaAs substrates.

Silicon and polycrystalline silicon can be etched by HF-$HNO_3$ mixtures while GaAs is etched by using bromine in methanol or hydrogen peroxide mixed with sulfuric acid in water. A number of selective etches have also been developed for the heterostructure technology.

While chemical etching is simple and inexpensive, it is not compatible with submicron technology or technology which demands deep anisotropic etching.

## 11.5.2　Plasma Etching

$\mathcal{R}$　Plasma etching resolves one of the main problems with wet chemical etching viz feature size control. It also provides efficient etching for a wide variety of films including those that are difficult to etch by wet chemicals. The plasma is produced by passing an rf electrical discharge through a gas at a low pressure. The rf discharge creates ions and electrons. The ions can be used to interact with the elements in the substrate and cause etching.

The ions, being charged particles, can be accelerated in the electric field and be made to bombard the substrate with controlled energy. If the ion energy is large, the ions simply sputter off atoms from the surface in a rather unselective manner. However, this provides extremely anisotropic etching with almost no undercutting effects. At low energies the ions can cause chemical reactions at the surface and cause removal of atoms selectively.

In typical plasmas one introduces fluorine or chlorine containing gases. The fluorocarbons (e.g., $CF_4$) or silicon tetrafluoride ($SiF_4$) and silicon tetrachloride ($SiCl_4$) are often used for the plasma. Once the plasma is formed fluorine (or chlorine) ions are produced which cause the etching process.

### 11.5.3   Reactive Ion Beam Etching (RIBE)

$\mathcal{R}$   An important tool in microelectronic technology is the ion-implantation which is widely used to dope semiconductors. The implanter accelerates ions to a prechosen energy and shoots them into the semiconductor by controlling the energy. The depth at which the ions are embedded can be controlled. The ion-implanter can also be used for etching by using appropriate ions and it provides focusing of the ion beam. At low energies, the ions can be used to selectively etch very small feature sizes. The advantages are the same as for plasma etching.

### 11.5.4   Ion Beam Milling

$\mathcal{R}$   The ion beam milling is another application of the ions in "chiseling" off material from a substrate. The ion milling requires a focused beam of ions of energy high enough that the ions can physically knock out atoms from the film. The focus is on removing the atoms physically rather than through a chemical reaction. The process is thus highly directional and can produce extremely anisotropically etched structures. Ion milling is particularly advantageous if the etching involves small patterns, very steep walls or materials which are relatively inert and cannot be etched by chemical reactions.

The ion milling is dominated by geometric effects as shown in Fig. 11.10. In Fig. 11.10a; we show a substrate with a resist in the process of being etched by a perpendicular ion-beam. The impact of the beam causes the substrate material to fly off randomly. Some of the material can get redeposited on the etched sides forming "ears." Ions bounding off the edges can cause "trenches" to be formed around the resist pattern. These effects are controlled by impinging the beam at a

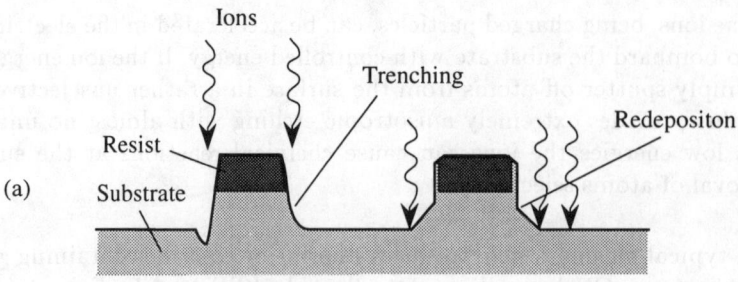

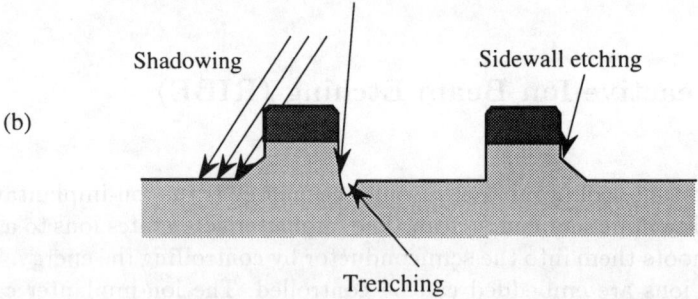

Figure 11.10: The importance of geometric effects in ion beam milling. In (a) the perpendicular incident beam can produce trenching effects as well as redeposition causing sidewall "ears;" (b) if the beam comes at an oblique angle and the substrate is rotated, the trenching and "ear" formation can be balanced.

slight angle and rotating the substrate.

The resolution of ion beam milling is controlled by the ion-beam spot size and can reach 0.1 $\mu$m or less.

**EXAMPLE 11.3** A 0.2 $\mu$m film of AlGaAs is to be etched. The etching rate is 1.0 $\mu$m/hour. Calculate the time needed for the etching. If the rate changes by -1%, calculate the number of monolayers of AlGaAs that will remain unetched.

The time needed for the etching is

$$t = \frac{0.2}{1.0} = 0.2 \text{ hour}$$

A -1% error in the etch rate means that the thickness etched is

$$d = 0.2 \times 0.99 = 0.198 \ \mu m$$

The unetched film remaining is

$$\Delta d = 0.2 - 0.198 \quad = \quad 0.002 \ \mu m$$
$$= \quad 20 \ \mathring{A}$$

A monolayer of AlGaAs is 2.83 Å, so that seven monolayers of film will remain. Such thickness errors may appear small, but are unacceptable in many heterostructure devices.

## 11.6  EPITAXIAL REGROWTH

$\mathcal{R}$ The spectacular growth of semiconductor microelectronics owes a great deal to the concept of the integrated circuit. The ability to fabricate transistors, resistors, and capacitors on the same wafer is critical to the low cost and high reliability we have come to expect from microelectronics. It is natural to expect similar dividents from the concept of the optoelectronic integrated circuit (OEIC). In the OEIC, the optoelectronic device (the laser or detector or modulator) would be integrated on the same wafer with an amplifier or logic gates.

One of the key issues in OEICs involves etching and regrowth. As we have seen in Chapters 6, 7, 8 and 9, the optoelectronic devices have a structure that is usually not compatible with the structure of an electronic device. The optimum layout then involves growing one of the device structures epitaxially and then masking the region to be used as, say, the optoelectronic device and etching away the epitaxial region. Next a regrowth is done to grow the electronic device with a different structure. The process is shown schematically in Fig. 11.11. While this process looks simple conceptually, there are serious problems associated with etching and regrowth.

A critical issue in the epitaxial growth of a semiconductor layer is the quality of the semiconductor-vacuum interface. This semiconductor surface must be "clean," i.e., there should be no impurity layers (e.g., an oxide layer) on the surface. Even if a fraction of a monolayer of the surface atoms have impurities bonded to them, the quality of the epitaxial layer suffers drastically. The growth may occur to produce microcrystalline regions separated by grain boundaries or may be amorphous in nature. In either case, the special properties arising from the crystalline nature of the material are then lost.

The issue of surface cleanliness and the nature of the surface can be addressed when one is doing a single epitaxial growth. For example, a clean wafer can be loaded into the growth chamber and the remaining impurities on the surface can be removed by heating the substrate. The proper arrangement of surface

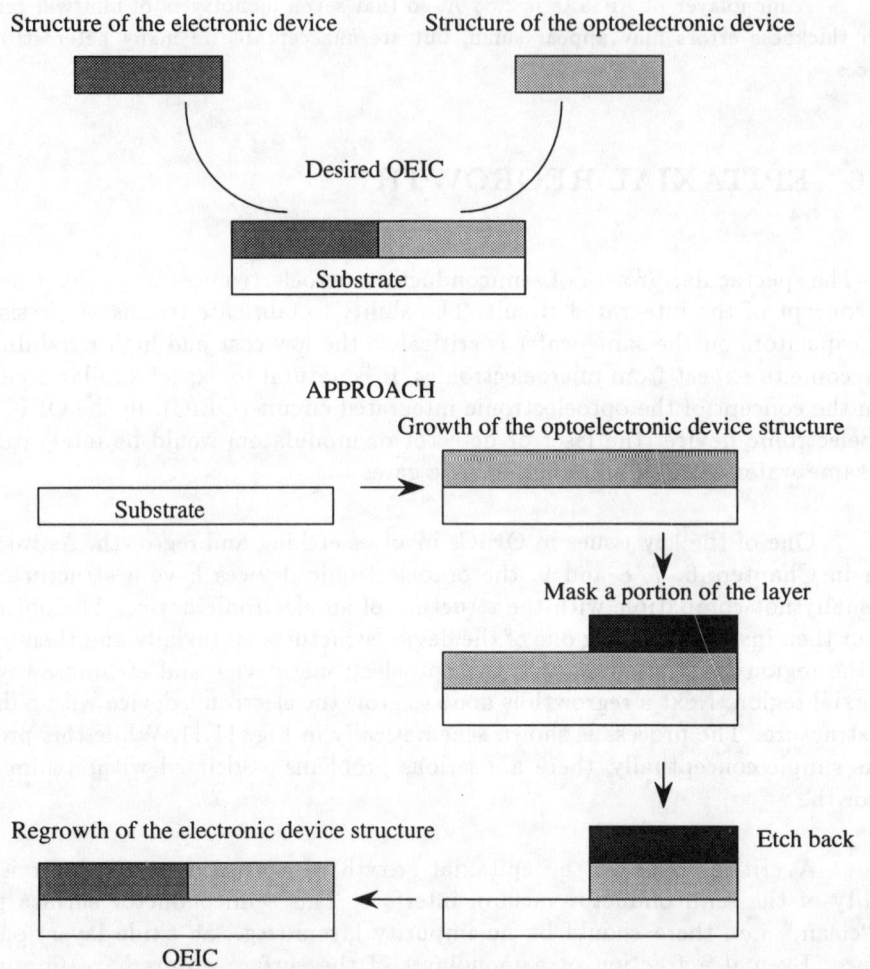

Figure 11.11: The importance of regrowth is clear when one examines the difference in the structure of electronic and optoelectronic devices. Etching and regrowth is essential for fabrication of optoelectronic integrated circuits (OEIC).

atoms (which can be monitored by electron diffraction techniques) can be ensured by adjusting the substrate temperature and specy overpressure. Now consider the problems associated with etching after the first epitaxial growth has occurred. As the etching starts, foreign atoms or molecules are introduced on the wafer as the semiconductor is etched. The etching process is quite damaging and as it ends, the surface of the etched wafer is quite rough and damaged. In addition, in most growth techniques the wafer has to be physically moved from the high purity growth chamber to the etching system. During this transportation, the surface of the wafer may collect some "dirt." During the etching process this "dirt" may not be etched off and may remain on the wafer. As a result of impurities and surface damage, when the second epitaxial layer is grown after etching, the quality of the layer suffers.

A great deal of processing research in OEICs focuses on improving the etching/regrowth process. So far the OEICs fabricated in various laboratories have performances barely approaching the performance of hybrid circuits. Clearly the problem of etching/regrowth is hampering the progress in OEIC technology.

## 11.7  FABRICATION OF OPTICAL FIBERS

$\longrightarrow$  As discussed in Chapter 3, optical fibers are light waveguides which are composed of a core region where the optical field is confined and a cladding region which is responsible for the confinement. The overall diameter of the fiber is $\sim 100-200\ \mu$m while the core can range from a few microns for the single mode fiber to fifty microns or more for a multimode fiber. The refractive index of the core and the cladding region may change abruptly at their interface or may gradually change. A fabrication process should be able to respond to these variations in fiber design and be able to produce uniform fibers hundreds of kilometers in length. The fiber purity has to be maintained as well to produce fibers with losses which are below 1 dB/km. Such fibers are now routinely produced using ultrapure glass and advanced fiber drawing techniques.

### 11.7.1  The Preform Production and Fiber Pulling

$\longrightarrow$  The fabrication of the optical fibers starts with the preparation of a "preform" which is a glass rod or tube in which the refractive index varies from the center to the outside. As noted in Chapter 1, Section 1.6, the refractive index can be altered by doping a glass network with specific dopants.

During the early days of fiber fabrication, a double crucible method was used. In this method two crucibles, one containing the cladding layer and the other the core layer, are arranged. As the crucibles are heated, the glasses flow through an aperture and the composite glass is pulled to form a fiber. This method is not very reliable and is difficult to control with great precision.

Most fibers are made by the chemical vapor deposition (CVD) technique. In the CVD process shown in Fig. 11.12, gases are introduced in a controlled manner through values into a growth chamber. The growth chamber has a high purity fused quartz glass tube on which a temperature gradient can be applied. As the gases flow through the tube, they have a chemical reaction, depositing a glass layer. The composition and rate of the deposited layer can be carefully controlled.

The CVD process can be used to deposit glass either inside a glass tube or on the outside of a glass rod. In both cases a gradient in the index of the preform is produced.

The preform that is produced by CVD has a high density of OH ions ($\sim$ 30 parts per million). The OH ions play a key role in the transmission losses in a fiber and, therefore, need to be removed. This is done by dehydrating the preform in an atmosphere of $SOCl_2$ at a temperature of $\sim$ 1700 K for a time of $\sim$ 5 hours. This process reduces the OH ion density to $\sim$ 0.5 parts per million.

The preform that is the outcome of the CVD process is about a meter long and has a diameter of $\sim$ 2 cm. This rod has to be now pulled to produce the optical fiber. A flow chart of a typical fiber pulling system is shown in Fig. 11.13. In this method the index profile is simply determined by the profile of the preform. It is essential that there be a high degree of control on the pulling speed and the feed speed of the preform. From the equation of continuity of mass we have

$$A_p v_p = A_f v_f \qquad (11.4)$$

where $A_p$ and $A_f$ are the areas of the preform and the fiber, respectively, and $v_p$ and $v_f$ are the velocities of the feed of the preform and the pull of the fiber. In-situ monitoring of the fiber being pulled is essential to ensure a high quality fiber.

In addition to glass fibers, plastic clad fibers are also being produced for applications where attenuation is not the only concern. Plastic fibers have a core of $SiO_2$ and a cladding of plastic. The attenuation is quite high in these fibers ($\sim$ 2-5 dB/km) and, therefore, they cannot be used for long haul applications. However, they are useful in short distance communication where the environment is harsh (high temperature, radiation, etc.).

The optical fiber fabrication has undergone tremendous improvements over the last decade. This is reflected in the cost of optical fibers. One meter of a high quality single mode fiber costs about 20 cents. This is to be compared to about 2 cents for a meter of telephone copper wire. When one considers the enormous

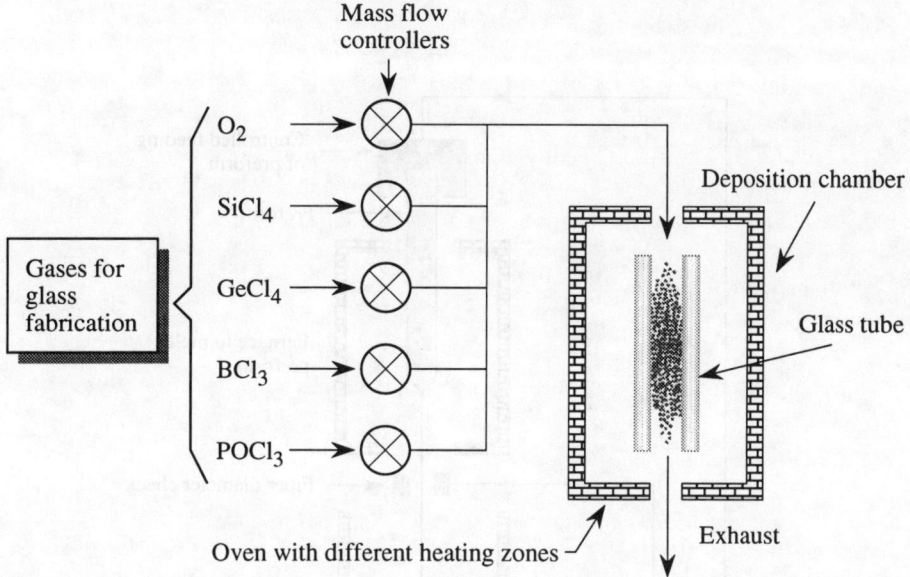

Figure 11.12: The CVD process used to produce the preforms used to pull the fibers. (Based on *Fiber Optics, Devices, and Systems*, P.K. Cheo, Prentice-Hall, New Jersey (1985).)

bandwidth potential of an optical fiber, the fiber turns out to be a lot cheaper for application where a large data transmission is to occur.

The optical fiber, being so thin and fragile, must be encased in protective jackets for use. Optical cables are manufactured and, depending upon the needs, they could contain anywhere from one to a couple of dozen optical fibers. Individual optical fibers are encased in the cable using protective jackets made from polyurethane. The optical cables have to satisfy the requirements of low weight (mass densities of cables are $\sim 5$ to $6$ kg/km), high flexibility (it should be possible to bend them in circles of radius as small as $\sim 2.5$ cm), resistance to kinks and crushing.

A variety of optical cables are available and can be selected for specific applications. For applications involving single optical fiber links such as between offices, computers, etc., a single-fiber strengthened optical-fiber cable is used. This cable can be used in conduits, cable ducts, etc., and consists of a single optical fiber surrounded by strength members. These are then encapsulated in a polyurethane jacket.

For applications where dual or duplex fiber-optic links are to be created between buildings or computers, the two-fiber strengthened optical cable is used. The cable has two single-fiber cables, each with a central fiber which is surrounded

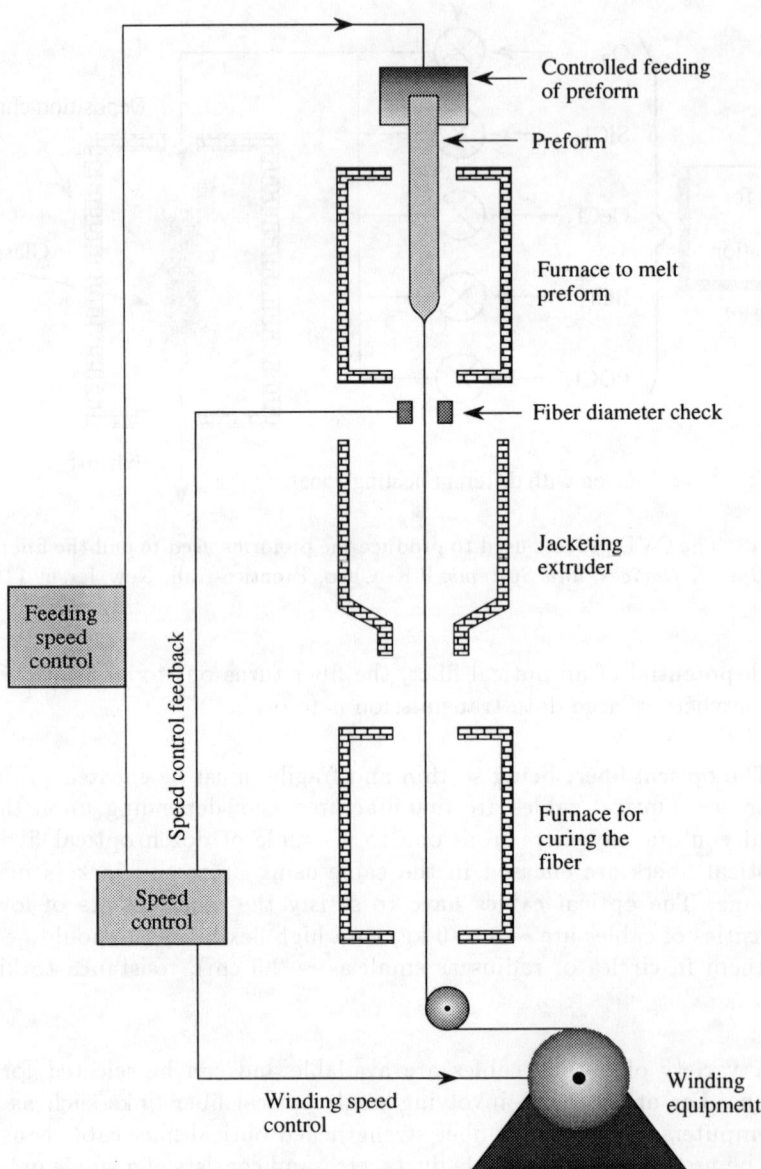

Figure 11.13: A schematic by a fiber pulling apparatus. The preform produced by the CVD process is collapsed by heat and pulled into fibers. (Based on *Fiber Optics, Devices, and Systems*, P.K. Cheo, Prentice-Hall, New Jersey (1985).)

by strength members and a protective jacket.

The extra-strength-member heavy duty optical fiber cable is designed for single-fiber per channel transmission systems with very high cable strength and crush resistance.

## 11.8  FIBER COUPLING AND SPLICING

$\longrightarrow$  An important challenge in optical communication systems is the connecting and disconnecting of optical fibers. This seemingly simple task can cause serious delays in the installation of a fiber system. Splicing and connecting optical fibers is an art considering that the core of a fiber could be only a few microns. Any misalignment can cause a serious loss rendering the system inoperable. This is a special problem for single mode fibers (human hair thickness is about 100 $\mu$m). It has to be kept in mind that these alignments have to be done in the field, not in a controlled laboratory environment of optical tables and vibration free structures.

Several problems arise in the coupling of optical fibers leading to losses. Some of the losses occur due to: i) rough finish of the fiber ends leading to a diffuse scattering of light at the interfaces; ii) sand or dust at the finished edges causing light to ooze out of the coupled joint; iii) misalignment of the cores due to core irregularity, lateral misalignment, gaps between the coupled ends and angular misalignment. The misalignment losses are the most serious losses and are schematically shown in Fig. 11.14. We will briefly discuss these here.

### Misalignment Due to Core Irregularity
This is a problem where the manufacturing process is not of high quality leading to fibers with a core which is not in the center of the fiber as shown in Fig. 11.14a.

### Lateral Misalignment
In this case, the centers of the cores are misaligned. An approximate loss for this misalignment is shown in Fig. 11.14b. Alignment of better than a micron is needed, especially for single mode fibers.

### Gap Loss
When the polished mirror ends of the two fibers are not butted together well, a gap loss can occur. The gap loss depends upon the numerical aperture of the fibers. Higher the numerical aperture, greater is the gap loss, as shown in Fig. 11.14c.

### Angular Loss
The angular loss is due to misorientation in the angles of the two fibers as shown

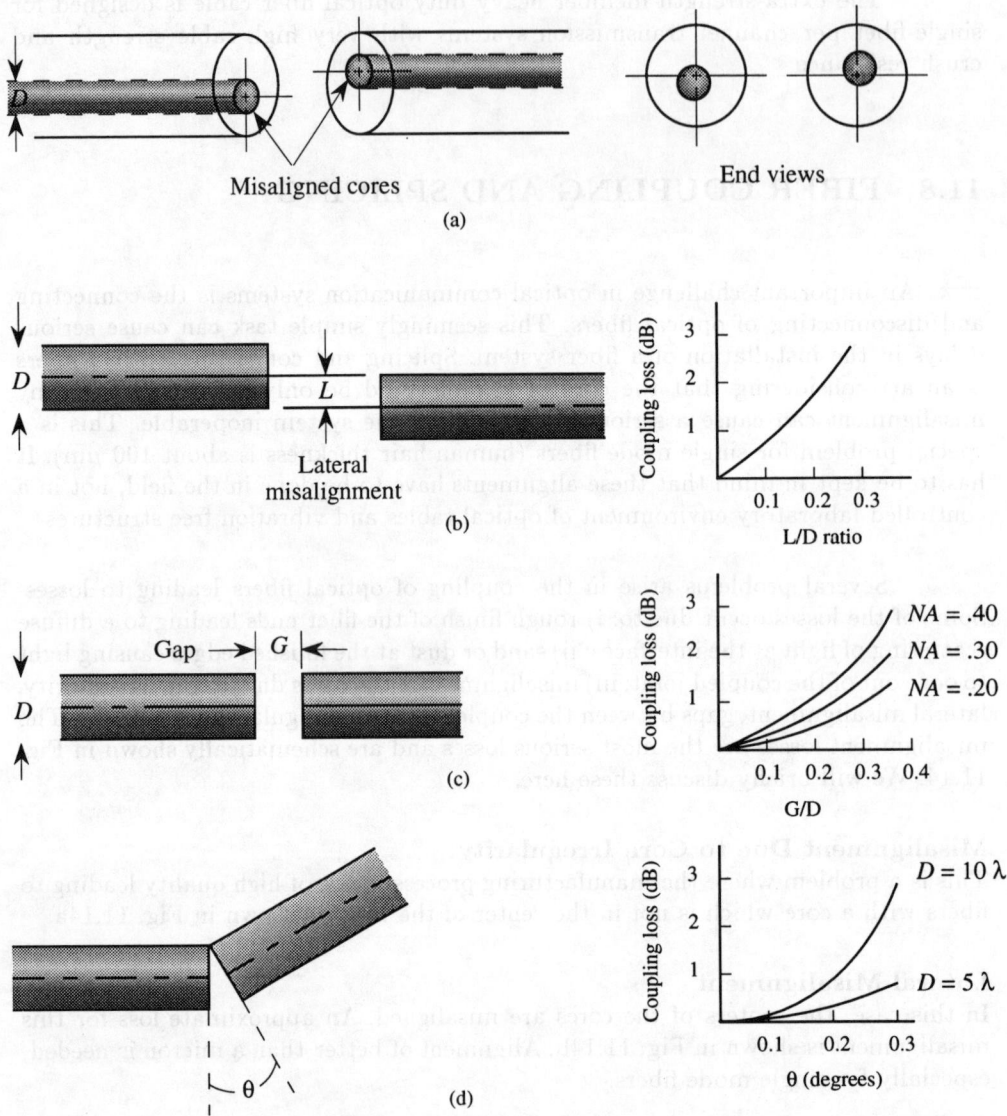

Figure 11.14: Misalignment losses in optical fibers. (a) Schematic of misaligned cores in two fibers. (b) Misalignment due to a lateral shift in the core axis. (c) Misalignment due to a gap in the end butt. The losses are shown for various numerical apertures. (d) Angular misalignment losses shown are approximate.

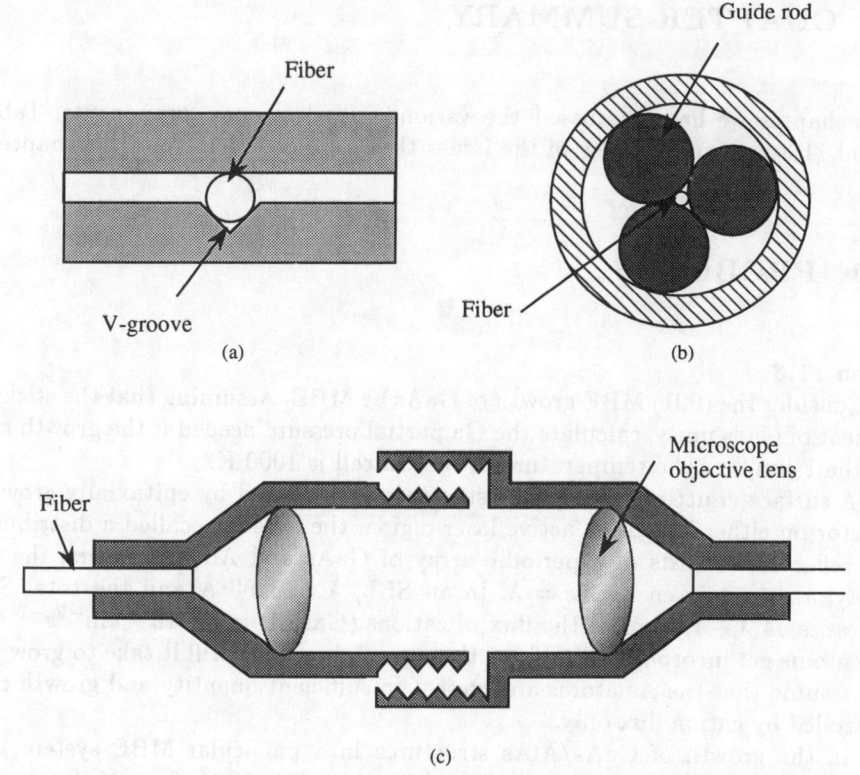

Figure 11.15: Various approaches used to produce fiber joints with low loss. (a) A V-groove is used to grip the fiber in a well defined region. (b) Using three guide rods, the fiber is gripped on three points. (c) Light is coupled from one fiber to another via lenses.

in Fig. 11.14d. This may arise from a cut that is not exactly perpendicular to the fiber axis.

    A number of techniques have been developed to produce fiber joints that are acceptable (losses are less than ∼ 0.2 dB). In Fig. 11.15 we show approaches that are used to produce demountable joints. The general philosophy is based on first enclosing the fiber in a container that has several surfaces touching the fiber so as to "grip" the fiber. To provide a firm grip, a V-groove structure is often used as shown. Microlens systems are also used, as shown in Fig. 11.15c, to couple light from one fiber to another.

## 11.9  CHAPTER SUMMARY

In this chapter we have discussed the various optoelectronic components. Tables 11.2 and 11.3 give an overview of the issues that have emerged from this chapter.

## 11.10  PROBLEMS

### Section 11.3

**11.1** Consider the (001) MBE growth of GaAs by MBE. Assuming that the sticking coefficient of Ga is unity, calculate the Ga partial pressure needed if the growth rate has to be 1 $\mu$m/hr. The temperature of the Ga cell is 1000 K.

**11.2** A surface emitting laser (SEL) structure is produced by epitaxially growing a reflector on either side of an active laser region. the reflector, called a distributed Bragg reflector, consists of a periodic array of GaAs and AlGaAs layers; the periodicity, $a$, being given by $2a = \lambda$. In an SEL, $\lambda = 8000$ Å and the total SEL thickness is $54\,\lambda + 1.0\ \mu$m. If the flux of cations (Ga, Al) is $5 \times 10^{14}$ cm$^{-2}$s$^{-1}$ and all the atoms get incorporated to form the crystal, how long will it take to grow the SEL? Assume that the As atoms are present in sufficient quantity and growth rate is controlled by cation flux only.

**11.3** In the growth of GaAs/AlAs structures in a particular MBE system, the background pressure of Ga when the Ga shutter is off is $10^{-7}$ Torr. If the growth rate of AlAs is 1 $\mu$m/hr, what fraction of Ga atoms are incorporated in the AlAs region? The Ga cell is at 1000 K.

### Section 11.4

**11.4** To produce lateral confinement of electronic states in a semiconductor, the feature sizes of the device have to be $\sim$100 Å. Calculate the energy of photons and electrons that can be used in such a lithography.

**11.5** In optoelectronics, the smallest feature sizes of typical lasers are $\sim$5 $\mu$m. However, for laser known as distributed feedback (DFB) lasers, the smallest feature size is $\sim$1000 Å. Discuss the kind of lithography systems that can be used for the two kinds of lasers.

**11.6** In a particular resist spinner, the initial deposit of resist on a 3 inch GaAs wafer is 0.2 cm high. After a spin at 4000 rpm, the resist thickness is 1 $\mu$m. How much resist has been wasted in this spin?

### Section 11.5

**11.7** Selective etchents are often used in microelectronic technology. Consider an etch that etches GaAs at the rate of 1 $\mu$m/hr and Al$_{0.3}$Ga$_{0.7}$As at 500 Å/hr. AlGaAs

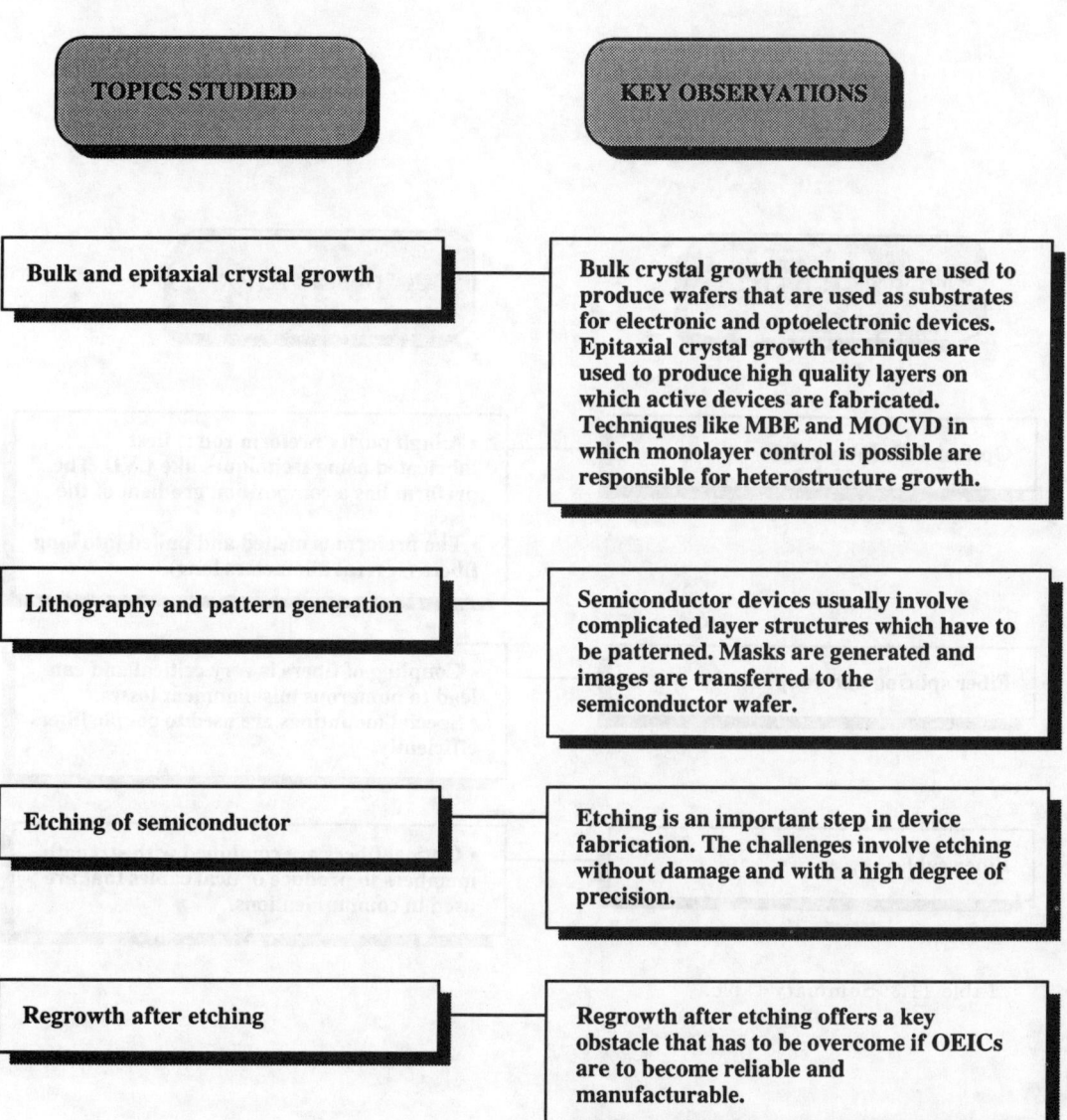

Table 11.2: Summary table.

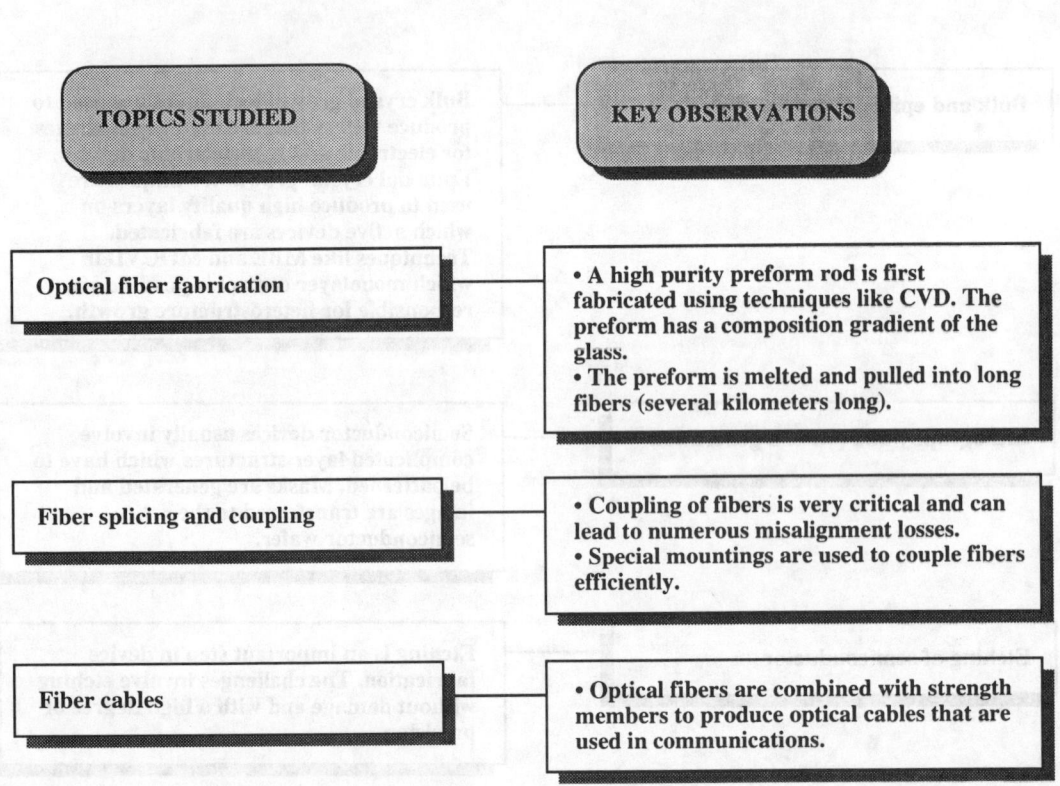

**TOPICS STUDIED**

**KEY OBSERVATIONS**

**Optical fiber fabrication**

• A high purity preform rod is first fabricated using techniques like CVD. The preform has a composition gradient of the glass.
• The preform is melted and pulled into long fibers (several kilometers long).

**Fiber splicing and coupling**

• Coupling of fibers is very critical and can lead to numerous misalignment losses.
• Special mountings are used to couple fibers efficiently.

**Fiber cables**

• Optical fibers are combined with strength members to produce optical cables that are used in communications.

Table 11.3: Summary table.

is used as an etch stop in a heterostructure where 0.5 $\mu$m of GaAs has to be etched. If the AlGaAs layer is 30 Å, what should be the tolerance in the etching time?

## 11.11  REFERENCES

- **Crystal Growth: Heterostructures**

  - A. Y. Cho, *Journal of Vacuum Science Technology*, vol. 8, 531 (1971).
  - F. Rosenberger, *Fundamentals of Crystal Growth*, Springer (1979).
  - G. Haas and M. H. Francombe, C. E. C. Wood in *Physics of Thin Films*, Academic Press (1980).

- **Microelectronic Processing**

  - J.W. Mayer and S.S. Lau, *Electronic Materials Science: For Integrated Circuits in Si and GaAs*, MacMillan, New York (1990).
  - W.S. Ruska, *Microelectronic Processing: An Introduction to the Manufacture of Integrated Circuits*, McGraw-Hill, New York (1988).
  - P.K.L. Yu and P.C. Chan, *Introduction to GaAs Technology*, ed. C. Wang, Wiley, New York (1988).

- **Optical Fibers**

  - P.K. Cheo, *Fiber Optics, Devices, and Systems*, Prentice-Hall, New Jersey (1985).
  - R.G. Seippel, *Optoelectronics*, Reston, Virginia (1981).

# APPENDIX
## A

# LIST OF SYMBOLS

| | |
|---|---|
| $a$ | lattice constant (edge of the cube for the semiconductor fcc lattice) |
| $AM$ | amplitude modulation |
| $ASK$ | amplitude shift keying |
| | |
| $B$ | bandwidth |
| $BER$ | bit error rate in a transmission stream |
| | |
| $c$ | velocity of light |
| $C_j, C_d$ | junction, diffusion capacitance in a $p$-$n$ diode |
| | |
| $d_{las}$ | active region thickness of a laser |
| | |
| $D_n$ | electron diffusion coefficient |
| $D_p$ | hole diffusion coefficient |
| $D, D^*$ | detectivity, specific detectivity of a detector |
| $DBR$ | distributed Bragg reflector |
| $DFB$ | distributed feedback (laser) |
| | |
| $e$ | magnitude of the electron charge |
| | |
| $\mathbf{E}$ | electric field |
| $E$ | energy of a particle |
| $E_F$ | Fermi level |
| $E_{Fi}$ | intrinsic Fermi level |

| | |
|---|---|
| $E_{Fn}$ | electron quasi-Fermi level |
| $E_{Fp}$ | hole quasi-Fermi level |
| $E^e(E^h)$ | energy of an electron (hole) in an optical absorption or emission measured from the bandedges |
| $E_c(E_v)$ | conduction (valence) bandedge |

| | |
|---|---|
| $f(E)$ | occupation probability of an electron state with energy $E$ at equilibrium. This is the Fermi-Dirac function |
| $f^e(E)$ | occupation function for an electron in non-equilibrium state. This is the quasi-Fermi function |
| $f^h(E)$ | occupation function for a hole $= 1 - f^e(E)$ |

| | |
|---|---|
| $F_{ext}$ | external force such as an electric or magnetic force |
| $F_f$ | fill factor of a solar cell |
| $FDM$ | frequency division multiplexing |
| $FM$ | frequency modulation |
| $FSK$ | frequency shift keying |

| | |
|---|---|
| $g, g_{th}$ | gain, threshold gain |

| | |
|---|---|
| $G_L$ | electron-hole generation rate due to a light beam |
| $G_{ph}$ | gain of a detector |

| | |
|---|---|
| $\hbar$ | Planck's constant divided by $2\pi$ |

| | |
|---|---|
| $H$ | magnetic field |

| | |
|---|---|
| $IMDD$ | intensity modulation direct detection |
| $I_{ph}$ | photon particle current |
| $I_E, I_B, I_C$ | emitter, base, and collector current in a BJT |
| $I_L$ | photocurrent due to an optical signal |
| $I_D$ | drain current in an FET |
| $I_o$ | reverse bias saturation current in a $p$-$n$ diode |
| $I_{GR}$ | generation recombination current in a diode |
| $I_{GR}^o$ | prefactor for the generation recombination current |
| $I_{sc}$ | short circuit current of a solar cell |

| | |
|---|---|
| $J$ | current density |
| $J_L$ | photocurrent density |

| | |
|---|---|
| $k_B$ | Boltzmann constant |

| | |
|---|---|
| $\ell$ | mean free path between successive collisions |
| $L_n$ | diffusion length for electron |
| $L_p$ | diffusion length for holes |
| | |
| $m_0$ | free electron mass |
| $m_e^*$ | electron mass |
| $m_h^*$ | hole mass |
| $m_{dos}^*$ | density of states mass |
| $m_\sigma^*$ | conductivity mass |
| $m_{hh}^*$ | mass of the heavy hole |
| $m_{\ell h}^*$ | mass of the light hole |
| $m_r^*$ | reduced mass of the electron-hole system |
| | |
| $M, M_e, M_h$ | multiplication factor, multiplication factor for electrons, mulitplication factor for holes |
| | |
| $n$ | electron concentration in the conduction band |
| $n_i$ | intrinsic electron concentration in the conduction band |
| $n_d$ | electrons bound to the donors |
| $n_r$ | refractive index |
| $n_{ph}(\hbar\omega)$ | photon density of photons with energy $\hbar\omega$ |
| $n_p(n_p)$ | equilibrium electron density in the $p$-side ($n$-side) of a $p$-n junction |
| $n_{th}$ | threshold carrier density for a laser |
| $NA$ | numerical aperture |
| $N_{cv}$ | joint density of states for electrons and holes |
| $N_e(E)$ | density of states of electrons in the conduction band |
| $N_h(E)$ | density of states of holes in the valence band |
| $N_c(E)$ | effective density of states in the conduction band |
| $N_v(E)$ | effective density of states in the valence band |
| $N_d$ | donor density |
| $N_a$ | acceptor density |
| $N_{2D}(E)$ | 2-dimensional density of states |
| $N_t$ | density of impurity states (trap states) |
| $NEP$ | noise equivalent power of a detector |
| | |
| $p$ | hole concentration in the valence band |
| $p_i$ | intrinsic hole concentration in the valence band |
| $p_{cv}$ | momentum matrix element for an optical transition between the valence and conduction band |
| $p_n(p_p)$ | equilibrium hole density in the $n$-side ($p$-side) of a $p$-n junction |

| | |
|---|---|
| $PM$ | phase modulation |
| $PSK$ | phase shift keying |
| $P_{op}$ | optical power density (energy flow/sec/area) |
| | |
| $QCSE$ | quantum confined Stark effect |
| | |
| $r_R$ | reflected amplitude to incident amplitude ratio |
| $R$ | reflection coefficient |
| $R_{ph}$ | responsivity of a detector material |
| $R_{spon}$ | total rate at which an electron-hole system recombines to emit photons by spontaneous recombination |
| $R_{st}$ | total rate at which $e$-$h$ recombine by stimulated emission of photons |
| $R_L$ | load resistance |
| $R^*$ | Richardson constant in a Schottky barrier |
| | |
| $SNR$ | signal to noise ratio |
| | |
| $t$ | transmitted amplitude to incident amplitude ratio |
| $t_d$ | delay time in large signal laser response |
| $t_{tr}$ | transit time of a carrier through a channel |
| | |
| $T$ | tunneling probability |
| | |
| $U(r)$ | position-dependent potential energy |
| | |
| $v$ | velocity of the electron |
| $v_s$ | saturation velocity of the carrier (electron, hole) |
| | |
| $V_{oc}$ | open circuit voltage of a solar cell |
| $V_{bi}$ | built-in voltage |
| $V_r(V_f)$ | reverse (forward) bias voltage in a diode |
| | |
| $W_{em}$ | spontaneous emission rate for an electron-hole radiative recombination |
| $W_{em}^{st}$ | stimulated emission rate for an electron-hole radiative recombination |
| $W_n(W_p)$ | depletion region edge on the $n$-side ($p$-side) of a $p$-$n$ junction |
| $W$ | depletion region width |
| $WDM$ | wavelength division multiplexing |
| $W_b, W_{bn}$ | base width, neutral base width of a bipolar transistor |

| | |
|---|---|
| $\alpha$ | optical absorption coefficient |
| $\alpha$ | current transfer ratio in a bipolar transistor |
| $\alpha_R$ | reflection loss coefficient in an optical cavity |
| $\alpha_{imp}$ | impact ionization coefficient for electrons |
| $\beta$ | base to collector current amplification factor in a BJT |
| $\beta_{imp}$ | impact ionization coefficient for holes |
| $\gamma_e$ | emitter efficiency of a bipolar transistor |
| $\gamma_{inj}$ | injection efficiency of a $p$-$n$ diode for electron (hole) current |
| $\Gamma$ | optical confinement factor in a semiconductor laser |
| $\Delta E_g$ | bandgap difference between two materials |
| $\Delta E_c, \Delta E_v$ | band discontinuity in the conduction, valence band in a heterostructure |
| $\epsilon_o$ | free space permittivity |
| $\epsilon$ | product of the relative dielectric constant and $\epsilon_o$ |
| $\psi$ | electron wavefunction |
| $\sigma_e(\sigma_n)$ | electron (hole) captive cross section for an impurity |
| $\sigma$ | conductivity of a material |
| $\mu$ | mobility of a material |
| $\mu_n(\mu_p)$ | electron (hole) mobility |
| $\tau_{sc}$ | scattering time between successive collisions. Also called relaxation time |
| $\omega$ | frequency |
| $\tau_o$ | rate at which an electron recombines radiatively with a hole at the same momentum value |
| $\tau_r$ | radiative recombination time for $e$-$h$ pair |
| $\tau_{nr}$ | non-radiative recombination time for a $e$-$h$ pair |
| $\tau_n$ | lifetime of an electron to recombine with a hole |
| $\tau_p$ | lifetime of a hole to recombine with an electron |
| $\tau_{sd}$ | storage delay time in a diode |
| $\tau_{ph}$ | photon lifetime in a laser cavity |
| $\delta n$ | excess electron density in a region. This is the density above the equilibrium density |
| $\delta p$ | excess hole density in a region |
| $\phi_m$ | metal work function |
| $\chi_s$ | electron affinity of a semiconductor |
| $\phi_s$ | work function of a semiconductor |
| $\phi_{ms}$ | difference between a metal and semiconductor work function |
| $\phi_b$ | barrier height seen by electrons coming from a metal towards a semiconductor |
| $\phi_{TE}$ | phase change produced by total internal reflection for a $TE$ wave |
| $\phi_{TM}$ | phase change produced by total internal reflection for a $TM$ wave |
| $\Phi_0$ | photon flux |
| $\lambda_c$ | cutoff wavelength of a detector |

| $\eta_Q$ | quantum efficiency of a detector |
|---|---|
| $\eta_{conv}$ | power conversion efficiency of a solar cell |
| $\eta_{det}$ | efficiency of a detector to convert an optical signal current to an electrical current |
| $\eta_{Qr}$ | radiative quantum efficiency of electron-hole recombination |
| $\eta_{fiber}$ | coupling efficiency for a optical fiber |
| $\theta_A$ | acceptance angle for an optical fiber |

# APPENDIX B

# IMPORTANT PROPERTIES OF SEMICONDUCTORS

The data and plots shown in this Appendix are extracted from a number of sources. A list of useful sources is given below.

- S. Adachi, *J. Appl. Phys.*, 58, R1 (1985).

- H.C. Casey, Jr. and M.B. Panish, *Heterostructure Lasers*, Part A, "Fundamental Principles;" Part B, "Materials and Operating Characteristics," Academic Press, N.Y. (1978).

- Landolt-Bornstein, *Numerical Data and Functional Relationship in Science and Technology*, Vol. 22, Eds. O. Madelung, M. Schulz, and H. Weiss, Springer-Verlog, N.Y. (1987).

- S.M. Sze, *Physics of Semiconductor Devices*, Wiley, N.Y. (1981). This is an excellent source of a variety of useful information on semiconductors.

| Material | Electron Mass $(m_0)$ | Hole Mass $(m_0)$ |
|---|---|---|
| AlAs | 0.1 | |
| AlSb | 0.12 | $m_{dos} = 0.98$ |
| GaN | 0.19 | $m_{dos} = 0.60$ |
| GaP | 0.82 | $m_{dos} = 0.60$ |
| GaAs | 0.067 | $m_{lh} = 0.082$ <br> $m_{hh} = 0.45$ |
| GaSb | 0.042 | $m_{dos} = 0.40$ |
| Ge | $m_l = 1.64$ <br> $m_t = 0.082$ | $m_{lh} = 0.044$ <br> $m_{hh} = 0.28$ |
| InP | 0.073 | $m_{dos} = 0.64$ |
| InAs | 0.027 | $m_{dos} = 0.4$ |
| InSb | 0.13 | $m_{dos} = 0.4$ |
| Si | $m_l = 0.98$ <br> $m_t = 0.19$ | $m_{lh} = 0.16$ <br> $m_{hh} = 0.49$ |

Table B.1: Electron and hole masses for several semiconductors. Some uncertainty remains in the value of hole masses for many semiconductors.

| Compound | Direct Energy Gap $E_g$ (eV) |
|---|---|
| $Al_xIn_{1-x}P$ | $1.351 + 2.23x$ |
| $Al_xGa_{1-x}As$ | $1.424 + 1.247x$ |
| $Al_xIn_{1-x}As$ | $0.360 + 2.012x + 0.698x^2$ |
| $Al_xGa_{1-x}Sb$ | $0.726 + 1.129x + 0.368x^2$ |
| $Al_xIn_{1-x}Sb$ | $0.172 + 1.621x + 0.43x^2$ |
| $Ga_xIn_{1-x}P$ | $1.351 + 0.643x + 0.786x^2$ |
| $Ga_xIn_{1-x}As$ | $0.36 + 1.064x$ |
| $Ga_xIn_{1-x}Sb$ | $0.172 + 0.139x + 0.415x^2$ |
| $GaP_xAs_{1-x}$ | $1.424 + 1.150x + 0.176x^2$ |
| $GaAs_xSb_{1-x}$ | $0.726 + 0.502x + 1.2x^2$ |
| $InP_xAs_{1-x}$ | $0.360 + 0.891x + 0.101x^2$ |
| $InAs_xSb_{1-x}$ | $0.18 + 0.41x + 0.58x^2$ |

Table B.2: Compositional dependence of the energy gaps of the binary III-V ternary alloys at 300 K. (After Casey and Panish (1978).)

| Semiconductor | Bandgap (eV) 300 K | Mobility at 300 K (cm$^2$/V-s) Electrons | Holes |
|---|---|---|---|
| C | 5.47 | 800 | 1200 |
| Ge | 0.66 | 3900 | 1900 |
| Si | 1.12 | 1500 | 450 |
| α-SiC | 2.996 | 400 | 50 |
| GaSb | 0.72 | 5000 | 850 |
| GaAs | 1.42 | 8500 | 400 |
| GaP | 2.26 | 110 | 75 |
| InSb | 0.17 | 8000 | 1250 |
| InAs | 0.36 | 33000 | 460 |
| InP | 1.35 | 4600 | 150 |
| CdTe | 1.56 | 1050 | 100 |
| PbTe | 0.31 | 6000 | 4000 |

Table B.3: Bandgaps, electron and hole mobilities of some semiconductors.

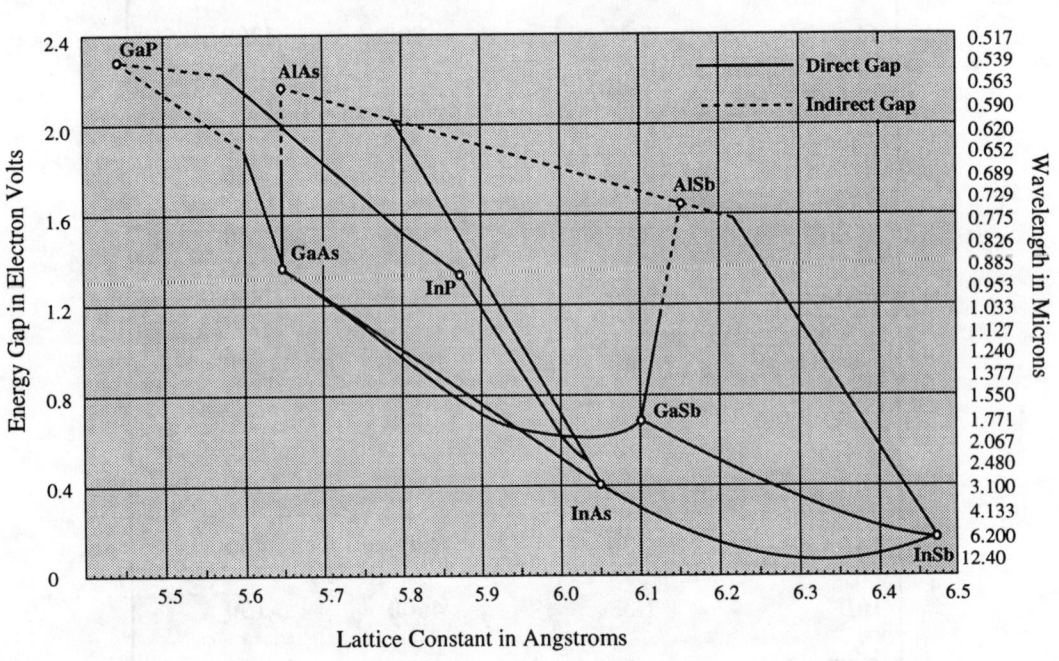

Figure B.1: Lattice constants and bandgaps of semiconductors at room temperature.

# APPENDIX
# C

# DENSITY OF STATES

## C.1  INTRODUCTION

An extremely important concept that has been used in this book is the concept of the density of states. The concepts of doping, optical absorption, mobility etc., in semiconductors are intimately linked to the density of states. In this appendix we will calculate the density of states of in a material in 3-, 2-, and 1-dimensions.

## C.2  THE ELECTRON WAVEFUNCTION and THE DENSITY OF STATES

A general solution for the electron in a semiconductor normalized to a volume $V$ has the form

$$\psi(r) = \frac{1}{\sqrt{V}}e^{\pm ik \cdot r} \qquad (C.1)$$

Note that we have suppressed the cell periodic part of the correct Bloch function since the following treatment is not affected by this. We will assume that the energy versus $k$ relation is parabolic and is

$$E = \frac{\hbar^2 k^2}{2m^*} \qquad (C.2)$$

where the factor $\frac{1}{\sqrt{V}}$ comes because we wish to have one electron per volume $V$ or

$$\int_V d^3r \mid \psi(r) \mid^2 = 1 \qquad (C.3)$$

We assume that the volume $V$ is a cube of side $L$.

In classical mechanics the energy momentum relation for the free electron is $E = p^2/2m^*$, and $p$ can be a *continuous variable*. The quantity $\hbar k$ appearing above is the effective momentum of the electron. Due to the wave nature of the electron in a finite volume, the quantity $k$ is not continuous but discrete. To correlate with physical conditions that we may want to describe, there are two kinds of boundary conditions that are imposed on the wavefunction. In the first one the wavefunction is considered to go to zero at the boundaries of the volume. In this case, the wave solutions are of the form $\sin(k_x x)$ or $\cos(k_x x)$, etc., and $k$-values are restricted to the positive values,

$$k_x = \frac{\pi}{L}, \frac{2\pi}{L}, \frac{3\pi}{L} \cdots \tag{C.4}$$

The standing wave solution is often used to describe stationary electrons confined in finite regions such as quantum wells discussed earlier. For describing moving electrons, the boundary condition used is known as a periodic boundary condition (Fig. C.1). Even though we focus our attention on a finite volume $V$, the wave can be considered to spread in all space as we conceive the entire space was made up of identical cubes of sides $L$. Then

$$\begin{aligned} \psi(x, y, z + L) &= \psi(x, y, z) \\ \psi(x, y + L, z) &= \psi(x, y, z) \\ \psi(x + L, y, z) &= \psi(x, y, z) \end{aligned} \tag{C.5}$$

Because of the boundary conditions the allowed values of $k$ are ($n$ are integers—positive and negative)

$$k_x = \frac{2\pi n_x}{L}; k_y = \frac{2\pi n_y}{L}; k_z = \frac{2\pi n_z}{L} \tag{C.6}$$

If $L$ is large, the spacing between the allowed $k$ values is very small. It is useful to discuss the *volume in k-space that each electronic state occupies*. As can be seen from Fig. C.2, this volume is (in 3-dimensions)

$$\left(\frac{2\pi}{L}\right)^3 = \frac{8\pi^3}{V} \tag{C.7}$$

If $\Omega$ is a volume of $k$-space, the number of electronic states in this volume are

$$\boxed{\frac{\Omega V}{8\pi^3}} \tag{C.8}$$

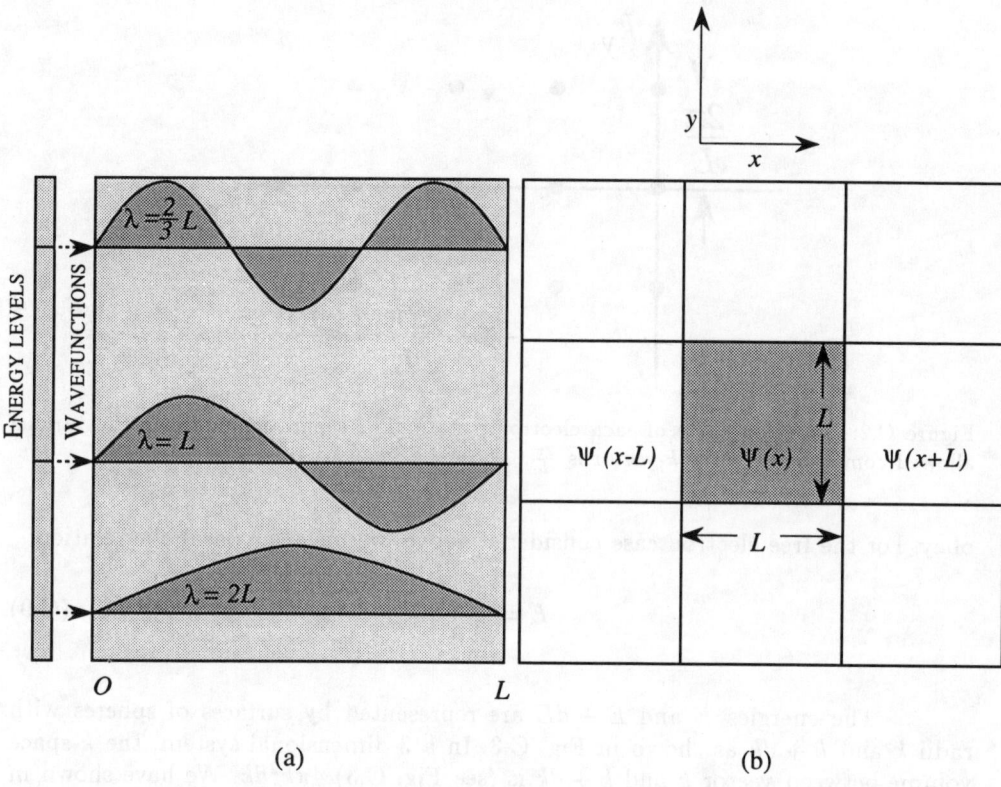

Figure C.1: A schematic showing (a) the stationary boundary conditions leading to standing waves and (b) the periodic boundary conditions leading to exponential solutions with the electron probability equal in all regions of space.

## C.2.1  Density of States for a 3-Dimensional System

We will now use the discussion of the previous subsection to derive the extremely important concept of density of states. Although we will use the periodic boundary conditions to obtain the density of states, the stationary conditions lead to the same result.

The concept of density of states is extremely powerful, and important physical properties such as optical absorption, transport, etc., are intimately dependent upon this concept. Density of states is the number of available electronic states *per unit volume per unit energy* around an energy $E$. If we denote the density of states by $N(E)$, the number of states in a unit volume in an energy interval $dE$ around an energy $E$ is $N(E)dE$. To calculate the density of states, we need to know the dimensionality of the system and the energy vs. $k$ relation that the electrons

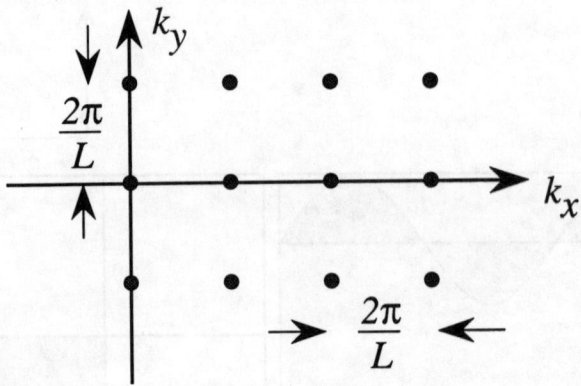

Figure C.2: $k$-space volume of each electronic state. The separation between the various allowed components of the $k$-vector is $\frac{2\pi}{L}$.

obey. For the free electron case considered above, we have the parabolic relation:

$$E = \frac{\hbar^2 k^2}{2m^*} \tag{C.9}$$

The energies $E$ and $E + dE$ are represented by surfaces of spheres with radii $k$ and $k + dk$ as shown in Fig. C.3. In a 3-dimensional system, the $k$-space volume between vector $k$ and $k + dk$ is (see Fig. C.3) $4\pi k^2 dk$. We have shown in Eqn. C.8 that the $k$-space volume per electron state is $\left(\frac{2\pi}{L}\right)^3$. Therefore, the number of electron states in the region between $k$ and $k + dk$ is

$$\frac{4\pi k^2 dk}{8\pi^3} V = \frac{k^2 dk}{2\pi^2} V \tag{C.10}$$

Denoting the energy and energy interval corresponding to $k$ and $dk$ as $E$ and $dE$, we see that the number of electron states between $E$ and $E + dE$ per unit volume are

$$N(E)dE = \frac{k^2 dk}{2\pi^2} \tag{C.11}$$

and since

$$E = \frac{\hbar^2 k^2}{2m^*} \tag{C.12}$$

$$k^2 dk = \frac{\sqrt{2} m^{*3/2} E^{1/2} dE}{\hbar^3} \tag{C.13}$$

and

$$N(E)dE = \frac{m^{*3/2} E^{1/2} dE}{\sqrt{2}\pi^2 \hbar^3} \tag{C.14}$$

In the next section we develop the concept that an electron is a "Fermion" and can have two possible states with a given energy. These states are called the spin states. The electron can have a spin state $\hbar/2$ or $-\hbar/2$. Accounting for spin, the density of states obtained above is simply multiplied by 2

$$N(E) = \frac{\sqrt{2}m^{*3/2}E^{1/2}}{\pi^2\hbar^3} \qquad (C.15)$$

## C.2.2   Density of States in Sub-3-Dimensional Systems

Let us now consider a 2-D system, a concept that has become a reality with use of quantum wells. Similar arguments tell us that the density of states for a parabolic band is (including spin)

$$N(E) = \frac{m^*}{\pi\hbar^2} \qquad (C.16)$$

Note that in a quantum well, a series of subbands is produced as discussed in Chapter 4. Each subband produces a constant density of states. As a result, the *total density of states has a staircase like shape.*

Finally, in a 1-D system or a "quantum wire" the density of states is (including spin)

$$N(E) = \frac{\sqrt{2}m^{*1/2}}{\pi\hbar}E^{-1/2} \qquad (C.17)$$

We note that as the dimensionality of the system changes, the energy dependence of the density of states also changes. As shown in Fig. C.4, for a 3-dimensional system we have a $E^{1/2}$ dependence; for a 2-dimensional system we have no energy dependence; and for a 1-dimensional system we have an $E^{-1/2}$ dependence.

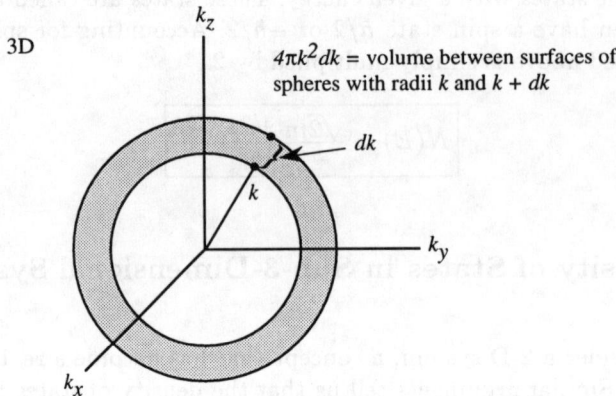

3D

$4\pi k^2 dk$ = volume between surfaces of spheres with radii $k$ and $k + dk$

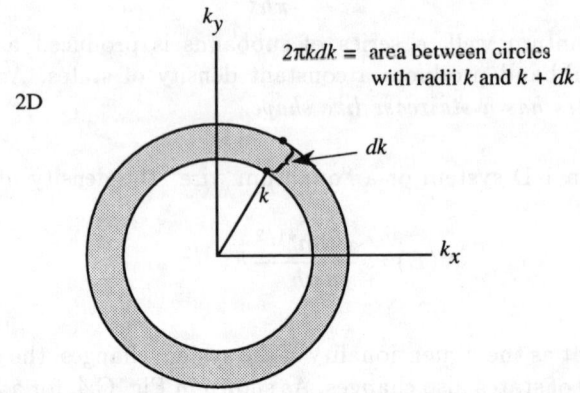

2D

$2\pi k dk$ = area between circles with radii $k$ and $k + dk$

1D

$2dk = k$ space between $k$ and $k + dk$

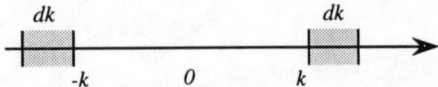

Figure C.3: Geometry used to calculate density of states in 3, 2, and 1-dimensions. By finding the $k$-space volume in an energy interval between $E$ and $E + dE$, one can find out how many allowed states there are.

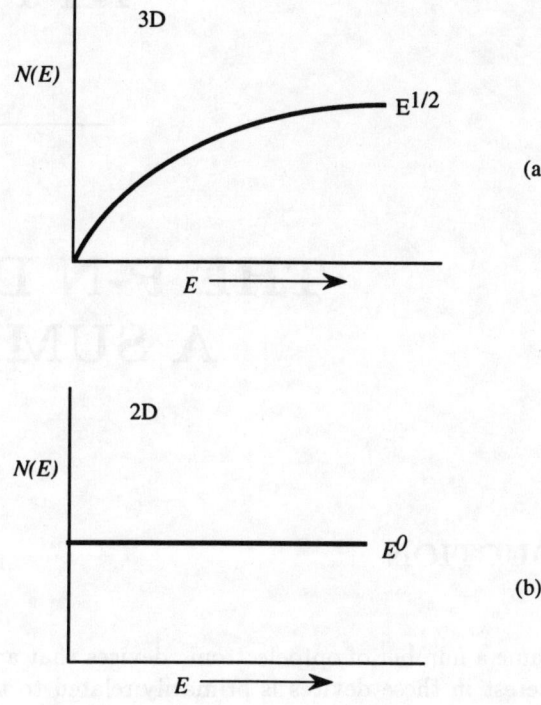

(a)

(b)

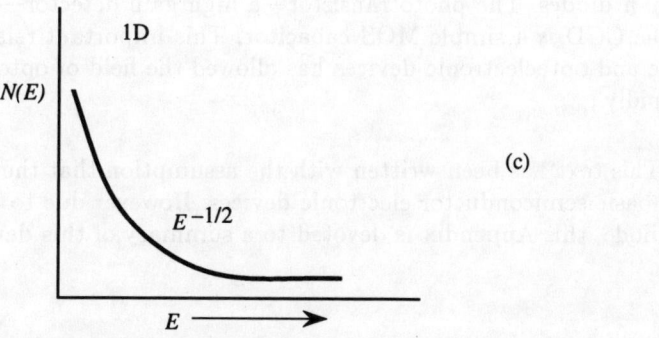

(c)

Figure C.4: Variation in the energy dependence of the density of states in (a) 3-dimensional, (b) 2-dimensional, and (c) 1-dimensional systems. The energy dependence of the density of states is determined by the dimensionality of the system. This feature is exploited in advanced semiconductor devices.

# APPENDIX D

# THE P-N DIODE: A SUMMARY

## D.1  INTRODUCTION

In this text we examine a number of optoelectronic devices that are based on semiconductors. Our interest in these devices is primarily related to the interaction of light with these devices. However these devices are based on principles very similar to electronic devices such as diodes, field effect transistors and bipolar transistors. For example most detectors are basically *p-n* diodes. The LED and the laser diode are also *p-n* diodes. The phototransistor—a high gain detector—is a bipolar transistor. The CCD is a simple MOS capacitor. This important relationship between electronic and optoelectronic devices has allowed the field of optoelectronics to advance rapidly.

This text has been written with the assumption that the reader is familiar with the basic semiconductor electronic devices. However due to the importance of the *p-n* diode, this Appendix is devoted to a summary of this device.

## D.2  THE P-N JUNCTION

As noted above the *p-n* diode is one of the most important optoelectronic devices. It forms the basis of most detectors and light emitting devices. We will review some of the salient features of this device.

## D.2.1 The Unbiased P-N Junction

$\mathcal{R}$  The *p-n* junction is one of the most important junctions in solid-state electronics. The fabrication techniques used to form *p*- and *n*-type regions involve i) epitaxial procedures where the dopant species are simply switched at a particular instant in time; ii) ion-implantation in which the dopant ions are implanted at high energies into the semiconductor (the junction is obviously not as abrupt as in the case of epitaxial techniques) and; iii) diffusion of dopants into an oppositely doped semiconductor.

We will assume in our analysis that the *p-n* junction is abrupt, even though this is really only true for epitaxially grown junctions. Let us first discuss the properties of the junction in the absence of any external bias where there is no current flowing in the diode.

What happens when the *p*- and *n*-type materials are made to form a junction and there is no externally applied field? We know that in absence of any applied bias, there is no current in the system and the Fermi level is uniform throughout the structure.

This gives the schematic view of the junction shown in Fig. D.1a. Three regions can be identified:

i) The *p*-type region at the far left where the material is neutral and the bands are flat. The density of acceptors exactly balances the density of holes;

ii) The *n*-type region in the far right where again the material is neutral and the density of immobile donors exactly balances the free electron density;

iii) The depletion region where the bands are bent and a field exists which has swept out the mobile carriers leaving behind negatively charged acceptors in the *p*-region and positively charged donors in the *n*-region as shown in Fig. D.1a.

In the depletion region, which extends a distance $W_p$ in the *p*-region and a distance $W_n$ in the *n*-region, an electric field exists. Any electrons or holes in the depletion region are swept away by this field. Thus a drift current exists which counterbalances the diffusion current which arises because of the difference in electron and hole densities across the junction.

In the absence of any applied bias, there is a built-in potential between the *n* and the *p* side as shown in Fig. D.2. Denoting $p_p$ and $p_n$ as the hole densities in

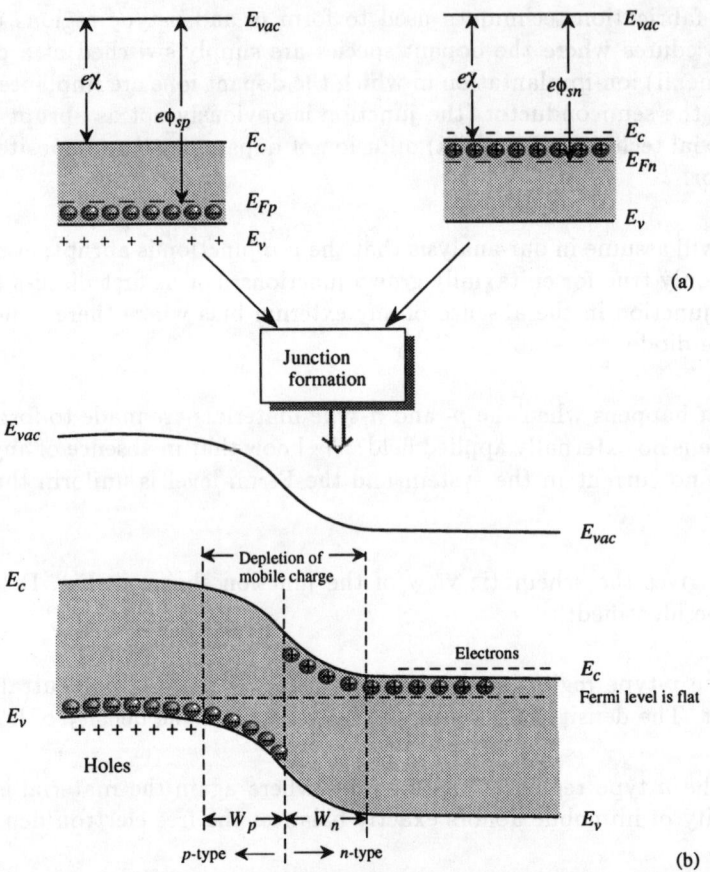

Figure D.1: (a) An idealized model of the $p$-$n$ junction without bias showing the neutral and the depletion areas. (b) A schematic showing various current and particle flow components in the $p$-$n$ diode at equilibrium. For electrons, the current flow is in the direction opposite to that of the particle flow. Electrons that enter the depletion region from the $p$-side and holes that enter the depletion region from the $n$-side are swept away and are the source of the drift components.

the $p$-type and $n$-type neutral regions the built-in potential is

$$V_{bi} = \frac{k_B T}{e} \; \ell n \; \frac{p_p}{p_n} \tag{D.1}$$

The built-in potential can also be written as

$$V_{bi} = \frac{k_B T}{e} \; \ell n \; \frac{n_n}{n_p} \tag{D.2}$$

where $n_n$ and $n_p$ are the electron densities in the $n$-type and $p$-type regions. Remember that the law of mass action tells us that

$$n_n p_n = n_p p_p = n_i^2 \tag{D.3}$$

We can thus write the following equivalent expressions

$$\frac{p_p}{p_n} = e^{eV_{bi}/k_B T} = \frac{n_n}{n_p} \tag{D.4}$$

We will now give the widths of the depletion regions on the $n$ and $p$ side in absence of an external bias. The values in presence of a bias are simply given by replacing the built-in bias by the total bias across the $p$ and $n$ regions.

$$W_p(V_{bi}) = \left\{ \frac{2\epsilon V_{bi}}{e} \left[ \frac{N_d}{N_a(N_a + N_d)} \right] \right\}^{1/2} \tag{D.5}$$

$$W_n(V_{bi}) = \left\{ \frac{2\epsilon V_{bi}}{e} \left[ \frac{N_a}{N_d(N_a + N_d)} \right] \right\}^{1/2} \tag{D.6}$$

$$W(V_{bi}) = W_p(V_{bi}) + W_n(V_{bi}) = \left( W_n^2(V_{bi}) + W_p^2(V_{bi}) + 2W_n(V_{bi})W_p(V_{bi}) \right)^{1/2}$$

$$W(V_{bi}) = \left[ \frac{2\epsilon V_{bi}}{e} \left( \frac{N_a + N_d}{N_a N_d} \right) \right]^{1/2} \tag{D.7}$$

From these discussions, we can draw the following important conclusions about the diode:

i) The electric field in the depletion region peaks at the junction and decreases linearly towards the depletion region edges.

ii) The potential drop in the depletion region has a quadratic form.

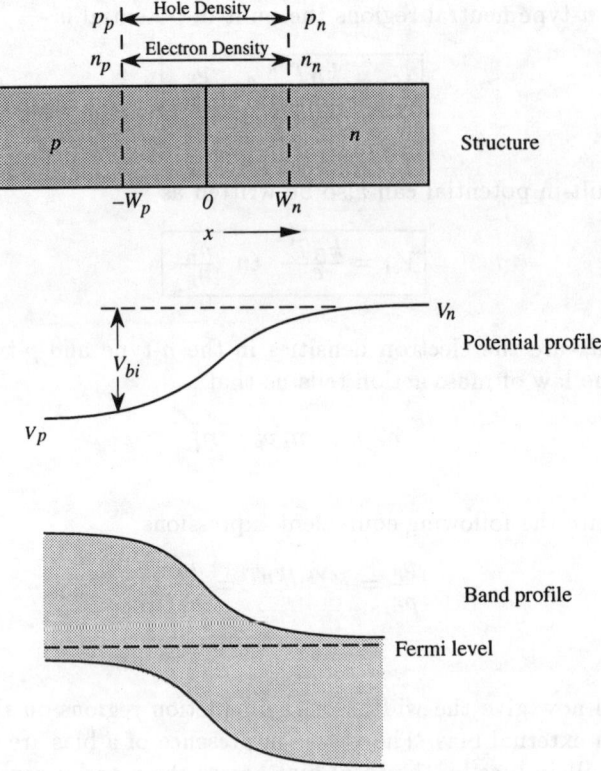

Figure D.2: A schematic showing the $p$-$n$ diode and the potential and band profiles. The voltage $V_{bi}$ is the built-in potential at equilibrium. The expressions derived in the text can be extended to the cases where an external potential is added to $V_{bi}$.

*We remind ourselves that this procedure can be extended to find the electric fields, potential and depletion widths for arbitrary values of $V_p$ and $V_n$ under certain approximations to be discussed next. Thus we can directly use these equations when the diode is under external bias $V$, by simply replacing $V_{bi}$ by $V_{bi} + V$. The applied bias can increase the total potential or decrease it as will be discussed later.*

The results of the calculations carried out above are schematically shown in Fig. D.3. Shown are the charge density and the electric field profiles. Notice that the electric field is nonuniform in the depletion region, peaking at the junction with a peak value (the sign of the field simply reflects the fact that in our study the field is pointing towards the negative $x$-axis)

$$F_m = -\frac{eN_dW_n}{\epsilon} = -\frac{eN_aW_p}{\epsilon} \qquad (D.8)$$

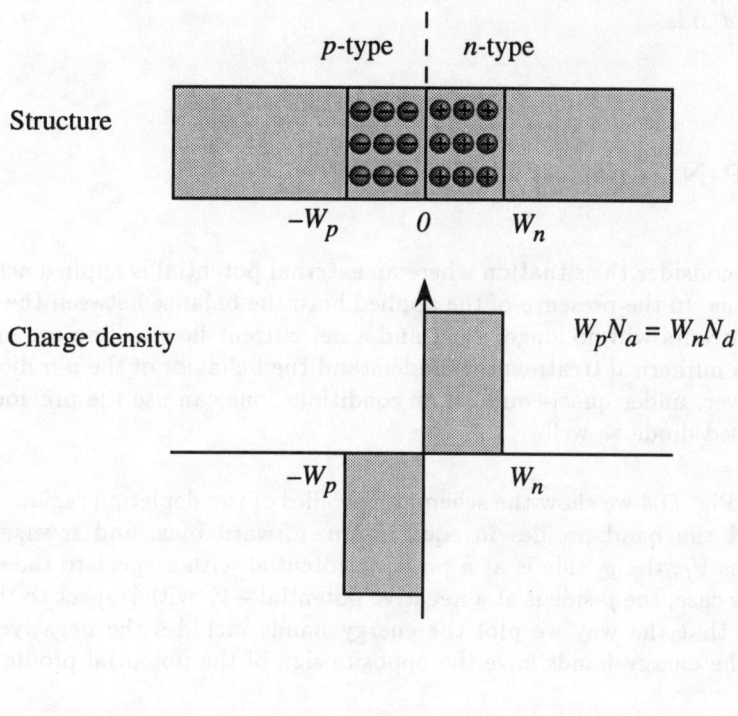

Structure

$p$-type | $n$-type

$-W_p$    $0$    $W_n$

Charge density

$-W_p$    $W_n$

$$W_p N_a = W_n N_d$$

Electric field

$-W_p$    $W_n$

$F_m$

Figure D.3: The $p$-$n$ structure, with the charge and the electric field profile in the depletion region. Note that in the depletion approximation there is no charge or electric field outside the depletion region. The electric field peaks at the junction as shown.

Notice that the depletion in the p- and n-sides can be quite different. If $N_a \gg N_d$, the depletion width $W_p$ is much smaller than $W_n$. Thus a very strong field exists over a very narrow region in the heavily doped side of the junction. *In such abrupt junction ($p^+n$ or $n^+p$) the depletion region exists primarily on the lightly doped side.*

## D.2.2   P-N Junction Under Bias

Let us now consider the situation where an external potential is applied across the p and n regions. In the presence of the applied bias, the balance between the drift and diffusion currents will no longer exist and a net current flow will occur. In general, one needs a numerical treatment to understand the behavior of the p-n diode under bias. However, under quasi-equilibrium conditions, one can use the previous results for the biased diode as well.

In Fig. D.4 we show the schematic profiles of the depletion region, potential profile, and the band profiles in equilibrium, forward bias, and reverse bias. In forward bias $V_f$, the p- side is at a positive potential with respect to the n-side. In reverse bias case, the p-side is at a negative potential $-V_r$ with respect to the n-side. Remember that the way we plot the energy bands includes the negative electron charge so the energy bands have the opposite sign of the potential profile.

In the forward bias case, the potential difference between the n- and p-side is ($V_f$ is taken as having a positive value)

$$V_{Tot} = V_{bi} - V_f \qquad (D.9)$$

while for the reverse biased case it is ($V_r$ is taken as having a positive value)

$$V_{Tot} = V_{bi} + V_r \qquad (D.10)$$

*Under the quasi-equilibrium approximations, the equations for electric field profile, potential profile, and depletion widths we have calculated in the previous section are directly applicable except that $V_{bi}$ is replaced by $V_{Tot}$. Thus the depletion width and the peak electric field at the junction decrease under forward bias, while they increase under reverse bias, as can be seen from Eqns. D.5 and D.6 if $V_{bi}$ is replaced by $V_{Tot}$.*

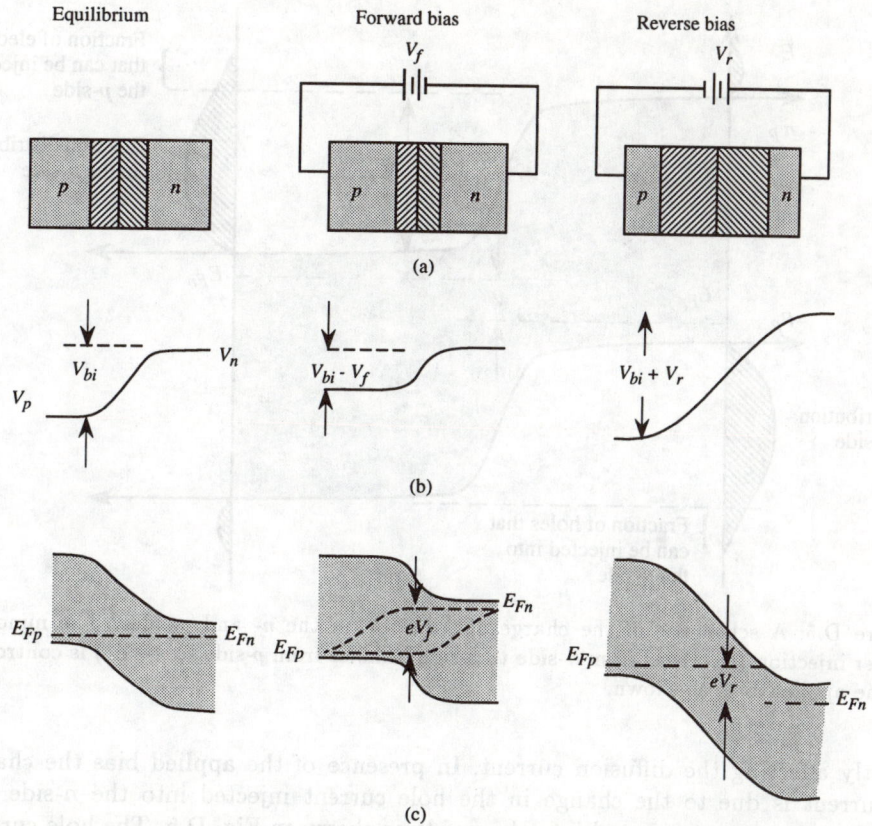

Figure D.4: A schematic showing (a) the biasing of a *p-n* diode in the equilibrium, forward and reverse bias; (b) the voltage profile, and (c) the energy band profiles. In the forward bias, the potential across the junction decreases, while in reverse bias it increases. The quasi-Fermi levels are shown in the depletion region.

## D.2.3   Charge Injection and Current Flow

We will now discuss the current flow in presence of an applied bias . The presence of the bias increases or decreases the electric field in the depletion region. However, under moderate external bias, the *electric field in the depletion region is always higher than the field for carrier velocity saturation* ($F \gtrsim 10 kV cm^{-1}$). *Thus the change in electric field does not alter the drift part of the electron or hole current in the depletion region. Regardless of the bias, electrons or holes that come into the depletion region are swept out and contribute to the same current independent of the field.* The situation is quite different for the diffusion current. Remember that the diffusion current depends upon the gradient of the carrier density. As the potential profile is greatly altered by the applied bias, the carrier profile changes accordingly,

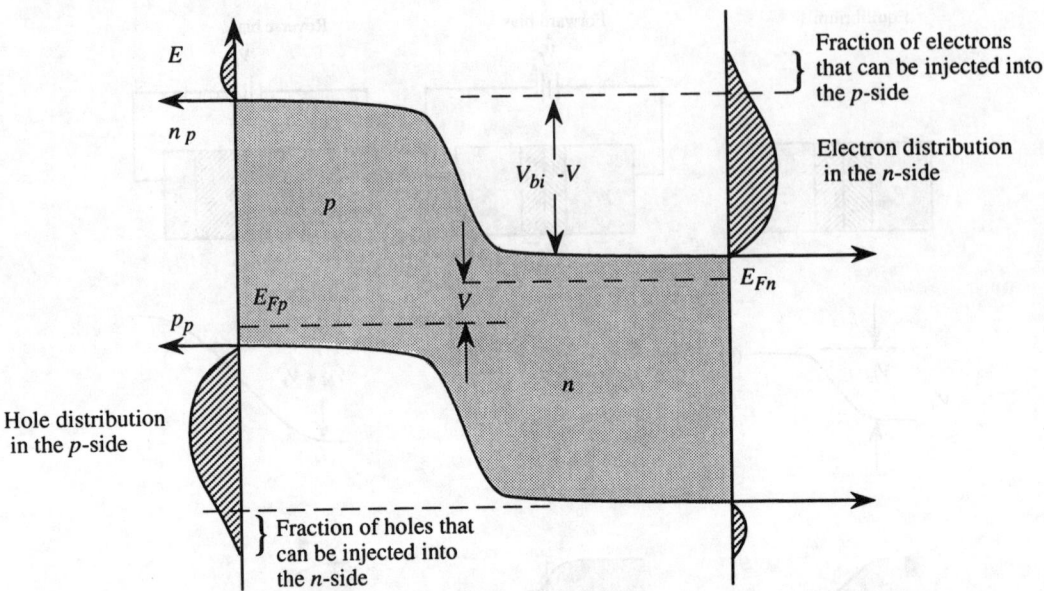

Figure D.5: A schematic of the charge distribution in the $n$- and $p$-sides. The minority carrier injection (electrons from $n$-side to $p$-side or holes from $p$-side to $n$-side) is controlled by the applied bias as shown.

greatly affecting the diffusion current. In presence of the applied bias the change in current is due to the change in the hole current injected into the $n$-side and the electron current injected into the $p$-side as shown in Fig. D.5. The hole current injected into the $n$-side is given by

$$I_p(W_n) = e \, \frac{AD_p}{L_p} \, p_n \left( e^{eV/k_BT} - 1 \right) \qquad (D.11)$$

Similarly the total electron current injected into the $p$-side region is given by

$$I_n(-W_p) = \frac{eAD_n}{L_n} \, n_p \left( e^{eV/k_BT} - 1 \right) \qquad (D.12)$$

We assume initially that in the ideal diode there is no recombination of the electron and hole injected currents in the depletion region. Thus the total current can be simply obtained by adding the hole current injected across $W_n$ and electron current injected across $-W_p$. The diode current is then

$$
\begin{aligned}
I(V) &= I_p(W_n) + I_n(-W_p) \\[2mm]
&= eA \left[ \frac{D_p}{L_p} p_n + \frac{D_n}{L_n} n_p \right] \left( e^{eV/k_BT} - 1 \right)
\end{aligned}
$$

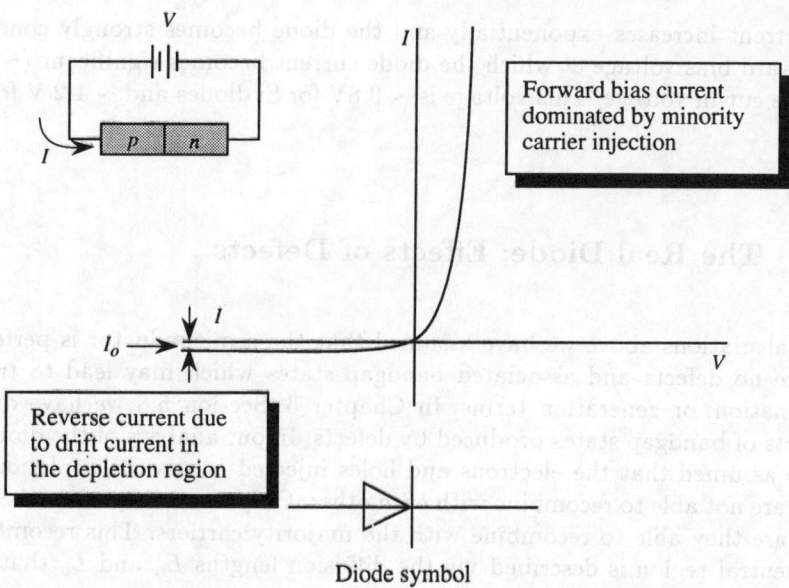

Figure D.6: The highly nonlinear and rectifying I-V current of the $p$-$n$ diode. The strong nonlinear response makes the diode a very important device for a number of applications.

$$I(V) \;=\; I_0 \left( e^{eV/k_B T} - 1 \right) \tag{D.13}$$

This equation, called the diode equation, gives us the current through a $p$-$n$ junction under forward ($V > 0$) and reverse bias ($V < 0$). Under reverse bias, the current simply goes towards the value $-I_0$, where

$$\boxed{I_0 = eA \left( \frac{D_p p_n}{L_p} + \frac{D_n n_p}{L_n} \right)} \tag{D.14}$$

Under forward bias the current increases exponentially with the applied forward bias. This strong asymmetry in the diode current is what makes the $p$-$n$ diode attractive for many applications.

We see from the discussions of this section that the current flow through the simple $p$-$n$ diode has some very interesting properties. We do not have the simple linear Ohm's law type behavior, but a strongly nonlinear and rectifying behavior. The current as shown in Fig. D.6 saturates to a value $I_0$ given by Eqn. 6.44 when a reverse bias is applied. Since this value is quite small, the diode is essentially nonconducting. On the other hand, when a positive bias is applied, the diode current increases exponentially and the diode becomes strongly conducting. The forward bias voltage at which the diode current becomes significant ($\sim$ mA) is called the cut-in voltage. This voltage is $\sim$ 0.8V for Si diodes and $\sim$ 1.2 V for GaAs diodes.

## D.2.4    The Real Diode: Effects of Defects

In the calculations above we have assumed that the semiconductor is perfect, i.e., there are no defects and associated bandgap states which may lead to trapping, recombination, or generation terms. In Chapter 5, Section 5.8 we have discussed the effects of bandgap states produced by defects. In our analysis of the diode ideal, we have assumed that the electrons and holes injected across the depletion region barrier, are not able to recombine with each other. Only when they enter the neutral regions are they able to recombine with the majority carriers. This recombination in the neutral region is described via the diffusion lengths $L_n$ and $L_p$ that appear in the expression for $I_0$.

In a real diode, a number of sources may lead to bandgap states. The states may arise if the material quality is not very pure so that there are chemical impurities present. The doping process itself can cause defects such as vacancies, interstitials, etc. Let us assume that the deep level states lead to a recombination time $\tau$.

The recombination current is now simply (current is equal to charge times volume times rate)

$$I_R = eAWR_t = \frac{eAWn_i}{2\tau} \, exp\left(\frac{eV}{2k_BT}\right)$$

$$= I_{GR}^o \, exp\left(\frac{eV}{2k_BT}\right) \tag{D.15}$$

where $W$ is the depletion width. *At zero applied bias, a generation current of $I_G$ balances out the recombination current.*

The generation-recombination current has an exponential dependence on the voltage as well, but the exponent is different. The generation-recombination current is

$$I_{GR} = I_R - I_G = I_R - I_R(V = 0)$$

$$= I_{GR}^o \left[exp\left(\frac{eV}{2k_BT}\right) - 1\right] \tag{D.16}$$

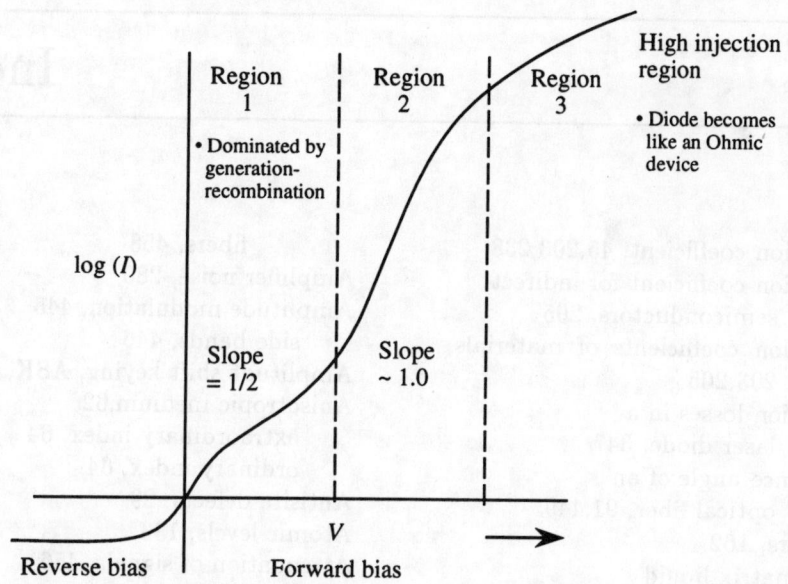

Figure D.7: The I-V characteristics of a real diode. At low biases, the recombination effects are quite pronounced leading to a curve with slope 1/2. At higher biases the slope becomes closer to unity. At still higher biases the behavior becomes more ohmic.

The total device current now becomes

$$I = I_0 \left[ exp \left( \frac{eV}{k_B T} \right) - 1 \right] + I_{GR}^o \left[ exp \left( \frac{eV}{2 k_B T} \right) - 1 \right]$$

or

$$I \cong I'_o \left[ exp \left( \frac{eV}{m k_B T} \right) - 1 \right] \tag{D.17}$$

The prefactor $I_{GR}^o$ can be much larger than $I_0$ for real devices. Thus at low applied voltages the diode current is often dominated by the second term. However, as the applied bias increases, the injection current starts to dominate. We thus have two regions in the forward I- V characteristics of the diode as shown in Fig. D.7.

At low applied bias the plot of $\frac{eV}{k_B T}$ and $\log(I)$ has a slope of 1/2 which turns over to 1.0 at higher voltages. The parameter $m$ of Eqn. D.17 is called the *diode ideality factor*. If the diode is of high quality, $m$ is close to unity, otherwise it approaches a value of 2.

# Index

Peter Anderson
19, Ashbrook Drive

Belfast   BT4  2FG

Tel  01232  654223